VOLUME ONE HUNDRED AND FIFTY ONE

ADVANCES IN CANCER RESEARCH

Novel Approaches to Colorectal Cancer

VOLUME ONE HUNDRED AND FIFTY ONE

ADVANCES IN CANCER RESEARCH

Novel Approaches to Colorectal Cancer

Edited by

FRANKLIN G. BERGER
Research & Outreach Director,
Colorectal Cancer Prevention Network,
University of South Carolina,
SC, United States

C. RICHARD BOLAND
Professor of Medicine,
University of California San Diego School of Medicine,
CA, United States

ELSEVIER

ACADEMIC PRESS
An imprint of Elsevier

Academic Press is an imprint of Elsevier
50 Hampshire Street, 5th Floor, Cambridge, MA 02139, United States
525 B Street, Suite 1650, San Diego, CA 92101, United States
The Boulevard, Langford Lane, Kidlington, Oxford OX5 1GB, United Kingdom
125 London Wall, London, EC2Y 5AS, United Kingdom

First edition 2021

ISBN: 978-0-12-824078-6
ISSN: 0065-230X

For information on all Academic Press publications
visit our website at https://www.elsevier.com/books-and-journals

Publisher: Zoe Kruze
Developmental Editor: Naiza Ermin Mendoza
Production Project Manager: James Selvam
Cover Designer: Alan Studholme

Typeset by SPi Global, India

Contents

Contributors

Parul Agarwal
Sidney Kimmel Cancer Center, Johns Hopkins University, Baltimore, MD, United States

Ludmil B. Alexandrov
Department of Cellular and Molecular Medicine; Department of Bioengineering; Moores Cancer Center, UC San Diego, La Jolla, CA, United States

Patrick M. Boland
Rutgers Cancer Institute of New Jersey, New Brunswick, NJ, United States

John M. Carethers
Division of Gastroenterology and Hepatology, Department of Internal Medicine, and Department of Human Genetics and Rogel Cancer Center, University of Michigan, Ann Arbor, MI, United States

Matthew Custodio
Department of Internal Medicine, Boston University Medical Center, Boston, MA, United States

Soumita Das
Department of Pathology; Rebecca and John Moore Comprehensive Cancer Center; HUMANOID Center of Research Excellence (CoRE), University of California, San Diego, CA, United States

Marcos Díaz-Gay
Department of Cellular and Molecular Medicine; Department of Bioengineering; Moores Cancer Center, UC San Diego, La Jolla, CA, United States

Amer Ali Abd El-Hafeez
Department of Cellular and Molecular Medicine, University of California, San Diego, CA, United States

Stacey Fedewa
Screening and Risk Factors Research, Surveillance and Health Equity Science Department, American Cancer Society, Atlanta, GA, United States

Seth I. Felder
Department of Gastrointestinal Oncology, H. Lee Moffitt Cancer Center and Research Institute, Tampa, FL, United States

Jason B. Fleming
Department of Gastrointestinal Oncology, H. Lee Moffitt Cancer Center and Research Institute, Tampa, FL, United States

Sanil Gandhi
Boston University, Boston, MA, United States

Robert A. Gatenby
Integrated Mathematical Oncology Department, H. Lee Moffitt Cancer Center and Research Institute, Tampa, FL, United States

Pradipta Ghosh
Department of Cellular and Molecular Medicine; Rebecca and John Moore Comprehensive Cancer Center; Department of Medicine; HUMANOID Center of Research Excellence (CoRE), University of California; Veterans Affairs Medical Center, San Diego, CA, United States

Ajay Goel
Department of Molecular Diagnostics and Experimental Therapeutics, Beckman Research Institute of City of Hope Comprehensive Cancer Center, Duarte, CA, United States

William M. Grady
Clinical Research Division, Fred Hutchinson Cancer Research Center; Division of Gastroenterology, University of Washington School of Medicine, Seattle, WA, United States

Heather Hampel
Division of Human Genetics, Biospecimen Research, The Ohio State University Comprehensive Cancer Center, Columbus, OH, United States

Dung T. Le
Sidney Kimmel Cancer Center, Johns Hopkins University, Baltimore, MD, United States

Christopher H. Lieu
Division of Medical Oncology, Gastrointestinal Medical Oncology, University of Colorado Anschutz Medical Campus, Aurora, CO, United States

Priti P. Maity
Department of Cellular and Molecular Medicine, University of California, San Diego, CA, United States

Caitlin C. Murphy
Division of Epidemiology, Department of Population & Data Sciences, University of Texas Southwestern Medical Center, Dallas, TX, United States

Swati G. Patel
Division of Gastroenterology & Hepatology, University of Colorado Anschutz Medical Campus, Rocky Mountain Regional Veterans Affairs Medical Center, Aurora, CO, United States

Dionne Rebello
Department of Gastroenterology, Boston University Medical Center, Boston, MA, United States

Elliott Rebello
Warren Alpert Medical School of Brown University, Providence, RI, United States

Hemant K. Roy
Department of Gastroenterology, Boston University Medical Center, Boston, MA; Department of Internal Medicine, Baylor College of Medicine, Houston, TX, United States

Juan Ruiz-Bañobre
Medical Oncology Department; Translational Medical Oncology Group (Oncomet), Health Research Institute of Santiago (IDIS), University Clinical Hospital of Santiago de Compostela, University of Santiago de Compostela (USC), CIBERONC, Santiago de Compostela, Spain

Ibrahim M. Sayed
Department of Pathology, University of California, San Diego, CA, United States

Rebecca Siegel
Surveillance Research, Surveillance and Health Equity Science Department, American Cancer Society, Atlanta, GA, United States

Robert A. Smith
Cancer Prevention and Early Detection Department, American Cancer Society, Atlanta, GA, United States

Mingyang Song
Department of Epidemiology; Department of Nutrition, Harvard T.H. Chan School of Public Health; Clinical and Translational Epidemiology Unit; Division of Gastroenterology, Massachusetts General Hospital and Harvard Medical School, Boston, MA, United States

Kai Wang
Department of Epidemiology; Department of Nutrition, Harvard T.H. Chan School of Public Health, Boston, MA, United States

Xixi Xu
Department of Internal Medicine, Boston University Medical Center, Boston, MA, United States

Preface

Colorectal cancer (CRC) remains a major source of cancer incidence and mortality. For example, it is expected that in 2021, there will be 149,500 new cases of CRC and 52,980 deaths in the United States. These sobering numbers have driven extensive research on the disease, and considerable progress has been made regarding its etiology, prevention, diagnosis, and treatment. We submit Volume 151 of *Advances in Cancer Research* to highlight the wide range of developments in our understanding of CRC. We have selected highly relevant topics collectively representing the significant progress made in the past few years. The specific topics covered in 11 chapters were carefully chosen to be relatively broad, written by leaders in the respective fields, and reflecting the most recent concepts driving CRC research.

A review of early-onset colorectal cancer by Swati G. Patel, Caitlin C. Murphy, Christopher H. Lieu, and Heather Hampel describes the recent unexplained rise in CRC among those under age 50, which is occurring in the face of the falling incidence and mortality in people over age 50. Seth Felder, Jason Fleming, and Robert Gatenby present a creative discussion of the evolutionary dynamics of tumor growth in locally advanced non-metastatic rectal cancer that may stimulate current thinking about treatment modalities in this subset of CRCs. Robert A. Smith, Stacey Fedewa, and Rebecca Siegel present a comprehensive review of the history and current status of CRC screening, including discussion of the challenges associated with applying research findings to policy and practice. Kai Wang and Mingyang Song review the topic of personalized nutrition and the impact of various lifestyle issues on the gut microbiota, and ultimately the risk of CRC and survivorship. Parul Agarwal, Dung T. Le, and Patrick M. Boland have composed an authoritative review of immunotherapy of CRC, a topic that has undergone a renaissance in recent years due to its application to new therapies. John M. Carethers presents an outstanding review of the critical issue of racial and ethnic disparities in CRC incidence and mortality, including detailed explanations of the mechanisms underlying such disparities. Juan Ruiz-Bañobre and Ajay Goel have written a highly detailed summary of the genomic and epigenomic biomarkers in CRC, which may become the standard review on the topic for quite some time. Dionne Rebello, Elliott Rebello, Matthew Custodio, Xixi Xu, Sunil Gandhi and Hemant

K. Roy describe the important concept of field carcinogenesis in the context of CRC, providing insight into the origin, occurrence, and recurrence of colorectal neoplasia over time, and offering important new insights into risk factor assessment. Ibrahim M. Sayed, Amer Ali Abd El-Hafeez, Priti P. Maity, Soumita Das, and Pradipta Ghosh have written a highly stimulating chapter on the use of multidimensional organoids to advance the study of CRC, including novel methods to examine the molecular basis of neoplastic progression, and new approaches for the design and testing of drug therapies. Marcos Díaz-Gay and Ludmil B. Alexandrov have written a definitive review of the use of mutational signatures to more precisely characterize CRC, particularly with regard to the mechanistic basis of the mutagenic events that drive the disease. Finally, William A. Grady has written a fantastic review of the nature and importance of epigenetic alterations in CRC and their use as cancer biomarkers, making a complex topic both understandable and interesting to all audiences.

We feel honored to have the accumulated expertise and engagement of the research leaders who have written the reviews incorporated into this volume of *Advances in Cancer Research*. We are pleased with the relevance and quality of their contributions, and hope that they stimulate discussion within and among the research, clinical, and educational communities.

Franklin G. Berger
C. Richard Boland

CHAPTER ONE

Early age onset colorectal cancer

Swati G. Patel[a,*], Caitlin C. Murphy[b], Christopher H. Lieu[c], and Heather Hampel[d]

[a]Division of Gastroenterology & Hepatology, University of Colorado Anschutz Medical Campus, Rocky Mountain Regional Veterans Affairs Medical Center, Aurora, CO, United States
[b]Division of Epidemiology, Department of Population & Data Sciences, University of Texas Southwestern Medical Center, Dallas, TX, United States
[c]Division of Medical Oncology, Gastrointestinal Medical Oncology, University of Colorado Anschutz Medical Campus, Aurora, CO, United States
[d]Division of Human Genetics, Biospecimen Research, The Ohio State University Comprehensive Cancer Center, Columbus, OH, United States
*Corresponding author: e-mail address: swati.patel@cuanschutz.edu

Contents

Advances in Cancer Research, Volume 151
ISSN 0065-230X
https://doi.org/10.1016/bs.acr.2021.03.001

Abstract

The incidence and mortality associated with colorectal cancer (CRC) diagnosed in patients under the age of 50 have been steadily increasing. The exact etiology of these epidemiologic trends is unclear. This chapter will provide a comprehensive review on the topic of early age onset colorectal cancer (EAO-CRC), defined as colorectal cancer (CRC) diagnosed in patients under the age of 50. Topics reviewed will include the epidemiology of EAO-CRC around the world, clinical and pathological features of EAO-CRC in contrast to later age onset CRC (CRC diagnosed on those over the age of 50) and the observed molecular and somatic characteristics. This chapter will review the etiologies to EAO-CRC and the established, as well as proposed risk factors for disease. Evidence-based approaches to prevention, early detection, treatment and survivorship will be presented.

Abbreviations

EAO-CRC	early age onset colorectal cancer
MSTF	Multi-Society Task Force
NCCN	National Comprehensive Cancer Network
USPSTF	United States Preventive Services Task Force
SEER	Surveillance, Epidemiology and End Results
FDR	first-degree relative
SDR	second-degree relative
NSAID	non-steroidal anti-inflammatory drug

1. Introduction

Over the last several decades, incidence and mortality rates of colorectal cancer (CRC) have decreased in the United States (US) (Siegel, Miller, Goding Sauer, et al., 2020). Reasons for this decline include increasing uptake of CRC screening and colonoscopic polypectomy in those over the age of 50 (Edwards, Noone, Mariotto, et al., 2014), as well as changing risk factors (e.g., decreased smoking, increased aspirin use) (Brenner & Chen, 2018).

Contrary to this overall trend, the incidence and mortality rates of CRC in persons under the age of 50 is increasing (Siegel, Fedewa, Anderson, et al., 2017). The term "Early Age Onset CRC" (EAO-CRC) includes a heterogeneous population of patients. Most studies have defined EAO-CRC as CRC occurring in patients under ages ranging from 40 to 55, often combining patients at increased risk due to family history of CRC or hereditary cancer syndromes with patients with no predisposing conditions. Although a specific cut-off age does not necessarily separate biologically or pathophysiologically different disease entities, 50 is an intuitive cut off based on long-standing recommendations to initiate average-risk screening at this age (NCCN, 2013; Rex, Boland, Dominitz, et al., 2017; U.S. Preventive Services Task Force, Bibbins-Domingo, Grossman, Curry, et al., 2016). Though professional societies are currently considering decreasing the average-risk screening age to 45 to address the increasing burden of CRC in younger patients (U.S. Preventive Services Task Force, 2020; Wolf, Fontham, Church, et al., 2018), for the purposes of this chapter, EAO-CRC will refer to CRC diagnosed in those under the age of 50, unless otherwise specified. We will review the epidemiology, clinicopathological features, molecular characteristics, somatic landscape, etiology and evidence-based management approach to EAO-CRC. This chapter will focus on sporadic EAO-CRC. More detailed reviews of hereditary colorectal cancer syndromes can be found elsewhere (Ballester, Rashtak, & Boardman, 2016; Patel & Ahnen, 2012; Syngal, Brand, Church, et al., 2015; Stoffel, Mangu, Limburg, et al., 2015).

2. Epidemiology

2.1 Overall incidence and mortality

In the US, CRC is the second most common cancer and the third leading cause of cancer-related death in men and women under the age of 50 (Bhandari, Woodhouse, & Gupta, 2017). In 2020, 11% of all colon cancer and 15% of all rectal cancer diagnoses are estimated to occur in patients under the age of 50 (Siegel et al., 2020). The median age of diagnosis for those under age 50 is 44 with 75.2% of all colorectal cancers occurring between the age of 40 and 49 (You, Xing, Feig, et al., 2012).

CRC incidence has been steadily increasing in younger patients for the last several decades, with the sharpest rise seen in incidence of rectal cancer (Fig. 1). Based on data from the North American Association of Central Cancer Registries (NAACR), which includes 47 states and the District of Columbia (DC), there has been a 1.1% increase per year (95% confidence interval [CI] 0.3–2.0%) from 2006 to 2015. This includes an increase of 0.7% per year (95% CI 0.5–0.9%) for colon tumors and 1.7% per year (95% CI ¼ 1.4–2.0%) increase for rectal tumors (Siegel, Medhanie, Fedewa, et al., 2019). By the year 2030, 22% of all rectal cancers in the US are expected to be diagnosed in patients under the age of 50 (Bailey, Hu, You, et al., 2015).

Fig. 1 Age-adjusted Surveillance, Epidemiology and End Results (SEER) incidence rates of colon and rectal cancer from 1975 to 2016 by age. Data plotted by accessing SEER*Explorer (https://seer.cancer.gov/explorer/index.html).

Bhandari et al. reported age specific incidence rates per 100,000 for ages 30–34 (men 4.9 vs women 4.2), ages 35–39 (men 9.9 vs women 7.6), ages 40–44 (men 16.4 vs women 15.3) and ages 45–49 (men 30.8 vs women 25.9) (Bhandari et al., 2017).

Mortality rates of CRC in the US in those age 20–54 declined from 6.3/100,000 in 1970 to 3.9/100,000 in 2004 but increased 13% to 4.3/100,000 in 2014 (Siegel, Miller, & Jemal, 2017). Specific to those ages 40–49, the mortality rate increased from 5.4/100,000 in 1992 to 6.5/100,000 in 2015.

2.2 Geographic variations

There are significant regional variations in the burden of EAO-CRC within the US. Abualkhair et al. analyzed data from 18 population-based Surveillance, Epidemiology, and End Results (SEER) registries from 2000 to 2015 and found that the incidence at age 45 was 26.8/100,000 in the South compared to 19.7/100,000 in the West ($P < 0.0001$) (Abualkhair, Zhou, Ochoa, et al., 2020). Although western states have a lower absolute incidence, these states have seen the sharpest relative increase in incidence. NAACR data showed that of the 10 states with the highest average annual percent change in incidence, six were located in the western US. Comparing 1995–96 to 2014–15, Washington State's EAO-CRC incidence increased from 6.7 to 11.5/100,000. Over the same period, EAO-CRC incidence in Colorado increased from 6.0 to 9.5/100,000 (Siegel, Medhanie, et al., 2019). Though incidence rates are increasing the fastest in the western states, Appalachian and southeastern states still have the highest absolute incidence, raising the possibility of shared risk factors such as environmental exposures and lifestyle factors that lead to more prevalent co-morbidities in these regions.

Zahnd, Gomez, Steck, et al. (2021) examined SEER data from 2000 to 2016 and reported a sharper increase in EAO-CRC incidence in rural populations (10.4–14.1 per 100,000, overall 35.0% change) compared to urban populations (9.4–11.2 per 100,000, overall 19.5% change).

Internationally, there is significant variability in EAO-CRC incidence (12.9/100,000 in Korea vs 3.5/100,000 in India). Of 42 countries included in a recent analysis, EAO-CRC incidence has been increasing in 19 countries across North America, Europe, Asia and Australia. In nine of these countries, EAO-CRC has been increasing as later age onset CRC is stable or has been decreasing (Siegel, Torre, Soerjomataram, et al., 2019).

2.3 Racial differences in EAO-CRC

Between the year 2000 and 2013, EAO-CRC incidence increased 2.5% in American Indian/Alaskan Natives, 2.3% in non-Hispanic Whites, 1.0% in non-Hispanic Blacks and 0.2% in Asian/Pacific Islanders (Siegel, Miller, Fedewa, et al., 2017). In an analysis of SEER data, Murphy et al. reported that from 1992 to 1996 to 2010–14, CRC incidence increased from 7.5 to 11.0 per 100,000 in White persons and from 11.7 to 12.7 per 100,000 in Black persons. The increase in rectal cancer was larger in White (from 2.7 to 4.5 per 100,000) compared to Black (from 3.4 to 4.0 per 100,000) persons (Murphy, Wallace, Sandler, et al., 2019).

The recent increase in mortality rates is limited to White persons, among whom there has been a 1.4% increase per year from 2004 to 2014 (3.6/100,000 to 4.1/100,000). Among Black persons, mortality rates declined by 0.4–1.1% annually (Siegel, Miller, & Jemal, 2017), however, black patients still had a higher risk of cancer-related death (colon cancer: hazard ratio 1.36, 95% CI 1.27–1.45; rectal cancer: 1.52, 95% CI 1.38–1.68) from 2000 to 2009, when compared to White patients (Holowatyj, Ruterbusch, Rozek, et al., 2016). Five-year relative survival was 54.9% in Black compared to 68.1% in White patients (Holowatyj et al., 2016). The racial disparities are likely multifactorial, stemming from a combination of socioeconomic, access to care, environmental and biologic factors (Ashktorab, Kupfer, Brim, et al., 2017). Several patient-, health system-, and policy-level factors may contribute to racial/ethnic disparities in survival, including delays in and advanced stage at diagnosis, poor access to and/or adherence to care, and availability of treatment. Many of these factors can be traced back to patient experiences of racism or may be a result of structural racism (May, Yang, Corona, et al., 2020).

3. Clinical and pathologic features

The majority of CRCs in young patients are identified due to signs and symptoms rather than incidentally or through screening (Table 1). In a series of over 1000 EAO-CRCs, the most common presenting symptom was rectal bleeding (50.8%), followed by abdominal pain (32.5%) and change in bowel habits (18.0%) (Dozois et al., 2008). When compared to symptomatic older patients with CRC, Chen et al. reported that young patients are more likely to present with symptoms of hematochezia (28.8% vs 23.2%) and abdominal pain (41.2% vs 27.2%).

Table 1 Clinical and pathologic features of colorectal cancer diagnosed in patient under and over the age of 50.

Clinical and pathological features		EAO-CRC (<age 50)	LAO-CRC (≥age 50)
Presenting with symptoms (%)		86.4 (Dozois, Boardman, Suwanthanma, et al., 2008)–95.6 (Dharwadkar, Greenan, Singal, et al., 2021; Myers, Feingold, Forde, et al., 2013)	33.9–79.0 (Dharwadkar et al., 2021; Rho, Gilabert, Polom, et al., 2017)
Incidental or screen detected (%)		1.6–5.2 (Chen, Sundaram, Chew, et al., 2017a; Dharwadkar et al., 2021)	3.4–14.6 (Chen et al., 2017a; Dharwadkar et al., 2021)
Duration of symptoms (days)		243 (Chen et al., 2017a)	154 (Chen et al., 2017a)
Time to diagnosis (days)		152–217 (Chen et al., 2017a; Scott, Rangel, Osler, et al., 2016)	29.5–87 (Chen et al., 2017a; Scott et al., 2016)
Family history of CRC (%)		13.8–33.5 (Chen et al., 2017a; Rho et al., 2017; Strum & Boland, 2019)	8.3–19.3 (Chen et al., 2017a; Rho et al., 2017; Strum & Boland, 2019)
Location (%)	Right colon	16.2–35.2 (Chen et al., 2017a; Dharwadkar et al., 2021; Dozois et al., 2008; Georgiou, Khakoo, Edwards, et al., 2019; Myers et al., 2013; Rho et al., 2017; Strum & Boland, 2019; Teng, Lee, Cai, et al., 2016; Willauer, Liu, Pereira, et al., 2019)	28.5–51.5 (Chen et al., 2017a; Rho et al., 2017; Strum & Boland, 2019; Willauer et al., 2019)
	Left colon	29.1–53.0 (Chen et al., 2017a; Dozois et al., 2008; Georgiou et al., 2019; Myers et al., 2013; Teng et al., 2016; Willauer et al., 2019)	28.9–48.5 (Chen et al., 2017a; Dharwadkar et al., 2021; Strum & Boland, 2019; Willauer et al., 2019)
	Rectal	25.4–49.1 (Chen et al., 2017a; Dharwadkar et al., 2021; Dozois et al., 2008; Georgiou et al., 2019; Myers et al., 2013; Rho et al., 2017; Teng et al., 2016)	20.0–35.2 (Chen et al., 2017a; Dharwadkar et al., 2021; Rho et al., 2017; Willauer et al., 2019)

Continued

Table 1 Clinical and pathologic features of colorectal cancer diagnosed in patient under and over the age of 50.—cont'd

Clinical and pathological features		EAO-CRC (<age 50)	LAO-CRC (≥age 50)
Histology (%)	Mucinous	10.0–15.0 (Dharwadkar et al., 2021; Georgiou et al., 2019; Liang, Huang, Cheng, et al., 2003; Teng et al., 2016; Willauer et al., 2019)	4.7–16.0 (Dharwadkar et al., 2021; Liang et al., 2003; Willauer et al., 2019)
	Signet ring	1.0–13.0 (Chang, Pai, Rybicki, et al., 2012; Dharwadkar et al., 2021; Georgiou et al., 2019; Teng et al., 2016; Willauer et al., 2019)	0.9–4.0 (Dharwadkar et al., 2021; Wang, Wang, & Ping, 2015; Willauer et al., 2019; You et al., 2012)
	Poor or no differentiation	7.2–27.9 (Dharwadkar et al., 2021; Georgiou et al., 2019; Liang et al., 2003; Teng et al., 2016)	3.2–18.0 (Dharwadkar et al., 2021; Liang et al., 2003; You et al., 2012)
Stage (%)	Early	11.0–47.0 (Dharwadkar et al., 2021; Dozois et al., 2008; Georgiou et al., 2019; Myers et al., 2013; Strum & Boland, 2019; Teng et al., 2016)	37.5–69.7 (Chen et al., 2017a; Dharwadkar et al., 2021; Myers et al., 2013; Strum & Boland, 2019)
	Late	61.2–89.0 (Chen et al., 2017a; Dharwadkar et al., 2021; Dozois et al., 2008; Georgiou et al., 2019; Myers et al., 2013; Rho et al., 2017; Strum & Boland, 2019; Teng et al., 2016)	30.3–62.5 (Chen et al., 2017a; Dharwadkar et al., 2021; Myers et al., 2013; Rho et al., 2017; Strum & Boland, 2019)

EAO-CRC, early age onset colorectal cancer; LAO-CRC, later age onset colorectal cancer; CRC, colorectal cancer.

EAO-CRC patients experience symptoms for longer (243 vs 154 days) and have a longer delay to diagnosis (152 vs 87 days), compared to later-age onset (LAO) CRC patients (Table 1) (Chen et al., 2017a).

Younger patients more commonly present with left sided cancers. LAO-CRC patients are more likely to have right sided cancers (31.1% vs 20.0%, $P < 0.001$) and EAO-CRC patients are more likely to have rectal cancer (31.2% vs 22.4%, $P < 0.001$) (Table 1) (Abdelsattar, Wong, Regenbogen, et al., 2016).

EAO-CRC also appears to have more aggressive histopathology than those in older patients. Overall, mucinous and signet-ring histologies were seen in 10.0–14.5% (Chang et al., 2012; Georgiou et al., 2019; Liang et al., 2003; Teng et al., 2016) and 2.0–13.0% (Chang et al., 2012; Georgiou et al., 2019; Teng et al., 2016) of EAO-CRCs, respectively, with up to 27.9% of cancers being poorly differentiated or undifferentiated (Chen et al., 2017a). Data from the National Cancer Database showed that EAO-CRC was modestly, but significantly, more likely to have a mucinous and/or signet-ring histology compared to older onset CRCs (12.6% vs 10.8%, $P < 0.01$) and poor or no differentiation (20.4% vs 18%; $P < 0.01$) (You et al., 2012).

EAO-CRC also presents at a more advanced stage (Table 1). Abdelsattar et al. reported a relative risk of 1.37 (1.33–1.41) and 1.58 (1.53–1.63) for younger patients to present with regional or distant metastasis, respectively, compared to older patients (Abdelsattar et al., 2016). Chen et al. found that the difference in stage at presentation between EAO-CRC and later-onset CRCs could not be explained simply by a longer time between the onset of symptoms and diagnosis because younger patients with stage III or IV disease had shorter symptom and workup periods compared to those with stage I or II disease (Chen et al., 2017a). Thus, advanced stage at diagnosis is not likely explained by longer dwell time or time to diagnosis.

The clinical characteristics LAO-CRC patients in these studies include screened patients, thus may not reflect the true differences between EAO-CRC and LAO-CRC. To address this issue, Dharwadkar et al. (2021) compared 319 EAO and 905 LAO-CRC patients in a predominantly unscreened population. For both groups, the majority presented with symptoms. In contrast to prior studies, this group found no significant difference in anatomic tumor site (69.0% of tumors in the EAO-CRC group were left-sided vs 64.1% in the LAO-CRC group, $P = 0.25$). Similar to previous studies, this group did find that EAO patients were more likely to have poorly or undifferentiated tumors compared to LAO patients (24.3% vs 14.2%, $P < 0.01$) and stage III or IV disease at diagnosis (75.9% vs 67.3%, $P = 0.01$) (Dharwadkar et al., 2021).

Despite seemingly later stage and more aggressive histology at presentation, EAO-CRC patients appear to have equivalent, if not improved survival. In one large report of SEER data, the stage-adjusted, cancer-specific survival was better in younger patients compared to those diagnosed over the age of 50 (local: 95.1% vs 91.9%; regional: 76% vs 70.3%; distant: 21.3% vs 14.1%) (Abdelsattar et al., 2016).

It is important to note that the literature on EAO-CRC clinical and pathologic features is drawn from retrospective series that have not consistently separated sporadic cancers from those occurring in patients with hereditary cancer syndromes. Two recent studies characterized cohorts of EAO-CRC patients with multi-gene panel testing. Although these are small cohorts (450 and 430 patients, each), Pearlman, Frankel, Swanson, et al. (2017) reported that left sided cancers (including rectal) comprised a larger proportion of sporadic EAO-CRCs compared to hereditary EAO-CRCs (74.9% vs 58.3%). Similarly, Stoffel, Koeppe, Everett, et al. (2018) reported that 72.6% of sporadic EAO-CRCs were left sided vs 38.0% of hereditary EAO-CRCs.

4. Genetic and familial EAO-CRC

4.1 Spectrum of germline variants in EAO-CRC

A series of studies have shed light on the spectrum of germline variants in EAO-CRC patients (Table 2). In 2015, Mork et al. (2015) performed germline genetic testing in 193 patients ≤35 years old referred for genetic counseling between 2009 and 2013. 34% (66/193) had a germline variant consistent with a hereditary cancer syndrome; 22 had LS, 16 had FAP, two had biallelic *MUTYH*, two had biallelic MMR variants (constitutional mismatch repair deficiency) and one had a pathogenic variant in *TP53*. Although this was a highly selected, risk-enriched population, this study highlighted that relying on syndrome phenotypes for genetic risk assessment can miss hereditary syndromes in EAO-CRC patients.

A series of subsequent studies have demonstrated that EAO-CRC patients have a wide spectrum of germline variants, including in genes not typically associated with CRC. In 2017, Yurgelun et al. (2017) reported results of 25-gene panel germline testing in 1058 consecutive, unselected CRC patients including 336 under age 50. 14% (47/336) of the EAO-CRC patients had a pathogenic variant, including 22 Lynch Syndrome variants, three *APC* variants, three biallelic *MUTYH* variants, four *APC* I1307K variants, three monoallelic *MUTYH* variants, six *BRCA1/2* variants, six *ATM* variants, one *CHEK2* variant and one *BRIP1* variant.

Table 2 Pathological germline variants in early age onset colorectal cancer.

Study	Cohort characteristics	Genes tested	Pathogenic or likely pathogenic variants N (%)	Germline variants found (N)	Variants of uncertain significance N (%)
Mork, You, Ying, et al. (2015) 2009–13 Single-center US	<35 y/o: $n = 193$ Mean age 29	Phenotype driven; all patients had been referred for genetic counseling	44/193 (22.8)	*APC* (13), biallelic *MUTYH* (2), CMMRD (2), *TP53* (1), monoallelic *MUTYH* (1), *MLH1* (6), *MSH2* (5), *MSH6* (2), *PMS2* (1), *EPCAM* (1)	6/193 (3.1)
Yurgelun, Kulke, Fuchs, et al. (2017) 2008–14 Single-center US	<50 y/o: $n = 336$ Median age 56 (from 1058 consecutive CRC patients of all ages)	25-gene panel	52 variants in 47 patients/336 (14.0)	*MLH1* (13), *MSH2* (4), *MSH6* (3), *PMS2* (2), *APC* (3), *BRCA1* (2), *BRCA2* (4), *APC* I1307K (4), biallelic *MUTYH* (3), monoallelic *MUTYH* (6), *BRIP1* (1), *ATM* (6), *CHEK2* (1)	479/1260 (38.0)[a]
Pearlman et al. (2017) 2013–16 Multi-center US	<50 y/o: $n = 450$ Mean age 43	25-gene panel	75 variants in 72 patients/450 (16.0)	*MLH1* (13), *MSH2* (17), *MSH6* (2), *PMS2* (6), *APC* (6), *APC* I1307K (4), biallelic *MUTYH* (4), monoallelic *MUTYH* (8), *SMAD4* (1), *BRCA1* (2), *BRCA2* (4), *ATM* (4), *CHEK2* (1), *PALB2* (2), *CDKN2A* (1)	145/450 (32.2)[b]
Stoffel et al. (2018) 1998–2015 Single-center US	<50 y/o: $n = 430$ Mean age 40; (retrospectively selected from genetic counseling service)	124 or 67 gene panel	79/315 (25.1)	*MSH2* (25), *MLH1* (24), *MSH6* (5), *PMS2* (2), *APC* (10), *MUTYH* (8), *SMAD4* (2), *BRCA1* (1), *TP53* (1), *CHEK2* (1)	21/315 (6.7)
Toh, Chiang, Chong, et al. (2018) 2014–16 Single-center Singapore	<50 y/o: $n = 88$ Mean age 41	64-gene panel, excluding MMR genes	12/88 (13.6)	*APC* (4), monoallelic *MUTYH* (2), *ATM* (1), *BRCA2* (1), *NTHL1* (1), *PALB2* (1), *WRN* (2)	236 VUS in 88 patients[b]

[a]Includes total cohort of patients (data on only <50 cohort not available).
[b]Patients had more than one variant. Study did not report the number of patients with at least 1 variant.
MSI-H, microsatellite unstable; MMR, mismatch repair; US, United States; CMMRD, constitutional mismatch repair deficiency.

Pearlman et al. (2017) conducted similar multi-gene panel testing in 450 EAO-CRC patients from a consortium of 51 Ohio hospitals. 16% (72/450) of patients had a pathogenic germline variant. 38 (8.4%) had a LS gene variant, 11 patients had polyposis gene variants (six with familial adenomatous polyposis, four with biallelic *MUTYH* and one with *SMAD4*). Phenotypically unexpected results included six *BRCA1/2*, four *ATM*, two *PALB2* and one each of *CHEK2* and *CDKN2A* variants. Fig. 2 summarizes the pathogenic variants by CRC risk found in EAO-CRC patients from the Yurgelun et al. and Pearlman et al. cohorts.

Stoffel et al. reported that 25% (79/315) EAO-CRC patients who had been selected for germline testing had a pathogenic variant. Twenty patients had polyposis variants and one each had *BRCA1*, *TP53* and *CHEK2* variants. In 2018, Toh et al. reported whole exome sequencing analyzed for targeted cancer-associated genes from 88 EAO-CRC patients, however, MMR variant carriers were excluded (Toh et al., 2018). In addition to six polyposis patients, they found six additional patients with variants in *ATM*, *BRCA2*, *PALB2*, *NTHL1* and *WRN* (2). These studies demonstrate that EAO-CRC patients have a wide spectrum of pathogenic germline variants, often without manifesting the expected phenotype or expected family history.

In addition to traditionally non-CRC risk genes, it is important to note that some of the pathogenic variants identified in these EAO-CRC patients were "moderate" risk genes, such as *MUTYH* heterozygotes and the I1307K allele of *APC*. For instance, Yurgelun et al. reported that 15% (16/106) of the variants for their entire (all age) cohort were in moderate penetrance genes. 21% (15/72) of the variants found in the Pearlman et al. study and were in moderate penetrance genes. It is unclear if these germline variants drove the EAO-CRC or were coincidental findings. There is limited data on the long-term colonic and extra-colonic cancer risks associated with these moderate penetrance genes, and accordingly, professional guidelines are evolving on how best to minimize metachronous cancer risk for these patients (NCCN, 2020a).

These studies also demonstrated high rates of variants of uncertain significance (VUS), or variants where there is limited or conflicting information regarding pathogenicity. Yurgelun et al. reported that 38% of their entire cohort had at least one VUS and Pearlman et al. found a VUS in 32% of their patients.

Another important finding from these studies was that a substantial proportion of patients with variants did not meet the expected family history

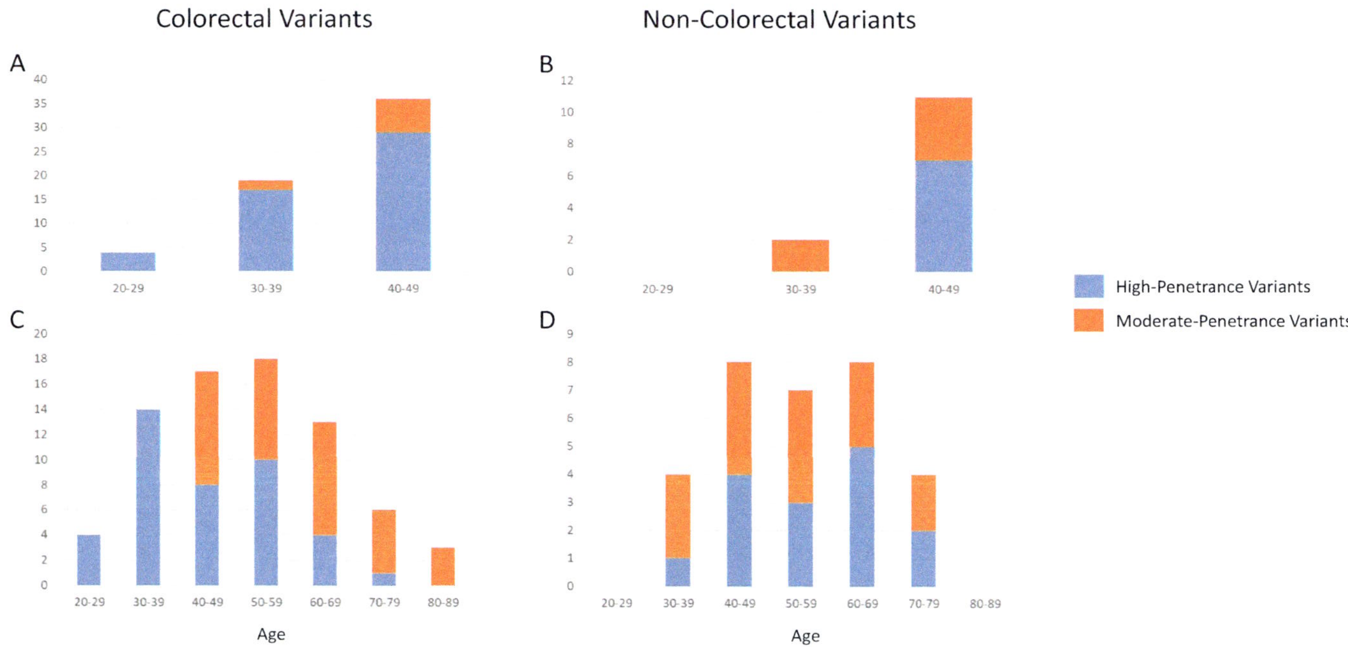

Fig. 2 Germline Variants (n) by age and penetrance. (A) Colorectal variants from Pearlman et al. (2017). (B) Non-colorectal variants from Pearlman et al. (2017). (C) Colorectal variants from Yurgelun et al. (2017). (D) Non-colorectal variants from Yurgelun et al. (2017). High-penetrance colorectal genes included *MLH1*, *MSH2*, *MSH6*, *PMS2*, *APC*, biallelic *MUTYH*, *SMAD4*. Moderate-penetrance colorectal genes included *APC* I1307K, monoallelic *MUTYH*. High-penetrance non-colorectal genes included *BRCA1*, *BRCA2*, *CDKN2A*, *PALB2*, *TP53*. Moderate penetrance non-colorectal genes included *ATM*, *CHEK2*, *BARD1*, *BRIP1*, *NBN*. Note: The Y-axis represents number of patients with a variant. Some patients had more than one variant; the higher-penetrance variant is represented. *Adapted from Patel, S. G., Boland, C. R. (2020). Genetic testing use and expectations in early onset colorectal cancer.* Current Treatment Options in Gastroenterology, 18, *589–603 (In Press).*

criteria. Mork et al. reported that 19% of the patients diagnosed with a hereditary syndrome did not have a family history consistent with the identified syndrome. Similarly, 19% and 26% of patients with a pathogenic variant in the Yurgelun and Pearlman cohorts, respectively, had no family history of cancer at all.

4.2 Familial CRC

Up to 33.5% of patients diagnosed with EAO-CRC have some family history of CRC (Chen et al., 2017a; Rho et al., 2017; Strum & Boland, 2019) (Table 1). Chen, Sundaram, Chew, et al. (2017b) reported that EAO-CRC patients are more likely to have a family history of CRC than older patients (24.9% vs 16.8%). When definitively excluding patients with germline hereditary syndromes, 14.0% (Pearlman et al., 2017) to 16.3% (Stoffel et al., 2018) of patients with EAO-CRC still have a first-degree relative (FDR) with CRC.

Based on these risks, the US Multi-Society Task Force (MSTF) on Colorectal Cancer (Rex et al., 2017) recommends that individuals with an FDR with CRC diagnosed under the age of 60, or more than one FDR with CRC (any age), should undergo screening colonoscopy starting at age 40 or 10 years earlier than the earliest cancer in the family and repeat screening every 5 years. Those with an FDR diagnosed with CRC over the age of 60 should begin screening at age 40 with any of several acceptable screening strategies (annual high sensitivity fecal occult blood test, Multi-Target FIT/DNA, flexible sigmoidoscopy, CT colonography, colonoscopy) but can proceed at average-risk intervals (Rex et al., 2017). More broadly, the National Comprehensive Cancer Network (NCCN) recommends colonoscopy screening at age 40, or 10 years prior to CRC diagnosis and repeated every 5 years, for any individual who has one or more FDRs with colorectal cancer diagnosed at any age (NCCN, 2020b).

Despite this increased risk, adherence to screening recommendations is low among persons with a family history of CRC. Lowery et al. summarized 17 original studies where adherence to start age and appropriate interval ranged between 31% and 47% (Lowery, Ahnen, Schroy 3rd, et al., 2016). A meta-analysis of these 17 studies determined that only 40% (range 26–54%) of individuals with an FDR with CRC had undergone colonoscopy, and only 31% (range 12–51%) had undergone colonoscopy at 5 year intervals (Ait Ouakrim, Lockett, Boussioutas, et al., 2013). A recent comparative effectiveness study by Paskett et al. found that this can be improved to 78.6% with the addition of personalized risk messaging and patient navigation (Paskett, Bernardo, Young, et al., 2020).

4.3 Familial adenomas

Colorectal adenomas are known precursors to adenocarcinoma. An advanced adenoma is defined as an adenoma $\geq$1 cm, the presence of villous architecture, or with high-grade dysplasia. A family history of advanced adenomas increases risk of CRC. An individual with one or more FDR with any advanced adenoma has a twofold increased risk of developing CRC (Tuohy, Rowe, Mineau, et al., 2014; Winawer, Zauber, Gerdes, et al., 1996). Cottet, Pariente, Nalet, et al. (2007) reported a 2.27 (1.09–5.09) increased odds of developing CRC or a large adenoma in those with a FDR with a large adenoma.

Based on this increased risk, the MSTF recommends that those with an FDR with an advanced adenoma at any age should initiate CRC screening at age 40 and continue screening according to average-risk intervals. Those who have an FDR diagnosed with an advanced adenoma under the age of 60 should start CRC screening with colonoscopy at age 40 or 10 years earlier than the earliest advanced adenoma in the family and then every 5 years or earlier depending on the colonoscopic findings (Rex et al., 2017). The NCCN recommends that FDRs of those with confirmed advanced adenomas should undergo colonoscopy at age 40 or at the age of advanced adenoma diagnosis, with repeat colonoscopy every 5–10 years (NCCN, 2020b).

Unfortunately, patients with a personal history of adenomas have poor recollection of these findings. Among 3986 participants in the Prostate, Lung, Colorectal and Ovarian (PLCO) Screening trial surveyed about their colonoscopy findings, only 44.4% accurately reported whether they had an adenoma on colonoscopy (Wiseman, Silver, Klabunde, et al., 2020). Although the data are sparse among patients with advanced adenomas, there is a similar deficiency in patient knowledge of personal findings and familial CRC risk. In a survey of 137 patients diagnosed with advanced adenomas under age 60, only 28.5% were aware that their polyp was precancerous, only 40.1% of respondents reported that their FDRs are at increased risk of CRC and only 38.0% were aware that their FDRs should get earlier CRC screening (Patel, Ahnen, Gumidyala, et al., 2020).

As reviewed above, current data suggests that 14–16% of unselected EAO-CRC is associated with a pathogenic germline variant in a cancer predisposition gene and an additional 14.3–16% of EAO-CRC patients have a FDR with CRC. It is unclearly how many EAO-CRC patients have a FDR with an advanced adenoma. Thus, patients without a germline variant and without an FDR with CRC are considered sporadic.

5. EAO-CRC pathogenesis

5.1 Somatic alterations in EAO-CRC

In 2007, Hong et al. published an analysis of gene expression using microarrays in EAO-CRCs from Singapore (Hong, Ho, Eu, et al., 2007). They studied 12 cases of EAO-CRC, and reported consistent upregulation of 7 genes in the tumor tissue compared to normal colons from healthy controls, including: *CYR61*, *UCHL1*, *FOS*, *FOSB*, *EGR1*, *VIP* and *KRT24*, and validated these findings in 6 additional cases. In 2014, Kirzin et al. compared molecular and clinical findings from 39 cases of sporadic CRC in 39 patients ≤45 years old vs 36 CRC patients >60 years old (Kirzin, Marisa, Guimbaud, et al., 2014). After excluding the cases with MSI (Lynch syndrome, Lynch-like syndrome and *MLH1* hypermethylation cases), EAO-CRC was more distally located, more likely to have synchronous metastases, to have a positive family history for CRC, to have a wild-type *BRAF* gene and the absence of a methylator phenotype (because *MHL1* hypermethylation cases were excluded), and to have evidence of chromosomal instability (CIN).

In 2017, Perea et al. from Madrid reported the frequent somatic deletion of the *NOMO-1* gene in EAO-CRC and homozygous losses of *NOMO-1* in CRCs that showed microsatellite stability (MSS) (Perea, Garcia, Perez, et al., 2017). A follow-up clinical and molecular analysis from the Perea group in 2019 reported that right-sided EAO-CRCs tended to show clinical features of Lynch syndrome and less CIN, whereas the later-onset CRCs showed a higher degree of chromosomal copy number variation (CNV) and CIN. In the EAO-CRC group, 76% were in the left colon or rectum. Interestingly, none of the rectal cancers in the early-onset group showed MSI, so presumably none of those represented Lynch syndrome (Alvaro, Cano, Garcia, et al., 2019).

In 2019, a multicenter study of 18,218 CRC cases were subjected to targeted next-generation genomic sequencing of 3769 exons from 403 cancer-related genes, and of 47 introns commonly rearranged in cancer tissues. The patient groups were divided into an early-onset group (1420 cases <40 years old), an intermediate-aged group (3248 cases 40–49 years old) and an older-onset group (13,550 cases ≥50 years old) (Lieu, Golemis, Serebriiskii, et al., 2019). Generally, the genomic alterations were similar between the younger and older cohorts, with a few notable differences.

TP53 and *CTNNB1* alterations were more common in the youngest cohort, whereas *APC*, *KRAS*, *BRAF* and *FAM123B* variants more frequent in the older cohort.

5.2 Molecular characteristics of EAO-CRC

Guinney, Dienstmann, Wang, et al. (2015) described four consensus molecular subtypes (CMS) of colorectal cancer: CMS1 (microsatellite instability immune), hypermutated, microsatellite unstable and strong immune activation; CMS2 (canonical), epithelial, marked WNT and MYC signaling activation; CMS3 (metabolic), epithelial and evident metabolic dysregulation; and CMS4 (mesenchymal), prominent transforming growth factor–β activation, stromal invasion and angiogenesis. Willauer et al. (2019) described the CMSs of 626 CRCs and found that the subtypes differed significantly by age group. Patients with CRC diagnosed under age 40 that the highest proportion of CMS1 (46%), which is characterized by tumors containing MSI-high and inflammatory or immunogenic markers. CMS2 was similar in all age groups. CMS3 and CMS4 were uncommon in patients younger than 40 years (4% and 13%, respectively).

"Microsatellite and chromosomally stable" or MACS tumors are near-diploid tumors without the aneuploidy that characterizes chromosomal instability and without microsatellite instability. Chan et al. first reported that 14/22 EAO-CRCs (64%, mean age 33 years) exhibited the MACS phenotype vs 2/16 late-onset CRCs (13%, mean age 68 years) (Chan, Curtis, Leung, et al., 2001). The higher frequency of MACS CRCs was confirmed by Boardman et al. in a prospectively collected series of 84 microsatellite-stable (MSS) EAO-CRCs and compared with 90 MSS later-onset CRCs (>65 years old). The MSS tumors in the EAO-CRC group were more frequently diploid (46%) than the older-onset group (26%, $P < 0.01$), and were more frequently late-stage (42% vs 18%, $P = 0.02$) and located in the rectum (50% vs 23%, $P = 0.04$) (Boardman, Johnson, Petersen, et al., 2007). It is unclear what pathological process drives MACS, or whether this is reflective of the distal colorectal location of these tumors vs a characteristic unique to EAO-CRCs.

5.3 Epigenetic alterations in EAO-CRC

Non-coding DNA of the human genome is made up of tandem repeats including long terminal repeats (LTRs), long and short interspersed nuclear elements (LINEs and SINEs) and shorter tandem repeats, such as satellites and microsatellites (Birney, Stamatoyannopoulos, Dutta, et al., 2007; Casa & Gabellini, 2012; Lander, 2011).

LINE-1 are repetitive sequences that comprise approximately 18% of the human genome (Feng, Moran, Kazazian Jr., et al., 1996; Hohjoh & Singer, 1997). *LINE-1* sequences are retrotransposons and mobile elements that, if expressed, permit our genome to rearrange itself (Kazazian Jr, 2004). *LINE-1* is comprised of a 5′-untranslated region (UTR), two open reading frames, and a 3′-UTR. The 5′-UTR contains a CpG island (like many promoters), and this is usually heavily methylated in normal cells, preventing transcription of the DNA. The frequency of *LINE-1* methylation in human DNA makes it a useful surrogate for estimating global hypomethylation, particularly in cancer.

Goel et al. reported significantly lower levels of LINE-1 methylation in EAO-CRC compared to sporadic MSS later age onset CRCs, sporadic MSI CRCs, Lynch Syndrome CRCs and normal colonic mucosa (Antelo, Balaguer, Shia, et al., 2012; Goel, Xicola, Nguyen, et al., 2010). Survival was significantly worse in the groups with *LINE-1* methylation $<65\%$ ($P = 0.039$) (Antelo et al., 2012).

De novo constitutional *MLH1* epimutations have been described rarely in patients with EAO-CRC (Goel, Nguyen, Leung, et al., 2011). In these instances, neither parent shows methylation of *MLH1*, but this epimutation is found in all tissues of the affected patient (i.e., blood, hair follicles, cheek swabs), with deletion of the wild-type allele in the tumor, leading to MSI. There has been a report of a cancer family with two linked single nucleotide variants in the *MLH1* gene [(one in the promoter, c.−27C>A, and one near the transcription start site) (85G>T)]. This combination leads to constitutional methylation of *MLH1*, which is erased in spermatozoa, but then recurs in the somatic cells (Hitchins, Rapkins, Kwok, et al., 2011).

6. Etiology and risk factors for EAO-CRC

6.1 Unrecognized hereditary or familial risk

Although hereditary syndromes contribute to 14–16% (Pearlman et al., 2017; Yurgelun et al., 2017) of all EAO-CRCs and they are grossly under-recognized (Giardiello, Allen, Axilbund, et al., 2014; Hampel & de la Chapelle, 2011), there is no reason to suspect that the incidence of hereditary syndromes is increasing or that penetrance is changing. Thus, although there is much work to be done to improve recognition of hereditary syndromes, they likely have not contributed to the rise in EAO-CRC that has been observed over the past several decades. Similarly, although there are gross deficiencies in acquisition and follow-up of accurate cancer family history (Fletcher, Lobb, Bauer, et al., 2007; Lowery et al., 2016), no data suggest that this deficiency is worsening such that it would contribute to the rise in EAO-CRC.

6.2 Birth cohort effect

Siegel, Fedewa, Anderson, et al. (2017) used age-period-cohort modeling to determine the influence of period effects (generally occur because of changes in clinical practice) vs birth cohort effects (due to generation-specific changes in risk factors) in the rising incidence of EAO-CRC. SEER incidence data from 1974 to 2013 was analyzed by age group.

Interestingly, the incidence curve for those age 50–54 is similar to the older age groups in the 1970s–80s, but then reflects the younger age group after the mid-1990s. Siegel et al. conclude that the younger birth cohorts are carrying the elevated risk with them as they age and that this supports a strong cohort effect in the data. Wolf et al. (2018) illustrate this cohort effect (Fig. 3). The inflection point for the birth cohort effect is for those born after 1960. This strong birth cohort effect suggests exposures increasingly prevalent in early life, or accumulated across the life course, may contribute to increasing incidence rates of EAO-CRC.

6.3 Risk factors

Increasing incidence rates of EAO-CRC point to risk factors that have increased in parallel over the same time period. For example, obesity, metabolic syndrome, sedentary lifestyle, smoking and western diets are established risk factors for later age onset CRC (Pan & DesMeules, 2009; Song, Garrett, & Chan, 2015; Zheng, Zhao, Li, et al., 2017). There has been a steady rise in sedentary lifestyles, diabetes mellitus, and obesity in the US and recent data suggest these risk factors now occur at a greater frequency in young adults compared to decades past (Bhupathiraju & Hu, 2016). The percentage of US adults (over the age of 20) reported to be obese in the US has steadily increased from 30.5% (1999–2002) to 34.9% (2007–10) to 36.5% (2011–14) (National Center for Health Statistics, 2017). Physical activity and obesity play are well-established risk factors of CRC in older adults (Zheng et al., 2017). It is estimated that for each 5-unit increase in body mass index, there is an associated 13–18% increased

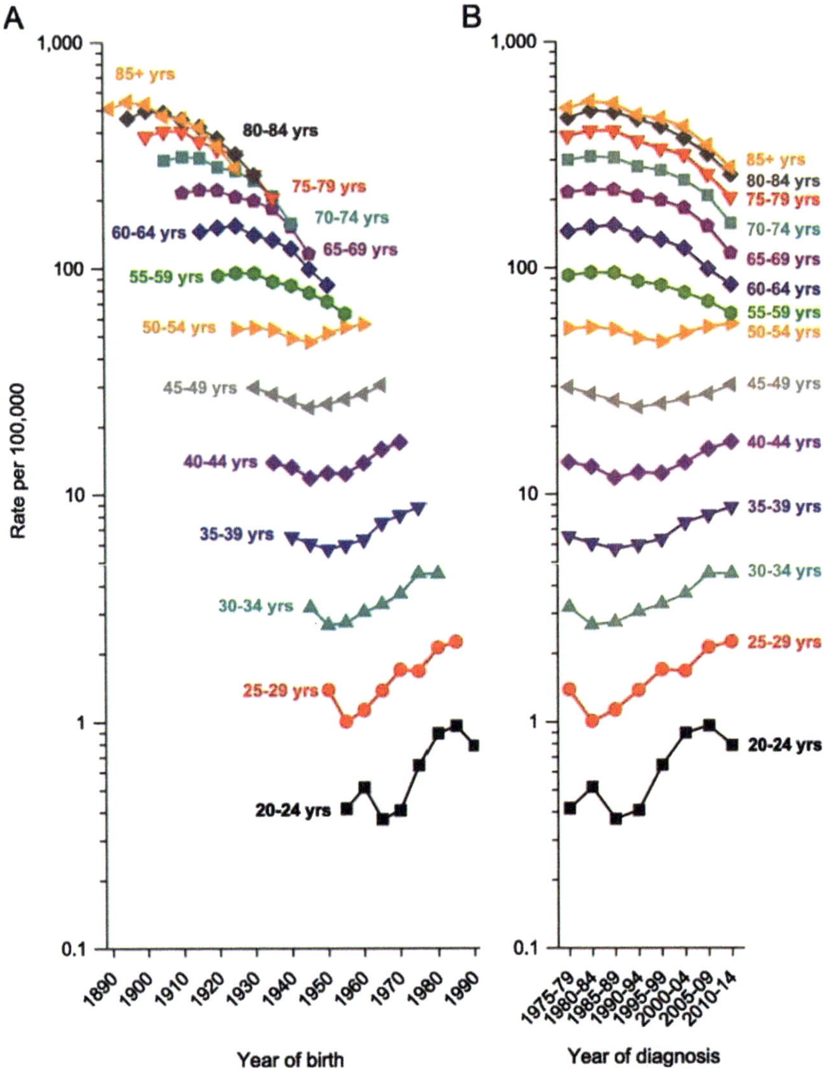

Fig. 3 Trends in colorectal cancer incidence rates by age and year of birth, and by age and year of diagnosis, United States, 1975–2014. Data source: Surveillance, Epidemiology, and End Results (SEER) program, SEER 9 registries, delayed adjusted rates, 1975–2014, National Cancer Institute. *With permission from Wolf, A. M. D., Fontham, E. T. H., Church, T. R., et al. (2018) Colorectal cancer screening for average-risk adults: 2018 guideline update from the American Cancer Society.* CA: A Cancer Journal for Clinicians, 68, *250–281.*

risk for CRC (Pan & DesMeules, 2009). Weight gain between early adulthood and mid-life has been shown to be associated with a significantly increased risk for CRC (Karahalios, English, & Simpson, 2015). On the other hand, regular physical activity is associated with a 24–31% reduction in CRC (Wolin, Yan, Colditz, et al., 2009). Diabetes mellitus has been associated with a 20–38% increased risk for CRC compared to non-diabetics, even when controlling for smoking, obesity, and physical activity (Yuhara, Steinmaus, Cohen, et al., 2011). Cigarette smoking (Botteri, Iodice, Bagnardi, et al., 2008) and moderate to heavy alcohol consumption (Cho, Smith-Warner, Ritz, et al., 2004; Fedirko, Tramacere, Bagnardi, et al., 2011; Mizoue, Inoue, Wakai, et al., 2008) are also associated with an increased risk of CRC. Although there has been a decline in White new smokers between 2006 and 2013, there has been an increase in new smokers among young Black and Hispanic men and a slower rate of decline among young Black and Hispanic women (Thompson, Mowery, Tebes, et al., 2017).

There are a growing body of data on the contribution of traditional risk factors to the rise in EAO-CRC. Risk factors that have been studied in later age onset CRC and EAO-CRC are summarized in Table 3. Liu et al. (2019) reported data from the Nurses' Health Study II showing a 1.93 (95% CI 1.15–3.25) adjusted relative risk for EAO-CRC in those with a body mass index of ≥30. From the same cohort, Zheng et al. reported that those who had the most components of a western diet (fifth quintile) had a 1.67 (95% CI 1.18–2.37) odds of early age onset advanced adenomas compared to those with the lowest western diet consumption (first quintile) (Zheng, Hur, Nguyen, et al., 2020). Nguyen et al. recently reported that those who were sedentary for more than 14 h per week had higher risk of (1.68, 95% CI 1.09–2.63) for EAO-CRC. The risk was more pronounced for rectal cancers (RR 2.62, 95% CI 1.15–6.00) (Nguyen, Liu, Zheng, et al., 2018). When comparing 329 cases of CRC diagnosed under age 45 to controls, Rosato et al. (2013) found that ≥14 alcoholic drinks per week (1.56, 95% CI 1.12–2.16) and the highest tertile of processed meat consumption (1.56, 95% CI 1.11–2.20) were associated with risk of EAO-CRC. Vegetable consumption was protective (RR 0.40, 95% CI 0.28–0.56). Jung et al. (2015) reported that ≥20 pack-year smoking history (30–39 year olds: 4.41, 95% CI 1.80–10.80; 40–49 year olds: 1.64 95% CI 1.14–2.37), hepatic steatosis (30–39 year olds: 1.71, 95% CI 1.10–2.68) and elevated

Table 3 Metabolic risk factors and supporting data in later age onset and early age onset CRC.

Clinical risk factor	Hypothesized mechanism(s)	Later age onset CRC risk	Early age onset CRC risk
Diabetes	• Insulin resistance; chronic hyperglycemia; therapeutic hyperinsulinemia (Larsson, Orsini, & Wolk, 2005)	Yes (Deng, Gui, Zhao, et al., 2012; Pan & DesMeules, 2009; Yuhara et al., 2011; Zheng et al., 2017)	Limited (Jung, Ryu, Chang, et al., 2015)
Prediabetes	• Chronic hyperglycemia; oxidative stress; accumulation of advanced glycation end-products (Scappaticcio, Maiorino, Bellastella, et al., 2017)	Yes (Huang, Cai, Qiu, et al., 2014)	No data
Non-alcoholic fatty liver disease	• Insulin resistance; altered levels of adipocytokines; chronic, low-grade inflammation (Muhidin, Magan, Osman, et al., 2012)	Yes (Hamaguchi, Hashimoto, Obora, et al., 2019; Kanwal, Kramer, Mapakshi, et al., 2018; Lee, Lim, & Park, 2012; Lee, Park, Yun, et al., 2020; Michelotti, Machado, & Diehl, 2013)	Limited (Rosato, Bosetti, Levi, et al., 2013)
Inflammatory bowel disease	• Chronic, low-grade inflammation; oxidative stress; altered immune response, mediated by microbiome (Kim & Chang, 2014)	Yes (Munkholm, 2003)	Yes (Jung et al., 2015)
Obesity	• Metabolic syndrome; insulin resistance; chronic, low-grade inflammation; altered levels of adipocytokines (Bardou, Barkun, & Martel, 2013)	Yes (Karahalios et al., 2015; Levi, Kark, Barchana, et al., 2011; Win, Dowty, English, et al., 2011; Zhang, Wu, Giovannucci, et al., 2015; Zheng et al., 2017)	Limited and mixed (Chen, Zheng, Zong, et al., 2020; Gausman, Dornblaser, Anand, et al., 2019; Liu, Wu, Ng, et al., 2019; Low, Demb, Liu, et al., 2020)
Dyslipidemia	• Metabolic syndrome; prolonged exposure to dietary cholesterol; indirect influence via hormonal or metabolic pathways (Cruse, Lewin, & Clark, 1979; McKeown-Eyssen, 1994)	Mixed (Järvinen, Knekt, Hakulinen, et al., 2001; Kreger, Anderson, Schatzkin, et al., 1992; Sandler, Lyles, Peipins, et al., 1993; Strohmaier, Edlinger, Manjer, et al., 2013)	Limited (Chen et al., 2020)
Hypertension	• Metabolic syndrome; prolonged exposure inhibits apoptosis (Hamet, 1997)	Mixed (Ahmed, Schmitz, Anderson, et al., 2006; Stocks, Lukanova, Johansson, et al., 2008; Trevisan, Liu, Muti, et al., 2001)	Limited (Chen et al., 2020)

Adapted, with permission, from Stoffel, E. M., Murphy, C. C. (2019) Epidemiology and mechanisms of the increasing incidence of colon and rectal cancers in young adults. *Gastroenterology*, *158*, 341–353.

triglyceride levels (30–39 year olds: 2.00, 95% CI 1.26–3.16) were risk factors for advanced colonic neoplasia (where the reference group was average risk women >50). Chen et al. (2020) performed an analysis of US claims data from 2006 to 2015 including 4673 cases of EAO-CRC. They found that metabolic syndrome (defined as presence of three or more of the following: obesity, hypertension, hyperlipidemia or hyperglycemia/type II diabetes) was associated with a 1.25 (95% CI 1.09–1.43) increased odds of EAO-CRC. Low et al. (2020) compared 651 EAO-CRC cases among veterans to age-matched controls. As expected, they identified older age (OR 1.05, 95% CI 1.03–1.07) and male sex (OR 2.21. 95% CI 1.68–2.91) as risk factors for EAO-CRC and aspirin use was protective (OR 0.66, 95% CI 0.52–0.84). Interestingly, overweight and obesity was associated with a lower odds of CRC in this cohort (0.69, 95% CI 0.55–0.87), perhaps reflecting weight loss prior to EAO-CRC diagnosis. Similarly, Gausman et al. compared 269 EAO-CRC patients to matched controls and reported no association between obesity and EAO-CRC (Gausman et al., 2019). These data are limited by cross sectional design and do not address the important question of when in a patient's lifetime exposure to a risk factor influences cancer development later in life (Nimptsch & Wu, 2018).

There are multiple additional risk factors currently under consideration as possible drivers of EAO-CRC. Alterations in gut microbiota can induce changes in mutational patterns, gene expression, metabolism, local and systemic immune responses (Song & Chan, 2019). Thus, risk factors such as antibiotic use (particularly in childhood), antibiotics in the food chain, mode of birth delivery, breast feeding or periodontal disease, and unrecognized dietary toxins may be contributing (Stoffel & Murphy, 2019). The role of the microbiome in CRC pathogenesis and prognosis is discussed in detail in the chapter by Dr. Kai Wang and Dr. Mingyang Song. Regional variations in disease burden raise the possibility of water supply or ambient exposures, though evidence for these risk factors is limited (Dwyer, Murphy, Boland, et al., 2019). Ongoing epidemiologic research is needed to better characterize these risk factors.

7. Prevention and early detection of EAO-CRC

7.1 Improving identification of high-risk populations

7.1.1 Family history of CRC and adenomas

At least 30% of the population likely has an FDR with CRC or an advanced adenoma. Based on data demonstrating that having a FDR with a large or histologically advanced adenoma or CRC increases the lifetime risk of CRC by up to fourfold (Butterworth, Higgins, & Pharoah, 2006; Fuchs, Giovannucci, Colditz, et al., 1994; Lieberman, Prindiville, Weiss, et al., 2003; Lynch, Ahnen, Byers, et al., 2003), national guidelines agree that these individuals should initiate CRC screening at age 40 (NCCN, 2013; Rex et al., 2017; U.S. Preventive Services Task Force, 2008). Unfortunately, adherence to this recommendation in the 40–49 year old group is low. Tsai, Xirasagar, Li, et al. (2015) analyzed the National Health Interview data from 2005 and 2010 and found that overall CRC screening had increased by two- to threefold between 2005 and 2010, and colonoscopy use was slightly higher in FDRs of patients with CRC had compared to those without a family history (70% vs 68%). However, colonoscopy in those aged 40–49 with a FDR with CRC remained very low (15% in 2005; 41% in 2010). Screening for those with a FDR who has had an advanced adenoma are undoubtedly even lower. Improving identification of and screening in this population is an immediate step to addressing the rising CRC rates in the young.

There are a multitude of cited barriers at the provider, patient and systems level. There is inadequate family history collection and documentation, inadequate knowledge of associated risk and insufficient action on family history information. Fletcher et al. (2007) reported that only 39% of patients under the age of 50 were asked about family history of CRC. CRC family history was only documented in 59% of patient charts, but only half of these records included relative's age of CRC diagnosis. Of those who had a strong family history of CRC, only 46% were aware that they needed to start CRC screening earlier than 50. Not surprisingly, only 45% of these patients had undergone appropriate screening.

There are multiple resources available to help clinicians gather and act upon cancer family history. The National Colorectal Cancer Roundtable recently published a family history toolkit which is designed to improve clinicians' ability to systematically collect, document and act on family history of CRC and neoplastic polyps. This comprehensive, step-by-step guide can be customized to various medical practice settings (NCCRT, 2019).

7.1.2 Universal tumor screening for lynch syndrome

Numerous multi-disciplinary professional societies endorse screening all colorectal tumors for Lynch Syndrome with either immunohistochemistry for loss of mismatch repair protein expression and/or microsatellite instability testing (Evaluation of Genomic Applications in Practice and Prevention

(EGAPP) Working Group, 2009; Giardiello et al., 2014; NCCN, 2016; Sepulveda, Hamilton, Allegra, et al., 2017). Despite this recommendation, many healthcare institutions face challenges to incorporating universal tumor testing of CRC patients to screen for LS. Even at National Cancer Institute designated Comprehensive Cancer Centers, only 71% reported that they routinely perform MSI/IHC testing on CRCs to screen for LS, whereas only 36% and 15%, respectively, perform LS tumor screening at American College of Surgeons–accredited Community Hospital Comprehensive Cancer Programs and Community Hospital Cancer Programs (Beamer, Grant, Espenschied, et al., 2012). There is a need for continued efforts to overcome current challenges, including securing resources for a large-scale screening program to ensure patients are appropriately navigated through the system, ensuring access to genetics expertise for interpretation and action on tumor testing and dispelling misconceptions about cost and potential delays in care.

7.1.3 Risk-based screening

The challenge with risk-based screening is that, as discussed above, the specific risk factors for EAO-CRC are not fully elucidated, thus it is not surprising that available risk models have relatively weak discriminatory power (Ma & Ladabaum, 2014). Furthermore, many of the existing models have not been specifically applied to the EAO-CRC populations. Emerging models which include incorporation of clinical risk factors as well as genetic variants may provide more precise risk assessment (McGeoch, Saunders, Griffin, et al., 2019), however are far from immediate clinical application.

7.2 Avoiding delay in diagnosis

Another immediate opportunity to address EAO-CRC is to diagnose CRC as early as possible in symptomatic patients. Bleeding, abdominal pain, anemia and change in bowel habits are the most commonly reported symptoms of EAO-CRC. Since these symptoms are often caused by benign etiologies in young patients (hemorrhoids, irritable bowel, etc.), there may be a low index of suspicion among clinicians for CRC in people under age 50. This can unfortunately lead to significant delays in the diagnosis (Table 1). Although this delay in care is likely multifactorial, including patient-related delay in seeking care or difficulty accessing care, it is important to acknowledge delays in care from dismissal of symptoms by health care providers. Scant hematochezia is often dismissed as a benign anorectal source such as hemorrhoids. The presence of hemorrhoids does not exclude more proximal pathology and bleeding hemorrhoids can obscure subtle symptoms suggestive of a more proximal source (Bat, Pines, Rabau, et al., 1985). In one series of 145 patients presenting with rectal bleeding, primary care providers predicted an anal source (hemorrhoids or fissure) in 63 patients based on their history and physical examination. Of these 63 patients, 11 (17.5%) had a more proximal rectal or colonic source on lower endoscopy (Goulston, Cook, & Dent, 1986). Based on this data showing how unreliable clinical history and physical examination can be and the observation that up to 15% of CRCs occur under the age of 50, the ASGE recommends endoscopic evaluation for all patients presenting with bleeding symptoms (ASGE Standards of Practice Committee, Pasha, Shergill, et al., 2014). Similarly, the AGA published a practice update in 2020 recommending a diagnostic evaluation of the colon and rectum for all patients who present with symptoms of rectal bleeding, weight loss, change in bowel habit, abdominal pain or iron deficiency anemia (Boardman, Vilar, You, et al., 2020).

7.3 Changing the screening age

In response to the rising burden of CRC in young patients, the American Cancer Society (ACS) published a qualified recommendation in 2018 to initiate screening in all average risk patients at age 45 instead of 50 (Wolf et al., 2018). This recommendation was based on results from two simulation models from the Cancer Intervention and Surveillance Modeling Network (CISNET), also used by the U.S. Preventive Services Task Force (USPSTF) for their 2016 recommendations (U.S. Preventive Services Task Force et al., 2016); however, the ACS made adjustments to CISNET models to account for recent increasing in incidence rates of EAO-CRC. These models showed a favorable balance of benefit to harm. An update of the modeling report by CISNET drafted in 2020 found that of the 57 screening strategies that were considered efficient, the majority (47) began average-risk screening at age 45 (U.S. Preventive Services Task Force, 2020). Per 1000 individuals screened starting at age 45, all three models showed a favorable balance of life-years gained compared to complications. Based on this data, the USPSTF issued a draft Grade B recommendation in 2020 to consider screening average risk patients at age 45. The MSTF and the NCCN are currently also revisiting the starting age for average-risk screening as well. There are still many unanswered questions about how best to approach screening for average risk individuals under the age of 50 (Imperiale, Kahi, & Rex, 2018; Liang, Allison, Ladabaum, et al., 2018), though a recent analysis suggests that starting average-risk screening at age 45 costs $7700 and $33,900 per quality adjusted life year, fecal immunochemical testing and colonoscopy screening, respectively (Ladabaum, Mannalithara, Meester, et al., 2019). A detailed discussion about the approach to CRC screening can be found in Dr. Robert Smith's chapter.

8. EAO-CRC treatment and survivorship

For patients with pathogenic germline variants consistent with hereditary cancer syndromes, the specific approach to surgical management, adjuvant therapy and risk-reduction efforts in non-colorectal organs is reviewed in detail elsewhere (Giardiello et al., 2014; Herzig, Hardiman, Weiser, et al., 2017; Syngal et al., 2015).

Although there appears to be a different pattern of molecular profile in EAO-CRCs, these cancers are still a heterogeneous group and there is currently no specific tailored approach to treatment based on age alone (Lieu et al., 2019). EAO-CRCs are more likely to be MSI-H given the higher likelihood of Lynch Syndrome in younger patients. MSI-H status may impact the use of chemotherapy in stage II EAO-CRC patients, as well as the use of immunotherapy for patients with stage IV disease (Le, Uram, Wang, et al., 2015; Sargent, Marsoni, Monges, et al., 2010). Secondary to their young age and presumed improved performance status compared to older adults, younger patients tend to be treated more aggressively. Matched for stage and tumor characteristics, younger patients are more likely to be treated with systemic chemotherapy and multi-agent regimens compared to older patients (Kneuertz, Chang, Hu, et al., 2015; Manjelievskaia, Brown, McGlynn, et al., 2017). This more aggressive approach to therapy has not been associated with survival gain in stage II patients (RR 0.90, 95% CI 0.69–1.17), but led to a marginal gain for stage III (0.89, 0.81–0.97) and IV (0.84, 0.79–0.90) patients (Kneuertz et al., 2015). Although the risk of metachronous cancer may be higher in a younger patient, Kozak et al. concluded that extended colon resection is not warranted in younger patients given the low rate of metachronous cancers seen in their series (Kozak, Kalady, Gamaleldin, et al., 2017). There does not appear to be a substantial benefit to over-treating younger patients. Another analysis of 6284 patients from nine phase III studies in stage IV CRC showed that young age was modestly associated with poorer progression free survival but not overall survival or response rate in treated patients with advanced

CRC, and young patients had more nausea, but less diarrhea and neutropenia with chemotherapy in general (Blanke, Bot, Thomas, et al., 2011).

Long-term sequelae of treatment need to be carefully considered in EAO-CRC patients given the treatment impact on quality of life and financial toxicity. Because EAO-CRC patients are more likely to be diagnosed at later stages requiring multimodal therapies, survivors can face long-term functional deficits, symptoms and have significantly increased reports of anxiety, poor body image and embarrassment associated with bowel habits (Bailey, Tran Cao, Hu, et al., 2015). Surgical resection does not likely impact fertility, however there is a moderate risk of impaired fertility with systemic chemotherapy and/or radiation, depending on the type, dose and duration (Shandley & McKenzie, 2019). Current clinical guidelines recommend offering cryopreservation of sperm to all patients undergoing gonadotoxic chemotherapy (Oktay, Harvey, Partridge, et al., 2018). Embryo cryopreservation is the most established option for women, however unfertilized oocytes can also be preserved for women without a partner or whose beliefs preclude preservation of embryos. Additional approaches to preserving fertility include translocation of ovaries away from radiation fields and the use of gonadotropin-releasing hormone agonists to prevent chemotherapy induced ovarian failure (Oktay et al., 2018).

Other special considerations for young patients include financial and career counseling, psychological support, family planning and nutritional counseling. The components are integral to the multidisciplinary management of patients with EAO-CRC.

9. Conclusions

An increase in the incidence of EAO-CRC in the US has been well-documented, as well as increases in mortality rates for this group. A major concern is that this trend will continue to increase and become a serious health burden in people under 50. Increasing EAO-CRC has also been seen in several non-US populations. The underlying cause of this problem is unknown. Germline mutations in genes known to cause familial cancer syndromes are responsible for a minority of these cases (16–25%). This incidence underscores the need to test all EAO-CRC patients for germline mutations, but this does not explain most of the problem. Young patients with lower abdominal symptoms or rectal bleeding should be fully evaluated for the possibility of CRC. Family histories need to be taken and recorded in a more systematic fashion, perhaps using on-line tools outside of the medical provider's office and referrals to genetics need to be made for high-risk individuals. Methods should be developed to record the presence of advanced adenomas and this information needs to be shared with first-degree relatives along with appropriate screening recommendations. Research should be directed at determining the biological basis of this problem, which may require an analysis of mutational signatures in these tumors and possibly tracing those patterns back to changes in the microbiome or the presence of unexpected mutagens in the diet or changes in the fecal metabolome. Finally, the medical community needs to determine the optimal age to initiate screening for CRC and the most efficient and cost-effective means to do this.

Disclosure statements

SGP: no conflicts of interest. CM: no conflicts of interest. CL: no conflicts of interest. HH: Scientific advisory board for Genome Medical, InVitae Genetics, and Promega, has conducted collaborative research with Myriad Genetics Laboratories, Inc., Ambry Genetics, and InVitae Genetics, and has stock in Genome Medical.

References

Abdelsattar, Z. M., Wong, S. L., Regenbogen, S. E., et al. (2016). Colorectal cancer outcomes and treatment patterns in patients too young for average-risk screening. *Cancer, 122*, 929–934.

Abualkhair, W. H., Zhou, M., Ochoa, C. O., et al. (2020). Geographic and intra-racial disparities in early-onset colorectal cancer in the SEER 18 registries of the United States. *Cancer Medicine, 9*, 9150–9159.

Ahmed, R. L., Schmitz, K. H., Anderson, K. E., et al. (2006). The metabolic syndrome and risk of incident colorectal cancer. *Cancer, 107*, 28–36.

Ait Ouakrim, D., Lockett, T., Boussioutas, A., et al. (2013). Screening participation for people at increased risk of colorectal cancer due to family history: A systematic review and meta-analysis. *Familial Cancer, 12*, 459–472.

Alvaro, E., Cano, J. M., Garcia, J. L., et al. (2019). Clinical and molecular comparative study of colorectal cancer based on age-of-onset and tumor location: Two main criteria for subclassifying colorectal cancer. *International Journal of Molecular Sciences, 20*, 968.

Antelo, M., Balaguer, F., Shia, J., et al. (2012). A high degree of LINE-1 hypomethylation is a unique feature of early-onset colorectal cancer. *PLoS One, 7*, e45357.

ASGE Standards of Practice Committee, Pasha, S. F., Shergill, A., et al. (2014). The role of endoscopy in the patient with lower GI bleeding. *Gastrointestinal Endoscopy, 79*, 875–885.

Ashktorab, H., Kupfer, S. S., Brim, H., et al. (2017). Racial disparity in gastrointestinal cancer risk. *Gastroenterology, 153*, 910–923.

Bailey, C. E., Hu, C. Y., You, Y. N., et al. (2015). Increasing disparities in the age-related incidences of colon and rectal cancers in the United States, 1975-2010. *JAMA Surgery, 150*, 17–22.

Bailey, C. E., Tran Cao, H. S., Hu, C. Y., et al. (2015). Functional deficits and symptoms of long-term survivors of colorectal cancer treated by multimodality therapy differ by age at diagnosis. *Journal of Gastrointestinal Surgery, 19*, 180–188. discussio 188.

Ballester, V., Rashtak, S., & Boardman, L. (2016). Clinical and molecular features of young-onset colorectal cancer. *World Journal of Gastroenterology*, *22*, 1736–1744.
Bardou, M., Barkun, A. N., & Martel, M. (2013). Obesity and colorectal cancer. *Gut*, *62*, 933–947.
Bat, L., Pines, A., Rabau, M., et al. (1985). Colonoscopic findings in patients with hemorrhoids, rectal bleeding and normal rectoscopy. *Israel Journal of Medical Sciences*, *21*, 139–141.
Beamer, L. C., Grant, M. L., Espenschied, C. R., et al. (2012). Reflex immunohistochemistry and microsatellite instability testing of colorectal tumors for Lynch syndrome among US cancer programs and follow-up of abnormal results. *Journal of Clinical Oncology*, *30*, 1058–1063.
Bhandari, A., Woodhouse, M., & Gupta, S. (2017). Colorectal cancer is a leading cause of cancer incidence and mortality among adults younger than 50 years in the USA: A SEER-based analysis with comparison to other young-onset cancers. *Journal of Investigative Medicine*, *65*, 311–315.
Bhupathiraju, S. N., & Hu, F. B. (2016). Epidemiology of obesity and diabetes and their cardiovascular complications. *Circulation Research*, *118*, 1723–1735.
Birney, E., Stamatoyannopoulos, J. A., Dutta, A., et al. (2007). Identification and analysis of functional elements in 1% of the human genome by the ENCODE pilot project. *Nature*, *447*, 799–816.
Blanke, C. D., Bot, B. M., Thomas, D. M., et al. (2011). Impact of young age on treatment efficacy and safety in advanced colorectal cancer: A pooled analysis of patients from nine first-line phase III chemotherapy trials. *Journal of Clinical Oncology*, *29*, 2781–2786.
Boardman, L. A., Johnson, R. A., Petersen, G. M., et al. (2007). Higher frequency of diploidy in young-onset microsatellite-stable colorectal cancer. *Clinical Cancer Research: An Official Journal of the American Association for Cancer Research*, *13*, 2323–2328.
Boardman, L. A., Vilar, E., You, Y. N., et al. (2020). AGA Clinical practice update on young adult-onset colorectal cancer diagnosis and management: Expert review. *Clinical Gastroenterology and Hepatology*, *18*, 2415–2424.
Botteri, E., Iodice, S., Bagnardi, V., et al. (2008). Smoking and colorectal cancer: A meta-analysis. *JAMA*, *300*, 2765–2778.
Brenner, H., & Chen, C. (2018). The colorectal cancer epidemic: Challenges and opportunities for primary, secondary and tertiary prevention. *British Journal of Cancer*, *119*, 785–792.
Butterworth, A. S., Higgins, J. P., & Pharoah, P. (2006). Relative and absolute risk of colorectal cancer for individuals with a family history: A meta-analysis. *European Journal of Cancer*, *42*, 216–227.
Casa, V., & Gabellini, D. (2012). A repetitive elements perspective in Polycomb epigenetics. *Frontiers in Genetics*, *3*, 199.
Chan, T. L., Curtis, L. C., Leung, S. Y., et al. (2001). Early-onset colorectal cancer with stable microsatellite DNA and near-diploid chromosomes. *Oncogene*, *20*, 4871–4876.
Chang, D. T., Pai, R. K., Rybicki, L. A., et al. (2012). Clinicopathologic and molecular features of sporadic early-onset colorectal adenocarcinoma: An adenocarcinoma with frequent signet ring cell differentiation, rectal and sigmoid involvement, and adverse morphologic features. *Modern Pathology*, *25*, 1128–1139.
Chen, F. W., Sundaram, V., Chew, T. A., et al. (2017a). Advanced-stage colorectal cancer in persons younger than 50 years not associated with longer duration of symptoms or time to diagnosis. *Clinical Gastroenterology and Hepatology*, *15*, 728–737 e3.
Chen, F. W., Sundaram, V., Chew, T. A., et al. (2017b). Low prevalence of criteria for early screening in young-onset colorectal cancer. *American Journal of Preventive Medicine*, *53*, 933–934.

Chen, H., Zheng, X., Zong, X., et al. (2020). Metabolic syndrome, metabolic comorbid conditions and risk of early-onset colorectal cancer. *Gut*. Online Ahead of Print.
Cho, E., Smith-Warner, S. A., Ritz, J., et al. (2004). Alcohol intake and colorectal cancer: A pooled analysis of 8 cohort studies. *Annals of Internal Medicine*, *140*, 603–613.
Cottet, V., Pariente, A., Nalet, B., et al. (2007). Colonoscopic screening of first-degree relatives of patients with large adenomas: Increased risk of colorectal tumors. *Gastroenterology*, *133*, 1086–1092.
Cruse, P., Lewin, M., & Clark, C. G. (1979). Dietary cholesterol is co-carcinogenic for human colon cancer. *Lancet*, *1*, 752–755.
Deng, L., Gui, Z., Zhao, L., et al. (2012). Diabetes mellitus and the incidence of colorectal cancer: An updated systematic review and meta-analysis. *Digestive Diseases and Sciences*, *57*, 1576–1585.
Dharwadkar, P., Greenan, G., Singal, A. G., et al. (2021). Is colorectal cancer in patients younger than 50 years of age the same disease as in older patients? *Clinical Gastroenterology and Hepatology*, *19*, 192–194.
Dozois, E. J., Boardman, L. A., Suwanthanma, W., et al. (2008). Young-onset colorectal cancer in patients with no known genetic predisposition: Can we increase early recognition and improve outcome? *Medicine (Baltimore)*, *87*, 259–263.
Dwyer, A. J., Murphy, C. C., Boland, C. R., et al. (2019). A summary of the fight colorectal cancer working meeting: Exploring risk factors and etiology of sporadic early-age onset colorectal cancer. *Gastroenterology*, *157*, 280–288.
Edwards, B. K., Noone, A. M., Mariotto, A. B., et al. (2014). Annual report to the nation on the status of cancer, 1975-2010, featuring prevalence of comorbidity and impact on survival among persons with lung, colorectal, breast, or prostate cancer. *Cancer*, *120*, 1290–1314.
Evaluation of Genomic Applications in Practice and Prevention (EGAPP) Working Group. (2009). Recommendations from the EGAPP working group: Genetic testing strategies in newly diagnosed individuals with colorectal cancer aimed at reducing morbidity and mortality from Lynch syndrome in relatives. *Genetics in Medicine*, *11*, 35–41.
Fedirko, V., Tramacere, I., Bagnardi, V., et al. (2011). Alcohol drinking and colorectal cancer risk: An overall and dose-response meta-analysis of published studies. *Annals of Oncology*, *22*, 1958–1972.
Feng, Q., Moran, J. V., Kazazian, H. H., Jr., et al. (1996). Human L1 retrotransposon encodes a conserved endonuclease required for retrotransposition. *Cell*, *87*, 905–916.
Fletcher, R. H., Lobb, R., Bauer, M. R., et al. (2007). Screening patients with a family history of colorectal cancer. *Journal of General Internal Medicine*, *22*, 508–513.
Fuchs, C. S., Giovannucci, E. L., Colditz, G. A., et al. (1994). A prospective study of family history and the risk of colorectal cancer. *The New England Journal of Medicine*, *331*, 1669–1674.
Gausman, V., Dornblaser, D., Anand, S., et al. (2019). Risk factors associated with early-onset colorectal cancer. *Clinical Gastroenterology and Hepatology*, *18*, 2752–2759.
Georgiou, A., Khakoo, S., Edwards, P., et al. (2019). Outcomes of patients with early onset colorectal cancer treated in a UK Specialist Cancer Center. *Cancers (Basel)*, *11*, 1558.
Giardiello, F. M., Allen, J. I., Axilbund, J. E., et al. (2014). Guidelines on genetic evaluation and management of Lynch syndrome: A consensus statement by the U.S. Multi-Society Task Force on colorectal cancer. *Gastrointestinal Endoscopy*, *80*, 197–220.
Goel, A., Nguyen, T. P., Leung, H. C., et al. (2011). De novo constitutional MLH1 epimutations confer early-onset colorectal cancer in two new sporadic Lynch syndrome cases, with derivation of the epimutation on the paternal allele in one. *International Journal of Cancer*, *128*, 869–878.

Goel, A., Xicola, R. M., Nguyen, T. P., et al. (2010). Aberrant DNA methylation in hereditary nonpolyposis colorectal cancer without mismatch repair deficiency. *Gastroenterology, 138*, 1854–1862.

Goulston, K. J., Cook, I., & Dent, O. F. (1986). How important is rectal bleeding in the diagnosis of bowel cancer and polyps? *Lancet, 2*, 261–265.

Guinney, J., Dienstmann, R., Wang, X., et al. (2015). The consensus molecular subtypes of colorectal cancer. *Nature Medicine, 21*, 1350–1356.

Hamaguchi, M., Hashimoto, Y., Obora, A., et al. (2019). Non-alcoholic fatty liver disease with obesity as an independent predictor for incident gastric and colorectal cancer: A population-based longitudinal study. *BMJ Open Gastroenterology, 6*, e000295.

Hamet, P. (1997). Cancer and hypertension: A potential for crosstalk? *Journal of Hypertension, 15*, 1573–1577.

Hampel, H., & de la Chapelle, A. (2011). The search for unaffected individuals with Lynch syndrome: Do the ends justify the means? *Cancer Prevention Research (Philadelphia, Pa.), 4*, 1–5.

Herzig, D., Hardiman, K., Weiser, M., et al. (2017). The American Society of Colon and Rectal Surgeons clinical practice guidelines for the management of inherited polyposis syndromes. *Diseases of the Colon and Rectum, 60*, 881–894.

Hitchins, M. P., Rapkins, R. W., Kwok, C. T., et al. (2011). Dominantly inherited constitutional epigenetic silencing of MLH1 in a cancer-affected family is linked to a single nucleotide variant within the 5'UTR. *Cancer Cell, 20*, 200–213.

Hohjoh, H., & Singer, M. F. (1997). Sequence-specific single-strand RNA binding protein encoded by the human LINE-1 retrotransposon. *The EMBO Journal, 16*, 6034–6043.

Holowatyj, A. N., Ruterbusch, J. J., Rozek, L. S., et al. (2016). Racial/ethnic disparities in survival among patients with young-onset colorectal cancer. *Journal of Clinical Oncology, 34*, 2148–2156.

Hong, Y., Ho, K. S., Eu, K. W., et al. (2007). A susceptibility gene set for early onset colorectal cancer that integrates diverse signaling pathways: Implication for tumorigenesis. *Clinical Cancer Research: An Official Journal of the American Association for Cancer Research, 13*, 1107–1114.

Huang, Y., Cai, X., Qiu, M., et al. (2014). Prediabetes and the risk of cancer: A meta-analysis. *Diabetologia, 57*, 2261–2269.

Imperiale, T. F., Kahi, C. J., & Rex, D. K. (2018). Lowering the starting age for colorectal cancer screening to 45 years: Who will come…and should they? *Clinical Gastroenterology and Hepatology, 16*, 1541–1544.

Järvinen, R., Knekt, P., Hakulinen, T., et al. (2001). Dietary fat, cholesterol and colorectal cancer in a prospective study. *British Journal of Cancer, 85*, 357–361.

Jung, Y. S., Ryu, S., Chang, Y., et al. (2015). Risk factors for colorectal neoplasia in persons aged 30 to 39 years and 40 to 49 years. *Gastrointestinal Endoscopy, 81*, 637–645.e7.

Kanwal, F., Kramer, J. R., Mapakshi, S., et al. (2018). Risk of hepatocellular cancer in patients with non-alcoholic fatty liver disease. *Gastroenterology, 155*, 1828–1837.e2.

Karahalios, A., English, D. R., & Simpson, J. A. (2015). Weight change and risk of colorectal cancer: A systematic review and meta-analysis. *American Journal of Epidemiology, 181*, 832–845.

Kazazian, H. H., Jr. (2004). Mobile elements: Drivers of genome evolution. *Science, 303*, 1626–1632.

Kim, E. R., & Chang, D. K. (2014). Colorectal cancer in inflammatory bowel disease: The risk, pathogenesis, prevention and diagnosis. *World Journal of Gastroenterology, 20*, 9872–9881.

Kirzin, S., Marisa, L., Guimbaud, R., et al. (2014). Sporadic early-onset colorectal cancer is a specific sub-type of cancer: A morphological, molecular and genetics study. *PloS one, 9*, e103159.

Kneuertz, P. J., Chang, G. J., Hu, C. Y., et al. (2015). Overtreatment of young adults with colon cancer: More intense treatments with unmatched survival gains. *JAMA Surgery*, *150*, 402–409.

Kozak, V. N., Kalady, M. F., Gamaleldin, M. M., et al. (2017). Colorectal surveillance after segmental resection for young onset colorectal cancer: Is there evidence for extended resection? *Colorectal Disease*, *19*, 386–392.

Kreger, B. E., Anderson, K. M., Schatzkin, A., et al. (1992). Serum cholesterol level, body mass index, and the risk of colon cancer. The Framingham Study. *Cancer*, *70*, 1038–1043.

Ladabaum, U., Mannalithara, A., Meester, R. G. S., et al. (2019). Cost-effectiveness and national effects of initiating colorectal cancer screening for average-risk persons at age 45 years instead of 50 years. *Gastroenterology*, *157*, 137–148.

Lander, E. S. (2011). Initial impact of the sequencing of the human genome. *Nature*, *470*, 187–197.

Larsson, S. C., Orsini, N., & Wolk, A. (2005). Diabetes mellitus and risk of colorectal cancer: A meta-analysis. *Journal of the National Cancer Institute*, *97*, 1679–1687.

Le, D. T., Uram, J. N., Wang, H., et al. (2015). PD-1 blockade in tumors with mismatch-repair deficiency. *The New England Journal of Medicine*, *372*, 2509–2520.

Lee, Y. I., Lim, Y. S., & Park, H. S. (2012). Colorectal neoplasms in relation to non-alcoholic fatty liver disease in Korean women: A retrospective cohort study. *Journal of Gastroenterology and Hepatology*, *27*, 91–95.

Lee, J.-M., Park, Y.-M., Yun, J.-S., et al. (2020). The association between nonalcoholic fatty liver disease and esophageal, stomach, or colorectal cancer: National population-based cohort study. *PLoS One*, *15*, e0226351.

Levi, Z., Kark, J. D., Barchana, M., et al. (2011). Measured body mass index in adolescence and the incidence of colorectal cancer in a cohort of 1.1 million males. *Cancer Epidemiology, Biomarkers & Prevention*, *20*, 2524–2531.

Liang, P. S., Allison, J., Ladabaum, U., et al. (2018). Potential intended and unintended consequences of recommending initiation of colorectal cancer screening at age 45 years. *Gastroenterology*, *155*, 950–954.

Liang, J. T., Huang, K. C., Cheng, A. L., et al. (2003). Clinicopathological and molecular biological features of colorectal cancer in patients less than 40 years of age. *The British Journal of Surgery*, *90*, 205–214.

Lieberman, D. A., Prindiville, S., Weiss, D. G., et al. (2003). Risk factors for advanced colonic neoplasia and hyperplastic polyps in asymptomatic individuals. *JAMA*, *290*, 2959–2967.

Lieu, C. H., Golemis, E. A., Serebriiskii, I. G., et al. (2019). Comprehensive genomic landscapes in early and later onset colorectal cancer. *Clinical Cancer Research: An Official Journal of the American Association for Cancer Research*, *25*, 5852–5858.

Liu, P. H., Wu, K., Ng, K., et al. (2019). Association of obesity with risk of early-onset colorectal cancer among women. *JAMA Oncology*, *5*, 37–44.

Low, E. E., Demb, J., Liu, L., et al. (2020). Risk factors for early-onset colorectal cancer. *Gastroenterology*, *159*, 492–501.

Lowery, J. T., Ahnen, D. J., Schroy, P. C., 3rd, et al. (2016). Understanding the contribution of family history to colorectal cancer risk and its clinical implications: A state-of-the-science review. *Cancer*, *122*, 2633–2645.

Lynch, K. L., Ahnen, D. J., Byers, T., et al. (2003). First-degree relatives of patients with advanced colorectal adenomas have an increased prevalence of colorectal cancer. *Clinical Gastroenterology and Hepatology*, *1*, 96–102.

Ma, G. K., & Ladabaum, U. (2014). Personalizing colorectal cancer screening: A systematic review of models to predict risk of colorectal neoplasia. *Clinical Gastroenterology and Hepatology*, *12*, 1624–34 e1.

Manjelievskaia, J., Brown, D., McGlynn, K. A., et al. (2017). Chemotherapy use and survival among young and middle-aged patients with colon cancer. *JAMA Surgery, 152*, 452–459.

May, F. P., Yang, L., Corona, E., et al. (2020). Disparities in colorectal cancer screening in the United States before and after implementation of the Affordable Care Act. *Clinical Gastroenterology and Hepatology, 18*, 1796–1804 e2.

McGeoch, L., Saunders, C. L., Griffin, S. J., et al. (2019). Risk prediction models for colorectal cancer incorporating common genetic variants: A systematic review. *Cancer Epidemiology, Biomarkers & Prevention, 28*, 1580–1593.

McKeown-Eyssen, G. (1994). Epidemiology of colorectal cancer revisited: Are serum triglycerides and/or plasma glucose associated with risk? *Cancer Epidemiology, Biomarkers & Prevention, 3*, 687–695.

Michelotti, G. A., Machado, M. V., & Diehl, A. M. (2013). NAFLD, NASH and liver cancer. *Nature Reviews Gastroenterology & Hepatology, 10*, 656–665.

Mizoue, T., Inoue, M., Wakai, K., et al. (2008). Alcohol drinking and colorectal cancer in Japanese: A pooled analysis of results from five cohort studies. *American Journal of Epidemiology, 167*, 1397–1406.

Mork, M. E., You, Y. N., Ying, J., et al. (2015). High prevalence of hereditary cancer syndromes in adolescents and young adults with colorectal cancer. *Journal of Clinical Oncology, 33*, 3544–3549.

Muhidin, S. O., Magan, A. A., Osman, K. A., et al. (2012). The relationship between nonalcoholic fatty liver disease and colorectal cancer: The future challenges and outcomes of the metabolic syndrome. *Journal of Obesity, 2012*, 637538.

Munkholm, P. (2003). The incidence and prevalence of colorectal cancer in inflammatory bowel disease. *Alimentary Pharmacology & Therapeutics, 18*, 1–5.

Murphy, C. C., Wallace, K., Sandler, R. S., et al. (2019). Racial disparities in incidence of Young-onset colorectal cancer and patient survival. *Gastroenterology, 156*, 958–965.

Myers, E. A., Feingold, D. L., Forde, K. A., et al. (2013). Colorectal cancer in patients under 50 years of age: A retrospective analysis of two institutions' experience. *World Journal of Gastroenterology, 19*, 5651–5657.

National Center for Health Statistics. (2017). *Health, United States, 2016: With Chartbook on long-term trends in health. Hyattsville, MD.* 2016.

NCCN. (2013). *NCCN clinical practice guidelines in oncology: Colorectal cancer screening. Volume 2014.* http://www.nccn.org/professionals/physician_gls/pdf/colorectal_screening.pdf.

NCCN. (2016). *Genetic/familial high-risk assessment: Colorectal.* National Comprehensive Cancer Network. https://www.nccn.org/professionals/physician_gls/default.aspx.

NCCN. (2020a). *Genetic/familial high-risk assessment: Colorectal.* National Comprehensive Cancer Network. https://www.nccn.org/professionals/physician_gls/default.aspx.

NCCN. (2020b). *Colorectal cancer screening guidelines.* National Comprehensive Cancer Network. https://www.nccn.org/professionals/physician_gls/default.aspx.

NCCRT. (2019). *Risk assessment and screening toolkit to detect familial, hereditary and early onset colorectal cancer.* https://nccrt.org/resource/risk-assessment-and-screening-toolkit-to-detect-familial-hereditary-and-early-onset-colorectal-cancer/.

Nguyen, L. H., Liu, P. H., Zheng, X., et al. (2018). Sedentary behaviors, TV viewing time, and risk of young-onset colorectal cancer. *JNCI Cancer Spectrum, 2*, pky073.

Nimptsch, K., & Wu, K. (2018). Is timing important? The role of diet and lifestyle during early life on colorectal neoplasia. *Current Colorectal Cancer Reports, 14*, 1–11.

Oktay, K., Harvey, B. E., Partridge, A. H., et al. (2018). Fertility preservation in patients with cancer: ASCO clinical practice guideline update. *Journal of Clinical Oncology, 36*, 1994–2001.

Pan, S. Y., & DesMeules, M. (2009). Energy intake, physical activity, energy balance, and cancer: Epidemiologic evidence. *Methods in Molecular Biology, 472*, 191–215.

Paskett, E. D., Bernardo, B. M., Young, G. S., et al. (2020). Comparative effectiveness of two interventions to increase colorectal cancer screening for those at increased risk based on family history: Results of a randomized trial. *Cancer Epidemiology, Biomarkers & Prevention*, *29*, 3–9.

Patel, S. G., & Ahnen, D. J. (2012). Familial colon cancer syndromes: An update of a rapidly evolving field. *Current Gastroenterology Reports*, *14*, 428–438.

Patel, S. G., Ahnen, D. J., Gumidyala, A., et al. (2020). Poor knowledge of personal and familial colorectal cancer risk and screening recommendations associated with advanced colorectal polyps. *Digestive Diseases and Sciences*, *65*, 2542–2550.

Pearlman, R., Frankel, W. L., Swanson, B., et al. (2017). Prevalence and spectrum of germline cancer susceptibility gene mutations among patients with early-onset colorectal cancer. *JAMA Oncology*, *3*, 464–471.

Perea, J., Garcia, J. L., Perez, J., et al. (2017). NOMO-1 gene is deleted in early-onset colorectal cancer. *Oncotarget*, *8*, 24429–24436.

Rex, D. K., Boland, C. R., Dominitz, J. A., et al. (2017). Colorectal cancer screening: Recommendations for physicians and patients from the U.S. multi-society task force on colorectal cancer. *Gastroenterology*, *153*, 307–323.

Rho, Y. S., Gilabert, M., Polom, K., et al. (2017). Comparing clinical characteristics and outcomes of young-onset and late-onset colorectal cancer: An international collaborative study. *Clinical Colorectal Cancer*, *16*, 334–342.

Rosato, V., Bosetti, C., Levi, F., et al. (2013). Risk factors for young-onset colorectal cancer. *Cancer Causes & Control*, *24*, 335–341.

Sandler, R. S., Lyles, C. M., Peipins, L. A., et al. (1993). Diet and risk of colorectal adenomas: Macronutrients, cholesterol, and fiber. *Journal of the National Cancer Institute*, *85*, 884–891.

Sargent, D. J., Marsoni, S., Monges, G., et al. (2010). Defective mismatch repair as a predictive marker for lack of efficacy of fluorouracil-based adjuvant therapy in colon cancer. *Journal of Clinical Oncology*, *28*, 3219–3226.

Scappaticcio, L., Maiorino, M. I., Bellastella, G., et al. (2017). Insights into the relationships between diabetes, prediabetes, and cancer. *Endocrine*, *56*, 231–239.

Scott, R. B., Rangel, L. E., Osler, T. M., et al. (2016). Rectal cancer in patients under the age of 50 years: The delayed diagnosis. *American Journal of Surgery*, *211*, 1014–1018.

Sepulveda, A. R., Hamilton, S. R., Allegra, C. J., et al. (2017). Molecular biomarkers for the evaluation of colorectal cancer. *American Journal of Clinical Pathology*, *147*, 221–260.

Shandley, L. M., & McKenzie, L. J. (2019). Recent advances in fertility preservation and counseling for reproductive-aged women with colorectal cancer: A systematic review. *Diseases of the Colon and Rectum*, *62*, 762–771.

Siegel, R. L., Fedewa, S. A., Anderson, W. F., et al. (2017). Colorectal cancer incidence patterns in the United States, 1974-2013. *Journal of the National Cancer Institute*, *109*, 01.

Siegel, R. L., Medhanie, G. A., Fedewa, S. A., et al. (2019). State variation in early-onset colorectal cancer in the United States, 1995-2015. *Journal of the National Cancer Institute*, *111*, 1104–1106.

Siegel, R. L., Miller, K. D., Fedewa, S. A., et al. (2017). Colorectal cancer statistics, 2017. *CA: A Cancer Journal for Clinicians*, *67*, 177–193.

Siegel, R. L., Miller, K. D., Goding Sauer, A., et al. (2020). Colorectal cancer statistics, 2020. *CA: A Cancer Journal for Clinicians*, *70*, 145–164.

Siegel, R. L., Miller, K. D., & Jemal, A. (2017). Colorectal cancer mortality rates in adults aged 20 to 54 years in the United States, 1970–2014. *JAMA*, *318*, 572–574.

Siegel, R. L., Torre, L. A., Soerjomataram, I., et al. (2019). Global patterns and trends in colorectal cancer incidence in young adults. *Gut*, *68*, 2179–2185.

Song, M., & Chan, A. T. (2019). Environmental factors, gut microbiota, and colorectal cancer prevention. *Clinical Gastroenterology and Hepatology*, *17*, 275–289.

Song, M., Garrett, W. S., & Chan, A. T. (2015). Nutrients, foods, and colorectal cancer prevention. *Gastroenterology, 148*, 1244–60 e16.

Stocks, T., Lukanova, A., Johansson, M., et al. (2008). Components of the metabolic syndrome and colorectal cancer risk; a prospective study. *International Journal of Obesity, 32*, 304–314.

Stoffel, E. M., Koeppe, E., Everett, J., et al. (2018). Germline genetic features of young individuals with colorectal cancer. *Gastroenterology, 154*, 897–905.e1.

Stoffel, E. M., Mangu, P. B., Limburg, P. J., et al. (2015). Hereditary colorectal cancer syndromes: American Society of Clinical Oncology clinical practice guideline endorsement of the familial risk-colorectal cancer: European Society for Medical Oncology clinical practice guidelines. *Journal of Oncology Practice/ American Society of Clinical Oncology, 11*, e437–e441.

Stoffel, E. M., & Murphy, C. C. (2019). Epidemiology and mechanisms of the increasing incidence of colon and rectal cancers in young adults. *Gastroenterology, 158*, 341–353.

Strohmaier, S., Edlinger, M., Manjer, J., et al. (2013). Total serum cholesterol and cancer incidence in the Metabolic syndrome and Cancer Project (Me-Can). *PLoS One, 8*, e54242.

Strum, W. B., & Boland, C. R. (2019). Clinical and genetic characteristics of colorectal cancer in persons under 50 years of age: A review. *Digestive Diseases and Sciences, 54*, 3059–3065.

Syngal, S., Brand, R. E., Church, J. M., et al. (2015). ACG clinical guideline: Genetic testing and management of hereditary gastrointestinal cancer syndromes. *The American Journal of Gastroenterology, 110*, 223–262.

Teng, A., Lee, D. Y., Cai, J., et al. (2016). Patterns and outcomes of colorectal cancer in adolescents and young adults. *The Journal of Surgical Research, 205*, 19–27.

Thompson, A. B., Mowery, P. D., Tebes, J. K., et al. (2017). Time trends in smoking onset by sex and race/ethnicity among adolescents and young adults: Findings from the 2006-2013 National Survey on Drug Use and Health. *Nicotine & Tobacco Research, 20*, 312–320.

Toh, M. R., Chiang, J. B., Chong, S. T., et al. (2018). Germline pathogenic variants in homologous recombination and DNA repair genes in an Asian cohort of young-onset colorectal cancer. *JNCI Cancer Spectrum, 2*, pky054.

Trevisan, M., Liu, J., Muti, P., et al. (2001). Markers of insulin resistance and colorectal cancer mortality. *Cancer Epidemiology, Biomarkers & Prevention, 10*, 937–941.

Tsai, M. H., Xirasagar, S., Li, Y. J., et al. (2015). Colonoscopy screening among US adults aged 40 or older with a family history of colorectal cancer. *Preventing Chronic Disease, 12*, E80.

Tuohy, T. M., Rowe, K. G., Mineau, G. P., et al. (2014). Risk of colorectal cancer and adenomas in the families of patients with adenomas: A population-based study in Utah. *Cancer, 120*, 35–42.

U.S. Preventive Services Task Force. (2008). Screening for colorectal cancer: U.S. preventive services task force recommendation statement. *Annals of Internal Medicine, 149*, 627–637.

U.S. Preventive Services Task Force. (2020). *Colorectal cancer screening: A decision analysis for the U.S. preventive services task force (Draft). Volume 2021.* https://uspreventiveservicestaskforce.org/uspstf/draft-recommendation/colorectal-cancer-screening3.

U.S. Preventive Services Task Force, Bibbins-Domingo, K., Grossman, D. C., Curry, S. J., et al. (2016). Screening for colorectal cancer: US preventive services task force recommendation statement. *JAMA, 315*, 2564–2575.

Wang, R., Wang, M. J., & Ping, J. (2015). Clinicopathological features and survival outcomes of colorectal cancer in young versus elderly: A population-based cohort study of SEER 9 registries data (1988-2011). *Medicine (Baltimore), 94*, e1402.

Willauer, A. N., Liu, Y., Pereira, A. A. L., et al. (2019). Clinical and molecular characterization of early-onset colorectal cancer. *Cancer*, *125*, 2002–2010.

Win, A. K., Dowty, J. G., English, D. R., et al. (2011). Body mass index in early adulthood and colorectal cancer risk for carriers and non-carriers of germline mutations in DNA mismatch repair genes. *British Journal of Cancer*, *105*, 162–169.

Winawer, S. J., Zauber, A. G., Gerdes, H., et al. (1996). Risk of colorectal cancer in the families of patients with adenomatous polyps. National Polyp Study Workgroup. *The New England Journal of Medicine*, *334*, 82–87.

Wiseman, K. P., Silver, M. I., Klabunde, C. N., et al. (2020). Accuracy of self-reported colonic polyps: Results from the prostate, lung, colorectal, and ovarian screening trial study of colonoscopy utilization. *Cancer Epidemiology, Biomarkers & Prevention*, *29*, 982–989.

Wolf, A. M. D., Fontham, E. T. H., Church, T. R., et al. (2018). Colorectal cancer screening for average-risk adults: 2018 guideline update from the American Cancer Society. *CA: A Cancer Journal for Clinicians*, *68*, 250–281.

Wolin, K. Y., Yan, Y., Colditz, G. A., et al. (2009). Physical activity and colon cancer prevention: A meta-analysis. *British Journal of Cancer*, *100*, 611–616.

You, Y. N., Xing, Y., Feig, B. W., et al. (2012). Young-onset colorectal cancer: Is it time to pay attention? *Archives of Internal Medicine*, *172*, 287–289.

Yuhara, H., Steinmaus, C., Cohen, S. E., et al. (2011). Is diabetes mellitus an independent risk factor for colon cancer and rectal cancer? *The American Journal of Gastroenterology*, *106*, 1911–1921. quiz 1922.

Yurgelun, M. B., Kulke, M. H., Fuchs, C. S., et al. (2017). Cancer susceptibility gene mutations in individuals with colorectal cancer. *Journal of Clinical Oncology*, *35*, 1086–1095.

Zahnd, W. E., Gomez, S. L., Steck, S. E., et al. (2021). Rural-urban and racial/ethnic trends and disparities in early-onset and average-onset colorectal cancer. *Cancer*, *127*, 239–248.

Zhang, X., Wu, K., Giovannucci, E. L., et al. (2015). Early life body fatness and risk of colorectal cancer in U.S. women and men-results from two large cohort studies. *Cancer Epidemiology, Biomarkers & Prevention*, *24*, 690–697.

Zheng, X., Hur, J., Nguyen, L. H., et al. (2020). Comprehensive assessment of diet quality and risk of precursors of early-onset colorectal cancer. *Journal of the National Cancer Institute*. Online Ahead of Print.

Zheng, J., Zhao, M., Li, J., et al. (2017). Obesity-associated digestive cancers: A review of mechanisms and interventions. *Tumour Biology*, *39*, 1010428317695020.

CHAPTER TWO

Treatment-induced evolutionary dynamics in nonmetastatic locally advanced rectal adenocarcinoma

Seth I. Felder[a,*], Jason B. Fleming[a], and Robert A. Gatenby[b]

[a]Department of Gastrointestinal Oncology, H. Lee Moffitt Cancer Center and Research Institute, Tampa, FL, United States
[b]Integrated Mathematical Oncology Department, H. Lee Moffitt Cancer Center and Research Institute, Tampa, FL, United States
*Corresponding author: e-mail address: seth.felder@moffitt.org

Contents

Advances in Cancer Research, Volume 151
ISSN 0065-230X
https://doi.org/10.1016/bs.acr.2021.02.003

Abstract

Multi-modal treatment of non-metastatic locally advanced rectal adenocarcinoma (LARC) includes chemotherapy, radiation, and life-altering surgery. Although highly effective for local cancer control, metastatic failure remains significant and drives rectal cancer-related mortality. A consistent observation of this tri-modality treatment paradigm is that histologic response of the primary tumor to neoadjuvant treatment(s), which varies across patients, predicts overall oncologic outcome. In this chapter, we will examine this treatment response heterogeneity in the context of evolutionary dynamics. We hypothesize that improved understanding of eco-evolutionary pressures rendering small cancer cell populations vulnerable to extinction may influence treatment strategies and improve patient outcomes. Applying effective treatment(s) to cancer populations causes a "race to extinction." We explore principles of eco-evolutionary extinction in the context of these small cancer cell populations, evaluating how treatment(s) aim to eradicate the cancer populations to ultimately result in cure. In this chapter, we provide an evolutionary rationale for limiting continuous treatment(s) with the same agent or combination of agents to avoid selection of resistant cancer subpopulation phenotypes, allowing "evolutionary rescue." We draw upon evidence from nature demonstrating species extinction rarely occurring as a single event phenomenon, but rather a series of events in the slide to extinction. We posit that eradicating small cancer populations, similar to small populations in natural extinctions, will usually require a sequence of different external perturbations that produce negative, synergistic dynamics termed the "extinction vortex." By exploiting these unique extinction vulnerabilities of small cancer populations, the optimal therapeutic sequences may be informed by evolution-informed strategies for patients with LARC.

1. Introduction

Curative cancer therapy represents an anthropogenic extinction. That is, a human-induced extinction of the cancer cell clade (population originating from one or a few individuals) that occupies the body of a patient. This can be a formidable task. The population is large—up to 1 billion cancer cells may be present in each cubic centimeter of tumor so that, for example, 10 g of tumor may contain more cancer cells than the current human population. Individual cancer cells are subject to mutational events and can epigenetically alter expression of the human genome resulting in considerable phenotypic heterogeneity. The cancer ecology can be highly diverse due to variations in blood flow, extracellular matrix, and immune response. These changes in microenvironmental selection forces can further increase phenotypic diversity. Furthermore, cancer cells can use niche construction strategies to, for example, promote angiogenesis or acidify the environment to promote invasion and reduce immune response. The clade may be geographically dispersed as individuals or small groups can travel locally or in the blood and lymph to form distant colonies, where they adapt to new environmental conditions.

In considering the events leading to extinction, the popular imagination typically focuses on the famous mass extinction of the mighty Dinosaur clade by the overwhelming evolutionary force generated by a large impact event. This concept of single-cause, single-event extinction is reflected in cancer therapy which has traditionally searched for "magic bullets" to kill cancer cells but not host cells, and oncologic practices that administer treatment agents at maximum dose density.

Many argue that a new mass extinction is underway—the Anthropocene era—in which large numbers of species are to become extinct due to human activities. Although extremely unfortunate, Anthropogenic extinctions have provided an opportunity for detailed investigations of the eco-evolutionary dynamics of "background extinctions." Over the history of earth, 99.9% of living species have become extinct, but only rarely as part of a mass extinction (Raup, 1991). In fact, background extinctions usually involve only a single species and the dynamics of extinction are often subtle and not discernable from the fossil record. Over geologic time, background extinctions have averaged about 5 per year but this has increased by at least an order of magnitude in the Anthropocene era (Otto, 2018). Investigation of well-documented extinctions such as the Passenger Pigeon, Heath Hen, and Galapagos goat have yielded new insights. As a result, the traditional concept of one-cause, one-event extinctions has given way to dynamics that are typically complex with multiple causes and multiple events. This provides a clear lesson for oncology: most species extinctions are the result of a series of perturbations, none of which, by itself, could have eradicated the population. Perhaps, magic bullets and maximum tolerated dose may need to be replaced by evolution-informed strategies.

Here we discuss current treatment of rectal adenocarcinoma in the context of emerging concepts in the eco-evolutionary dynamics of species extinction. We hypothesize that observations and principles obtained from recent investigations of background extinctions in the Anthropocene era can provide novel insights into current practices designed to cure localized rectal cancer and eradicate subclinical micro-metastases using adjuvant and/or neoadjuvant therapies. Specifically, Darwinian concepts provide an infrastructure to design treatment strategies and standardize a language for cancer treatment that is both descriptive and mechanistic.

2. Evolving treatment paradigms of non-metastatic locally advanced rectal adenocarcinoma

Treatment of non-metastatic locally advanced rectal adenocarcinoma (LARC) is multi-modality and highly effective, historically comprised of neoadjuvant (chemo)radiation, radical resection, and adjuvant chemotherapy (ACT). Local pelvic recurrence following this sequence of therapies is low, yet metastatic failure remains significant and responsible for most cancer-related mortality. Because this pattern of disease recurrence has been stable over the past two decades, many research efforts have focused on the sequences of treatment delivery. Total neoadjuvant treatment (TNT), consisting of (chemo)radiation and neoadjuvant chemotherapy prior to curative-intent surgery, is a recent shift in the treatment paradigm, favorably effecting metastasis-free survival rates. In addition, the proportion of patients treated with a TNT strategy resulting in a pathologic complete response (pCR), with no histologic viable tumor remaining in the rectum or perirectal lymph nodes, has markedly increased, ranging between 25% and 38%. These complete pathologic responders experience almost no risk of local recurrence, however, still possess a risk, though much attenuated, for distant recurrence. Consequently, the response of the primary tumor to neoadjuvant treatment(s), rather than the clinical primary stage, best predicts a patient's anticipated oncologic outcome. Tumor regression, however, varies substantially among patients, with no reliable predictor markers of a neoadjuvant treatment response. This close association between treatment effect and extent of downstaging to the risk of recurrence highlights the heterogeneity of rectal tumor biology. Examining the treatment of LARC in the context of evolutionary dynamics may better inform treatment strategies and potentially improve patient outcomes.

3. What is evolution?

To frame clinical observations into an eco-evolutionary context, we must first define Darwinian principles, which are often misunderstood in cancer biology. Briefly, evolution by natural selection requires three conditions: (1) A population of individuals with heritable phenotypic variation undergoing births and deaths, (2) A "struggle for existence" in which a population's capacity for growth exceeds ecological limits so that not all individuals can survive and proliferate, and (3) Heritable phenotypic properties of individuals within the population influencing survival and proliferation.

In cancer biology, it is commonly assumed that evolution is simply an extension of the conventional view of cancer as "a disease of the genes." Indeed, the title of the seminal article on cancer as evolution by Cairns (1975) is "Mutation, selection and the natural history of cancer." However, the mutation-selection sequence neglects the central roles of both environmental selection forces (Gatenby, Gillies, & Brown, 2011) and competitive interactions of individual cancer cells with each other and with normal host cells. Therefore, in the sense that mutations contribute to heritable phenotypic variation, the "evolution as mutation model" is accurate but incomplete. Cancer cells have access to the entire human genome so that epigenetic changes in expression of normal genes, e.g. those involved in fetal development (Ma et al., 2010) or xenobiotic metabolism (Dashwood, 2006), can also contribute to heritable phenotypic variation. Thus, in cancer evolution, genetic expression and mutations determine the set of evolutionarily available cellular traits, which is the "fuel" for evolution. However, the ecological context (i.e., microenvironmental properties), by imposing selection pressures, determines the direction and endpoints of natural selection. In this way, the fitness of any cancer cell is the result of its phenotypic properties interacting with local environmental conditions. From a Darwinian viewpoint, the molecular heterogeneity observed in cancers (Gerlinger et al., 2012) is the result of variation in environmental selection forces (Helmlinger, Yuan, Dellian, & Jain, 1997; Ibrahim-Hashim et al., 2017; Lloyd et al., 2016) caused by, for example, spatial and temporal heterogeneity of blood flow (Gillies, Brown, ARA, & Gatenby, 2018). In turn, phenotypic adaptations to specific environmental condition will favor gain- or loss-of-function mutations in different genes. To summarize, the observed mutations are the *result* of environmentally driven evolution rather than its *cause*.

4. Evolutionary dynamics in cancer treatment: Acclimation and adaptation

Cancer treatment imposes strong selection pressures on cancer cells which can adapt through two mechanisms, (1) *Acclimation* through epigenetic phenotypic alteration (phenotypic plasticity) and/or (2) *Adaptation* through evolutionary "creative destruction" (Hoyal Cuthill, Guttenberg, & Budd, 2020). Not all heritable phenotypes in the cancer population are equally capable of proliferating or surviving when treatment is applied. Via cell turnover (births and deaths), cells with heritable phenotypes that are resistant (i.e., maximally fit) will proliferate faster and replace cells that express less fit (treatment sensitive) phenotypes. When the cell environment contains a cytotoxic therapy, a cell that is resistant will have a high fitness value and proliferate at the expense of those that are sensitive (Gallaher, Enriquez-Navas, Luddy, Gatenby, & ARA, 2018). Importantly, however, in the absence of the cytotoxic agent, synthesis, maintenance, and operation of the molecular machinery of resistance may incur a cost (Gatenby et al., 2009a; Schneider et al., 1989; Shen et al., 2008), providing no benefit to the cell. This fitness expenditure trade-off is an important dynamic in evolution-based therapy principles (Kam et al., 2015; Orlando, Gatenby, & Brown, 2012; Park et al., 2019).

In viewing the evolutionary dynamics during cancer therapy, it is critical to recognize that evolution results from interactions between environmental selection forces and the phenotype (not genotype) of the cancer cells. In a clinical setting in which cytotoxic chemotherapy is applied, there is often concern that the resulting mutagenesis may result in additional cells with a genotype that confers resistance to treatment. However, that *genotype* is resistant because it, in some way, encodes the molecular machinery necessary to overcome the cytotoxic effects of therapy. As a result, the new resistant *phenotype* is still subject to the evolutionary cost/benefit trade-off that weighs the benefit of resistance against the resources necessary to synthesize, maintain, and operate the molecular apparatus of resistance. Thus, any cell that acquires resistance due to the mutagenic effect of therapy or randomly acquirs mutations during treatment will still be subject to the evolutionary limitations conferred by its phenotype.

In the context of treatment exposure, a cancer population that can acclimate will exhibit minimal, if any, evolutionary response, or significant change in population sizes. A response requiring adaptation begins with a decline in tumor volume due to death of sensitive subpopulations, however, proliferation of the resistant subpopulation(s) allows the cancer population to rebound (Reed et al., 2020). Clinically, this repopulation is consistent with cancer recurrence.

5. The "magic bullet" cancer treatment concept to an evolutionary cancer treatment paradigm

Since Ehrlich's pioneering work more than a century ago, the concept of a "magic bullet" (Strebhardt & Ullrich, 2008) drug that kills cancer cells while sparing normal host cells has dominated the ongoing quest for the cure of cancers. In the vast global drug development industry, the focus remains almost entirely on identifying new and "better" drugs. There has been little interest in investigating drug administration strategies beyond conventional Maximum Tolerated Dose (MTD) or integration of drugs from multiple manufacturers into broader treatment strategies that require agents to be combined either in parallel or in sequence. This prevailing oncology approach has been successful in developing new and innovative cancer treatment agents and has provided at least one effective treatment for nearly all metastatic cancers. However, this success has not greatly increased the overall cure rate for most common metastatic cancer types due to almost inevitable evolutionary resistance. While the continued search for magic bullets is critical, our approach accepts that, when none are available, extinction of cancer populations may still be possible by exploiting the eco-evolutionary dynamics of background extinctions.

6. Intentional anthropogenic extinction—The galapagos goat example

Most extinctions in the anthropocene era are inadvertent and the result of carelessness, indifference, or accident. An exception is the intentional anthropogenic extinction of the Galapagos goats (Guo, 2006). Non-native goats had been introduced into the Galapagos islands as food sources for sporadic visits from fishing boats and even pirates. The goats became feral and by 1990 had increased to a population of about 250,000 on the Pinta, Isabela, and Santiago islands causing significant damage to native species in this World Heritage Site. This resulted in Project Isabela to eradicate the goat population. The initial successful strategy employed mass slaughter using automatic weapons mounted in trucks and helicopters. Although initially successful in eradicating about 90% of the feral goats, this strategy became progressively less effective as "resistance" emerged. Individuals and small groups of goats became sensitized to the sound of approaching helicopters and trucks escaping into densely forested regions. (Carrion, Donlan, Campbell, Lavoie, & Cruz, 2011) To achieve its goal of total population extinction, Project Isabela was required to deploy a new strategy—"Judas goats." These were sterilized female goats sent into the forested areas equipped with a radio-tracking collar and coated with hormones. As social animals, the Judas goats quickly joined the surviving groups allowing hunters to locate and kill them. The Galapagos goat was declared extinct in 2005 (Lavoie, Harcourt-Carrasco, & Moya, 2007).

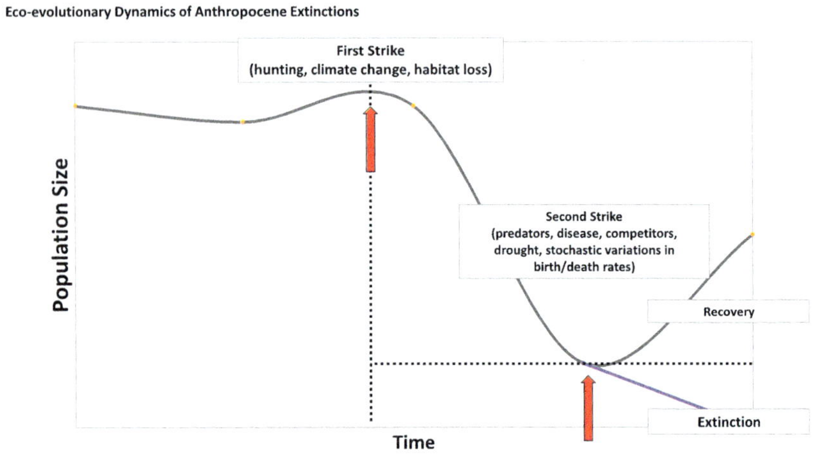

Fig. 1 The eco-evolutionary dynamics of Anthropocene extinctions. Unlike the famous one-cause, one-event extinction of the dinosaurs, most species are lost through less dramatic but more complex "background extinctions." The increase in background extinction in the Anthropocene Era has allowed these extinctions to be studied in detail. Typically, these are multi-cause, multi-event processes that begin with a "first strike" that causes a significant decline in the size, geographic distribution, and genetic diversity of the population. The small surviving populations is then subject to extinction due to Allee and stochastic effects "second strikes." Successful second strike perturbations are nearly always different from the first strike. If second strikes are insufficient or delayed, the small population can recover "evolutionary rescue." Thus, most natural extinctions are caused by a sequence of events none of which, in itself, could eradicate the population.

Recent recognition of extinction as a multi-step, multi-cause process is apparent in the eradication of the Galapagos goats (Fig. 1). The initial population decline is typically the result of a "first strike" (in this case hunting) that reduces the total population size but also decreases heterogeneity and causes geographic fragmentation of the surviving population. Importantly, however, the first strike rarely causes extinction. As in the Galapagos goats, the intense selection pressure caused by the rapid population decline produces an adaptive strategy that prevents extinction—termed "evolutionary rescue." Once a resistance strategy emerges, continued application of the initial perturbation (in this case hunting), does not result in extinction. A new strategy or strategies—collectively termed "second strike"—must then be deployed to produce extinction. Thus, the final descent into extinction is rarely caused by the same perturbation that initiated the population decline. Furthermore, successful second-strike agents usually exploit the unique eco-evolutionary properties of small populations that render them highly vulnerable to extinction. In the Galapagos goat extinction, for example, release of a small number of Judas goats pushed the survivors into extinction. This strategy would have been ineffective as a first strike because the population was too large just as the first strike strategy was unable to eradicate the small group of survivors.

7. Small cancer populations—"Second strike"

In viewing cancer cure as an intentional Anthropocene extinction, successful first line therapy is the equivalent of a first strike which reduces the size and heterogeneity of, for example, an initially large, metastatic tumor burden. However, a clear lesson from Anthropocene extinctions is that the first strike rarely eradicates the initial population. Instead, the first strike selects for resistance mechanisms producing small groups of survivors (Allee, 1931a; Odum & Allee, 1954; Otto, 2018) resulting in three possible outcomes: (1) a small stable population under threat of extinction, equivalent to "dormant disease" (2) proliferation ("evolutionary rescue") with re-population leading to treatment failure and patient death or (3) Extinction and cure (Fig. 2). These dynamics raise questions about the standard oncologic practice of "continuous application of treatment at maximum tolerated dose until progression." That is, continued administration of the first strike agent after the tumor decline is, in effect, only treating cancer cells that are resistant. Thus, during this time in which the treatment seems to be "working," the cancer cells are beginning to proliferate and occupy new microenvironmental niches increasing phenotypic heterogeneity. Applying a new treatment (in effect, a second-line therapy) too late, the surviving cancer population has already increased and is no longer vulnerable to small population extinction dynamics. This emphasizes the temporal importance of treatments, applying second strikes at a time-point when a population is under threat of extinction and not yet able to diversify to rebound. We hypothesize that these "two-strike" eco-evolutionary dynamics provides novel insights for the TNT approach to treating LARC patients, delivering sequential effective and different therapies to increase the probability of extinction and thus, cure.

The descent of small populations to extinction typically requires a temporal sequence of perturbations that are nearly always qualitatively different from the initial events that reduced the originally large population. Importantly, these perturbations are usually quantitatively smaller than the first strike agents because they, within the unique eco-evolutionary

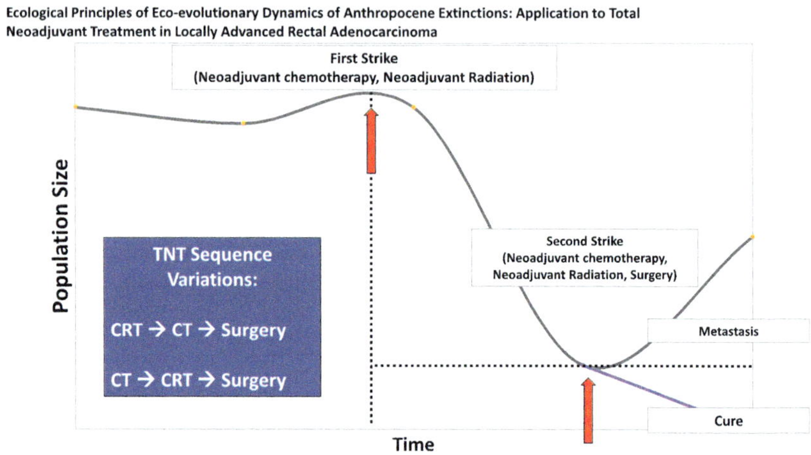

Fig. 2 The eco-evolutionary dynamics applied to non-metastatic locally advanced rectal cancer (LARC). The initial strike, either neoadjuvant chemotherapy (CT) or neoadjuvant chemoradiation (CRT) rarely eradicates the cancer population, but typically results in a surviving, vulnerable small cancer cell population. The surviving population may become net negative and cross an extinction threshold (minimal viable population), remain marginally positive compared to the critical density extinction threshold and narrowly persist in the absence of additional stressors (second-strikes), or the population may increase with diversification and rebound with time (evolutionary rescue). The status and probable outcome of the surviving population is not static, and experiences ongoing influence by dynamic Allee and stochastic effects. By sequencing LARC therapies, total neoadjuvant treatment (TNT) delivers sequential first and second-strikes during a window of potential maximum cellular proliferation (and smallest population volume), monopolizing on eco-evolutionary principles to limit the ability for rebound and rescue, and maximize the probability of cure.

dynamics of small populations results in destructive, reinforcing synergies (Raup, 1991) that is characterized as the "extinction vortex" (Brook, Sodhi, & Bradshaw, 2008; Fagan & Holmes, 2006). Unlike the single-cause, single-event concept, natural extinctions are driven by stressors that are not mutually exclusive, each exerting a relative influence on an increasingly vulnerable small population. The initial perturbation (i.e., hunting, chemotherapy) may be responsible for the initial decline toward an extinction threshold, however, the subsequent perturbations (i.e., predators, radiation) deepen the probability of completing the process. This suggests that treatment in a vacuum without reinforcing stressors incompletely exploits a vulnerable small cancer population, and therefore, risks evolutionary rescue and clinically macro-metastatic disease progression. Much like a population of a species declines then disappears, a complex of perturbations is responsible for cancer cell extinction, not an individual event. Although a second-strike treatment agent may not significantly impact a large cancer population, it may be lethal to a small cancer population. Similarly, the same ineffective perturbation (i.e., treatment) initially administered may be efficacious when applied to a small cancer population further along the extinction vortex, emphasizing the temporal importance secondary perturbations (i.e., treatment(s)).

8. Small populations and rectal cancer: Adjuvant chemotherapy, a second-strike too late?

Although large tumor volumes are the proximate cause of death in most cancer patients, the clinical course of each patient is often determined by the eco-evolutionary dynamics of small populations (Artzy-Randrup, Brown, Costa, Czerniecki, & Gatenby, 2021).This is evident in the 20% to 40% of localized rectal cancer patients receiving curative-intent treatment but then go on to develop macro-metastatic disease. This consistent outcome strongly suggests that a proportion of subclinical small cancer populations persist and then expand despite a complete course of therapeutic interventions. Although speculative, perhaps micro-metastases are present in all patients with LARC, some of which become extinct either spontaneously or secondarily to evolutionary pressures, (i.e., radiation, surgery, chemotherapy), while others expand and accelerate in growth (Hu & Curtis, 2020; Uchi et al., 2016). If a cancer subpopulation can acclimate with phenotypic plasticity to applied external pressures (i.e., treatments), the population size continues to expand. Similarly, if the subpopulation requires adaptation to applied stressors, tumor population declines initially due to the death of sensitive less-fit phenotypic subpopulations while higher fitness proliferate at their expense. As previously described, acclimation and adaptation govern evolving populations exposed to specific perturbations and partly describe the proliferation of one or more resistant small cancer subpopulations exposed to therapeutic agents resulting in disease recurrence (Gatenby & Brown, 2020a).

Currently, the most common multi-modality LARC paradigm consists of pelvic irradiation, often concurrent with a "radiosensitizing" dose-reduced chemotherapeutic agent, followed by surgical resection of the rectum and nodal drainage (mesorectum), and lastly, adjuvant chemotherapy (ACT). This treatment paradigm was established by the landmark German rectal cancer randomized trial, setting the sequence of tri-modality therapy for the next two decades (Sauer et al., 2004). What is notable from long-term follow-up of the German trial, as well as multiple international trials, is that pelvic irradiation impacts only local disease outcomes. The recurrence risks for LARC patients receiving either preoperative or postoperative pelvic radiation are almost identical regarding distant metastases and overall survival. Thus, adjuvant chemotherapy is typically considered the only systemic therapy, while the remaining components (i.e., radiation and surgery) are local treatments and not expected to significantly affect non-regional/distant disease recurrence. The explicit goal of systemic ACT is to eradicate subclinical metastatic cancer populations. The treatment paradigm established by the German trial- chemoradiation, surgery, followed by ACT sequence shifted the patterns of disease failure from local to distant.

Considered in an evolutionary context, ACT aims to produce a series of external perturbations within small cancer populations (i.e., micrometastatic or subclinical/occult metastasis) to promote extinction, as opposed to a single event, single cause perturbation associated with a large population extinction. Modern cancer therapy, similar to the meteor impact event as a single driver of mass dinosaur extinction, often relies on administering the maximal therapeutic doses of cytotoxic agents to exert massive external perturbations and evolutionary pressure. However, mass extinction events cumulatively account for no more than 5% of all extinctions, most species facing extinction through subtler, and more prolonged background extinctions (Leakey & Lewin, 1995; Raup, 1991). Although the majority of LARC patients treated do not experience disease recurrence, treatments fail to cause extinction in a significant proportion of patients. The small cancer populations within these patients treated with (chemo)radiation, surgery, and ACT either are eradicated, or they survive and then re-populate (evolutionary rescue). These treatment interactions of small cancer populations dictate a patient's overall outcome, that is, cure or death. Despite this, oncologic treatment regularly administers dose dense chemotherapy, ignoring evolutionary dynamics of small cancer populations, and instead predominantly relies on a one-cause, one-event extinction mechanism.

As noted above, the metaphorical "magic bullet," originating from the concept of a targeted antibiotic directed at a specific infectious disease, only loosely applies to cancer therapy. Partially realized over the past decades, chemotherapy agents have multiplied in number and specificity, intending to treat localized disease curatively but typically only delaying progression in the setting of metastatic disease. The unrelenting quest for more effective and less toxic chemotherapeutic agents continues, but it begs the question as to why even microscopic cancer populations often survive repeated exposures to highly cytotoxic therapies. Several hypotheses have been considered, including the restraints on maximum tolerated dose (MTD) to limit host toxicity as well as rapid evolutionary adaptation and resistance mechanisms within diverse small cancer populations.

In fact, the role and benefit of ACT for all patients with LARC remains uncertain and unresolved. Although ACT is a component of the guideline (NCCN 2020) recommended tri-modality treatment regimen in North America, limited data demonstrates a clear benefit in disease outcome. Indeed, ACT for LARC patients is extrapolated from randomized trials studying node-positive colon cancer. This appears reasonable based on The Cancer Genome Atlas (TCGA) data demonstrating biological similarity of left-sided colon cancers to rectal cancers (Cancer Genome Atlas Network, 2012). Collectively, however, the data for ACT in LARC has failed to improve disease-free survival. Despite disappointing results from several ACT rectal trials, certain patient sub-populations appear to derive benefit, again indicating that our understanding of heterogeneous tumor biology remains deficient. Pathologic node-positive (Stage 3) LARC patients treated with neoadjuvant (chemo)radiation, in distinction to pathologic Stage 0 through 2, appear to benefit from ACT. This observation suggests that adaptive ACT may be more appropriate, rather than empiric universal administration, to reduce the probability of metastatic disease occurrence. Although the data defining the role of ACT is confounded by inaccurate clinical staging, differences in tumor response to neoadjuvant (chemo)radiation, quality of the mesorectal surgical excision, low adherence to adjuvant systemic treatment (25–57%), and outdated chemotherapeutic regimens, it remains a guideline recommended treatment (Breugom et al., 2015a; Breugom, Swets, & Velde, 2015; Petersen, Harling, Kirkeby, Wille-Jorgensen, & Mocellin, 2012; Xu et al., 2017).

9. Small populations and rectal cancer: Total neoadjuvant treatment exploits the second-strike

If systemic chemotherapy is recommended and considered beneficial in patients with LARC, as currently accepted in North America, then why not give it before surgery? Over the past several years, the tri-modality approach to LARC has gradually shifted, delivering all therapies preoperativley prior to surgical consideration. In this way, a series of perturbations resulting from a TNT approach applies the unique ecological and evolutionary dynamics of small and declining populations to maximize "extinction therapy" (Gatenby & Brown, 2020a; Reed et al., 2020). Systemic chemotherapy is administered using two options, either as "induction" (before ((chemo)radiation) or as "consolidation" (after (chemo) radiation), both termed total neoadjuvant treatment (TNT). Presently, the rationale for this paradigm change is primarily driven by the distant metastatic risk which has remained constant despite advances in imaging, surgical technique, and chemotherapy agents. Not only is the first exposure to systemic dose ACT at least 14–20 weeks from time of initial diagnosis, the compliance and full receipt is universally poor. Multiple phase 2 and several phase 3 trials have demonstrated superior tolerability and compliance of neoadjuvant systemic chemotherapy, with modern data reporting an increased probability of a pathologic complete response (pCR), disease-free, and metastasis-free survival. No differences in local tumor control nor OS have been firmly established, however, this is likely due to lack of long-term outcome data. Of considerable importance, TNT has been shown to increase the odds of pCR by 40% when compared to standard neoadjuvant CRT alone, greatly expanding the possibility of non-operative strategies for select patients (Petrelli et al., 2020). Obtaining a clinical complete response (cCR) to neoadjuvant therapies, an assumed surrogate of pCR, without the need for rectal surgery, is a relatively new concept with immense patient implications. It has been consistently reported that radical rectal resection in the setting of a pCR imparts little, if any, oncologic benefit, yet subjects a patient to a risk of perioperative mortality and significant morbidity. A high likelihood of cure in LARC patients with a cCR suggests surgical resection is superfluous, rather than the mainstay of curative treatment. Therefore, rectal organ preservation may be a safe alternative to radical rectal resection for patients with a cCR following receipt of TNT.

Given promising results applying TNT as the standard platform for LARC moving forward, the selective evolutionary pressures of neoadjuvant therapies can be examined. TNT strategically exploits several effective, but typically individually non-curative therapies, to enable eradication of a subclinical metastatic population.

The preferred sequence of TNT (induction chemotherapy (INCT) then CRT or CRT then consolidation chemotherapy (CCT)) is presently unknown. Recent data has indicated DFS and distant-metastatic failure rates are better in TNT treated patients compared to the historical CRT-surgery-ACT sequence. Here, we further explore these neoadjuvant approaches in an evolutionary context, rather than compare the two strategies in terms of toxicity, compliance, or even, efficacy.

10. Alternative strategies—Small population vulnerabilities to extinction: Stochastic and Allee effects

Within the Anthropocene extinction framework, stochastic and Allee effects exert selection pressures, largely dictating the fate of small populations, including cancers. Two main stochastic elements influence a small populations' risk for extinction: demographic and environmental stochasticity (Beissinger, 2000; Saether, Engen, Islam, McCleery, & Perrins, 1998). Although acting in different density-dependent ways, both demographic and environmental stochasticity account for the probability that once a critical population threshold is crossed, the slide to extinction becomes inevitable, consistent with the conservation biology minimal viable populations (MVP) concept (Brook, Bradshaw, Traill, & Frankham, 2011; Gilpin, 1986; Saether et al., 1998; Shoemaker et al., 2020; Tier & Hanson, 1981).

Demographic stochasticity describes the likelihood of extinction among an increasingly smaller population due to a larger relative fluctuation in population size. The random variation at a demographic level of an individual cell surviving or dying results in a population size that deviates from the expected deterministic level. This demographic stochasticity minimally affects large size populations, where population growth remains close to its deterministic rate. However, once smaller, the random variation between individuals of small populations exert a greater relative influence in population growth (Artzy-Randrup et al., 2021).

Environmental stochasticity describes the fluctuations of environmental conditions temporally on a population's demographic rate. Environmental variance presumably affects an entire population collectively, contrary to demographic stochasticity which is intimately impacted by population size. To mitigate the blanketing effect of temporal environmental fluctuations, natural populations such as a growing cancer, are comprised of multiple subpopulations. These subpopulations experience environmental micro-niches such that no single, small subpopulation is entirely exposed to all environmental fluctuations. Cancer cells can engineer micro-environments by altering angiogenesis and acidification of the local environment and promote their survival at the expense of less adept cells. These eco-evolutionary niche-construction strategies imply an element of ecological inheritance passed down to the next generation of cells. When therapy is applied, these environmental conditions might continue sufficiently so as to protect the local cells from the immediate effects of therapy, allowing time to acclimate then evolve favorable adaptations to resist treatments (Mullon & Ohtsuki, 2020). For a rectal cancer, environmental stochasticity may be governed by spatial and temporal heterogeneity of tumor vascularity delivering substrate and growth factors, cytotoxic treatment agents, and radiation (Gillies et al., 2018). The observed phenotypic adaptations of subpopulations resulting from mutations or epigenetic alterations are therefore environmentally driven, rather than its cause. In the same way, cytotoxic therapies impose intense selection pressures resulting in the development and evolution of resistance mechanisms (Gatenby et al., 2009b).

11. Alternative approaches—Stochastic effects and total neoadjuvant treatment

Neoadjuvant chemotherapy in non-metastatic localized cancers is not new and has been previously explored in other disease sites, including head and neck and cervical cancers. For LARC, earlier treatment of subclinical micrometastatic disease to reduce the chance for metastatic dissemination remains the primary rationale. A delay in exposure to systemic therapy may allow growth of small volume disease outside of the pelvis. In addition to this, the local effects of neoadjuvant chemotherapy are of importance given the amplified pCR rate seen using a TNT approach. In fact, primary neoadjuvant chemotherapy with selective omission of pelvic irradiation has been explored due to the significant anticipated acute and late toxicities and concern for over-treatment when universally applying pelvic radiation. The Alliance N1048 Phase II/III PROSPECT Trial has completed accrual to study the role of neoadjuvant FOLFOX chemotherapy alone in non-metastatic LARC patients. This treatment platform is based on data showing no compromise in negative margin rectal resection, local pelvic control, or disease-free survival after 53 months of follow up. Unexpectedly, a 25% pCR rate was observed indicating marked activity of preoperative systemic chemotherapy alone in these patients yet was higher than similar studies reporting the role of primary chemotherapy (Deng et al., 2019; Schrag et al., 2014; Ueki et al., 2016).

Neoadjuvant chemotherapy utilizes embryonic tumor blood supply and reliably decreases a tumor's size which may translate into enhanced radiation susceptibility due to better tumor oxygenation (Atkin et al., 2006). Neoadjuvant treatment is hypothetically administered when surgically induced hypoxia secondary to fibrosis may be lowest, possibly improving intra-tumoral access, exposure, and effectiveness of cytotoxic agents. Although speculative, these factors may enhance tumor sensitivity to these therapeutic interventions. Because neoadjuvant chemotherapy or neoadjuvant CRT does not typically eradicate a primary tumor completely (i.e., pCR) and accelerated re-population of resistant surviving tumor clones may take place, a TNT strategy then administers a second therapeutic intervention (CRT or consolidation chemotherapy depending on the management sequence), to reduce the potential for cellular proliferation. Therefore, a TNT strategy addresses the time factor concern, that is, accelerated re-population of tumor clones surviving the initial therapeutic intervention. It remains possible, however, that surviving cells may favor selection of radio- or chemo-resistant clones, reducing the efficacy of the subsequent therapeutic intervention. Alternatively, the smaller proportion of surviving tumor cells may be partially damaged while proliferating, increasing susceptibility to the subsequent therapy. Notably, randomized trials in other disease sites including cervical and head and neck cancers comparing INCT followed by radiation to radiation alone have failed to show DFS or OS improvement, with some reporting worse outcomes (da Costa et al., 2019; Hong et al., 2018). The explanations for these findings are speculative but revolve around the development of resistant clones and remain an area of concern.

Naive rectal tumors exhibit complex clonal architectural and significant intra-tumoral heterogeneity that are spatially disparate within regions of a tumor (Bettoni et al., 2019, 2017). This genetic variability may allow tumors to access a "genetic reservoir" in their evolutionary response to therapeutic perturbations (Hardiman et al., 2016). Whole exome sequencing of irradiated normal adjacent rectal mucosa has not been shown to introduce novel somatic mutations, however, intratumoral genetic heterogeneity increases within the residual rectal tumor, suggesting evolutionary pressures and positive clonal selection following treatment exposure. Although the same selection pressures on non-malignant mucosa is imposed, non-neoplastic cells are unable to evolve resistance strategies since cell division, death, and survival is predetermined by local signals and therefore is not subject to Darwinian dynamics (Ma, Poole, Goyette, & Gaus, 2017). Higher proportions of potentially deleterious somatic mutations in post-treatment, irradiated rectal tumor have been described. The enriched mutations in the irradiated tumor occurs among genes involved in DNA damage repair, genetic instability, cell cycle regulation, proliferation, survival, and differentiation (Bettoni et al., 2019). Similar alterations in intra-tumoral genetic heterogeneity have been reported for esophageal adenocarcinoma after exposure to neoadjuvant chemotherapy (Findlay et al., 2016; Murugaesu et al., 2015). These studies suggest that neoadjuvant therapies sculpt the mutational landscape and clonal architecture, promoting changes in frequencies of somatic mutations due to repopulation of residual clones in the residual tumor. Tumors not completely responding to therapies may acquire novel mutations among the surviving tumor cell subpopulations compared to their baseline counterparts, leading to increased intratumoral heterogeneity, and potentially possess selective growth advantages risking more diverse residual tumors that could be biologically more aggressive (Bettoni et al., 2019). More effective therapies, optimizing therapeutic sequences, or applying variable agents rather than the same drug(s) are needed to overcome the resistance of small cancer populations resulting in residual disease responsible for metastatic failure.

12. Alternative strategies—Allee and aggregation effects and total neoadjuvant treatment

Allee dynamics, as opposed to stochastic effects, represent a natural phenomenon in which an individual organism's proliferation potential decreases within small populations (Bottger et al., 2015; Boukal & Berec, 2002; Neufeld et al., 2017; Tobin, Berec, & Liebhold, 2011). This observation was described in the 1930s by Warder Clyde Allee (Allee, 1931a; Odum & Allee, 1954), an American ecologist, who showed that the fitness of an individual decreased in smaller populations, and correspondingly increased its risk of extinction (Allee, 1931a, 1931b; Odum & Allee, 1954). Allee effects are evident in common laboratory experience in which many cancer cells must be injected into an experimental animal or placed in a culture dish to establish a colony. More recently, Allee effects have been explicitly identified in pre-clinical and clinical studies (Bottger et al., 2015; Johnson et al., 2019; Konstorum, Hillen, & Lowengrub, 2016; Neufeld et al., 2017; Sewalt, Harley, Heijster, & Balasuriya, 2016). There is a statistical possibility that a small population becomes extinct by chance fluctuations alone, randomly falling to zero. This probability is low, although the risk of chance extinction increases as a population declines, such that spontaneous regression of cancer has been previously observed (Cole, 1981; Cui et al., 2003). Although rare, these events should not be discounted. We recognize these cancer populations to be 'small,' however, rare events happen all the time within a cellular system operating on orders of significant magnitude, where each cubic centimeter of tumor can contain up to one billion cancer cells (Cairns, 1975; Diaz et al., 2012). Most LARC tumors experience reduction in tumor volume when exposed to neoadjuvant therapies, even those without significant pathologic downstaging. This is observed after both neoadjuvant (chemo)radiation or primary chemotherapy and magnified when administered sequentially as TNT. Although speculative, the greater degree of tumor regression associated with TNT implies a higher likelihood of spontaneous tumor regression due to an overall smaller total population of residual small cancer populations, or at least, a greater Allee influence (Nougaret et al., 2012). However, the extent of tumor volume reduction is highly variable as well as time dependent (Cusumano et al., 2020; Yeo et al., 2012). The complete or partial protection of residual subpopulations provides time to develop a resistant phenotype and subsequently proliferate (Gatenby & Brown, 2020a).

In the context of small cancer cell populations, Allee dynamics result in "aggregation effects" (Alllee & Bowen, 1932; Parrish & Edelstein-Keshet, 1999), the tendency to experience enhanced vulnerability among isolated, rather than larger populations, who are better equipped to mitigate an external threat through defensive mutual collective interactions. These Allee and aggregation effects of small cancer cell populations are expected to negatively impact the ability to adapt to external perturbations (i.e., treatment(s)) or host immune responses (Brown, Cunningham, & Gatenby, 2017). However, as population size increases, favorable interactions among individuals produce increased proliferation rates, growth, and population size. Due to growing species-specific competition for space and resources, the proliferation rate would be expected to decline secondary to these evolutionary dynamics. However, Charles Darwin observed that these competitive threats from an enlarging population size unexpectedly result in increased proliferation rates among large populations. The benefits of large populations and survival are well recognized in cancer, perhaps by allowing dilutional effects of cytotoxic agents (i.e., reduction in drug to cell ratio), increased diversity of resistance mechanisms, and collective responses to immune-mediated interactions (Brown et al., 2017).

The strength of the Allee effect correlates with the risk of extinction of small populations. In the context of a single or a small population of cancer cells, both stochastic and Allee effects impart vulnerability and probability of survival. The Allee interactions result in inter-cell competition and fitness stressors through external variables such as angiogenesis, matrix remodeling, acidification, and metabolic cooperation (Gillies et al., 2018). Depending on the strength of the Allee effect within a small population, growth may become net negative and cross an extinction threshold (Minimum viable population), growth may remain marginally positive compared to the critical density extinction threshold and narrowly persist in the absence of additional external perturbations (i.e., treatments), or the population may increase and accelerate with time ("evolutionary rescue"). These small cancer populations subjected to Allee effects, however, must always be considered temporally, since extinction risk is dynamic and changes with time in respect to size of the population. It follows that once a population begins to rebound, the probability of extinction increasingly diminishes at that time point (Artzy-Randrup et al., 2021; Bell, 2009, 2017; Gonzalez, Ronce, Ferriere, & Hochberg, 2013). By sequencing LARC therapies, TNT delivers treatment during a window of potential maximum cellular proliferation after the initial insult attempting to negate the ability to rebound and rescue the small cancer subpopulation.

13. Alternative strategies—Summary

Any cancer treatment, or "magic bullet," generates selection pressures. In the setting of LARC, these agents are targeted at potential occult metastatic cancer populations. Though small, these cancer populations may possess phenotypic heterogeneity and an ability to rapidly proliferate, such that proportions survive due to resistance adaptation imposed by selection forces. Current cancer treatment practices MTD delivery, balancing a precarious therapeutic ratio, attributing favorable outcomes (i.e., cancer extinction/cure) to large population mass extinction dynamics. Although tumor extinction is a possibility following MTD treatment, it negates the more likely probability of evolutionary rescue. Microscopic cancer populations can evolve effective survival mechanisms through resistance and avoid extinction. The underpinnings delineating the mechanisms of drug resistance or immune evasion is pharmacologically important, but arguably of more importance, is questioning the repetitious delivery of the same therapeutic agents that will not and do not achieve extinction of a cancer population. Chemotherapeutic treatment course regimens, often empiric in frequency and number, are deep-seated oncologic management strategies. Despite a growing level of recognition of Darwinian dynamics in cancer biology and its critical role in treatment failure, $<1\%$ of clinical trials discuss the role of evolution in treatment resistance (Aktipis, Kwan, Johnson, Neuberg, & Maley, 2011). From the perspective of Allee and stochastic dynamics, these therapies often accomplish reduction in the cancer populations, but among a significant proportion of patients, do not eradicate the remaining cancer population.

An alternative cancer treatment paradigm considers the small population dynamics (Artzy-Randrup et al., 2021) as an intentional Anthropocene extinction (Gatenby, Artzy-Randrup, Epstein, Reed, & Brown, 2020; Gatenby, Zhang, & Brown, 2019). Pelvic irradiation to a LARC represents a "first strike," followed by surgical removal and ACT as "second strikes" driving the small remaining cancer populations down the extinction vortex. Alternatively, a TNT management approach sequences either neoadjuvant chemotherapy (induction) as a "first strike," followed by (chemo)radiation and surgery as "second strikes." Currently, the optimal sequence of neoadjuvant therapies for LARC remains unclear but is highly relevant to establishing a treatment strategy most effective in achieving small cancer population extinction. Since successful "second strikes" typically require sequential perturbations to the disproportionally vulnerable small cancer populations, in comparison to large populations, TNT is a well-positioned approach to maximizing the probability of cancer cure.

A TNT paradigm for LARC integrates extinction vortex principles. For example, a primary tumor and regional lymph nodes are irradiated and, in most patients, results in a reduction of tumor burden. This is evident in the resected specimens, histologically displaying variable degrees of fibrosis and viable tumor characterized as tumor regression grades (Fokas et al., 2017). For LARC patients approached with a TNT strategy, the anticipated cancer cell decline is then immediately consolidated with a "second-strike" of cytotoxic preoperative systemic chemotherapy subsequently followed by short-interval surgical resection. As seen in nature, when small invasive populations rebound, their phenotypic diversity is magnified. In small cancer populations permitted to proliferate due to delays or ineffective treatment(s), the population expands and genetically diversifies by occupying new niches. Assuming 1 cm^3 of cancer contains up to one billion cells, becoming clinically detectable requires generations of cancer cells, driving the repeated division-death process of cellular evolution (Gatenby & Brown, 2020a, 2020b). Timing is therefore critical, as an opportunity for eradication/extinction may be lost if secondary strikes are not delivered when the cancer population is declining or at its nadir (Artzy-Randrup et al., 2021). Multi-drug chemotherapy regimens, rather than single agent treatment, is the standard approach for LARC. And consistent with routine oncologic practice, these agents are administered concomitantly over the course of typically eight cycles. Applying small population dynamics principles, however, sequential agent administration may be a more effective approach. If exposure of the same combination of agents does not result in extinction, the strategy allows for adaptive responses of a heterogenous small cancer population. Consequently, resistant survivors expand in size and greater genetic diversity emerges among evolutionary subpopulations. Perhaps the treatment periods (i.e., cycles) might be too extended, missing the temporal window of opportunity to deliver an optimal second strike. Or, another agent or alteration in the sequences of cytotoxic agent exposure over shorter time intervals may optimally exploit the stochasticity of evolutionary forces.

Using small cancer populations as a guiding principle, LARC treatment might be conceptualized as an evolutionary game between the treating oncologists and the cancer (Stankova, Brown, Dalton, & Gatenby, 2018). Since the cancer can only react to selection forces, the oncologist possesses initial advantages, and should aim to anticipate and steer the cancer's eco-evolutionary dynamics by applying treatments. Traditional approaches have not utilized this dynamic, usually prescribing a regimen to the maximal allowable therapeutic ratio of patient tolerance, until clinical evidence of disease progression. In this way, cancer cells evolve an adaptive response to therapy and recover ecologically, better positioned to counteract future therapies. By awaiting clinically measurable disease, the unique vulnerabilities of the window comprised of smaller volume disease, probabilistically teetering toward extinction, is incompletely exploited (Gatenby & Brown, 2020b).

14. Future directions

Although the magic bullet conceptual approach, reflecting the conventional single-cause, single-event model of extinction, dominates cancer drug development and delivery, evolutionary principles suggest the probability of cure is more likely when small cancer subpopulations vulnerabilities can be maximally exploited. Departing from standard oncologic practices, sequential treatment using variable agents, may push towards an extinction vortex from repeated environmental perturbations, mitigating the chances that the remaining small cancer populations are sufficiently ecologically or demographically heterogenous to allow evolutionary rescue and clinically recur. Considered in this context, the dynamics of drug delivery should be carefully examined along with drug targets and agents. A mix of demographic (i.e., net negative populations size from cytotoxicity), ecologic (i.e., substrate effectors such as angiogenic inhibitors), and predatorial (i.e., immune modulators) perturbations may maximize the probability of extinction. The currently evolving treatment paradigms of TNT for LARC employ cytotoxic multi-drug agents sequentially with radiation. The outcomes of these approaches have observed high complete pathologic response rates, directly translating to a greater than 90% cure for most of these complete responders. Future studies might further investigate an evolutionary-informed approach, condensing the treatment periods between administration, which may decrease the time for resistant populations to increase in size and heterogeneity, alternating varying drug agents over short time periods, reducing the emergence of resistant subpopulations, or delivering treatments both prior and following curative-intent resection, extending sequential perturbations when the cancer subpopulation is in rapid decline or at its nadir and less able to rebound with genetic diversity.

Acknowledgment

This work is supported by grants U54 CA143970 and R01 CA077575.

Conflicts of interest

The authors declared no conflict of interest.

References

Aktipis, C. A., Kwan, V. S., Johnson, K. A., Neuberg, S. L., & Maley, C. C. (2011). Overlooking evolution: A systematic analysis of cancer relapse and therapeutic resistance research. *PLoS One*, *6*(11), e26100. Epub 2011/11/30 https://doi.org/10.1371/journal.pone.0026100. 22125594. PMC3219640.

Allee, W. C. (1931a). *Animal aggregations, a study in general sociology. Vol ix* (p. 431). Chicago: The University of Chicago Press.

Allee, M. H. (1931b). *Jane's island. Vol. vii*. Boston, New York: Houghton Mifflin Company. 1 p., 2 l., 235 p. 1 p.

Alllee, W. C., & Bowen, E. S. (1932). Studies in animal aggregations: Mass protection against colloidal silver among goldfishes. *Journal of Experimental Zoology*, *61*(2), 185–207. https://doi.org/10.1002/jez.1400610202.

Artzy-Randrup, Y. E. T., Brown, J. S., Costa, R. L. B., Czerniecki, B. J., & Gatenby, R. A. (2021). Novel evolutionary dynamics of small populations in breast cancer adjuvant and neoadjuvant therapy. *NPJ Breast Cancer*. In Press.

Atkin, G. K., Daley, F. M., Bourne, S., Glynne-Jones, R., Northover, J. M., & Wilson, G. D. (2006). The impact of surgically induced ischaemia on protein levels in patients undergoing rectal cancer surgery. *British Journal of Cancer*, *95*(7), 928–933. Epub 2006/10/04 https://doi.org/10.1038/sj.bjc.6603362. 17016487. PMC2360543.

Beissinger, S. R. (2000). Ecological mechanisms of extinction. *Proceedings of the National Academy of Sciences of the United States of America*, *97*(22), 11688–11689. Epub 2000/10/26 https://doi.org/10.1073/pnas.97.22.11688. 11050195. PMC34335.

Bell, G. G. A. (2009). Evolutionary rescue can prevent extinction following environmental change. *Ecology Letters, 12*(9), 942–948.

Bell, G. (2017). Evolutionary rescue. *Annual Review of Ecology, Evolution, and Systematics, 48*(1), 605–627.

Bettoni, F., Masotti, C., Correa, B. R., Donnard, E., Dos Santos, F. F., Sao Juliao, G. P., et al. (2019). The effects of neoadjuvant chemoradiation in locally advanced rectal cancer-the impact in intratumoral heterogeneity. *Frontiers in Oncology, 9*, 974. Epub 2019/10/16 https://doi.org/10.3389/fonc.2019.00974. 31612112. PMC6776613.

Bettoni, F., Masotti, C., Habr-Gama, A., Correa, B. R., Gama-Rodrigues, J., Vianna, M. R., et al. (2017). Intratumoral genetic heterogeneity in rectal cancer: Are single biopsies representative of the entirety of the tumor? *Annals of Surgery, 265*(1), e4–e6. Epub 2016/08/02 https://doi.org/10.1097/SLA.0000000000001937. 27479130.

Bottger, K., Hatzikirou, H., Voss-Bohme, A., Cavalcanti-Adam, E. A., Herrero, M. A., & Deutsch, A. (2015). An emerging allee effect is critical for tumor initiation and persistence. *PLoS Computational Biology, 11*(9), e1004366. Epub 2015/09/04 https://doi.org/10.1371/journal.pcbi.1004366. 26335202. PMC4559422.

Boukal, D. S., & Berec, L. (2002). Single-species models of the Allee effect: Extinction boundaries, sex ratios and mate encounters. *Journal of Theoretical Biology, 218*(3), 375–394. Epub 2002/10/17 12381437.

Breugom, A. J., Swets, M., Bosset, J. F., Collette, L., Sainato, A., Cionini, L., et al. (2015). Adjuvant chemotherapy after preoperative (chemo)radiotherapy and surgery for patients with rectal cancer: A systematic review and meta-analysis of individual patient data. *The Lancet Oncology, 16*(2), 200–207. Epub 2015/01/16 https://doi.org/10.1016/S1470-2045(14)71199-4. 25589192.

Breugom, A. J., Swets, M., & van de Velde, C. J. (2015). Adjuvant chemotherapy for rectal cancer—Authors' reply. *The Lancet Oncology, 16*(4), e155. Epub 2015/04/08 https://doi.org/10.1016/S1470-2045(15)70110-5. 25846091.

Brook, B. W., Bradshaw, C. J., Traill, L. W., & Frankham, R. (2011). Minimum viable population size: Not magic, but necessary. *Trends in Ecology & Evolution, 26*(12), 619–620. author reply 20-2. Epub 2011/10/18 https://doi.org/10.1016/j.tree.2011.09.006. 22001527.

Brook, B. W., Sodhi, N. S., & Bradshaw, C. J. (2008). Synergies among extinction drivers under global change. *Trends in Ecology & Evolution, 23*(8), 453–460. Epub 2008/06/28 https://doi.org/10.1016/j.tree.2008.03.011. 18582986.

Brown, J. S., Cunningham, J. J., & Gatenby, R. A. (2017). Aggregation effects and population-based dynamics as a source of therapy resistance in cancer. *IEEE Transactions on Biomedical Engineering, 64*(3), 512–518. Epub 2017/01/24 https://doi.org/10.1109/TBME.2016.2623564. 28113286. PMC5438296.

Cairns, J. (1975). Mutation selection and the natural history of cancer. *Nature, 255*(5505), 197–200. 1143315.

Cancer Genome Atlas Network. (2012). Comprehensive molecular characterization of human colon and rectal cancer. *Nature, 487*(7407), 330–337. Epub 2012/07/20 https://doi.org/10.1038/nature11252. 22810696. PMC3401966.

Carrion, V., Donlan, C. J., Campbell, K. J., Lavoie, C., & Cruz, F. (2011). Archipelago-wide island restoration in the Galapagos Islands: Reducing costs of invasive mammal eradication programs and reinvasion risk. *PLoS One, 6*(5), e18835. Epub 2011/05/19 https://doi.org/10.1371/journal.pone.0018835. 21589656. PMC3092746.

Cole, W. H. (1981). Efforts to explain spontaneous regression of cancer. *Journal of Surgical Oncology, 17*(3), 201–209. Epub 1981/01/01 https://doi.org/10.1002/jso.2930170302. 6166811.

Cui, Z., Willingham, M. C., Hicks, A. M., Alexander-Miller, M. A., Howard, T. D., Hawkins, G. A., et al. (2003). Spontaneous regression of advanced cancer:

Identification of a unique genetically determined, age-dependent trait in mice. *Proceedings of the National Academy of Sciences of the United States of America, 100*(11), 6682–6687. https://doi.org/10.1073/pnas.1031601100. 12724523. PMC164507.

Cusumano, D., Boldrini, L., Yadav, P., Yu, G., Musurunu, B., Chiloiro, G., et al. (2020). External validation of early regression index (ERITCP) as predictor of pathologic complete response in rectal cancer using magnetic resonance-guided radiation therapy. *International Journal of Radiation Oncology, Biology, Physics, 108*(5), 1347–1356. Epub 2020/08/08 https://doi.org/10.1016/j.ijrobp.2020.07.2323. 32758641.

da Costa, S. C. S., Bonadio, R. C., Gabrielli, F. C. G., Aranha, A. S., Dias Genta, M. L. N., Miranda, V. C., et al. (2019). Neoadjuvant chemotherapy with cisplatin and gemcitabine followed by chemoradiation versus chemoradiation for locally advanced cervical cancer: A randomized phase II trial. *Journal of Clinical Oncology, 37*(33), 3124–3131. Epub 2019/08/27 https://doi.org/10.1200/JCO.19.00674. 31449470.

Dashwood, R. H. (2006). Xenobiotic metabolism relevance to cancer. *The Journal of Nutrition, 136*(10), 2681S–2682S. Epub 2006/09/22 https://doi.org/10.1093/jn/136.10.2681S. 16988148.

Deng, Y., Chi, P., Lan, P., Wang, L., Chen, W., Cui, L., et al. (2019). Neoadjuvant modified FOLFOX6 with or without radiation versus fluorouracil plus radiation for locally advanced rectal cancer: Final results of the chinese FOWARC trial. *Journal of Clinical Oncology, 37*(34), 3223–3233. Epub 2019/09/27 https://doi.org/10.1200/JCO.18.02309. 31557064. PMC6881102.

Diaz, L. A., Jr., Williams, R. T., Wu, J., Kinde, I., Hecht, J. R., Berlin, J., et al. (2012). The molecular evolution of acquired resistance to targeted EGFR blockade in colorectal cancers. *Nature, 486*(7404), 537–540. Epub 2012/06/23 https://doi.org/10.1038/nature11219. 22722843. PMC3436069.

Fagan, W. F., & Holmes, E. E. (2006). Quantifying the extinction vortex. *Ecology Letters, 9*(1), 51–60. Epub 2006/09/09 https://doi.org/10.1111/j.1461-0248.2005.00845.x. 16958868.

Findlay, J. M., Castro-Giner, F., Makino, S., Rayner, E., Kartsonaki, C., Cross, W., et al. (2016). Differential clonal evolution in oesophageal cancers in response to neo-adjuvant chemotherapy. *Nature Communications*, 7, 11111. Epub 2016/04/06 https://doi.org/10.1038/ncomms11111. 27045317. PMC4822033.

Fokas, E., Strobel, P., Fietkau, R., Ghadimi, M., Liersch, T., Grabenbauer, G. G., et al. (2017). Tumor regression grading after preoperative chemoradiotherapy as a prognostic factor and individual-level surrogate for disease-free survival in rectal cancer. *Journal of the National Cancer Institute, 109*(12). https://doi.org/10.1093/jnci/djx095. Epub 2017/12/06 29206996.

Gallaher, J. A., Enriquez-Navas, P. M., Luddy, K. A., Gatenby, R. A., & ARA, A. (2018). Spatial heterogeneity and evolutionary dynamics modulate time to recurrence in continuous and adaptive cancer therapies. *Cancer Research, 78*(8), 2127–2139. Epub 2018/02/01 https://doi.org/10.1158/0008-5472.CAN-17-2649. 29382708. PMC5899666.

Gatenby, R. A., Artzy-Randrup, Y., Epstein, T., Reed, D. R., & Brown, J. S. (2020). Eradicating metastatic cancer and the eco-evolutionary dynamics of Anthropocene extinctions. *Cancer Research, 80*, 613–623. https://doi.org/10.1158/0008-5472.CAN-19-1941. Epub 2019/11/28 31772037.

Gatenby, R. A., & Brown, J. S. (2020a). Integrating evolutionary dynamics into cancer therapy. *Nature Reviews. Clinical Oncology*. https://doi.org/10.1038/s41571-020-0411-1. Epub 2020/07/24 32699310.

Gatenby, R. A., & Brown, J. S. (2020b). The evolution and ecology of resistance in cancer therapy. *Cold Spring Harbor Perspectives in Medicine, 10*(11), a040972. https://doi.org/10.1101/cshperspect.a040972. Epub 2020/11/04 33139405. PMC7605238.

Gatenby, R. A., Brown, J., & Vincent, T. (2009). Lessons from applied ecology: Cancer control using an evolutionary double bind. *Cancer Research*, *69*(19), 7499–7502. Epub 2009/09/16 https://doi.org/10.1158/0008-5472.can-09-1354. 19752088.
Gatenby, R. A., Gillies, R. J., & Brown, J. S. (2011). Of cancer and cave fish. *Nature Reviews. Cancer*, *11*(4), 237–238. Epub 2011/05/10 https://doi.org/10.1038/nrc3036. 21548400. PMC3971416.
Gatenby, R. A., Silva, A. S., Gillies, R. J., & Frieden, B. R. (2009). Adaptive therapy. *Cancer Research*, *69*(11), 4894–4903. https://doi.org/10.1158/0008-5472.CAN-08-3658. 19487300. PMC3728826.
Gatenby, R. A., Zhang, J., & Brown, J. S. (2019, https://doi.org/10.1158/0008-5472.CAN-19-0807). First strike—Second strike strategies in metastatic cancer: Lessons from the evolutionary dynamics of extinction. *Cancer Research*, *79*(13), 3174–3177.
Gerlinger, M., Rowan, A. J., Horswell, S., Math, M., Larkin, J., Endesfelder, D., et al. (2012). Intratumor heterogeneity and branched evolution revealed by multiregion sequencing. *The New England Journal of Medicine*, *366*(10), 883–892. Epub 2012/03/09 https://doi.org/10.1056/NEJMoa1113205. 22397650. PMC4878653.
Gillies, R. J., Brown, J. S., ARA, A., & Gatenby, R. A. (2018). Eco-evolutionary causes and consequences of temporal changes in intratumoural blood flow. *Nature Reviews. Cancer*. https://doi.org/10.1038/s41568-018-0030-7. Epub 2018/06/13 29891961.
Gilpin, M. E. S. M. (1986). Minimum viable populations: Proceses of species extinction. In M. E. Soule (Ed.), *Conservation biology: The science of scarcity and diversity sunderland* (pp. 19–34). Massachusetts: Sinauer.
Gonzalez, A., Ronce, O., Ferriere, R., & Hochberg, M. E. (2013). Evolutionary rescue: An emerging focus at the intersection between ecology and evolution. *Philosophical Transactions of the Royal Society of London. Series B, Biological Sciences*, *368*(1610), 20120404. Epub 2012/12/05 https://doi.org/10.1098/rstb.2012.0404. 23209175. PMC3538460.
Guo, J. (2006). Invasive species. The Galapagos Islands kiss their goat problem goodbye. *Science*, *313*(5793), 1567. Epub 2006/09/16 https://doi.org/10.1126/science.313.5793.1567. 16973856.
Hardiman, K. M., Ulintz, P. J., Kuick, R. D., Hovelson, D. H., Gates, C. M., Bhasi, A., et al. (2016). Intra-tumor genetic heterogeneity in rectal cancer. *Laboratory Investigation*, *96*(1), 4–15. Epub 2015/11/17 https://doi.org/10.1038/labinvest.2015.131. 26568296. PMC4695247.
Helmlinger, G., Yuan, F., Dellian, M., & Jain, R. K. (1997). Interstitial pH and pO2 gradients in solid tumors in vivo: High-resolution measurements reveal a lack of correlation. *Nature Medicine*, *3*(2), 177–182. Epub 1997/02/01 https://doi.org/10.1038/nm0297-177. 9018236.
Hong, R. L., Hsiao, C. F., Ting, L. L., Ko, J. Y., Wang, C. W., Chang, J. T. C., et al. (2018). Final results of a randomized phase III trial of induction chemotherapy followed by concurrent chemoradiotherapy versus concurrent chemoradiotherapy alone in patients with stage IVA and IVB nasopharyngeal carcinoma-Taiwan Cooperative Oncology Group (TCOG) 1303 Study. *Annals of Oncology*, *29*(9), 1972–1979. Epub 2018/07/18 https://doi.org/10.1093/annonc/mdy249. 30016391.
Hoyal Cuthill, J. F., Guttenberg, N., & Budd, G. E. (2020). Impacts of speciation and extinction measured by an evolutionary decay clock. *Nature*, *588*(7839), 636–641. Epub 2020/12/11 https://doi.org/10.1038/s41586-020-3003-4. 33299185.
Hu, Z., & Curtis, C. (2020). Looking backward in time to define the chronology of metastasis. *Nature Communications*, *11*(1), 3213. Epub 2020/06/27 https://doi.org/10.1038/s41467-020-16995-y. 32587245. PMC7316862.
Ibrahim-Hashim, A., Robertson-Tessi, M., Enriquez-Navas, P. M., Damaghi, M., Balagurunathan, Y., Wojtkowiak, J. W., et al. (2017). Defining cancer subpopulations

by adaptive strategies rather than molecular properties provides novel insights into intratumoral evolution. *Cancer Research*, *77*(9), 2242–2254. https://doi.org/10.1158/0008-5472.CAN-16-2844. 28249898.

Johnson, K. E., Howard, G., Mo, W., Strasser, M. K., Lima, E., Huang, S., et al. (2019). Cancer cell population growth kinetics at low densities deviate from the exponential growth model and suggest an Allee effect. *PLoS Biology*, *17*(8), e3000399. Epub 2019/08/06 https://doi.org/10.1371/journal.pbio.3000399. 31381560. PMC6695196.

Kam, Y., Das, T., Tian, H., Foroutan, P., Ruiz, E., Martinez, G., et al. (2015). Sweat but no gain: Inhibiting proliferation of multidrug resistant cancer cells with "ersatzdroges". *International Journal of Cancer*, *136*(4), E188–E196. Epub 2014/08/27 https://doi.org/10.1002/ijc.29158. 25156304. 4262548.

Konstorum, A., Hillen, T., & Lowengrub, J. (2016). Feedback regulation in a cancer stem cell model can cause an allee effect. *Bulletin of Mathematical Biology*, *78*(4), 754–785. Epub 2016/04/27 https://doi.org/10.1007/s11538-016-0161-5. 27113934. PMC4946570.

Lavoie, C. C. F., Harcourt-Carrasco, S., & Moya, M. (2007). *The atlas of Isabela Project: An illustrative document of the biggests successful goat eradication porject on island*. Charles Darwin Foundation.

Leakey, R. E., & Lewin, R. (1995). *The sixth extinction: Patterns of life and the future of humankind* (1st ed., p. 271). New York: Doubleday.

Lloyd, M. C., Cunningham, J. J., Bui, M. M., Gillies, R. J., Brown, J. S., & Gatenby, R. A. (2016). Darwinian dynamics of intratumoral heterogeneity: Not solely random mutations but also variable environmental selection forces. *Cancer Research*, *76*(11), 3136–3144. https://doi.org/10.1158/0008-5472.CAN-15-2962. 27009166. PMC5384728.

Ma, Y., Poole, K., Goyette, J., & Gaus, K. (2017). Introducing membrane charge and membrane potential to t cell signaling. *Frontiers in Immunology*, *8*, 1513. Epub 2017/11/25 https://doi.org/10.3389/fimmu.2017.01513. 29170669. PMC5684113.

Ma, Y., Zhang, P., Wang, F., Yang, J., Yang, Z., & Qin, H. (2010). The relationship between early embryo development and tumourigenesis. *Journal of Cellular and Molecular Medicine*, *14*(12), 2697–2701. Epub 2010/10/30 https://doi.org/10.1111/j.1582-4934.2010.01191.x. 21029369. PMC3822720.

Mullon, C. D. W. J., & Ohtsuki, H. (2020). Evolutionary feedbacks between genes and their extended effects under limited gene flow and biased ecological inheritance. *BioRxiv*.

Murugaesu, N., Wilson, G. A., Birkbak, N. J., Watkins, T., McGranahan, N., Kumar, S., et al. (2015). Tracking the genomic evolution of esophageal adenocarcinoma through neoadjuvant chemotherapy. *Cancer Discovery*, *5*(8), 821–831. Epub 2015/05/25 https://doi.org/10.1158/2159-8290.CD-15-0412. 26003801. PMC4529488.

Neufeld, Z., von Witt, W., Lakatos, D., Wang, J., Hegedus, B., & Czirok, A. (2017). The role of Allee effect in modelling post resection recurrence of glioblastoma. *PLoS Computational Biology*, *13*(11), e1005818. Epub 2017/11/18 https://doi.org/10.1371/journal.pcbi.1005818. 29149169. PMC5711030.

Nougaret, S., Rouanet, P., Molinari, N., Pierredon, M. A., Bibeau, F., Azria, D., et al. (2012). MR volumetric measurement of low rectal cancer helps predict tumor response and outcome after combined chemotherapy and radiation therapy. *Radiology*, *263*(2), 409–418. Epub 2012/03/23 https://doi.org/10.1148/radiol.12111263. 22438364.

Odum, H. T., & Allee, W. C. (1954). Note on the stable point of populations showing both intraspecific cooperation and disoperation. *Ecology*, *35*(1), 95–97. https://doi.org/10.2307/1931412.

Orlando, P. A., Gatenby, R. A., & Brown, J. S. (2012). Cancer treatment as a game: Integrating evolutionary game theory into the optimal control of chemotherapy. *Physical Biology*, *9*(6), 065007. Epub 2012/12/01 https://doi.org/10.1088/1478-3975/9/6/065007. 23197192. PMC3653600.

Otto, S. P. (2018). Adaptation, speciation and extinction in the anthropocene. *Proceedings of the Biological Sciences*, *285*(1891). https://doi.org/10.1098/rspb.2018.2047, 20182047. Epub 2018/11/16 30429309. PMC6253383.

Park, D. S., Robertson-Tessi, M., Luddy, K. A., Maini, P. K., Bonsall, M. B., Gatenby, R. A., et al. (2019). The goldilocks window of personalized chemotherapy: Getting the immune response just right. *Cancer Research*, *79*(20), 5302–5315. https://doi.org/10.1158/0008-5472.Can-18-3712. canres.3712.2018.

Parrish, J. K., & Edelstein-Keshet, L. (1999). Complexity, pattern and evolutionary trade-offs in animal aggregation. *Science*, *284*, 99–100. https://doi.org/10.1126/science.284.5411.99.

Petersen, S. H., Harling, H., Kirkeby, L. T., Wille-Jorgensen, P., & Mocellin, S. (2012). Postoperative adjuvant chemotherapy in rectal cancer operated for cure. *Cochrane Database of Systematic Reviews*, *2012*(3), CD004078. Epub 2012/03/16 https://doi.org/10.1002/14651858.CD004078.pub2. 22419291. PMC6599875.

Petrelli, F., Trevisan, F., Cabiddu, M., Sgroi, G., Bruschieri, L., Rausa, E., et al. (2020). Total neoadjuvant therapy in rectal cancer: A systematic review and meta-analysis of treatment outcomes. *Annals of Surgery*, *271*(3), 440–448. Epub 2019/07/19 https://doi.org/10.1097/SLA.0000000000003471. 31318794.

Raup, D. M. (1991). *Extinction: Bad genes or bad luck?. Vol. xvii* (p. 210). New York: W.W. Norton.

Reed, D. R., Metts, J., Pressley, M., Fridley, B. L., Hayashi, M., Isakoff, M. S., et al. (2020). An evolutionary framework for treating pediatric sarcomas. *Cancer*, *126*, 2577–2587. https://doi.org/10.1002/cncr.32777. Epub 2020/03/17 32176331.

Saether, B. E., Engen, S., Islam, A., McCleery, R., & Perrins, C. (1998). Environmental stochasticity and extinction risk in a population of a small songbird, the great tit. *The American Naturalist*, *151*(5), 441–450. Epub 2008/09/25 https://doi.org/10.1086/286131. 18811318.

Sauer, R., Becker, H., Hohenberger, W., Rodel, C., Wittekind, C., Fietkau, R., et al. (2004). Preoperative versus postoperative chemoradiotherapy for rectal cancer. *The New England Journal of Medicine*, *351*(17), 1731–1740. Epub 2004/10/22 https://doi.org/10.1056/NEJMoa040694. 15496622.

Schneider, J., Bak, M., Efferth, T., Kaufmann, M., Mattern, J., & Volm, M. (1989). P-glycoprotein expression in treated and untreated human breast cancer. *British Journal of Cancer*, *60*(6), 815–818. 2574987. 2247284.

Schrag, D., Weiser, M. R., Goodman, K. A., Gonen, M., Hollywood, E., Cercek, A., et al. (2014). Neoadjuvant chemotherapy without routine use of radiation therapy for patients with locally advanced rectal cancer: A pilot trial. *Journal of Clinical Oncology*, *32*(6), 513–518. Epub 2014/01/15 https://doi.org/10.1200/JCO.2013.51.7904. 24419115. PMC5795691.

Sewalt, L., Harley, K., van Heijster, P., & Balasuriya, S. (2016). Influences of Allee effects in the spreading of malignant tumours. *Journal of Theoretical Biology*, *394*, 77–92. Epub 2016/01/24 https://doi.org/10.1016/j.jtbi.2015.12.024. 26802481.

Shen, F., Chu, S., Bence, A. K., Bailey, B., Xue, X., Erickson, P. A., et al. (2008). Quantitation of doxorubicin uptake, efflux, and modulation of multidrug resistance (MDR) in MDR human cancer cells. *The Journal of Pharmacology and Experimental Therapeutics*, *324*(1), 95–102. https://doi.org/10.1124/jpet.107.127704. 17947497.

Shoemaker, L. G., Sullivan, L. L., Donohue, I., Cabral, J. S., Williams, R. J., Mayfield, M. M., et al. (2020). Integrating the underlying structure of stochasticity into community ecology. *Ecology*, *101*(2), e02922. Epub 2019/10/28 https://doi.org/10.1002/ecy.2922. 31652337. PMC7027466.

Stankova, K., Brown, J. S., Dalton, W. S., & Gatenby, R. A. (2018). Optimizing cancer treatment using game theory: A review. *JAMA Oncology*. https://doi.org/10.1001/jamaoncol.2018.3395. Epub 2018/08/12 30098166.

Strebhardt, K., & Ullrich, A. (2008). Paul Ehrlich's magic bullet concept: 100 years of progress. *Nature Reviews. Cancer*, *8*(6), 473–480. Epub 2008/05/13 https://doi.org/10.1038/nrc2394. 18469827.

Tier, C., & Hanson, F. B. (1981). Persistence in density dependent stochastic populations. *Mathematical Biosciences*, *53*, 89–117. https://doi.org/10.1016/0025-5564(81)90041-9.

Tobin, P. C., Berec, L., & Liebhold, A. M. (2011). Exploiting Allee effects for managing biological invasions. *Ecology Letters*, *14*, 615–624. https://doi.org/10.1111/j.1461-0248.2011.01614.x.

Uchi, R., Takahashi, Y., Niida, A., Shimamura, T., Hirata, H., Sugimachi, K., et al. (2016). Integrated multiregional analysis proposing a new model of colorectal cancer evolution. *PLoS Genetics*, *12*(2), e1005778. Epub 2016/02/20 https://doi.org/10.1371/journal.pgen.1005778. 26890883. PMC4758664.

Ueki, T., Manabe, T., Inoue, S., Ienaga, J., Yamanaka, N., Egami, T., et al. (2016). A feasibility study of neoadjuvant XELOX without radiotherapy for locally advanced lower rectal cancer. *Anticancer Research*, *36*(2), 741–747. Epub 2016/02/07 26851033.

Xu, Z., Mohile, S. G., Tejani, M. A., Becerra, A. Z., Probst, C. P., Aquina, C. T., et al. (2017). Poor compliance with adjuvant chemotherapy use associated with poorer survival in patients with rectal cancer: An NCDB analysis. *Cancer*, *123*(1), 52–61. Epub 2016/08/26 https://doi.org/10.1002/cncr.30261. 27560162.

Yeo, S. G., Kim, D. Y., Park, J. W., Oh, J. H., Kim, S. Y., Chang, H. J., et al. (2012). Tumor volume reduction rate after preoperative chemoradiotherapy as a prognostic factor in locally advanced rectal cancer. *International Journal of Radiation Oncology, Biology, Physics*, *82*(2), e193–e199. Epub 2011/05/25 https://doi.org/10.1016/j.ijrobp.2011.03.022. 21605941.

CHAPTER THREE

Early colorectal cancer detection—Current and evolving challenges in evidence, guidelines, policy, and practices

Robert A. Smith[a,*], Stacey Fedewa[b], and Rebecca Siegel[c]
[a]Cancer Prevention and Early Detection Department, American Cancer Society, Atlanta, GA, United States
[b]Screening and Risk Factors Research, Surveillance and Health Equity Science Department, American Cancer Society, Atlanta, GA, United States
[c]Surveillance Research, Surveillance and Health Equity Science Department, American Cancer Society, Atlanta, GA, United States
*Corresponding author: e-mail address: robert.smith@cancer.org

Contents

Synopsis

The understanding at the beginning of the last century that colorectal cancer began as a localized disease that progressed and became systemic, and that most colorectal cancer arose from adenomatous polyps gave rise to aggressive attempts at curative treatment and eventually attempts to detect advanced lesions before they progressed to invasive

Advances in Cancer Research, Volume 151
ISSN 0065-230X
https://doi.org/10.1016/bs.acr.2021.03.005

disease. In the last four decades, steadily greater uptake of screening has led to reductions in colorectal cancer incidence and mortality. However, the fullest potential of screening is not being met due to the lack of organized screening, where a systems approach could lead to higher rates of screening of average and high risk groups, higher quality screening, and prompt followup of adults with positive screening tests.

Abstract

Since the beginning of the 20th century, there has been a general understanding that colorectal cancer is a clonal disease that progresses from a localized stage with a favorable prognosis through progressively more advanced stages which have progressively worse prognosis. That understanding led first to determined efforts to detect and treat early stage symptomatic disease, and then to detect pre-symptomatic colorectal cancer and precursor lesions, where there was hope that the natural history of the disease could be arrested and the incidence and premature mortality of colorectal cancer averted. Toward the end of the last century, guidelines for colorectal cancer screening, growth in the number of technical options for screening, and a steady increase in the proportion of the adult population who attended screening contributed to the beginning of a significant decline in colorectal cancer incidence and mortality. Despite this progress, colorectal cancer remains the third leading cause of death among men and women in the United States. Screening for early detection of precursor lesions and localized cancer offers the single most productive opportunity to further reduce the burden of disease, and yet nearly four in five deaths from colorectal cancer are associated with having never been screened, not recently screened, or not followed up for an abnormal screening test. This simple observation is a call to action in all communities to apply existing knowledge to fulfill the potential to prevent avertable incidence and mortality.

Key points

- Colorectal cancer is the second leading cause of cancer death worldwide and in the United States.
- The clinical and scientific foundation for understanding the natural history of colorectal cancer, and the potential for prevention and early detection of localized cancers through screening examinations was established nearly a century ago.
- Although considerable progress has been made in reducing colorectal cancer incidence and mortality through screening, greater progress could be made through prioritizing the application of existing knowledge in public health programs and routine clinical care, and improving access to care.

1. Introduction

In the past century, colorectal cancer (CRC) has been a leading cause of cancer death in higher income countries, and over this period it has remained among the top three causes of cancer death in the United States

(US) among men and women combined (Wingo et al., 2003). Although formal recommendations for CRC screening were not promoted until 1980 (Eddy, 1980a, 1980b), the essential requirements for screening outlined by Wilson and Jungner in 1968 accumulated over many decades beginning in the late 19th century, most notably the knowledge of the natural history of the disease (Smith, 2020; Wilson & Jungner, 1968). In this article, we will describe the descriptive epidemiology of CRC in the US, the evolution of knowledge and technical development that has led to approaches to prevention and early detection, current utilization of screening, and enduring challenges and barriers that stand in the way of achieving the fullest potential to prevent CRC and avert avoidable deaths.

2. The descriptive epidemiology of colorectal cancer

As shown in Table 1, approximately 147,950 new cases of CRC were diagnosed in the United States in 2020, including 104,610 people with colon tumors and 43,340 people with rectal tumors (Siegel et al., 2020). Although most of these individuals were aged 50 years and older, 17,930 (12%) were younger than 50 years.

In addition, there were an estimated 53,200 CRC deaths in 2020, including 3640 decedents (7%) aged younger than 50 years. Deaths specific to anatomic subsite of incident cancer are not reportable because studies have shown that more than one-third of deaths from rectal cancer are misclassified as colon cancer on death certificates (Yin, Morris, Bates, & German, 2011). This misclassification hinders colorectal cancer research and is partly due to the use of the term "colon cancer" as a catch-all to refer to rectal cancer as well as colon cancer in educational messaging.

Table 1 Estimated numbers[a] of new colorectal cancer cases and deaths by age, United States, 2020.

	Cases			Deaths[a]
Age	**Colorectum**	**Colon**	**Rectum**	**Colorectum**
0–49 years	17,930	11,540	6390	3640
50–64 years	50,010	32,290	17,720	13,380
65 + years	80,010	60,780	19,230	36,180
All ages	147,950	104,610	43,340	53,200

[a]Estimates are rounded to the nearest 10 and exclude in situ carcinoma. Deaths for colon and rectal cancers are combined because a large proportion of rectal cancer deaths are misclassified as colon.

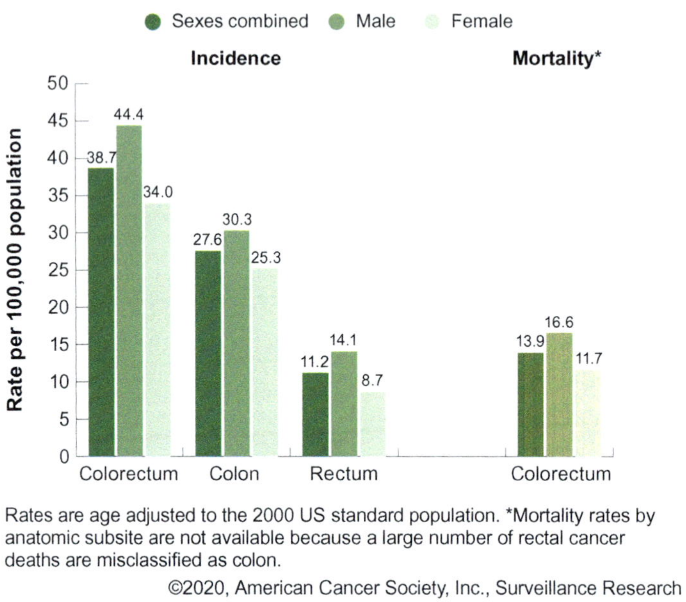

Fig. 1 Colorectal cancer incidence and mortality rates by subsite and sex, U.S., 2012–17.

The average annual age-standardized CRC incidence rate was 38.7 per 100,000 persons in the United States during 2012–16 and was about 30% higher in males than in females (Fig. 1).

As shown in Fig. 2, risk escalates rapidly with age; rates approximately double every 5 years before age 50 and increase by 30% from 55 years upward. For example, the rate increases from 90.2 (per 100,000 population) in individuals aged 60–64 years to 121.4 in ages 65–69 years, peaking at 258.8 in those 85 years and older.

Among people in their early 50s, the natural pattern of age-specific incidence is disrupted by lead time associated with first-time CRC screening and the detection of prevalent, asymptomatic cancers in the subset of the population who begins screening at or close to age 50. For example, incidence rates are higher in individuals aged 50–51 years than in those 52–55 years (Fig. 2 inset). However, although guidelines have recommended that CRC screening should begin at age 50, in a large study in Minnesota Sinicrope et al. (2012) observed that the mean age of initiating CRC screening was 55, meaning some adults do not begin screening until they are closer to age 60.

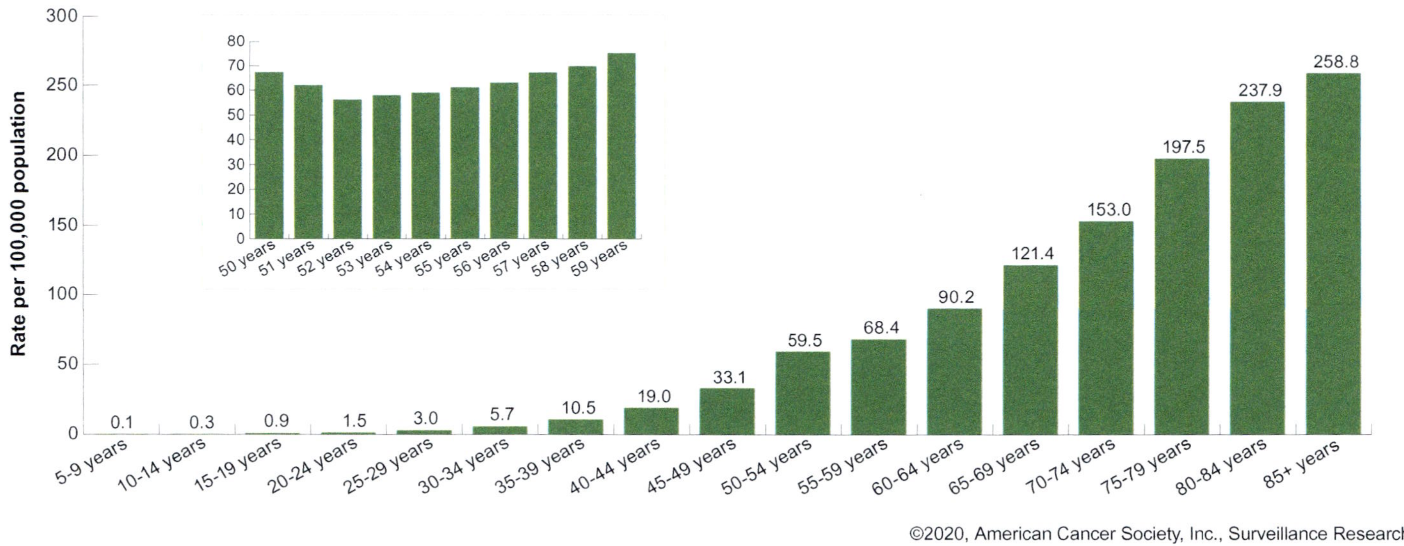

Fig. 2 Age-specific colorectal cancer incidence rates, U.S., 2012–16.

2.1 Temporal trends in incidence

CRC incidence (1975–2016) and mortality (1930–2017) trends are shown by sex in Fig. 3. CRC incidence rates increased in the United States from 1975 through the mid-1980s but have since generally decreased.

The decline before 2000 is attributed equally to changing patterns in risk factors (e.g., reductions in smoking) and increasing use of CRC screening (Edwards et al., 2010), whereas the accelerated decline during the 2000s is thought to reflect a period of rapid uptake of first-time colonoscopy screening. Colonoscopy has a larger capacity for cancer prevention than the fecal occult blood test (FOBT), which was previously the most common

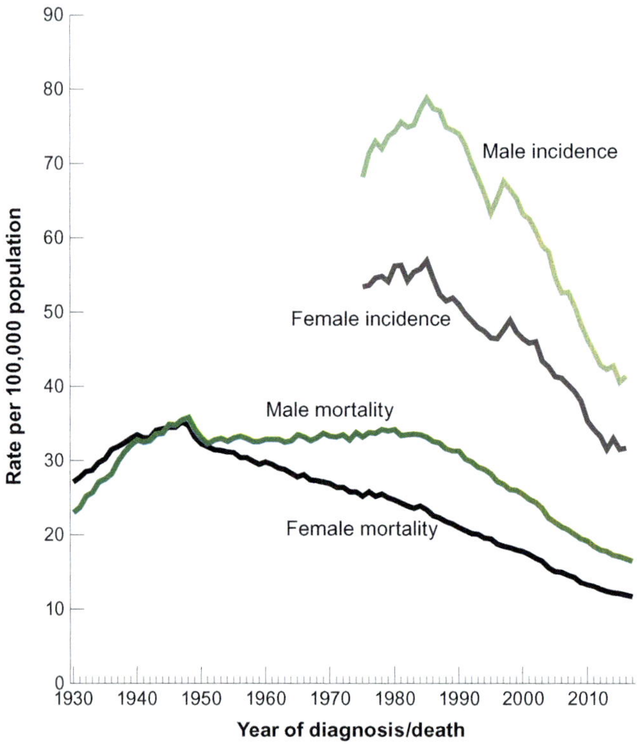

Rates are age adjusted to the 2000 US standard population. Incidence rates are adjusted for delays in reporting and exclude appendix. Due to changes in International Classification of Diseases (ICD) coding, numerator information for mortality has changed over time.

Fig. 3 Trends in colorectal cancer incidence (1975–2016) and mortality (1930–2017) rates by sex, U.S.

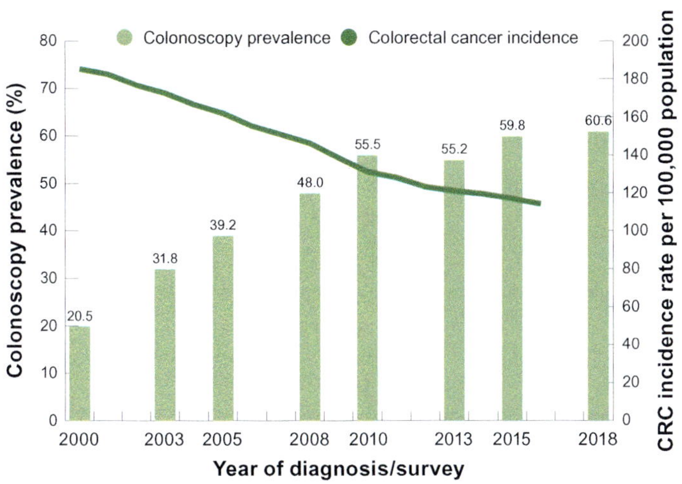

Incidence rates exclude appendix and are age-adjusted to the 2000 US standard population. Rates are adjusted to account for reporting delays. Colonoscopy screening prevalence indicates a colonoscopy within the past 10 years at the time of survey.

Fig. 4 Trends in colonoscopy use (2000–18) and colorectal cancer incidence rates (2000–16) among individuals aged 50 years and older, U.S.

CRC screening test (Schenck et al., 2009). As shown in Fig. 4, among adults aged 50 years and older, colonoscopy prevalence tripled from 20% in 2000 to 61% in 2018 (National Center for Health Statistics & Division of Health Interview Statistics, 2019). The surge is strongly associated with Medicare expansion of colonoscopy screening coverage in 2001 from only those at high-risk to all individuals 50 and older (Schenck et al., 2009).

Fig. 5 shows trends in CRC incidence and mortality in three age groups. Since 2011, declines in incidence are confined to individuals 65 years and older, with rates in ages 50–64 years increasing by 1.0% per year from 2011 to 2016 following a decade of decline (Siegel et al., 2020).

The reversal in the direction of the trend mirrors that which occurred in individuals younger than 50 years in the mid-1990s and likely reflects elevated disease risk in generations born since 1950 being carried forward by aging generations, a phenomenon referred to as a birth cohort effect (Holford, 1991). A strong birth cohort effect has been demonstrated in CRC incidence patterns in the United States (Siegel, Fedewa, et al., 2017) and in other high-income countries (Araghi et al., 2019), and also described by Patel et al. in this volume. Similar to early patterns among

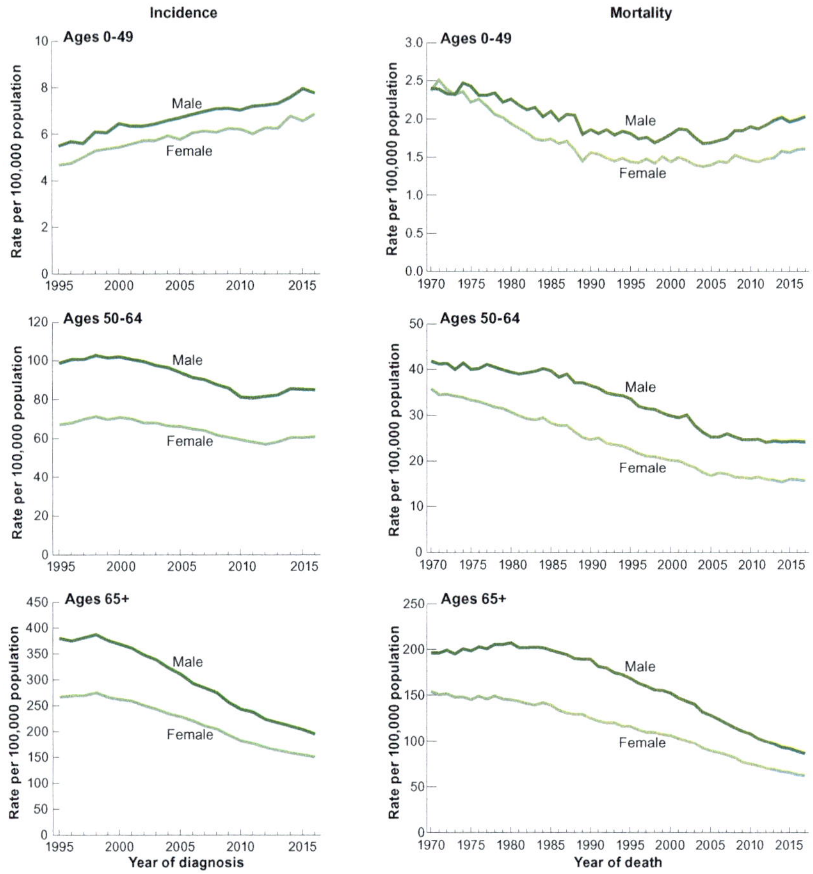

Rates are age adjusted to the 2000 US standard population. Incidence rates are adjusted for reporting delays and exclude appendix.

©2020, American Cancer Society, Inc., Surveillance Research

Fig. 5 Trends in colorectal cancer incidence (1995–2016) and mortality (1970–2017) rates by age and sex, U.S.

younger adults, the increasing trend in individuals 50–64 years excludes malignancies in the proximal colon, for which rates are stable.

During 2012–16, incidence rates in individuals aged younger than 50 years rose similarly for tumors in the proximal and distal colon as well as in the rectum by about 2% annually (Siegel et al., 2020). The trend is confined to advanced-stage disease diagnosis and is steepest among people who are non-Hispanic White or Native American. Incidence rates are also increasing slightly (by 0.4–0.5% annually) among young Black and Asian/Pacific Islander adults, driven exclusively by rectal tumors. Similar patterns have been observed globally (Siegel et al., 2019).

2.2 Stage at diagnosis and survival

As of January 1, 2019, there were more than 1.5 million Americans living with a history of CRC (Miller et al., 2019). As shown in Fig. 6, the most important predictor of survival after a cancer diagnosis is the stage of disease. 5-year relative survival rates for CRC range from 90% for patients with localized stage disease to 14% for those diagnosed with distant stage.

Individuals with rectal tumors are slightly more likely than those with colon tumors to be diagnosed with localized disease, probably because of the earlier appearance of symptoms. Despite the rapid uptake of screening since 2000, the proportion of cases diagnosed at a localized stage has only increased slightly, from 34% in the mid-1990s to 36% during 2012 through 2016, because colonoscopy predominantly prevents cancer through the removal of slow-growing premalignant polyps (Lieberman et al., 2000).

The 5-year relative survival rate for CRC has increased from 50% in the mid-1970s to 64% during 2009 through 2015 (Howlader et al., 2020). These gains reflect improvements in earlier detection through screening, advances in imaging techniques (e.g., positron emission tomography) that improve staging, and treatment. Headway in the treatment of metastatic disease, including improved surgical techniques and increased cancer-directed surgery (Wancata, Banerjee, Muenz, Haymart, & Wong, 2016), advances in the treatment of liver metastases (Chua, Saxena, Chu, Zhao, & Morris, 2011; Ruers et al., 2017; Zampino et al., 2016), and the development of targeted therapies (Kopetz et al., 2009; Murphy, Harlan, Lund, Lynch, & Geiger, 2015; Piawah & Venook, 2019) is evident in population-based survival rates in recent decades. For example, the 2-year relative survival rate for patients diagnosed with distant-stage disease increased from 21% during the mid-1990s to 37% during 2009 through 2015, with a larger improvement for rectal cancer (from 22% to 41%) than for colon cancer (from 21% to 36%) (Siegel et al., 2020). This progress is evident across race and age, although gains are most prominent among patients who are white and non-elderly (Sineshaw, Robbins, & Jemal, 2014).

Fig. 7 shows the CRC stage distribution among U.S. adults over the period 2012–16 by age and race/ethnicity. Stage at diagnosis, and cancer survival independent of stage, are associated with socioeconomic status, race, and age (Andrew et al., 2018; Ward et al., 2004).

Individuals who are Black are the most likely of any racial or ethnic group to be diagnosed with distant-stage CRC (25% vs 20% of NHWs and APIs) and also have the lowest overall 5-year survival rate (60% vs 68% among APIs and 66% among NHWs). These disparities are largely

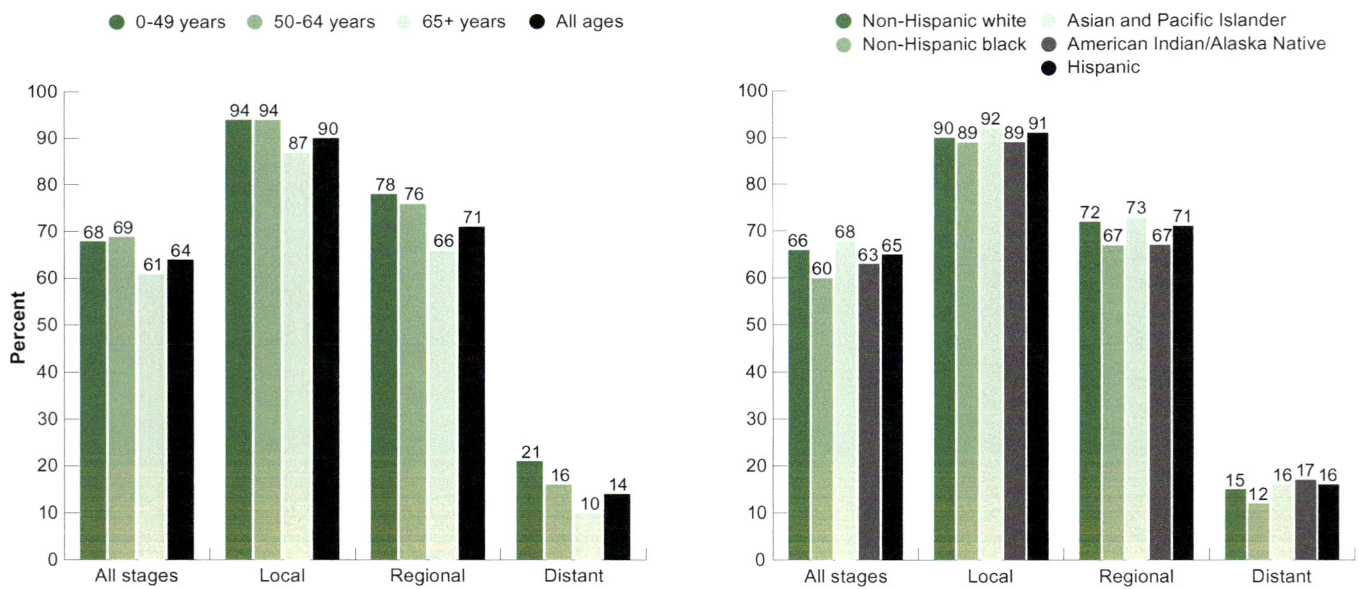

*Cause-specific survival rates are the probability of not dying from colorectal cancer within 5 years of diagnosis. Rates are based on cases diagnosed from 2009 to 2015, all followed through 2016. Rates for American Indians/Alaska Natives are based on small case numbers, particularly for distant-stage disease.

Fig. 6 Colorectal cancer 5-year relative survival rates (%) by age and race/ethnicity, U.S., 2009–15.

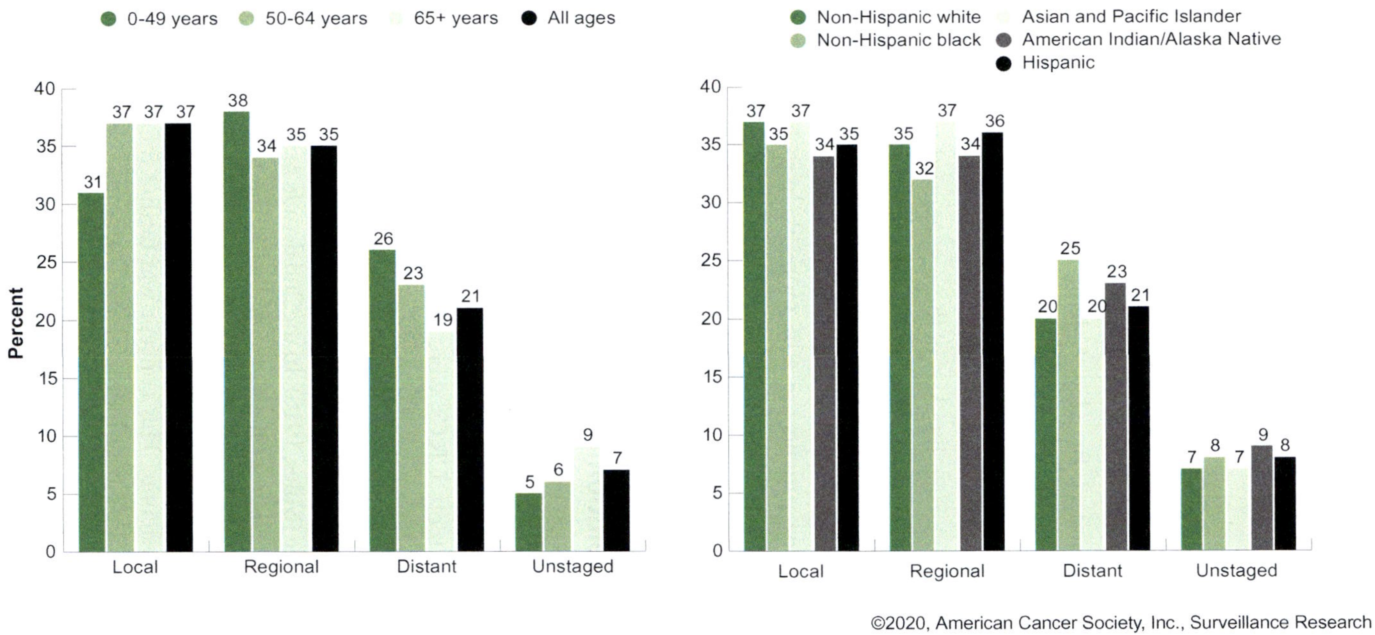

Fig. 7 Colorectal cancer stage distribution (%) by age and race/ethnicity, 2012–16.

driven by longstanding socioeconomic inequalities stemming from systemic racism that result in differences in access to early detection and the receipt of timely, high-quality treatment (Bach et al., 2002; Ward et al., 2004). Access to care is directly related to stage at diagnosis, which plays the largest role in racial/ethnic survival disparities (Ellis et al., 2018). A recent nationwide study found that greater than one-half of the Black-White survival disparity is explained by differences in insurance status, and one-quarter is because of differences in tumor characteristics (e.g., grade, anatomic location) (Sineshaw, Ng, Flanders, Brawley, & Jemal, 2018). Notably, when CRC is diagnosed at a localized stage, 5-year survival is comparable (range, 89–92%) across racial/ethnic groups.

As Fig. 7 shows, early-onset cancers are much more likely to be diagnosed at an advanced stage. Just 31% of those younger than 50 are diagnosed with localized stage disease vs 37% of those 50 and older; conversely, 26% of young adults are diagnosed with metastatic disease compared to 23% of those aged 50–64 years and 19% of those 65 years and older. Even after accounting for screening, young patients are 40% more likely to be diagnosed with advanced disease, largely because of avoidable delays in diagnosis (Chen, Sundaram, Chew, & Ladabaum, 2017).

3. The evolution of colorectal cancer screening

Prior to the final decades of the 19th century, treating colorectal cancer was judged to be futile and mostly palliative, owing to long-standing deterministic concepts about the inherently systemic nature of carcinogenesis, and because most CRCs were advanced at the time of diagnosis (Ballantyne, 1988). However, improvements in surgical techniques for rectal cancer in the early 20th century, with excision of zones associated with recurrence, led to greater confidence that treating rectal cancer with surgical excision was feasible (Miles, 1910). In the 2nd decade of the 20th century the prevailing view of carcinogenesis changed from the belief that cancer was inherently systemic at inception to the understanding that cancer began as a single cell and progressed from a localized disease to more advanced stages when it became systemic. This concept was further refined by Lockhart-Mummery and Duke's description of the adenoma-carcinoma sequence in the natural history of colorectal cancer (Lockhart-Mummery & Dukes, 1927), providing greater impetus to pursue curative intent therapy for earlier stage disease, and to also to detect and remove benign polyps (Ballantyne, 1988).

The understanding that "earlier is better" naturally led to the pursuit of strategies to diagnosis CRC at the first indication of symptoms, typically rectal bleeding, and even before patients became symptomatic since it was presumed that occult bleeding must precede visible bleeding. However, early testing for occult blood in the laboratory was complex and unreliable. The earliest approach to earlier diagnosis was rigid proctosigmoidoscopy, which was developed in 1895, and represented the first attempts at opportunistic screening examinations for early detection of rectal cancer (Winawer, 2015). On-going work at St. Marks Hospital in London in the 1930s demonstrated that patients diagnosed at an earlier stage had better survival (Winawer, 2015). With growing acceptance that CRC arose from polyps, adults with polyps detected with sigmoidoscopy often would undergo a barium enema to determine if there were additional lesions outside of the reach of the sigmoidoscope, and if so, surgery was undertaken to remove them. Winawer described the morbidity and mortality associated with the procedures in this early period as "significant"(Winawer, 2015).

As data accumulated on the use of sigmoidoscopy in the examination of symptomatic and asymptomatic adults in the first half of the twentieth century, it had become well established that improvements in surgical technique and the opportunity to intervene when the disease was diagnosed early had resulted in an increase in the rate of operability and a decline in mortality. Although sigmoidoscopy was regarded as a surgical procedure, as early as 1930, Paulson argued that internists should regard proctosigmoidoscopy as any other endoscopic method and utilize it during examinations "when there is a history of even the slightest deviation from the usual bowel habits" (Paulson, 1930), and in 1948 Kirsner et al. lamented that too little progress was being made in the early detection of the early occult disease due to "infrequent use of the examining finger and the proctoscope" (Kirsner, Palmer, et al., 1948). In their review of gastroenterological literature from 1948 to 49, they cited studies showing a long duration of symptoms before medical attention had been sought, that nearly all patients reported at least vague symptoms during the period prior to diagnosis, and that in the examination of asymptomatic adults, a significant fraction of occult cancers was within the reach of the examining finger and sigmoidoscope (Kirsner et al., 1948).

By the early 1960s, a growing number of reports in the literature described case series of screening asymptomatic adults with sigmoidoscopy for rectal and colon cancer with favorable findings, leading to "expert opinions" in the medical literature endorsing office-based efforts to diagnose CRC early (Gilbertsen et al., 1961). In a major development that opened the door to a simple approach to wide-spread CRC screening, in 1967 Greegor reported that occult, early-stage colorectal cancer could be detected utilizing a commercially available guaiac-impregnated card at home (Hemoccult®) to detect occult blood, (Greegor, 1967) and in 1969 he endorsed routine testing for the presence of fecal occult blood with the guaiac-impregnated cards (Greegor, 1969).

In the 1970s, the long-standing tradition of practicing medicine based on expert consensus began to be replaced by the idea that clinical practice should be supported by evidence from well-designed studies, principally randomized clinical trials (RCTs) (Smith & Rennie, 2014). Although expert opinion regularly reinforced the value of sigmoidoscopy and stool testing for the early detection of CRC, neither sigmoidoscopy (rigid or flexible) nor fecal occult blood tests (FOBT) had been evaluated in a prospective RCT. As observed by Winawer, the introduction of the colonoscope into clinical practice in the early 1970s provided an opportunity for a diagnostic workup of positive CRC screening tests and the ability to remove polyps (Winawer, 2015). In the mid-1970s and early 1980s, five RCTs of CRC screening with FOBT were launched in the US and Europe, four of which demonstrated statistically significant reductions in colorectal cancer mortality associated with an invitation to screening (IARC Working Group on the Evaluation of Cancer-Preventive Interventions, 2019). In a later analysis, the US Minnesota Trial demonstrated a 20% reduction in the incidence of CRC associated with an invitation to screening, largely attributed to the removal of polyps by colonoscopy after a positive FOBT (Mandel et al., 2000).

Further confirmation of the importance of the adenoma-carcinoma sequence and the potential to reduce incidence and mortality through the early detection of cancer and polypectomy was provided by a case-control study of sigmoidoscopy by Selby, Friedman, Quesenberry Jr., and Weiss (1992) which demonstrated reduced CRC mortality attributable to removal of lesions within the reach of the sigmoidoscope, and through a report from the National Polyp Study in 1993, which demonstrated reduced CRC incidence associated with colonoscopic polypectomy (Winawer et al., 1993).

In 1980, the American Cancer Society (ACS) issued the first evidence-based guideline for CRC screening (Eddy, 1980b). As of 1977, the ACS had endorsed that all adults aged 40+ have an annual sigmoidoscopy, digital rectal exam (DRE), and FOBT as part of a period health checkup. The 1980 update recommended annual DRE after age 40, annual FOBT beginning at age 50, and sigmoidoscopy every 3–5 year after two initial sigmoidoscopies with normal findings 1 year apart (Eddy, 1980a). In 1992, an update of the ACS guideline included some notable changes (Levin & Murphy, 1992), including recommending adults begin screening at age 50, undergo sigmoidoscopy (preferably flexible) every 3–5 years (based on advice from a physician), *or* annual guaiac-based FOBT (gFOBT) (Levin & Murphy, 1992). In 1997, the American Gastroenterological Association (AGA) and the ACS issued new guidelines that were largely reflective of current recommendations insofar as options included both structural examinations (sigmoidoscopy, colonoscopy, or barium enema) and stool testing with gFOBT (Byers, Levin, Rothenberger, Dodd, & Smith, 1997; Winawer et al., 1997).

The AGA guideline endorsed regular CRC screening for adults aged 50 years and older with annual gFOBT, flexible sigmoidoscopy every 5 years, combination of annual FOBT and flexible sigmoidoscopy every 5 years, barium enema every 5–10 years, or colonoscopy every 10 years (Winawer et al., 1997). In the same year, the ACS updated their 1992 guideline, recommending a combination of annual FOBT and sigmoidoscopy every 5 years, or total colon examination with either colonoscopy every 10 years, or double-contrast barium enema every 5–10 years (Byers et al., 1997).

On-going technological development has led to additional CRC screening tests being included in modern guidelines, including more effective and acceptable stool testing with fecal immunochemical tests (FIT) (Robertson et al., 2017) and multi-target stool DNA tests (mt-sDNA), which combine an immunochemical test with a test for exfoliated DNA and other molecular markers (Imperiale et al., 2014), and colorectal cancer screening with computed tomography (CT colonography) (Johnson et al., 2008). More recently, there are a growing number of candidate serum tests in development that could lead to CRC screening with a single blood draw (Loktionov, 2020), including one with FDA approval conditional on specific circumstances for use (Lamb & Dhillon, 2017), although serum tests currently are not recommended for screening by any of the major guideline issuing organizations.

CRC screening to detect occult disease evolved based on observational study designs and inference, and subsequently the efficacy of screening with rehydrated gFOBT, and ironically, much later with sigmoidoscopy (Atkin et al., 2010; Schoen et al., 2012), was confirmed with RCTs. New randomized trials of colonoscopy have been initiated (Dominitz et al., 2017; Quintero et al., 2012), although they are prospective, parallel comparison trials of colonoscopy and FIT since a study with a control group not invited to screening is neither ethical nor feasible. Likewise, head-to-head comparisons of screening outcomes utilizing a candidate test with or without a standard test for comparison in a cohort with post-test evaluation with colonoscopy are common, acceptable approaches to evaluation because the efficacy of screening has been demonstrated in prospective RCTs.

4. Colorectal cancer screening guidelines

Table 2 shows the latest guidelines and recommendations recommended by the ACS (Wolf et al., 2018), the United States Preventive Services Task Force (USPSTF) (Bibbins-Domingo et al., 2016), and the United States Multi-Society Task Force (USMSTF), a group of experts representing the AGA, American College of Gastroenterology (ACG), and the American Society of Gastrointestinal Endoscopy (ASGE) (Rex et al., 2017).

For all practical purposes, the recommendations are the same, with only subtle differences that pertain to flexible sigmoidoscopy, which has very low

Table 2 ACS, USPSTF, USMSTF recommended colorectal cancer screening tests for average risk adults.

Recommendation/Most recent update	ACS 2018	USPSTF 2016	USMSTF 2017
Age to begin screening	45	50	50
Age to stop screening			
Visual CRC screening tests			
Colonoscopy	10 yrs.	10 yrs.	10 yrs. (1st tier)
Sigmoidoscopy	5 yrs.	5/10 yrs.	5–10 yrs. (2nd tier)
CT Colonography	5 yrs.	5 yrs.	5 yrs. (2nd tier)
Capsule endoscopy	NR	NR	5 yrs. (3rd tier)
Stool-based CRC screening tests			
High sensitivity guaiac FOBT (Hs-gFOBT)	1 yr.		NR
Fecal immunochemical test (FIT)	1 yr.		1 yr. (1st tier)
Multi-target stool DNA test (MtsDNA)	3 yrs.		3 yrs. (2nd tier)

ACS, American Cancer Society; UPSTF, United States Preventive Services Task Force; USMSTF, United States Multi-Society Task Force.

availability and uptake, and guidance related to the age to stop screening. Also, as will be described below, the ACS and USPSTF has not ever ranked or prioritized tests, although in 2017 the USMSTF prioritized tests into three tiers based on features and performance characteristics (see Table 2), in which, for all practical purposes, tier 1 tests are the two most commonly used tests in the US (Rex et al., 2017). The main difference between the guidelines at this time is the age to begin screening. The ACS recommends that average risk adults should begin screening at age 45, the USPSTF has proposed to reduce the age to begin screening to age 45, for which a final decision will be reached later in 2021 (United States Preventive Services Task Force, 2020), and the USMSTF recommends beginning screening at age 50 (Rex et al., 2017). The timing of the next USMSTF update is not known at this time.

5. Current CRC screening rates

Although screening with sigmoidoscopy and FOBT were promoted as tests that could find CRC cancer early before formal guidelines were issued in 1980 (Eddy, 1980a, 1980b), even after practice guidelines were in place, the uptake of CRC screening was slow into the 1980s and 1990s. In 1987, when the National Health Interview Survey (NHIS), an annual, in-person household survey representative of the non-institutionalized population in the United States, began to track CRC screening, 38% of women and men reported having ever had a FOBT, but only 15% of women and 12% of men reported having had a FOBT in the past year, and only 5% of women and 7% of men reported a proctosigmoidoscopy in the past 3 years (Centers for Disease Control and Prevention, 1996).

5.1 Cancer screening data sources

In this article, we present colorectal cancer screening estimates from the NHIS. Questions on cancer screening are asked periodically and definitions of being up-to-date are detailed in figure footnotes. In this article, cancer screening data from 2000 through 2018 are presented. Overall estimates and those according to race/ethnicity and education were age-adjusted, and all estimates were weighted to account for the complex sampling design. Estimates are based on self-report and do not distinguish between testing performed for asymptomatic and symptomatic reasons.

5.2 Participation in colorectal cancer screening

Colorectal cancer screening steeply rose in the early 2000s mostly with colonoscopy where the proportion of adults reporting a colonoscopy in the past 10 years nearly doubled from 20.5% to 39.2% between 2000 and 2005. Colonoscopy use continued to increase until 2015 and remained stable between 2015 and 2018. Stool-testing in the past year declined by more than half from 17.3% to 7.2% between 2000 and 2015 before rising to 10.8% in the recent time period (2015–18). Sigmoidoscopy in the past 5 years was reported by 9.1% of respondents in 2000 and currently <3% of people aged ≥50 years report having this test.

The proportion of up-to-date adults ≥50 years increased from 37.6% in 2000 to 46.8% in 2005, further increasing to 59.1% in 2010, then leveling off between 2010 and 2013, before steadily climbing to 65.6% in 2018. Despite impressive gains in colorectal cancer screening, disparities according to race/ethnicity and education continue (Fig. 8).

In 2018, <60% of Hispanics were up to date compared to 64.9% of blacks and 67.6% of whites. Educational disparities were even wider where about half (51.7%) of those with <HS degree were up to date compared to 72.6% of college graduates. The age-related disparities for CRC screening are persistent and somewhat unique. CRC screening prevalence continues to increase among people in their 60s and 70s but remain stunted among people in their 50s (Fig. 9). In 2018, 56.1% of people 50–59 were up-to-date with CRC screening compared to 76.4% among people 70–79 years.

More than 30 years since the NHIS began to measure CRC screening rates, the most recent data show a recent uptick in CRC screening rates, as noted above, likely due in large part to an nationwide, collaborative effort to increase CRC screening rates to 80% by 2018 led by the ACS National Colorectal Cancer Roundtable (NCCRT) (Wender, Brooks, Sharpe, & Doroshenk, 2020; Wender, Doroshenk, Brooks, Hotz, & Smith, 2018), but rates remain considerably lower overall than is optimal, and still lower in persistently underserved groups.

ACS lowered the recommended age for screening from 50 to 45 years in mid-2018 (Wolf et al., 2018). In a nationally representative study, the proportion of individuals aged 45–49 years reporting a CRC screening in the past year doubled during 2018, coinciding with the release of ACS guidelines (Fedewa, Siegel, Goding Sauer, Bandi, & Jemal, 2020). By the end of

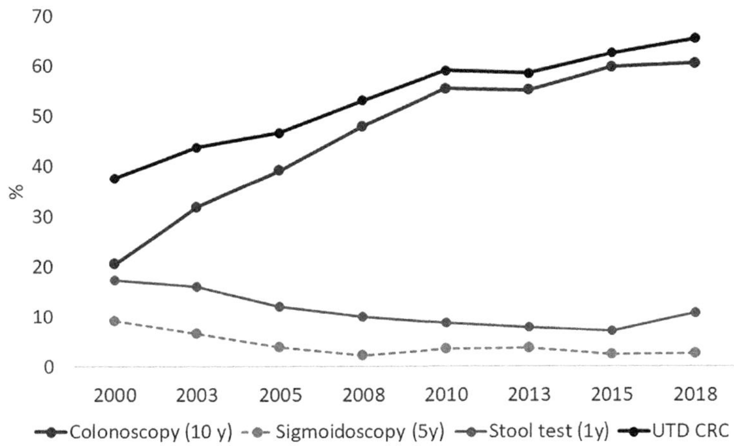

Abbreviations: National Health Interview Survey (NHIS); up to date (UTD); year (y)
a. Up to date colorectal cancer screening defined as colonoscopy in the past 10 years, sigmoidoscopy in the past 5 years, and/or stool-testing in the past year. All estimates are age-standardized to the 2000 population.

Fig. 8 Trends in colorectal cancer screening, NHIS 2000–18[a].

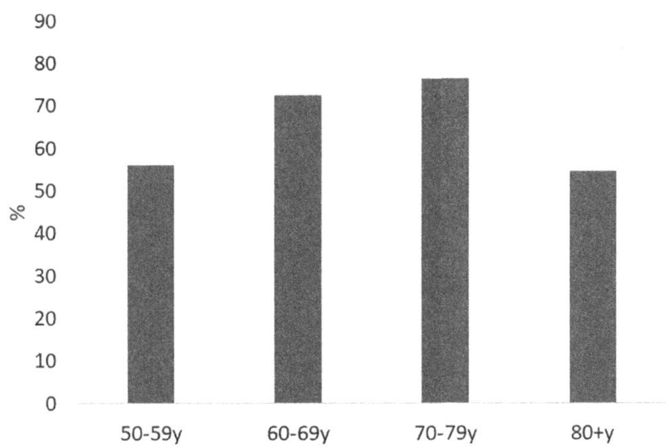

Abbreviations: National Health Interview Survey (NHIS); up to date (UTD); year (y)
a. Up to date colorectal cancer screening defined as colonoscopy in the past 10 years, sigmoidoscopy in the past 5 years, and/or stool-testing in the past year. All estimates are age-standardized to the 2000 population.

Fig. 9 Up to date colorectal cancer screening according to age, NHIS 2000–18[a].

2018, approximately 12% of adults 45–49 years of age reported having had a CRC screening test in the past year, and during 2018 approximately 21% of adults 45–49 years of age were up-to-date with ACS CRC screening recommendations.

6. Fulfilling the potential of colorectal cancer screening—Challenges and opportunities for improvement

6.1 Offering options for CRC screening

In the past, one common lament about the CRC screening guidelines was that the growing number of options for CRC screening were not ranked or prioritized, resulting in confusion over which test to recommend to patients. Jones, Vernon, and Woolf (2010) reported that adults who had discussed two or more screening options were more likely to report confusion and to be nonadherent with screening recommendations compared with adults who only discussed one option. The ACS and the USPTF have never stated a preference for one recommended test over another, since ranking is not as straightforward as it might seem, but have provided guidance about the features, benefits, limitations, and requirements individuals will face for test completion. Because microsimulation modeling showed very similar long term benefits associated with lifetime adherence to the recommended tests (Knudsen et al., 2016), the provision of a menu of options accompanied by details and features seems to have been a reasonable approach, especially since data has shown high variability among adults in the preference for one test over another that may not be influenced by priority ranking (Christy, Monahan, Stump, Rawl, & Champion, 2020; Inadomi et al., 2012; Moreno et al., 2016). Although the study by Jones et al. reported that discussing multiple testing options led to confusion and nonadherence in some patients (Jones et al., 2010), Inadomi et al. reported that only offering patients a colonoscopy resulted in substantially lower uptake of CRC screening compared with offering a choice of colonoscopy or a stool test and having a discussion about test features (Inadomi et al., 2012). In response to these findings, the 2018 ACS guideline included a recommendation that clinicians offer patients at least one stool testing option and one visual testing option (Wolf et al., 2018), to enhance screening uptake, and based on the patient's preference, available options within the chosen category can then be discussed further. The ACS also provided shared decision-making tools that clinicians and patients could use in the office to help patients who were uncertain about test choice to reach a decision (Volk et al., 2018). With respect to ranking multiple options, the 2017 USMSTF guideline update organized recommended tests into three priority tiers, with colonoscopy or FIT recommended as first-tier "tests of choice" when multiple tests are offered as alternatives (see Table 2) (Rex et al., 2017).

6.2 Adherence with screening guidelines

Although screening for any cancer represents a complex chain of interrelated events, at the most basic level successful screening has three core imperatives. First, high rates of *regular* attendance to screening are essential. In a retrospective cohort study of patients aged 55–90 years who died of CRC and were enrolled in the Kaiser Permanente Northern and Southern California systems for at least 5 years before diagnosis, two thirds of the deaths were attributable to having never been screened or not being up to date with screening (Doubeni et al., 2019). Second, screening needs to achieve high sensitivity, otherwise a screening test that fails to detect advanced neoplasia due to poor quality has been an empty exercise and may lead to a missed opportunity to prevent CRC incidence and mortality. The sensitivity of the test may be influenced by host factors, but more commonly is influenced by range of quality assurance factors associated with false negatives. Third, positive, non-colonoscopy tests must be promptly followed up with colonoscopy. In the study of CRC deaths in the Kaiser system mentioned above, 8.1% of deaths were attributed to failure to followup abnormal results (Doubeni et al., 2019). In addition, there is a need to address the challenge of identifying and appropriately screening groups at high risk for CRC due to personal or family history of advanced neoplasia, individuals with a family history of CRC, and known or suspected hereditary syndromes, and individuals with inflammatory bowel disease of significant duration. Although the incidence and mortality of CRC have declined in recent years largely due to screening, (Siegel, Fedewa, et al., 2017; Siegel, Miller, & Jemal, 2017) there is a substantial body of evidence showing that we fail to consistently and completely apply existing knowledge associated with the prevention and early detection of CRC, and this failing accounts for tens of thousands of new cases and deaths each year that otherwise could have been prevented.

Apart from a few large integrated health systems, the U.S. does not have a system of organized screening. Screening in the U.S. is opportunistic, i.e., it depends on a reminder from a health care clinician during an encounter for a checkup or during visits for acute or chronic conditions. Organized screening is distinguished from opportunistic screening primarily based on how the target population is invited to screening, and how outcomes are tracked, which typically is from a central population register. Thus individuals, and their providers can depend not only on regular notification that screening is due, but monitoring for nonadherence with screening recommendations or failure to be followed up after a positive screening test. Organized programs typically have other features, such as quality assurance programs, and ongoing evaluation to monitor screening and programmatic outcomes and provide feedback where appropriate to improve performance (Miles, Cockburn, Smith, & Wardle, 2004).

In populations where screening is opportunistic, some features of organized screening may exist, but they usually underperform compared with an organized system. Some electronic health records can aid clinicians to provide reminders during health care encounters, and may be used to notify patients that screening is due, but often these notifications are in on-line patient portals where they may not be seen, and followup for adherence after a referral to screening will be consigned to the next encounter, which leaves a great deal to chance. It has been clear for many years that opportunistic screening achieves some value. Indeed, self-reported screening rates among adults ages 50–74 in the US under an opportunistic model were higher in 2014 than the majority of 27 European Union (EU) Member States (Eurostat, 2014). However, there is no question that opportunistic screening is hopelessly inefficient as a strategy to achieve high target rates of ever screening and adherence to regular screening. Further, as noted by Doubeni, "… current practices of opportunistic screening in most healthcare systems deprive many individuals of their benefits of screening either because they do not receive regular healthcare or because their preventive care needs are crowded out by competing economic, social, medical and administrative needs of both the patient and/or provider" (Doubeni, 2014). These factors contribute to even lower CRC screening rates in minority, lower-income, and uninsured populations, and overall, they account for the historically slow uptake of screening over time, and still lower than optimal adherence to screening recommendations in the population (see also Carethers in this volume).

Expecting primary care to initiate CRC screening at the age at which screening is recommended to begin and to achieve high rates of regular CRC screening through an opportunistic model of health care is unrealistic, increasingly so as time pressures on primary care have risen and payment systems still prioritize acute care and procedures over prevention and early detection (Sarfaty, Wender, & Smith, 2011). First, we are long overdue to build on decades of accumulated literature documenting successful, evidence-based interventions associated with increasing and maintaining high CRC screening rates (Dougherty et al., 2018), recommendations from the Centers for Disease Control Community Guide (Centers for Disease Control Community Guide, 2006, 2010), and the opportunity electronic systems should have provided to have all eligible adults and their providers supported by reminder and follow-up systems. Second, the evidence shows that clinicians should stop equating CRC screening with colonoscopy, and instead, discuss options and ascertain patient preference. Screening rates are higher when patients are offered a choice between colonoscopy and a stool test (Inadomi et al., 2012). Third, patients who are non-adherent need focused efforts to better understand non-adherence (Dressler, Johnsen, Madsen, Rasmussen, & Jorgensen, 2021). Fourth, primary care practices and health systems in every community should be incentivized to achieve 80% screening or more, which is the current goal of the updated national campaign of the NCCRT (National Colorectal Cancer Roundtable, 2021).

6.3 Underutilization of stool testing

6.3.1 The shift to colonoscopy as the most common screening test

Inclusion of colonoscopy in screening recommendations in 1997 and coverage by Medicare in 2001 led to a dramatic increase in referral to colonoscopy for CRC screening by primary care physicians, resulting in rising rates of CRC screening with colonoscopy and declining rates of screening with FOBT and sigmoidoscopy (Zapka et al., 2012). In 1999, over 80% of physicians reported that they mostly recommended their patients undergo FOBT or flexible sigmoidoscopy for CRC screening (Klabunde et al., 2003). However, in 2006–07, a nationally representative survey of primary care physicians revealed that the large majority had shifted away from FOBT and sigmoidoscopy and increased utilization of colonoscopy because it was regarded as the best test and it was readily available to patients. Two-thirds of physicians also reported concerns about medical liability if they did not offer the test (Zapka et al., 2012). From 2000 to 2015, the proportion of adults screened with stool testing and flexible sigmoidoscopy would continue to decline, although a small increase in reported use of FOBT was observed in 2018 (see Fig. 8).

The decline in stool testing, but also continued utilization of low-sensitivity stool testing, has resulted in missed opportunities to adequately screen adults who are averse to CRC screening with an invasive procedure, or adults, who for various reasons, do not have access to screening colonoscopy, but could be adequately screened with a high-sensitivity stool test. Over the past two decades, the transition to colonoscopy as the most frequently utilized CRC screening test has led to a general lack of awareness about the availability of high-sensitivity stool tests, the higher sensitivity of FIT over gFOBT (Shapiro et al., 2017), the greater patient acceptance of 1-sample FIT tests compared with stool tests that require multiple samples and dietary restrictions (Chubak, Bogart, Fuller, Laing, & Green, 2013), and the success of organized screening programs with FIT as measured by screening uptake, annual testing, and reduced incidence and mortality (Levin et al., 2018). A recent summit on mailed FIT sponsored by the Centers for Disease Control and Prevention (CDC) concluded that there was a considerable body of research showing that mailed FIT outreach had been shown to successfully address a range of barriers to CRC screening and increase CRC screening rates in multiple settings (Gupta, Coronado, et al., 2020).

6.3.2 In-office FOBT

In 2005, Collins, Lieberman, Durbin, Weiss, and Veterans Affairs Cooperative Study #380 Group (2005) reported the comparative sensitivity utilizing Hemoccult® for advanced neoplasia (tubular adenomas 10 mm or greater, adenomas with villous histology or high-grade dysplasia, or invasive cancer) in single sample in-office stool testing (digital FOBT) following a DRE and 6-sample in-home FOBT testing in 2665 patients. The sensitivity for advanced neoplasia with in-home testing was 23.9% compared with 4.9% with digital FOBT. In the same issue of the *Annals of Internal Medicine*, Nadel et al. reported findings from the National Survey of Colorectal Cancer Screening Practices in Health Care Organizations, which asked physicians about FOBT practices, and the 2000 NHIS, which also included questions on FOBT use (Nadel et al., 2005). Although neither national guidelines, nor manufacturers, recommended digital FOBT, nearly one-third (32.5%) of physicians reported in-office digital FOBT as their only method of FOBT testing, and another 41.2% reported using both in-office and take-home testing. Nearly one-third of adults in the NHIS reported having an in-office test. In the same issue of the *Annals*, the editor, Dr. Hal Sox, described the results as shocking, that the common practice of in-office digital FOBT was thoroughly discredited, and that physicians should reexamine their CRC screening practices (Sox, 2005). Despite clear evidence that in-office FOBT is useless as a screening test, and a blistering editorial admonishing clinicians to adhere to screening recommendations 10 years earlier, in the 2015 NHIS, 10.4% of adults reported having undergone stool testing in the past year, among which 28.3% reported having undergone an in-office test (Chido-Amajuoyi, Sharma, Talluri, Tami-Maury, & Shete, 2019). Although rates of stool testing are low, limiting the adverse impact on a population basis of digital-FOBT, the potential for digital-FIT is being discussed on the basis of FITs greater sensitivity, and the appeal of completing a screening test in the office. Based on low return rates of in-home FIT tests in their institution, Naidu and Jacobson (Naidu & Jacobson, 2018) surveyed patients and clinicians and found that a majority of patients (68%) and greater majority of clinicians (88%) regarded digital-FIT acceptable if sensitivity were comparable to the take home method. If this level of accuracy could be achieved, there would be little reason to oppose in-office testing. However, in the absence of evidence, apart from one small study showing considerable divergence in results and lower predictive value with a digital FIT vs the take home method, there is little reason to be optimistic that digital-FIT has a future as an alternative to the take home method (Nakama, Zhang, Abdul Fattah, & Kamijo, 2001).

There is one approach to in-office stool testing that can be endorsed. Some programs are giving patients the opportunity to provide a spontaneous stool sample for testing while they are at the clinic (poop on demand), an alternative approach to home testing that has been reported to significantly boost FIT testing rates (National Colorectal Cancer Roundtable, 2019a).

As described above, given that only 66% of US adults report having had a recent CRC screening test within a time period that represents adherence to recommended guidelines, that the evidence suggesting underutilization of high-sensitivity stool tests, mainly FIT, in adults who would choose FIT over colonoscopy if given the opportunity, and that uptake of CRC screening in underserved, especially uninsured adults is very low, there is a need for an collective, mission-oriented approach to promoting greater utilization of one-sample FIT and mt-sDNA to increase CRC screening rates. The successful mailed FIT program supported by Kaiser Permanente of North California, and conclusions of the CDC summit on mailed FIT provide strong support for a strategy to increase utilization of high-sensitivity stool testing in all settings (Gupta, Coronado, et al., 2020; Levin et al., 2018).

One has to wonder if there is any strategy to put an end to in-office, digital FOBT. Despite numerous, repeated recommendations against digital-FOBT, a concerning proportion of primary care clinicians and patients still report utilizing this screening method. The FDA could require visible product labels on gFOBT cards, FIT tests, and containers that clearly state that these tests should not be used for CRC screening in the office after DRE. Finally, it probably is time that we educated the public that they should reject the offer of a digital-FOBT CRC screening in favor of a recommended stool test.

6.4 Followup of positive stool tests

Approximately 3–15% of asymptomatic adults undergoing CRC screening with gFOBT or FIT will have a positive finding depending on patient factors and screening history, but all abnormal stool tests should receive prompt followup with colonoscopy to prevent disease progression in those with true positive findings. Followup colonoscopy is commonly described as diagnostic colonoscopy, but it is more properly regarded as part of the continuum of screening, i.e., if there is a positive finding on a stool test, the screening test is not complete until the patient has undergone a followup colonoscopy. For many years, failure to properly followup a positive stool test with colonoscopy, or follow it up at all, has been a persistent shortcoming of stool testing

Few standards exist for the timing of followup colonoscopy after an abnormal FOBT. After a Veterans Health Administration (VHA) study showed only 44% of patients with positive FOBT findings had been followed up within 12 months, and that mean time from abnormal FOBT to colonoscopy in average risk patients was 199 days, the VHA established a performance standard of followup within 60 days (Department of Veterans Affairs Office of Inspector General, 2013). Most studies examining followup of positive FOBT findings use a 6-month period as quality benchmark, perhaps based on the assumption of slow growth and allowing for a degree of flexibility when establishing a benchmark, but for some patients a period of 6 months could result in a missed opportunity to arrest the transition of advanced neoplasia to CRC or advanced disease. Corley et al. evaluated the association of followup time after a positive FIT result with risk of colorectal cancer and advanced-stage disease in 70,124 patients aged 50 through 70 years from 2010 to 2014 in a large integrated health system. Compared with followup of an abnormal FOBT within 8–30 days, the investigators observed a nonsignificant increased risk of any and late stage CRC after 6 months that became significant after 9 months (Corley et al., 2017). Rutter and Inadomi raised concerns about potential bias from small numbers of patients who were followed up after 6 months and the large number of patients (30%) who were never followed up, and the inability to measure the presence of symptoms that may have developed during the followup period, prompting earlier evaluation than otherwise might have occurred (Rutter & Inadomi, 2017). Doubeni et al. citing limited empirical evidence, but noting that microsimulation modeling studies are consistent with worsening outcomes after 180 days, recommended 90 days as a reasonable target for quality improvement for colonoscopy followup after a positive FOBT (Doubeni et al., 2018), which should be understood as an upper threshold. In a previous publication he and colleagues suggested follow-up should take place within 1–3 months (Doubeni, Corley, & Levin, 2017), and that wait times longer than 180 days should be avoided (Doubeni et al., 2018).

The evidence shows that lack of followup and timely followup remain challenges for CRC screening with FOBT. In a review of followup colonoscopy in four health systems among 62,384 adults with positive FOBT from 2011 to 2012, Chubak et al. found that most adults who received followup colonoscopy underwent colonoscopy within 6 months, but across the four health systems, followup by 12 months ranged from 58.1% to 83.8% (Chubak et al., 2016). Bharti et al. reported followup of abnormal FIT from eight Federally Qualified Health Centers (FQHCs) from 2015 to 2016, where they observed both incomplete referral for followup, ranging from 73% to 96%, and completion of colonoscopy within 6 months ranging from 18% to 57% (Bharti et al., 2019). More recently, Green and colleagues reported positive FIT followup rates among Medicaid patients in two health plans within 12 months ranging from 32.8% to 35.9% (Green, Baldwin, West, Schwartz, & Coronado, 2020).

Factors associated with not receiving followup colonoscopy have been identified, and not surprisingly include patient factors, clinician factors and system factors. Cusumano et al. surveyed patients with no documented followup colonoscopy after an abnormal FIT and found that 26% were not aware that their test results were positive.(Cusumano et al., 2020) In a retrospective analysis (2016–18) of patients in an academic center with a positive stool tests, Cooper et al. reviewed the electronic medical records of all patients who did not complete followup colonoscopy to identify barriers, which included referral to the gastroenterology clinic rather than ordering a colonoscopy (with all patients failing to keep their appointment); the primary care provider attributing the positive finding to reasons other than colorectal neoplasia (23.8%) and not ordering a colonoscopy; 14.6% of patients had a colonoscopy in the past year and were judged to not need another colonoscopy (which begs the question of why the patient had a FIT test in the year after a colonoscopy); no reason for not following up the positive FIT in the patient record (16%); and patient refusing colonoscopy (7.4%) (Cooper et al., 2021). Anecdotal reports of long wait times for uninsured and Medicaid insured patients are common, but documentation in the literature is scarce. Patel et al. reported results of a secret-shopper survey in Connecticut on wait times for screening colonoscopy and found fewer than half of Connecticut practices accepted Medicaid patients, and participation varied in municipalities across the state (Patel, Nahar, Murray, & Salner, 2013). Counties with the fewest practices accepting Medicaid patients had the shortest wait times for a colonoscopy (10 days) compared with counties with a higher percentage of practices accepting Medicaid patients (53 days). One practice in Hartford had a wait time of 159 days. For some patients, a followup exam may be difficult to schedule, or not available, in their community due to a high proportion of gastroenterologists who do not accept Medicaid, which may increase wait times in the smaller number of facilities that accept Medicaid patients.

In a review of studies through June 2017 on interventions to improve followup of positive stool tests, Selby et al. reported that patient navigators, provider reminders, and feedback on performance data appeared to improve followup of positive findings in asymptomatic adults, but concluded overall evidence supporting useful system-level interventions was "scant and insufficient" (Selby et al., 2017) In a study of system-level strategies to improve timely followup colonoscopy after positive FOBT, the investigators established (1) a goal of followup within 30 day of a positive FIT, (2) tracking patients with positive FIT, (3) early telephone contact to schedule colonoscopy, (4) assigning followup responsibility to gastroenterology departments rather than primary care, and (5) increasing colonoscopy capacity (Selby et al., 2019). The investigators observed followup within 180 days between 72% and 88% across service areas, observing no significant difference between followup rates between higher vs lower service areas.

A novel program designed to increase CRC screening rates in FQHCs by increasing access to followup colonoscopy was initiated by the ACS NCCRT in 2014 and focused on two strategies, a medical neighborhood strategy and a screening navigation strategy (National Colorectal Cancer Roundtable, 2019b). The medical neighborhood strategy sought to access no-cost of low-cost colonoscopies and surgical services from local providers, emphasizing the humanitarian case and the business case, with the latter stressing that failing to prevent CRC or intervene early could result in a more advanced cancer that would be more expensive to treat in the future. The navigation strategy addressed concerns that FQHC patients would fail keep appointments and would be poorly prepped for the exam, by ensuring that appointments were kept and that patients were adequately prepared for the examination. At the end of the 4-year project, all three sites had a substantial increase of rates of annual screening with stool testing, and timely followup colonoscopies for patients with abnormal FOBT.

Although stool testing for CRC screening is utilized far less than colonoscopy, there is growing appreciation that screening with high sensitivity tests compares favorably over a lifetime of screening with colonoscopy, and that even single test sensitivity is respectable. It also is clear that many adults prefer stool testing, especially single sample testing that is available with some FITS and with mt-sDNA, and, although adults should have access all recommended tests, many adults of screening age will only have access to timely screening if they undergo an annual high sensitivity gFOBT or FIT. For these reasons and more, there is an urgent need to advocate for universal access to high-sensitivity, single sample stool tests, and promote uniform protocols for uptake and followup of abnormal tests with timely colonoscopy.

7. Identifying patients with a family history

Persons with a family history of colorectal cancer and adenomas are at increased risk of CRC, and represent a larger prevalent group compared with adults at elevated risk due to hereditary syndromes. Approximately 10% of US adults report having a first degree relative (FDR) diagnosed with CRC, and closer to 30% of adults report an affected first or second degree relative (Lowery et al., 2016). Risk is approximately two times higher with a single affected FDR, increases with increasing numbers of first and second degree relatives and relatives diagnosed before age 50, which also increases risk of an earlier age at diagnosis (Butterworth, Higgins, & Pharoah, 2006). Guidelines generally recommend that adults with a family history of a FDR diagnosed before age 60 should begin screening with colonoscopy at a younger age (Leddin et al., 2018; Lowery et al., 2016). Knowledge of the patient's family history is a simple and powerful tool to identify someone at higher risk who would benefit from tailored screening, typically regular, shorter-interval colonoscopy beginning at an earlier age. However, the opportunity to identify and appropriately screen adults with a family history of CRC is commonly unfulfilled. In a review of the contribution of family history to CRC risk and its clinical implications, Lowery et al. cited evidence that clinicians report insufficient knowledge and awareness of guidelines to implement recommendations, have inadequate communication with patients about risk, commonly fail to take and update the patient's family history, and act on the findings; adults with a family history tend to have higher rates of regular screening, but have low adherence to guidelines recommending beginning screening before age 50 (Lowery et al., 2016).

The trend in earlier onset CRC (EOCRC) has raised new awareness about risk before age 50 and opportunities for detecting CRC earlier. Gupta et al. assessed the effectiveness of screening recommendations based on having a family history of CRC using data from the Colon Cancer Family Registry from 1998 through 2007 (Gupta, Bharti, et al., 2020). Among adults with a family history meeting early screening criteria, 98% would have been recommended to begin screening at an age younger than their diagnosis age, with the potential for their CRC to have been diagnosed at an earlier stage, or even prevented. Although these findings suggest a simple opportunity to identify a significant fraction of EOCRC patients through adherence with recommended guidelines, Zazove et al. reviewed the effectiveness of family history prompts in the electronic medical record and observed "an almost total lack of discussion, testing, or referral for pertinent patients based on chart reviews" (Zazove, Plegue, Uhlmann, & Ruffin, 2015). Patient at high risk for EOCRC also contribute to missed opportunities to detect CRC earlier. Skinner et al. found that only 5.3% of adults aged 25–49 for whom colonoscopy had been recommended had been screened within 6 months of referral (Skinner et al., 2017). Missed opportunities for earlier diagnosis in EOCRC also occur when patients and clinicians attribute symptoms of CRC to other causes based on the patient's age, commonly resulting in much longer delay in diagnosis (>6 months) than would have occurred if the patient had been above an age when CRC screening is recommended (Taggarshe, Rehil, Sharma, Flynn, & Damadi, 2013).

Missed opportunities to identify patients at risk, refer high risk patients to screening, respond to persistent symptoms in a timely fashion, and act on screening recommendations are common and indicate the need for interventions at the system, physician and patient levels.

8. Conclusion

Over a century after the earliest insights that deaths from colorectal cancer could be avoided by detecting and treating localized disease, and potentially prevented by detection and removal of adenomatous polyps, considerable progress has been made in the prevention and early detection of CRC. However, too many people are diagnosed with colorectal cancer and there are too many deaths from colorectal cancer that could have been prevented. A glaring shortcoming in our opportunistic model of care is unequal access to care, and disparities in access to care and outcomes are well documented. For all adults, the lack of supportive systems, and more broadly as a part of a system, well understood rules, roles, relationships, and accountability in an organized multi-disciplinary care model would contribute to averting many of the diagnoses and deaths from colorectal cancer that occur each year. The modern emphasis on evidence-based medicine should extend from *what we should do in health care* to *what we do in health care.*

Disclosures

All authors are employed by the American Cancer Society, which receives grants from private and corporate foundations, including foundations associated with companies in the health sector for research outside of the submitted work. The authors are not funded by or key personnel for any of these grants and their salary is solely funded through American Cancer Society funds.

References

Andrew, A. S., Parker, S., Anderson, J. C., Rees, J. R., Robinson, C., Riddle, B., et al. (2018). Risk factors for diagnosis of colorectal cancer at a late stage: A population-based study. *Journal of General Internal Medicine, 33*(12), 2100–2105. https://doi.org/10.1007/s11606-018-4648-7.

Araghi, M., Soerjomataram, I., Bardot, A., Ferlay, J., Cabasag, C. J., Morrison, D. S., et al. (2019). Changes in colorectal cancer incidence in seven high-income countries: A population-based study. *The Lancet Gastroenterology & Hepatology, 4*(7), 511–518. https://doi.org/10.1016/S2468-1253(19)30147-5.

Atkin, W. S., Edwards, R., Kralj-Hans, I., Wooldrage, K., Hart, A. R., Northover, J. M., et al. (2010). Once-only flexible sigmoidoscopy screening in prevention of colorectal cancer: A multicentre randomised controlled trial. *Lancet, 375*(9726), 1624–1633. https://doi.org/10.1016/S0140-6736(10)60551-X.

Bach, P. B., Schrag, D., Brawley, O. W., Galaznik, A., Yakren, S., & Begg, C. B. (2002). Survival of blacks and whites after a cancer diagnosis. *JAMA, 287*(16), 2106–2113. https://doi.org/10.1001/jama.287.16.2106.

Ballantyne, G. H. (1988). Theories of carcinogenesis and their impact on surgical treatment of colorectal cancer. A historical review. *Diseases of the Colon and Rectum, 31*(7), 513–517. https://doi.org/10.1007/BF02553723.

Bharti, B., May, F. F. P., Nodora, J., Martinez, M. E., Moyano, K., Davis, S. L., et al. (2019). Diagnostic colonoscopy completion after abnormal fecal immunochemical testing and quality of tests used at 8 Federally Qualified Health Centers in Southern California: Opportunities for improving screening outcomes. *Cancer, 125*(23), 4203–4209. https://doi.org/10.1002/cncr.32440.

Bibbins-Domingo, K., Grossman, D. C., Curry, S. J., Davidson, K. W., Epling, J. W., Jr., Garcia, F. A., et al. (2016). Screening for colorectal cancer: US Preventive Services Task Force recommendation statement. *JAMA, 315*(23), 2564–2575. https://doi.org/10.1001/jama.2016.5989.

Butterworth, A. S., Higgins, J. P., & Pharoah, P. (2006). Relative and absolute risk of colorectal cancer for individuals with a family history: A meta-analysis. *European Journal of Cancer, 42*(2), 216–227. https://doi.org/10.1016/j.ejca.2005.09.023.

Byers, T., Levin, B., Rothenberger, D., Dodd, G. D., & Smith, R. A. (1997). American Cancer Society guidelines for screening and surveillance for early detection of colorectal polyps and cancer: Update 1997. American Cancer Society Detection and Treatment

Advisory Group on Colorectal Cancer. *CA: A Cancer Journal for Clinicians, 47*(3), 154–160. Retrieved from http://www.ncbi.nlm.nih.gov/pubmed/9152173.

Centers for Disease Control and Prevention. (1996). Trends in cancer screening—United States, 1987 and 1992. *MMWR. Morbidity and Mortality Weekly Report, 45*(3), 57–61. Retrieved from https://www.ncbi.nlm.nih.gov/pubmed/8544808.

Centers for Disease Control Community Guide. (2006). *Cancer screening: Provider reminder and recall systems—Colorectal cancer.* Retrieved from (February, 2006) https://www.thecommunityguide.org/findings/cancer-screening-provider-reminder-and-recall-systems-colorectal-cancer.

Centers for Disease Control Community Guide. (2010). *Cancer screening: Client reminders—Colorectal cancer.* Retrieved from (July, 2010) https://www.thecommunityguide.org/findings/cancer-screening-client-reminders-colorectal-cancer.

Chen, F. W., Sundaram, V., Chew, T. A., & Ladabaum, U. (2017). Advanced-stage colorectal cancer in persons younger than 50 years not associated with longer duration of symptoms or time to diagnosis. *Clinical Gastroenterology and Hepatology, 15*(5), 728–737.e723. https://doi.org/10.1016/j.cgh.2016.10.038.

Chido-Amajuoyi, O. G., Sharma, A., Talluri, R., Tami-Maury, I., & Shete, S. (2019). Physician-office vs home uptake of colorectal cancer screening using FOBT/FIT among screening-eligible US adults. *Cancer Medicine, 8*(17), 7408–7418. https://doi.org/10.1002/cam4.2604.

Christy, S. M., Monahan, P. O., Stump, T. E., Rawl, S. M., & Champion, V. L. (2020). Impact of tailored interventions on receipt of a preference-concordant colorectal cancer screening test. *Medical Decision Making, 40*(1), 29–41. https://doi.org/10.1177/0272989X19890603.

Chua, T. C., Saxena, A., Chu, F., Zhao, J., & Morris, D. L. (2011). Predictors of cure after hepatic resection of colorectal liver metastases: An analysis of actual 5- and 10-year survivors. *Journal of Surgical Oncology, 103*(8), 796–800. https://doi.org/10.1002/jso.21864.

Chubak, J., Bogart, A., Fuller, S., Laing, S. S., & Green, B. B. (2013). Uptake and positive predictive value of fecal occult blood tests: A randomized controlled trial. *Preventive Medicine, 57*(5), 671–678. https://doi.org/10.1016/j.ypmed.2013.08.032.

Chubak, J., Garcia, M. P., Burnett-Hartman, A. N., Zheng, Y., Corley, D. A., Halm, E. A., et al. (2016). Time to colonoscopy after positive fecal blood test in four U.S. health care systems. *Cancer Epidemiology, Biomarkers & Prevention, 25*(2), 344–350. https://doi.org/10.1158/1055-9965.EPI-15-0470.

Collins, J. F., Lieberman, D. A., Durbin, T. E., Weiss, D. G., & Veterans Affairs Cooperative Study #380 Group. (2005). Accuracy of screening for fecal occult blood on a single stool sample obtained by digital rectal examination: A comparison with recommended sampling practice. *Annals of Internal Medicine, 142*(2), 81–85. Retrieved from http://www.ncbi.nlm.nih.gov/pubmed/15657155.

Cooper, G. S., Grimes, A., Werner, J., Cao, S., Fu, P., & Stange, K. C. (2021). Barriers to follow-up colonoscopy after positive FIT or multitarget stool DNA testing. *Journal of American Board of Family Medicine, 34*(1), 61–69. https://doi.org/10.3122/jabfm.2021.01.200345.

Corley, D. A., Jensen, C. D., Quinn, V. P., Doubeni, C. A., Zauber, A. G., Lee, J. K., et al. (2017). Association between time to colonoscopy after a positive fecal test result and risk of colorectal cancer and cancer stage at diagnosis. *JAMA, 317*(16), 1631–1641. https://doi.org/10.1001/jama.2017.3634.

Cusumano, V. T., Corona, E., Partida, D., Yang, L., Yu, C., & May, F. P. (2020). Patients without colonoscopic follow-up after abnormal fecal immunochemical tests are often unaware of the abnormal result and report several barriers to colonoscopy. *BMC Gastroenterology, 20*(1), 115. https://doi.org/10.1186/s12876-020-01262-7.

Department of Veterans Affairs Office of Inspector General. (2013). *Evaluation of colorectal cancer screening and follow-up in veterans health administration facilities*. Retrieved from https://www.va.gov/oig/pubs/VAOIG-13-01741-215.pdf.

Dominitz, J. A., Robertson, D. J., Ahnen, D. J., Allison, J. E., Antonelli, M., Boardman, K. D., et al. (2017). Colonoscopy vs. fecal immunochemical test in reducing mortality from colorectal cancer (CONFIRM): Rationale for study design. *The American Journal of Gastroenterology*, *112*(11), 1736–1746. https://doi.org/10.1038/ajg.2017.286.

Doubeni, C. A. (2014). The impact of colorectal cancer screening on the US population: Is it time to celebrate? *Cancer*, *120*(18), 2810–2813. https://doi.org/10.1002/cncr.28789.

Doubeni, C. A., Corley, D. A., & Levin, T. R. (2017). Time to diagnostic testing after a positive colorectal cancer screening test. *JAMA*, *318*(5), 483. https://doi.org/10.1001/jama.2017.9312.

Doubeni, C. A., Fedewa, S. A., Levin, T. R., Jensen, C. D., Saia, C., Zebrowski, A. M., et al. (2019). Modifiable failures in the colorectal cancer screening process and their association with risk of death. *Gastroenterology*, *156*(1), 63–74.e66. https://doi.org/10.1053/j.gastro.2018.09.040.

Doubeni, C. A., Gabler, N. B., Wheeler, C. M., McCarthy, A. M., Castle, P. E., Halm, E. A., et al. (2018). Timely follow-up of positive cancer screening results: A systematic review and recommendations from the PROSPR Consortium. *CA: A Cancer Journal for Clinicians*, *68*(3), 199–216. https://doi.org/10.3322/caac.21452.

Dougherty, M. K., Brenner, A. T., Crockett, S. D., Gupta, S., Wheeler, S. B., Coker-Schwimmer, M., et al. (2018). Evaluation of interventions intended to increase colorectal cancer screening rates in the United States: A systematic review and meta-analysis. *JAMA Internal Medicine*, *178*(12), 1645–1658. https://doi.org/10.1001/jamainternmed.2018.4637.

Dressler, J., Johnsen, A. T., Madsen, L. J., Rasmussen, M., & Jorgensen, L. N. (2021). Factors affecting patient adherence to publicly funded colorectal cancer screening programmes: A systematic review. *Public Health*, *190*, 67–74. https://doi.org/10.1016/j.puhe.2020.10.025.

Eddy, D. (1980a). Cancer of the colon and rectum. *CA: A Cancer Journal for Clinicians*, *30*(4), 208–215.

Eddy, D. (1980b). Guidelines for the cancer related checkup: Recommendations and rationale. *CA: A Cancer Journal for Clinicians*, *30*(4), 194–240.

Edwards, B. K., Ward, E., Kohler, B. A., Eheman, C., Zauber, A. G., Anderson, R. N., et al. (2010). Annual report to the nation on the status of cancer, 1975–2006, featuring colorectal cancer trends and impact of interventions (risk factors, screening, and treatment) to reduce future rates. *Cancer*, *116*(3), 544–573. https://doi.org/10.1002/cncr.24760.

Ellis, L., Canchola, A. J., Spiegel, D., Ladabaum, U., Haile, R., & Gomez, S. L. (2018). Racial and ethnic disparities in cancer survival: The contribution of tumor, sociodemographic, institutional, and neighborhood characteristics. *Journal of Clinical Oncology*, *36*(1), 25–33. https://doi.org/10.1200/JCO.2017.74.2049.

Eurostat. (2014). *Colorectal cancer screening*. Retrieved from (September 8, 2020) https://ec.europa.eu/eurostat/statistics-explained/index.php/Healthcare_activities_statistics_-_preventive_services#Colorectal_cancer_screening.

Fedewa, S. A., Siegel, R. L., Goding Sauer, A., Bandi, P., & Jemal, A. (2020). Colorectal cancer screening patterns after the American Cancer Society's recommendation to initiate screening at age 45 years. *Cancer*, *126*(6), 1351–1353. https://doi.org/10.1002/cncr.32662.

Gilbertsen, V., Schultz, A., Cooper, J. P., Green, R. A., Nord, R., & Arnold, A. W. (1961). Cancer detection in office practice. *The Journal-Lancet*, *81*, 398–401. Retrieved from https://www.ncbi.nlm.nih.gov/pubmed/13898553.

Greegor, D. H. (1967). Diagnosis of large-bowel cancer in the asymptomatic patient. *JAMA, 201*(12), 943–945. Retrieved from https://www.ncbi.nlm.nih.gov/pubmed/6072632.
Greegor, D. H. (1969). Detection of silent colon cancer in routine examination. *CA: A Cancer Journal for Clinicians, 19*(6), 330–337. Retrieved from http://www.ncbi.nlm.nih.gov/pubmed/4983009.
Green, B. B., Baldwin, L. M., West, I. I., Schwartz, M., & Coronado, G. D. (2020). Low rates of colonoscopy follow-up after a positive fecal immunochemical test in a Medicaid health plan delivered mailed colorectal cancer screening program. *Journal of Primary Care and Community Health, 11*. https://doi.org/10.1177/2150132720958525, 2150132720958525.
Gupta, S., Bharti, B., Ahnen, D. J., Buchanan, D. D., Cheng, I. C., Cotterchio, M., et al. (2020). Potential impact of family history-based screening guidelines on the detection of early-onset colorectal cancer. *Cancer, 126*(13), 3013–3020. https://doi.org/10.1002/cncr.32851.
Gupta, S., Coronado, G. D., Argenbright, K., Brenner, A. T., Castaneda, S. F., Dominitz, J. A., et al. (2020). Mailed fecal immunochemical test outreach for colorectal cancer screening: Summary of a Centers for Disease Control and Prevention-sponsored Summit. *CA: A Cancer Journal for Clinicians, 70*(4), 283–298. https://doi.org/10.3322/caac.21615.
Holford, T. R. (1991). Understanding the effects of age, period, and cohort on incidence and mortality rates. *Annual Review of Public Health, 12*, 425–457. https://doi.org/10.1146/annurev.pu.12.050191.002233.
Howlader, N., Noone, A. M., Krapcho, M., Miller, D., Brest, A., Yu, M., et al. (2020). *SEER cancer statistics review, 1975–2017*. Retrieved from Bethesda, MD http://seer.cancer.gov/csr/1975_2015/.
IARC Working Group on the Evaluation of Cancer-Preventive Interventions. (2019). *IARC handbooks of cancer prevention; volume 17: Colorectal cancer screening*. Lyon: International Agency for Research on Cancer.
Imperiale, T. F., Ransohoff, D. F., Itzkowitz, S. H., Levin, T. R., Lavin, P., Lidgard, G. P., et al. (2014). Multitarget stool DNA testing for colorectal-cancer screening. *The New England Journal of Medicine, 370*(14), 1287–1297. https://doi.org/10.1056/NEJMoa1311194.
Inadomi, J. M., Vijan, S., Janz, N. K., Fagerlin, A., Thomas, J. P., Lin, Y. V., et al. (2012). Adherence to colorectal cancer screening: A randomized clinical trial of competing strategies. *Archives of Internal Medicine, 172*(7), 575–582. https://doi.org/10.1001/archinternmed.2012.332.
Johnson, C. D., Chen, M. H., Toledano, A. Y., Heiken, J. P., Dachman, A., Kuo, M. D., et al. (2008). Accuracy of CT colonography for detection of large adenomas and cancers. *The New England Journal of Medicine, 359*(12), 1207–1217. Retrieved from http://www.ncbi.nlm.nih.gov/entrez/query.fcgi?cmd=Retrieve&db=PubMed&dopt=Citation&list_uids=18799557.
Jones, R. M., Vernon, S. W., & Woolf, S. H. (2010). Is discussion of colorectal cancer screening options associated with heightened patient confusion? *Cancer Epidemiology, Biomarkers & Prevention, 19*(11), 2821–2825. https://doi.org/10.1158/1055-9965.EPI-10-0695.
Kirsner, J. B., Palmer, W. L., et al. (1948). Gastroenterology; a review of the literature from July 1946 to July 1947. *Archives of Internal Medicine (Chicago, Ill.), 81*(4), 534. passim. Retrieved from https://www.ncbi.nlm.nih.gov/pubmed/18888381.
Klabunde, C. N., Frame, P. S., Meadow, A., Jones, E., Nadel, M., & Vernon, S. W. (2003). A national survey of primary care physicians' colorectal cancer screening recommendations and practices. *Preventive Medicine, 36*(3), 352–362. https://doi.org/10.1016/s0091-7435(02)00066-x.

Knudsen, A. B., Zauber, A. G., Rutter, C. M., Naber, S. K., Doria-Rose, V. P., Pabiniak, C., et al. (2016). Estimation of benefits, burden, and harms of colorectal cancer screening strategies: Modeling study for the US Preventive Services Task Force. *JAMA*, *315*(23), 2595–2609. https://doi.org/10.1001/jama.2016.6828.

Kopetz, S., Chang, G. J., Overman, M. J., Eng, C., Sargent, D. J., Larson, D. W., et al. (2009). Improved survival in metastatic colorectal cancer is associated with adoption of hepatic resection and improved chemotherapy. *Journal of Clinical Oncology*, *27*(22), 3677–3683. https://doi.org/10.1200/JCO.2008.20.5278.

Lamb, Y. N., & Dhillon, S. (2017). Epi proColon(R) 2.0 CE: A blood-based screening test for colorectal cancer. *Molecular Diagnosis & Therapy*, *21*(2), 225–232. https://doi.org/10.1007/s40291-017-0259-y.

Leddin, D., Lieberman, D. A., Tse, F., Barkun, A. N., Abou-Setta, A. M., Marshall, J. K., et al. (2018). Clinical practice guideline on screening for colorectal cancer in individuals with a family history of nonhereditary colorectal cancer or adenoma: The Canadian Association of Gastroenterology Banff Consensus. *Gastroenterology*, *155*(5), 1325–1347. e1323. https://doi.org/10.1053/j.gastro.2018.08.017.

Levin, T. R., Corley, D. A., Jensen, C. D., Schottinger, J. E., Quinn, V. P., Zauber, A. G., et al. (2018). Effects of organized colorectal cancer screening on cancer incidence and mortality in a large community-based population. *Gastroenterology*, *155*(5), 1383–1391. e1385. https://doi.org/10.1053/j.gastro.2018.07.017.

Levin, B., & Murphy, G. P. (1992). Revision in American Cancer Society recommendations for the early detection of colorectal cancer. *CA: A Cancer Journal for Clinicians*, *42*(5), 296–299. Retrieved from https://www.ncbi.nlm.nih.gov/pubmed/1515968.

Lieberman, D. A., Weiss, D. G., Bond, J. H., Ahnen, D. J., Garewal, H., & Chejfec, G. (2000). Use of colonoscopy to screen asymptomatic adults for colorectal cancer. Veterans Affairs Cooperative Study Group 380. *The New England Journal of Medicine*, *343*(3), 162–168. https://doi.org/10.1056/NEJM200007203430301.

Lockhart-Mummery, J. P., & Dukes, C. (1927). The precancerous changes in the rectum and colon. *Surgery, Gynecology & Obstetrics*, *36*, 591–596.

Loktionov, A. (2020). Biomarkers for detecting colorectal cancer non-invasively: DNA, RNA or proteins? *World Journal of Gastrointestinal Oncology*, *12*(2), 124–148. https://doi.org/10.4251/wjgo.v12.i2.124.

Lowery, J. T., Ahnen, D. J., Schroy, P. C., 3rd, Hampel, H., Baxter, N., Boland, C. R., et al. (2016). Understanding the contribution of family history to colorectal cancer risk and its clinical implications: A state-of-the-science review. *Cancer*, *122*(17), 2633–2645. https://doi.org/10.1002/cncr.30080.

Mandel, J. S., Church, T. R., Bond, J. H., Ederer, F., Geisser, M. S., Mongin, S. J., et al. (2000). The effect of fecal occult-blood screening on the incidence of colorectal cancer. *The New England Journal of Medicine*, *343*(22), 1603–1607. Retrieved from http://www.ncbi.nlm.nih.gov/htbin-post/Entrez/query?db=m&form=6&dopt=r&uid=11096167.

Miles, W. E. (1910). The radical abdomino-perineal operation for cancer of the rectum and of the pelvic colon. *British Medical Journal*, *2*, 941–943.

Miles, A., Cockburn, J., Smith, R. A., & Wardle, J. (2004). A perspective from countries using organized screening programs. *Cancer*, *101*(Suppl. 5), 1201–1213. Retrieved from http://www.ncbi.nlm.nih.gov/entrez/query.fcgi?cmd=Retrieve&db=PubMed&dopt=Citation&list_uids=15316915.

Miller, K. D., Nogueira, L., Mariotto, A. B., Rowland, J. H., Yabroff, K. R., Alfano, C. M., et al. (2019). Cancer treatment and survivorship statistics, 2019. *CA: A Cancer Journal for Clinicians*, *69*(5), 363–385. https://doi.org/10.3322/caac.21565.

Moreno, C. C., Weiss, P. S., Jarrett, T. L., Roberts, D. L., Mittal, P. K., & Votaw, J. R. (2016). Patient preferences regarding colorectal cancer screening: Test features and cost

willing to pay out of pocket. *Current Problems in Diagnostic Radiology*, *45*(3), 189–192. https://doi.org/10.1067/j.cpradiol.2015.12.002.
Murphy, C. C., Harlan, L. C., Lund, J. L., Lynch, C. F., & Geiger, A. M. (2015). Patterns of colorectal cancer care in the United States: 1990–2010. *Journal of the National Cancer Institute*, *107*(10). https://doi.org/10.1093/jnci/djv198.
Nadel, M. R., Shapiro, J. A., Klabunde, C. N., Seeff, L. C., Uhler, R., Smith, R. A., et al. (2005). A national survey of primary care physicians' methods for screening for fecal occult blood. *Annals of Internal Medicine*, *142*(2), 86–94. Retrieved from http://www.ncbi.nlm.nih.gov/pubmed/15657156.
Naidu, H., & Jacobson, B. C. (2018). Patients and providers are amenable to fecal immunochemical testing by digital rectal exam. *Journal of Patient Experience*, *5*(3), 236–237. https://doi.org/10.1177/2374373517753176.
Nakama, H., Zhang, B., Abdul Fattah, A. S., & Kamijo, N. (2001). Does stool collection method affect outcomes in immunochemical fecal occult blood testing? *Diseases of the Colon and Rectum*, *44*(6), 871–875. https://doi.org/10.1007/BF02234712.
National Center for Health Statistics & Division of Health Interview Statistics. (2019). *National health interview survey public use data file 2018*. Retrieved from https://www.cdc.gov/nchs/nhis/nhis_2018_data_release.htm.
National Colorectal Cancer Roundtable. (2019a). *CRC news: July 16, 2019, poop on demand video*. Retrieved from https://nccrt.org/crc-news-july-16-2019/.
National Colorectal Cancer Roundtable. (2019b). *Summary report on links of care pilot to increase colorectal cancer screening for underserved patients*. Retrieved from Atlanta, GA https://nccrt.org/resource/summary-report-on-links-of-care-pilot-to-increase-colorectal-cancer-screening-for-underserved-patients/.
National Colorectal Cancer Roundtable. (2021). *Achieving 80% colorectal cancer screening rates in every community*. Retrieved from https://nccrt.org/80-in-every-community/.
Patel, V. B., Nahar, R., Murray, B., & Salner, A. L. (2013). Exploring implications of Medicaid participation and wait times for colorectal screening on early detection efforts in Connecticut—A secret-shopper survey. *Connecticut Medicine*, 77(4), 197–203. Retrieved from https://www.ncbi.nlm.nih.gov/pubmed/23691732.
Paulson, M. (1930). Proctosigmoidoscopy: A medical diagnostic procedure: A pleas for a wider and more frequent use of this method by the internist. *Annals of Internal Medicine*, *4*, 498–500.
Piawah, S., & Venook, A. P. (2019). Targeted therapy for colorectal cancer metastases: A review of current methods of molecularly targeted therapy and the use of tumor biomarkers in the treatment of metastatic colorectal cancer. *Cancer*, *125*(23), 4139–4147. https://doi.org/10.1002/cncr.32163.
Quintero, E., Castells, A., Bujanda, L., Cubiella, J., Salas, D., Lanas, A., et al. (2012). Colonoscopy versus fecal immunochemical testing in colorectal-cancer screening. *The New England Journal of Medicine*, *366*(8), 697–706. https://doi.org/10.1056/NEJMoa1108895.
Rex, D. K., Boland, C. R., Dominitz, J. A., Giardiello, F. M., Johnson, D. A., Kaltenbach, T., et al. (2017). Colorectal cancer screening: Recommendations for physicians and patients from the U.S. Multi-Society Task Force on colorectal cancer. *Gastroenterology*, *153*(1), 307–323. https://doi.org/10.1053/j.gastro.2017.05.013.
Robertson, D. J., Lee, J. K., Boland, C. R., Dominitz, J. A., Giardiello, F. M., Johnson, D. A., et al. (2017). Recommendations on fecal immunochemical testing to screen for colorectal neoplasia: A consensus statement by the US Multi-Society Task Force on colorectal cancer. *Gastroenterology*, *152*(5), 1217–1237.e1213. https://doi.org/10.1053/j.gastro.2016.08.053.
Ruers, T., Van Coevorden, F., Punt, C. J., Pierie, J. E., Borel-Rinkes, I., Ledermann, J. A., et al. (2017). Local treatment of unresectable colorectal liver metastases: Results of a

randomized phase II trial. *Journal of the National Cancer Institute, 109*(9), 1–10. https://doi.org/10.1093/jnci/djx015.

Rutter, C. M., & Inadomi, J. M. (2017). Time to diagnostic testing after a positive colorectal cancer screening test-reply. *JAMA, 318*(5), 483–484. https://doi.org/10.1001/jama.2017.9316.

Sarfaty, M., Wender, R., & Smith, R. (2011). Promoting cancer screening within the patient centered medical home. *CA: A Cancer Journal for Clinicians, 61*(6), 397–408. https://doi.org/10.3322/caac.20125.

Schenck, A. P., Peacock, S. C., Klabunde, C. N., Lapin, P., Coan, J. F., & Brown, M. L. (2009). Trends in colorectal cancer test use in the medicare population, 1998–2005. *American Journal of Preventive Medicine, 37*(1), 1–7. https://doi.org/10.1016/j.amepre.2009.03.009.

Schoen, R. E., Pinsky, P. F., Weissfeld, J. L., Yokochi, L. A., Church, T., Laiyemo, A. O., et al. (2012). Colorectal-cancer incidence and mortality with screening flexible sigmoidoscopy. *The New England Journal of Medicine, 366*(25), 2345–2357. https://doi.org/10.1056/NEJMoa1114635.

Selby, K., Baumgartner, C., Levin, T. R., Doubeni, C. A., Zauber, A. G., Schottinger, J., et al. (2017). Interventions to improve follow-up of positive results on fecal blood tests: A systematic review. *Annals of Internal Medicine, 167*(8), 565–575. https://doi.org/10.7326/M17-1361.

Selby, J. V., Friedman, G. D., Quesenberry, C. P., Jr., & Weiss, N. S. (1992). A case-control study of screening sigmoidoscopy and mortality from colorectal cancer. *The New England Journal of Medicine, 326*(10), 653–657. https://doi.org/10.1056/NEJM199203053261001.

Selby, K., Jensen, C. D., Zhao, W. K., Lee, J. K., Slam, A., Schottinger, J. E., et al. (2019). Strategies to improve follow-up after positive fecal immunochemical tests in a community-based setting: A mixed-methods study. *Clinical and Translational Gastroenterology, 10*(2), e00010. https://doi.org/10.14309/ctg.0000000000000010.

Shapiro, J. A., Bobo, J. K., Church, T. R., Rex, D. K., Chovnick, G., Thompson, T. D., et al. (2017). A comparison of fecal immunochemical and high-sensitivity guaiac tests for colorectal cancer screening. *The American Journal of Gastroenterology, 112*(11), 1728–1735. https://doi.org/10.1038/ajg.2017.285.

Siegel, R. L., Fedewa, S. A., Anderson, W. F., Miller, K. D., Ma, J., Rosenberg, P. S., et al. (2017). Colorectal cancer incidence patterns in the United States, 1974–2013. *Journal of the National Cancer Institute, 109*(8), 1–6. https://doi.org/10.1093/jnci/djw322.

Siegel, R. L., Miller, K. D., Goding Sauer, A., Fedewa, S. A., Butterly, L. F., Anderson, J. C., et al. (2020). Colorectal cancer statistics, 2020. *CA: A Cancer Journal for Clinicians, 70*(3), 145–164. https://doi.org/10.3322/caac.21601.

Siegel, R. L., Miller, K. D., & Jemal, A. (2017). Colorectal cancer mortality rates in adults aged 20 to 54 years in the United States, 1970–2014. *JAMA, 318*(6), 572–574. https://doi.org/10.1001/jama.2017.7630.

Siegel, R. L., Torre, L. A., Soerjomataram, I., Hayes, R. B., Bray, F., Weber, T. K., et al. (2019). Global patterns and trends in colorectal cancer incidence in young adults. *Gut, 68*(12), 2179–2185. https://doi.org/10.1136/gutjnl-2019-319511.

Sineshaw, H. M., Ng, K., Flanders, W. D., Brawley, O. W., & Jemal, A. (2018). Factors that contribute to differences in survival of black vs white patients with colorectal cancer. *Gastroenterology, 154*(4), 906–915.e907. https://doi.org/10.1053/j.gastro.2017.11.005.

Sineshaw, H. M., Robbins, A. S., & Jemal, A. (2014). Disparities in survival improvement for metastatic colorectal cancer by race/ethnicity and age in the United States. *Cancer Causes & Control, 25*(4), 419–423. https://doi.org/10.1007/s10552-014-0344-z.

Sinicrope, P. S., Goode, E. L., Limburg, P. J., Vernon, S. W., Wick, J. B., Patten, C. A., et al. (2012). A population-based study of prevalence and adherence trends in average risk colorectal cancer screening, 1997 to 2008. *Cancer Epidemiology, Biomarkers & Prevention, 21*(2), 347–350. https://doi.org/10.1158/1055-9965.EPI-11-0818.

Skinner, C. S., Ahn, C., Halm, E. A., Bishop, W. P., McCallister, K., Sanders, J. M., et al. (2017). Recommendation of colorectal cancer testing among primary care patients younger than 50 with elevated risk. *Preventive Medicine, 102*, 20–23. https://doi.org/10.1016/j.ypmed.2017.06.014.

Smith, R. A. (2020). The development of cancer screening guidelines. *The Medical Clinics of North America, 104*(6), 955–970. https://doi.org/10.1016/j.mcna.2020.08.009.

Smith, R., & Rennie, D. (2014). Evidence-based medicine—An oral history. *JAMA, 311*(4), 365–367. https://doi.org/10.1001/jama.2013.286182.

Sox, H. C. (2005). Office-based testing for fecal occult blood: Do only in case of emergency. *Annals of Internal Medicine, 142*(2), 146–148. Retrieved from http://www.ncbi.nlm.nih.gov/pubmed/15657163.

Taggarshe, D., Rehil, N., Sharma, S., Flynn, J. C., & Damadi, A. (2013). Colorectal cancer: Are the "young" being overlooked? *American Journal of Surgery, 205*(3), 312–316. discussion 316 https://doi.org/10.1016/j.amjsurg.2012.10.016.

United States Preventive Services Task Force. (2020). *Draft recommendation statement colorectal cancer: Screening*. Retrieved from https://uspreventiveservicestaskforce.org/uspstf/draft-recommendation/colorectal-cancer-screening3#citation9.

Volk, R. J., Leal, V. B., Jacobs, L. E., Wolf, A. M. D., Brooks, D. D., Wender, R. C., et al. (2018). From guideline to practice: New shared decision-making tools for colorectal cancer screening from the American Cancer Society. *CA: A Cancer Journal for Clinicians, 68*(4), 246–249. https://doi.org/10.3322/caac.21459.

Wancata, L. M., Banerjee, M., Muenz, D. G., Haymart, M. R., & Wong, S. L. (2016). Conditional survival in advanced colorectal cancer and surgery. *The Journal of Surgical Research, 201*(1), 196–201. https://doi.org/10.1016/j.jss.2015.10.021.

Ward, E., Jemal, A., Cokkinides, V., Singh, G. K., Cardinez, C., Ghafoor, A., et al. (2004). Cancer disparities by race/ethnicity and socioeconomic status. *CA: A Cancer Journal for Clinicians, 54*(2), 78–93. https://doi.org/10.3322/canjclin.54.2.78.

Wender, R., Brooks, D., Sharpe, K., & Doroshenk, M. (2020). The national colorectal cancer roundtable: Past performance, current and future goals. *Gastrointestinal Endoscopy Clinics of North America, 30*(3), 499–509. https://doi.org/10.1016/j.giec.2020.02.013.

Wender, R. C., Doroshenk, M., Brooks, D., Hotz, J., & Smith, R. A. (2018). Creating and implementing a National Public Health Campaign: The American Cancer Society's and National Colorectal Cancer Roundtable's 80% by 2018 initiative. *The American Journal of Gastroenterology, 113*(12), 1739–1741. https://doi.org/10.1038/s41395-018-0384-1.

Wilson, J., & Jungner, G. (1968). *Principles and practice of screening for disease*. Geneva, Switzerland: World Health Organization.

Winawer, S. J. (2015). The history of colorectal cancer screening: A personal perspective. *Digestive Diseases and Sciences, 60*(3), 596–608. https://doi.org/10.1007/s10620-014-3466-y.

Winawer, S. J., Fletcher, R. H., Miller, L., Godlee, F., Stolar, M. H., Mulrow, C. D., et al. (1997). Colorectal cancer screening: Clinical guidelines and rationale. *Gastroenterology, 112*(2), 594–642. https://doi.org/10.1053/gast.1997.v112.agast970594.

Winawer, S. J., Zauber, A. G., Ho, M. N., O'Brien, M. J., Gottlieb, L. S., Sternberg, S. S., et al. (1993). Prevention of colorectal cancer by colonoscopic polypectomy. The National Polyp Study Workgroup. *The New England Journal of Medicine, 329*(27), 1977–1981. https://doi.org/10.1056/NEJM199312303292701.

Wingo, P. A., Cardinez, C. J., Landis, S. H., Greenlee, R. T., Ries, L. A., Anderson, R. N., et al. (2003). Long-term trends in cancer mortality in the United States, 1930–1998. *Cancer, 97*(Suppl. 12), 3133–3275. https://doi.org/10.1002/cncr.11380.

Wolf, A. M. D., Fontham, E. T. H., Church, T. R., Flowers, C. R., Guerra, C. E., LaMonte, S. J., et al. (2018). Colorectal cancer screening for average-risk adults: 2018 Guideline update from the American Cancer Society. *CA: A Cancer Journal for Clinicians*, *68*(4), 250–281. https://doi.org/10.3322/caac.21457.

Yin, D., Morris, C. R., Bates, J. H., & German, R. R. (2011). Effect of misclassified underlying cause of death on survival estimates of colon and rectal cancer. *Journal of the National Cancer Institute*, *103*(14), 1130–1133. https://doi.org/10.1093/jnci/djr207.

Zampino, M. G., Magni, E., Ravenda, P. S., Cella, C. A., Bonomo, G., Della Vigna, P., et al. (2016). Treatments for colorectal liver metastases: A new focus on a familiar concept. *Critical Reviews in Oncology/Hematology*, *108*, 154–163. https://doi.org/10.1016/j.critrevonc.2016.11.005.

Zapka, J., Klabunde, C. N., Taplin, S., Yuan, G., Ransohoff, D., & Kobrin, S. (2012). Screening colonoscopy in the US: Attitudes and practices of primary care physicians. *Journal of General Internal Medicine*, *27*(9), 1150–1158. https://doi.org/10.1007/s11606-012-2051-3.

Zazove, P., Plegue, M. A., Uhlmann, W. R., & Ruffin, M. T. (2015). Prompting primary care providers about increased patient risk as a result of family history: Does it work? *Journal of American Board of Family Medicine*, *28*(3), 334–342. https://doi.org/10.3122/jabfm.2015.03.140149.

CHAPTER FOUR

Personalized nutrition for colorectal cancer

Kai Wang[a,b] and Mingyang Song[a,b,c,d,*]

[a]Department of Epidemiology, Harvard T.H. Chan School of Public Health, Boston, MA, United States
[b]Department of Nutrition, Harvard T.H. Chan School of Public Health, Boston, MA, United States
[c]Clinical and Translational Epidemiology Unit, Massachusetts General Hospital and Harvard Medical School, Boston, MA, United States
[d]Division of Gastroenterology, Massachusetts General Hospital and Harvard Medical School, Boston, MA, United States
*Corresponding author: e-mail address: mis911@mail.harvard.edu

Contents

Abstract

The substantial burden of colorectal cancer and its increasing trend in young adults highlight the importance of dietary and lifestyle modifications for improved cancer prevention and survivorship. In this chapter, we review the cutting-edge evidence for the interplay between diet/lifestyle and the gut microbiota in the incidence and prognosis of colorectal cancer. We focus on factors for which there are data supporting their importance for the gut microbiota and colorectal cancer, including excess body weight, fiber, red and processed meat, and coffee. We discuss the potential precision nutrition approaches for modifying and exploiting the gut microbiota for improved cancer prevention and survivorship.

Colorectal cancer (CRC) is the third leading cause of cancer death in both men and women in the United States, and ranks second when men and women are combined (Siegel, Miller, Fuchs, & Jemal, 2021). In 2021, approximately 149,500 individuals were estimated to be diagnosed with CRC and 52,980 to die from the disease (Siegel et al., 2021). The past several decades have witnessed declines in both CRC incidence rate, from a peak of 66.3 in 1985 to 38.7 per 100,000 persons in 2020, and CRC

Advances in Cancer Research, Volume 151
ISSN 0065-230X
https://doi.org/10.1016/bs.acr.2021.02.004

mortality rate, from a peak of 28.6 in 1976 to 13.9 per 100,000 in 2020 (Siegel et al., 2020). The decline before 2000 is thought to be equally attributed to changing patterns in risk factors (e.g., reductions in smoking) and increasing use of CRC screening (Edwards et al., 2010), whereas the accelerated decline since 2000 is believed to reflect the rapid dissemination of CRC screening, particularly colonoscopy screening (Siegel et al., 2020; Smith et al., 2019). To further increase the undertaking of CRC screening, in 2014, the National Colorectal Cancer Roundtable (NCCRT) launched a national campaign to increase CRC screening rates to 80% by 2018 (American Cancer Society, 2014). Although this goal has been achieved in a number of populations (Smith et al., 2019), the overall rate of screening, particularly colonoscopy screening, has reached a plateau since 2015 with a rate around 60% (Siegel et al., 2020).

CRC is most often diagnosed in elderly individuals. Despite the decline in the overall CRC incidence over the past decades, the incidence of CRC in younger adults has been increasing in the United States. According to the latest data of 2011 through 2016, the incidence rate of CRC among individuals aged 50 to 64 years and younger than 50 years increased by 1.0% and 2.0% annually, respectively (Siegel et al., 2020). Given these alarming increases, the US Preventive Services Task Force (USPSTF) released a draft recommendation in 2020 that average-risk adults may initiate routine screening at age 45 instead of 50 (US Preventive Services Task Force, 2020). The reasons for the increase in young-onset CRC remain largely to be elucidated, but a number of plausible hypotheses have been proposed, including greater exposure to potential risk factors, e.g., obesity, Western-style diet, physical inactivity, and antibiotic use, particularly during early life (Akimoto et al., 2020). Given the possibly decade-long latency in the transition from non-neoplastic cells to malignant neoplasms (Loeb, 2016; Tomasetti & Vogelstein, 2015) and that young-onset CRC is more likely to be diagnosed at an advanced-stage (Chen, Sundaram, Chew, & Ladabaum, 2017; Kneuertz et al., 2015; Myers et al., 2013), novel primary preventive strategies are urgently needed to prevent the initiation of young-onset CRC. The young-onset CRC has been extensively discussed in the chapter by Dr. Swati Patel.

CRC is potentially preventable, since more than one-half of all CRC cases and deaths are estimated to be attributable to modifiable environmental factors, e.g., excess body weight, an unhealthy diet, smoking, high alcohol consumption, and physical inactivity (Islami et al., 2018; Kamangar et al., 2020; Song & Giovannucci, 2016). Although the exact mechanisms through which each of these factors affects CRC are likely to differ, accumulating evidence suggests that the gut microbiota may represent a congregate mode of action linking these factors to CRC. The booming microbiota research in the last decade has allowed delineation of the composition and some functions of the gut microbiota and indicated its intricate interaction with environmental factors. On the one hand, the composition and function of the gut microbiota are largely shaped by diet and lifestyle factors (Rothschild et al., 2018); on the other hand, the gut microbiota plays important roles in nutrient processing and synthesis to sustain a balanced dialog between dietary metabolites and host response (Rooks & Garrett, 2016; Tremaroli & Bäckhed, 2012). For example, a healthy gut microbiota produces normal intestinal levels of secondary bile acids to help maintain a viable epithelial layer of cells by inducing apoptosis and promoting cell turnover, but the gut dysbiosis induced by factors such as colon inflammation may result in a diminished ability of the microbiota to produce secondary bile acids. Thereupon, if a high-fat-diet-induced overload of intestinal secondary bile acids is present simultaneously, this dysbiosis may promote selective growth of less differentiated cells that are resistant to apoptosis with a loss of farnesoid X receptor (FXR) function, contributing to colorectal carcinogenesis (Jia, Xie, & Jia, 2018). Appreciation of the interactions between the gut microbiota and dietary factors in impacting intestinal health not only helps explain the seemingly paradoxical relationships between secondary bile acids, intestinal inflammation, and CRC risk, but also sheds light on novel microbiota-based dietary interventional strategies to prevent CRC.

Often referred to as the "forgotten organ," the gut microbiota plays an important role in the host health. Although microbiome composition differs between individuals and at times fluctuates markedly within individuals, core features exist in the microbial communities that colonize the human body (Ding & Schloss, 2014). Although dysbiosis has been associated with CRC, whether this dysbiosis contributes to CRC or merely represents a bystander effect remains unclear. To establish causality, it will be necessary to decipher which specific taxa or metabolites drive CRC and how these taxa interact with the environmental factors in driving CRC. Enrichment of several putatively procarcinogenic microbial members in stool or cancer tissue, including *Fusobacterium nucleatum*, *Escherichia coli*, *Bacteroides fragilis*, *Enterococcus faecalis*, *Streptococcus gallolyticus*, and *Peptostreptococcus spp.*, has been identified to contribute to colorectal carcinogenesis by inducing tumor proliferation (Long et al., 2019; Yang et al., 2017), promoting inflammation (Chung et al., 2018), causing DNA damage (Rubinstein et al., 2013) and protecting tumor from immune attach (Long et al., 2019). In contrast, some so-called protective genera, including *Roseburia*, *Clostridium*, *Faecalibacterium*, and *Bifidobacterium*, are found to be depleted in CRC patients (Garrett, 2019; Saus, Iraola-Guzmán, Willis, Brunet-Vega, & Gabaldón, 2019; Ternes et al., 2020; Tilg, Adolph, Gerner, & Moschen, 2018; Wirbel et al., 2019; Wong & Yu, 2019). Several products of bacterial metabolism have been implicated in CRC, many of which are associated with dietary intake (Wu, Powers, Zhu, & Hannun, 2016), including products of protein fermentation, secondary bile acids from high-fat intake, and short-chain fatty acids (SCFAs) metabolized from carbohydrates and phytochemicals (Louis, Hold, & Flint, 2014). We also reported the substantial influence of commensal microbes in prognosis of cancer patients, e.g., *F. nucleatum* might be associated with a worse survival in CRC patients (Mima et al., 2016). Moreover, increasing evidence indicates that gut microbiota can modulate host responses to cancer immunotherapy (York, 2018). Studies continue to suggest potential modulating roles of individual microbial species and microbial communities in the effects of lifestyle and dietary factors on CRC. This would potentially reveal new avenues that can be exploited for microbiota-based personalized nutrition and lifestyle intervention for CRC prevention.

In this chapter, we review the evidence on how lifestyle and dietary factors may influence the risk and survivorship of CRC through the gut microbiota-related mechanisms. We focus on the emerging prospective evidence on how individuals' gut microbiota may modulate the effects of lifestyle and diet on CRC risk and related death. We review lifestyle and nutritional factors for which there are convincing data supporting their importance for CRC, including excess body weight, fiber, red and processed meat, and coffee. We first review the evidence for CRC incidence and then summarize the evidence for CRC survivorship.

1. Excess body weight

In the United States, 42.4% of adults are obese and an additional 32.5% are overweight (Centers for Disease Control and Prevention, 2020a, 2020b). Excess body weight is an established cause of CRC (World Cancer Research Fund/American Institute for Cancer Research, 2018). Each $5\,kg/m^2$ increase in body mass index was reported to be associated with 5% increased CRC risk (Table 1). This association appears to be stronger when body mass index was above $27\,kg/m^2$ (World Cancer Research Fund/American Institute for Cancer Research, 2018). Overall, excess body weight has been

Table 1 Summary of evidence and potential mechanisms underlying the relationship between lifestyle and dietary factors, gut microbiota, and colorectal cancer.

Risk factor	Estimated association with colorectal cancer (RR [95% CI])[a]	Associated microbes	Potential microbiota-related mechanisms
Overweight and obesity	1.05 (1.03–1.07) per 5 kg/m^2 increase in body mass index	Decreased *Bacteroidetes/Firmicutes* ratio; increased *Desulfovibrio* (Xu et al., 2017); decreased *Akkermansia muciniphila* (Depommier et al., 2019)	Increased LPS levels; epigenetic changes in the colonic epithelial cells; increased production of DNA-damaging bile acids
Dietary fiber	0.93 (0.87–1.00) per 10 g/day increase in dietary consumption	Increased SCFAs-producing bacteria	Epigenetic regulation and immunomodulatory benefits of SCFAs
Red and processed meat	1.12 (1.04–1.21) per 100 g/day increase in consumption	Increased mucin-degrading bacteria	Microbiota-dependent barrier dysfunction induced by heme iron; increased production of secondary bile acids and hydrogen sulfide by microbiota associated with high red meat consumption
Coffee	0.82 (0.69–0.96) comparing ≥2 cups/day of decaffeinated coffee to no decaffeinated coffee; no association observed for caffeinated coffee (Um et al., 2020)	Increased *Bifidobacterium spp.* (Jaquet, Rochat, Moulin, Cavin, & Bibiloni, 2009)	Given the roles of *Bifidobacterium* in the metabolism of SCFAs and anti-inflammation, it may suggest a synergistic beneficial effect of coffee with fiber against CRC

[a]Data are derived from the meta-analysis conducted by the World Cancer Research Fund/American Institute for Cancer Research, (WCRF/AICR, 2018) unless stated otherwise.

This table was adapted from our prior publication (DOI: https://doi.org/10.1016/j.cgh.2018.07.012; © 2019 by the AGA Institute) with minor modifications.

Abbreviations: CI, confidence interval; LPS, lipopolysaccharides; RR, relative risk; SCFA, short-chain fatty acid.

estimated to contribute to approximately 5% of incident CRC cases in the United States (Islami et al., 2018). Recently, it has been suggested that overweight and obesity may confer a particular propensity toward young-onset CRC (Hofseth et al., 2020).

There is convincing mechanistic evidence supporting a link between excess body weight and CRC, with the pathways including insulin/insulin-like growth factors, chronic inflammation, leptin, sex hormones, adipokines, and disrupted bile acid homeostasis. In recent years, the gut microbiota has been increasingly considered as a potential pathway underlying the link between excess body weight and CRC (Hofseth et al., 2020; Marchesi et al., 2016; Murphy, Jenab, & Gunter, 2018; Song & Chan, 2018; Song, Chan, & Sun, 2019; Wang & Jia, 2016; Ye, Xi, Huang, & Xu, 2020). Studies in both mice and humans have shown the effects of the gut microbiota on host metabolism by increasing energy harvest from food and modulating adiposity-related metabolic pathways. Lipopolysaccharide (LPS) is an endotoxin produced by gram-negative bacteria in the gut and participates in colorectal carcinogenesis through increased chronic inflammation (Maldonado, Sá-Correia, & Valvano, 2016). Circulating LPS was elevated in high-fat-diet-induced obesity due to gut microbiome remodeling (Tai, Wong, & Wen, 2015; Xu et al., 2017). Animal studies observed an increased abundance of the LPS-producing *Desulfovibrio* with obesity (Xu et al., 2017). Further studies are needed to confirm the role of *Desulfovibrio* in humans and its potential as a microbial target for modulation.

Studies have also reported a negative correlation of *Akkermansia muciniphila* abundance with overweight and obesity (Dao et al., 2016; Liu, Hong, et al., 2017; Yassour et al., 2016). These bacteria may protect against obesity through increasing intestinal endocannabinoids that reduce inflammation and enhance gut barrier function (Everard et al., 2013). In obese human volunteers, *Akkermansia* administration improved insulin sensitivity and inflammation, and mildly reduced body weight, compared to placebo (Depommier et al., 2019). *Akkermansia* is also a crucial player in gastrointestinal disorders. Treatment with *Akkermansia* inhibited dextran sodium sulfate-induced colitis in mice (Bian et al., 2019). *Akkermansia* blunted colitis and reduced tumorigenesis through regulation of macrophages and CD8+ cytotoxic T lymphocytes (Wang et al., 2020). A clinical trial has shown that the administration of *Akkermansia muciniphila* was safe and well tolerated and that microbial supplementation improved several metabolic parameters (Depommier et al., 2019).

Trimethylamine N-oxide (TMAO) is a microbial metabolite implicated in obesity (Koeth et al., 2013). As an oxidized form of trimethylamine (TMA), increasing evidence suggests that the level of TMAO is associated with an increased inflammation, oxidative stress, DNA damage, and disruption in protein folding (Chan et al., 2019). Plasma TMAO levels were found to be positively correlated with CRC risks, and subjects in the highest quartile of plasma TMAO concentration were at 3.4 times greater risk of rectal cancer than those in the lowest quartile (Bae et al., 2014). Another study demonstrated an increased serum level of TMAO among CRC patients, compared to healthy controls (Liu, Liu, et al., 2017). A number of bacterial species in the gut have been shown involved in TMAO production, and these bacteria largely belong to the *Firmicutes* and *Proteobacteria* phyla (Romano, Vivas, Amador-Noguez, & Rey, 2015). Of note, none of the bacterial species screened for TMAO production belongs to *Bacteroidetes* (Craciun & Balskus, 2012; Falony, Vieira-Silva, & Raes, 2015; Romano et al., 2015). These observations suggest that the level of TMAO production is closely influenced by changes in composition of the intestinal microbiota, such as the ratio of *Firmicutes* and *Bacteroidetes* abundances, which has been related to obesity.

2. Fiber

Dietary fiber is the part of plant foods or complex carbohydrates (e.g., cellulose and lignin) not digestible or absorbable by humans. Despite its minimal contribution to total energy, dietary fiber is believed to provide important health benefits. However, whether dietary fiber protects against CRC remains controversial because of the conflicting findings from epidemiologic studies. Numerous (Aune et al., 2011; Hansen et al., 2012; Murphy et al., 2012; Reynolds et al., 2019), but not all (He et al., 2019; Hullings et al., 2020; Oh et al., 2019), epidemiologic studies reported that a higher intake of dietary fiber was associated with a reduced CRC risk (Table 1). Given these data, the World Cancer Research Fund and the American Institute for Cancer Research (WCRF/AICR) concluded that there is "probable strong evidence" for foods containing dietary fiber to be CRC-preventive (World Cancer Research Fund/American Institute for Cancer Research, 2018). The Dietary Guidelines for Americans (2015–20) highlight dietary fiber as an under-consumed nutrient of public health concern and recommend consuming fiber-dense foods amounting to 38 and 25 g fibers per day for adult men and women, respectively (Scientific Advisory Committee on Nutrition, 2015; US Department of Health and Human Services & US Department of Agriculture, 2015). However, the proportion of the population managing to reach this recommended fiber intake is only 5% in the United States (Quagliani & Felt-Gunderson, 2017). Therefore, it has been indicated that a higher intake of total dietary fiber within the range of a typical American diet is unlikely to substantially reduce CRC risk (He et al., 2019).

Increasing data suggest a heterogeneity in the association between fiber intake and CRC. In some (He et al., 2019; Hullings et al., 2020; Oh et al., 2019), but not all studies (Schwingshackl et al., 2018), the protective association with CRC was only found for fiber from cereal or grains, but not other sources; and the protection was particularly strong for cancers at the distal colon and rectum (Hullings et al., 2020) or of particular molecular profile, e.g., tumors of microsatellite stable/microsatellite instability (MSI)-low, CpG island methylator phenotype (CIMP)-negative, *BRAF* gene-wildtype, and *KRAS* gene-wildtype (Carr et al., 2018; Hidaka et al., 2020). Despite the contributions of these findings to clarify the etiological pathways underlying the fiber-CRC association, one key question remaining unresolved is how these data may be applied to inform recommendations for fiber intake in CRC prevention.

Two general mechanisms underlying the benefit of dietary fiber against CRC have been proposed. First, insoluble fiber bulks luminal contents and may speed colonic transit to minimize the exposure of the colonic epithelium to ingested carcinogens, e.g., nitrosamines from charred meat. Second, bacteria in the lumen of the colon ferment soluble fiber into SCFAs and other metabolites with potentially beneficial properties. SCFAs, particularly butyrate, are transported into the colonic epithelium. Through β-oxidation, they provide ≥70% of the energy used by normal colonocytes (Fung, Cosgrove, Lockett, Head, & Topping, 2012). SCFAs also function as a histone deacetylase (HDAC) inhibitor inside the nucleus to epigenetically regulate gene expression, inhibit cell proliferation, and induce cell differentiation or apoptosis when added to tumor-derived cell lines, and have anti-inflammatory effects (Bultman, 2014; Fung et al., 2012; Hamer et al., 2008). However, very limited data have examined the heterogeneity by microbiota in the fiber-CRC association. We recently reported that higher intake of dietary fiber was more strongly associated with lower risk for *Fusobacterium nucleatum*-positive CRC but not *Fusobacterium nucleatum*-negative CRC (Mehta et al., 2017). Although the mechanism remains unclear, it is plausible that bacterial fermentation of fiber results in altered levels of SCFAs, leads to lower pH, increases transit time of gut contents, or changes the local immune surveillance, all of which make it less hospitable for certain commensal species, such as *F. nucleatum*, to establish in the colonic niche and potentiate colorectal carcinogenesis (Garrett, 2015; Smith et al., 2013; Sonnenburg et al., 2016). Indeed, a study using gnotobiotic mice models demonstrated that a high-fiber diet and a colonization of the butyrate-producing bacterium *Butyrivibrio fibrisolvens* were both needed to observe a significantly higher luminal butyrate and reduced CRC incidence (Donohoe et al., 2014). These data indicate that bacterial capability for SCFAs production may be an important determinant of the CRC-preventive effect of dietary fiber. Another mouse model study found that the administration of *Bifidobacterium longum* had a similar improving effect on colonic mucus layer defect as that observed after fiber administration (Schroeder et al., 2018). Given the roles of *Bifidobacterium* in metabolizing fiber into SCFAs (Papathanasopoulos & Camilleri, 2010; Slavin, 2013), it may suggest that the presence of distinct bacteria in the gut is also crucial for a healthy diet to confer benefit for the host. Our unpublished data showed that a higher fiber intake was associated with a lower inflammation level (assessed by C-reactive protein level) only in individuals without substantial *Prevotella copri* carriage in the gut. Although detailed mechanism remains unknown, given the role of *Prevotella copri* in carbohydrate metabolism (De Filippis et al., 2019), it may be important to consider *Prevotella copri* when examining the fiber-CRC association. These preliminary data suggest that dietary fiber may protect against colorectal tumorigenesis in a microbiota-dependent manner. Further studies exploring the fiber-microbiota interaction in CRC will be critical to develop novel probiotic/prebiotic strategies for CRC prevention.

3. Red and processed meat

The International Agency for Research on Cancer (IARC) reported conclusive evidence that processed meat causes CRC and limited evidence for a positive association between red meat and CRC (Bouvard et al., 2015). A meta-analysis of 10 studies by the WCRF/AICR showed that each 50 g of processed meat per day is associated with a 16% increased CRC risk, and each 100 g of red meat per day is associated, although not significantly, with a 12% increased CRC risk (WCRF/AICR, 2018). Based on these data, in

2015, the World Health Organization (WHO) classified processed meat as Group 1 carcinogen and red meat as group 2A carcinogen to humans (Bouvard et al., 2015).

Several underlying mechanisms have been proposed for the pro-CRC effect of processed and red meat, e.g., the heterocyclic amines and polycyclic aromatic hydrocarbons in red and processed meat induce DNA mutation, heme produces cytotoxic and genotoxic aldehydes through the lipid peroxidation process and the formation of N-nitroso compounds (Jeyakumar, Dissabandara, & Gopalan, 2017). However, inconsistent findings of red and processed meat in relation to CRC continue to be reported (Alexander, Weed, Cushing, & Lowe, 2011; Han et al., 2019). Some potential reasons may be methodology-related. For example, some meta-analyses of cohort studies often combined studies from different regions in the world, studies in both sexes, or studies on fresh red meat and processed meat with ambiguous classifications (Alexander et al., 2011; Han et al., 2019). These merges might have diluted the association.

In red and processed meat, in addition to the preservatives (e.g., nitrates and nitrites) and chemicals produced during meat processing and cooking (e.g., heme iron, heterocyclic amines, and polycyclic aromatic hydrocarbons), certain nutrients (e.g., saturated fat and sulfur) are increasingly appreciated to cause the carcinogenic effect (Song et al., 2019). Red meat, depending on the cut and origin of the meat, may contain more fat than white meat. The fat content in red meat has been implicated as a potential risk factor for CRC through an interaction with the gut microbiota to produce more secondary bile acids. In response to fat intake, primary bile acids are synthesized from cholesterol in the liver mainly as cholic acid (CA) and chenodeoxycholic acid (CDCA) (Portincasa & Calamita, 2012; van Erpecum et al., 2006). Before secretion of primary bile acids, liver cells conjugated them with one of the two amino acids, glycine or taurine, to form a total of 14 possible primary bile acids (Hang et al., 2019). Primary bile acids are released into the lumen of the intestine and undergo bacterial dihydroxylation to form 15 possible secondary bile acids (Hamilton et al., 2007; Hang et al., 2019; Ridlon, Kang, & Hylemon, 2006). Secondary bile acids, dominantly deoxycholic acid (DCA), lithocholic acid (LCA), and ursodeoxycholic acid (UDCA), are important gut bacterial metabolites and increasingly recognized as a pathway underlying the role of the gut microbiome in colorectal carcinogenesis (Ajouz, Mukherji, & Shamseddine, 2014; Cao et al., 2017; Feng & Chen, 2016; Long, Gahan, & Joyce, 2017; Ma et al., 2018; Scott et al., 2019; Song et al., 2019; Wong & Yu, 2019;

Yoshimoto et al., 2013). They may trigger a plethora of pro-cancer effects, including inflammation, oxidative DNA damage, and apoptosis resistance (Farhana et al., 2016; Gadaleta, Garcia-Irigoyen, & Moschetta, 2017; Hofmann, 1999; Ridlon et al., 2006; Zeng, Claycombe, & Reindl, 2015; Zeng, Umar, Rust, Lazarova, & Bordonaro, 2019). Recently, higher circulating levels of secondary bile acids were found to be associated with an increased risk of CRC (Cross et al., 2014; Kühn et al., 2020). Two major bile acid-sensing receptors, farnesoid X receptor (FXR) and G protein-coupled bile acid receptor 1 (TGR5), have been implicated in the CRC relationship (Jia et al., 2018). Major bacterial modifications in the biosynthesis of secondary bile acids include deconjugation of conjugated bile acids yielding free bile acids (mainly CA and CDCA) by bile salt hydrolase and $7\alpha/\beta$-dehydroxylation of CA and CDCA yielding DCA, LCA, and UDCA. Bile salt hydrolase is widely expressed in the gut bacterial community, including Gram-positive bacteria, e.g., *Clostridium* (Gopal-Srivastava & Hylemon, 1988; Masuda, 1981), *Enterococcus* (Wijaya et al., 2004), *Bifidobacterium* (Jarocki & Targonski, 2013; Tanaka, Hashiba, Kok, & Mierau, 2000), *Lactobacillus* (De Smet, Van Hoorde, Vande Woestyne, Christiaens, & Verstraete, 1995; Oh et al., 2008), *Eubacterium* (Martin et al., 2018; Song et al., 2017), *Escherichia* (Jones, Begley, Hill, Gahan, & Marchesi, 2008; Ridlon et al., 2006), and Gram-negative bacteria, e.g., *Bacteroides* (Stellwag & Hylemon, 1976). However, $7\alpha/\beta$-dehydroxylase, as the "gateway" enzyme in secondary bile acids biotransformation and encoded by the bile acids-inducible (*bai*) genes (Wirbel et al., 2019), appears restricted to a narrow bacterial phylogenetic group, e.g., *Clostridium* cluster XIV (Ridlon, Harris, Bhowmik, Kang, & Hylemon, 2016) and *Clostridium* sp. (Vital, Rud, Rath, Pieper, & Schluter, 2019; Wells & Hylemon, 2000). Thus, the abundances of the "gateway" secondary bile-producing bacteria in the gut, e.g., *Clostridium* cluster XIV and *Clostridium* sp., may be important in determining a personalized pro-CRC effect of the fat from red and processed meat.

Organic sulfur compounds are enriched in red meat and inorganic sulfur (sulfate and sulfite) is routinely used as preservatives in processed meat. These sulfur compounds are metabolized by sulfur-reducing or sulfur-fermenting bacteria into hydrogen sulfide in the gut (Song, Garrett, & Chan, 2015). Sulfur-metabolizing bacteria are a specialized group of phylogenetically diverse microbes with the capability to metabolize organic compounds for energy, while metabolizing dietary sulfur to hydrogen sulfide (Attene-Ramos, Wagner, Gaskins, & Plewa, 2007; Magee, Richardson, Hughes, &

Cummings, 2000; Ramasamy, Singh, Taniere, Langman, & Eggo, 2006). These bacteria and their resultant metabolites have been associated with CRC (Cai, Wang, Ju, Wang, & Zhu, 2010; Scanlan, Shanahan, & Marchesi, 2009), whereas hydrogen sulfide itself may cause epithelial DNA damage through production of free radicals (Attene-Ramos et al., 2007; Hildebrandt et al., 2009), and promote alterations in immune cell populations associated with inflammation (Cai et al., 2010; Ramasamy et al., 2006) and CRC (Attene-Ramos et al., 2007; Roediger, Duncan, Kapaniris, & Millard, 1993; Zeng & Combs Jr., 2008). Furthermore, gut-derived hydrogen sulfide may fragment the mucus bilayer of the gastrointestinal tract that is typically held together by disulfide bonds. This breach may precede tumorigenesis by exposing gut epithelium to immunogenic luminal bacteria (Ijssennagger et al., 2015; Ijssennagger, van der Meer, & van Mil, 2016; Schroeder et al., 2018). Several sulfidogenic bacteria were found to be enriched in tissue samples from patients with CRC, including *Fusobacterium*, *Bilophila wadsworthia*, and the genera *Lactococcus*, *Porphyromonas*, *Odoribacter*, *Bilophila*, and *Pyramidobacter* (Yazici et al., 2015, 2017). Hydrogen sulfide-producing pathways are increased in fecal samples from patients with CRC (Yachida et al., 2019). Recently, we found that a long-term adherence to a dietary pattern associated with higher abundance of sulfur-metabolizing bacteria in stool was associated with an increased risk of distal CRC (Nguyen et al., 2020).

4. Coffee

Coffee is one of the most commonly consumed beverages in the world (Butt & Sultan, 2011). Concerns have long existed that coffee may increase the risk of cancer and cardiovascular diseases. However, the emerging evidence has shown the health benefits of coffee (Ludwig, Clifford, Lean, Ashihara, & Crozier, 2014; van Dam, Hu, & Willett, 2020). An umbrella review including 218 meta-analyses of observational and interventional studies reported that intakes of three to four cups of coffee a day were associated with 17%, 15%, and 18% lower risk of all-cause mortality, incident cardiovascular disease, and incident cancer, respectively (Poole et al., 2017). Besides caffeine, coffee also contains a complex mixture of hundreds of biologically active phytochemicals, including polyphenols (mainly chlorogenic acids and lignans), alkaloid trigonelline, melanoidins formed during roasting, and modest amounts of magnesium, potassium, and vitamin B3 (van Dam et al., 2020). Many of the components, e.g., polyphenols, melanoidins, and diterpenes, are beneficial for gut health due to their anti-inflammatory and insulin-sensitizing properties, as well as their effects on reduction of bile acid secretion, modification of microbiome composition, and enhancement of critical bowel functions, e.g., motility and stool output, to decrease the exposure of epithelial cells to potential carcinogens in the colon (Alicandro, Tavani, & La Vecchia, 2017; Higdon & Frei, 2006; Ludwig et al., 2014; Schmit, Rennert, Rennert, & Gruber, 2016; Vitaglione, Fogliano, & Pellegrini, 2012). Based on the potential benefits of coffee's biochemical components, coffee has been suggested to protect against CRC (Cross & Gunter, 2018; Giovannucci, 1998; Je, Liu, & Giovannucci, 2009).

Despite the biologic plausibility for coffee's chemopreventive properties against CRC, only some (Arab, 2010; Galeone, Turati, La Vecchia, & Tavani, 2010; Li, Ma, Zhang, Zheng, & Wang, 2013; Schmit et al., 2016; Tian, Wang, Hong, & Zhang, 2013; Um et al., 2020; Yu, Bao, Zou, & Dong, 2011), but not all (Je et al., 2009; Kashino et al., 2018; Micek, Gniadek, Kawalec, & Brzostek, 2019; Sartini et al., 2019; Yamada et al., 2014; Zhang et al., 2010), epidemiologic evidence supported an inverse association between coffee intake and CRC risk. Heterogeneity has been reported in the coffee-CRC association according to study design (case-control vs cohort), coffee type (caffeinated vs decaffeinated), cancer site (colon vs rectum), sex, and ethnicity. A recent meta-analysis of 26 prospective studies with over 3 million subjects observed an inverse association only for decaffeinated coffee (vs caffeinated coffee), for colon cancer (vs rectal cancer), and in the United States subjects (vs Europe and Asia) (Sartini et al., 2019). Some studies attributed the observed null or heterogeneous associations to the possibility that caffeine has genotoxic and mutagenic properties (Higdon & Frei, 2006) and can lower insulin sensitivity (Keijzers, De Galan, Tack, & Smits, 2002) at high concentrations, leading to an increased CRC risk and offsetting the beneficial effects. However, consumption of caffeinated coffee (4–5 cups per day) for up to 6 months has been shown not to affect insulin resistance (Alperet et al., 2020; Ohnaka et al., 2012; Wedick et al., 2011). Some other studies attributed the heterogeneity to the variation in coffee cup size, brewing method, type of coffee bean, and the amounts of sugar and milk or cream added to coffee, which are generally not captured in epidemiologic studies. However, it has been indicated that within many populations, variation in cup sizes and brew strength is likely to be modest in comparison with the large variation in frequency of consumption (van Dam et al., 2020). Some hypotheses for the inconsistency have also been suggested. First, because tolerance effects can develop (Robertson, Wade, Workman, Woosley, & Oates, 1981), the observed association of coffee intake with CRC may not reflect long-term effect of coffee, particularly given the long-term latency of CRC initiation. Second, the activity of caffeine-metabolizing enzyme is partly inherited (Perera, Gross, & McLachlan, 2012). Individuals with genetically determined slower coffee metabolism tend to compensate by having lower habitual coffee intake than persons without this genetic predisposition (Cornelis et al., 2015). These data highlight the importance of personalized coffee intake for health benefit.

There has been an increasing interest to study the interplay between coffee intake and the gut microbiota. A clinical trial among 16 healthy adults reported that a daily dose of 3 cups of coffee during 3 weeks significantly increased the abundance and activity of *Bifidobacterium spp.* (Jaquet et al., 2009). Given the roles of *Bifidobacterium* in the metabolism of SCFAs and reduction of inflammation (Papathanasopoulos & Camilleri, 2010; Slavin, 2013), it may suggest a synergistic beneficial effect of coffee with other nutrients, e.g., fiber, against CRC. This is in line with a recent viewpoint that coffee as a plant food has similar beneficial properties to many vegetables and fruits, with one potential mechanism being the modulation of the gut microbiota by non-digested prebiotic constituents, e.g., phenolic acids and polyphenols (Kolb, Kempf, & Martin, 2020). Another recent trial among 30 healthy adults observed an increase in the gut microbial diversity after a single-dose consumption of sugar-free black coffee, and the microbial response was substantially modified by sex (Chong et al., 2020). In summary, although polyphenols and other nondigestible constituents of coffee like polysaccharides and melanoidins have the ability to modify the composition and metabolic function of gut microbiota like components in other plant foods (Kolb et al., 2020), whether these effects contribute to the beneficial properties of coffee and how their interactions with the gut microbiota may be exploited for CRC prevention remain to be studied.

5. CRC survivorship

With the tremendous advances in early detection and treatment of CRC, the number of individuals living with CRC is ever-increasing. As of 2016, more than 1.4 million Americans are living with CRC, of which 65% survive more than 5 years and 58% over 10 years, and this number is projected to be approximately 1.8 million as of 2026 (Miller et al., 2016). These data highlight the importance to develop effective adjuvant strategies to further improve patients' survivorship. Over the years, how healthy lifestyle and diet may be associated with CRC survivorship have been reported in a number of studies (Morales-Oyarvide et al., 2019; Song & Chan, 2018; Song, Wu, Meyerhardt, Ogino, et al., 2018; Song, Wu, Meyerhardt, Yilmaz, et al., 2018; Van Blarigan et al., 2018). With increasing appreciation of the gut dysbiosis in colorectal carcinogenesis, revealing how the microbial features may modify the survival benefits of healthy lifestyle and diet will shed light on personalized approaches to improved CRC treatment.

We recently reported that higher postdiagnosis intake of dietary fiber was associated with a better CRC survival (Song, Wu, Meyerhardt, Ogino, et al., 2018). Each 5 g per day increase in fiber intake was associated with 22% lower risk for CRC-specific mortality and 14% lower risk for all-cause mortality, and these benefits were even higher for the fiber from cereals and whole grains (Song, Wu, Meyerhardt, Ogino, et al., 2018). In line with the immune benefit of SCFAs, patients with higher abundance of SCFA-producing bacteria, e.g., *Bifidobacterium* and *Faecalibacterium*, appear to be more responsive to immunotherapy and have a better survivorship (Gopalakrishnan et al., 2018; Matson et al., 2018; Sivan et al., 2015). These findings do not only support the contribution of dietary fiber in improving CRC survivorship, but also suggest that the presence of SCFA-producing bacteria may be important for fiber's beneficial effect. Additionally, *F. nucleatum* in CRC tissue was found to be associated with a shorter survival in CRC patients (Mima et al., 2016). It remains unknown how lifestyle and diet may modify the effect of *F. nucleatum* on CRC survival.

Limited evidence exists on the role of red and processed meat intake in survival after CRC diagnosis (van Meer et al., 2013; van Zutphen, Kampman, Giovannucci, & van Duijnhoven, 2017). Overall, the existing studies suggest a null association of red and processed meat intake with survival among CRC patients.

Coffee contains various anti-carcinogenic substances, including polyphenols, diterpenes, melanoidins, and antioxidants (Alicandro et al., 2017). These substances may be beneficial for CRC survival by ameliorating systemic disturbances due to metabolic reprogramming of cancer or by promoting an anti-carcinogenic microenvironment that slows down tumor progression. Recently, we and another group reported in three prospective studies that a higher consumption of coffee after CRC diagnosis was associated with lower CRC recurrence and improved survival (Guercio et al., 2015; Hu et al., 2018). In our study (Hu et al., 2018), compared with nondrinkers, patients who consumed at least 4 cups of coffee per day had a 52% lower risk of CRC-specific and 30% reduced risk of all-cause death. We found this benefit was not modified by coffee type (caffeinated vs decaffeinated). We also found that a consistently high consumption of coffee in both pre- and post-diagnosis periods was associated with even lower risks of overall and CRC-specific death. Further studies are needed to better understand how the gut microbiota may modify the effect of coffee on CRC survival.

6. Future directions

With the increasing knowledge of how the gut microbiota influences carcinogenesis and treatment outcomes, microbial modulation has become a potential strategy for CRC prevention and treatment. The strategies to restore microbial dysbiosis include probiotics (living microorganisms conferring health benefit on the host), prebiotics (nondigestible food ingredient selectively stimulating the growth and activity of specific bacteria in the gut), postbiotics (soluble byproducts and metabolites secreted by the gut microbiota exerting biological benefit to the host), antibiotics (to deplete deleterious bacteria and reverse the detrimental dysbiosis), and fecal microbiota transplantation (to administrate fecal transplants from healthy donors to patients' gastrointestinal tract to restore microbial homeostasis). Advantages, disadvantages, and putative mechanisms of these different approaches have been discussed in other review (Fong, Li, & Yu, 2020). The current review highlights the potential of precision nutrition approaches in modifying and exploiting the gut microbiota for improved cancer prevention and survivorship. At present, the evidence in that regard remains very limited. Integrated epidemiologic and clinical studies are needed to prospectively interrogate the interplay between diet and the gut microbiota in CRC development and progression and to assess the microbiota-tailored dietary approaches for intervention. Investigations in these areas will lead to the ultimate realization of the promises of precision cancer prevention.

References

Ajouz, H., Mukherji, D., & Shamseddine, A. (2014). Secondary bile acids: An under-recognized cause of colon cancer. *World Journal of Surgical Oncology*, *12*, 164. https://doi.org/10.1186/1477-7819-12-164.

Akimoto, N., Ugai, T., Zhong, R., Hamada, T., Fujiyoshi, K., Giannakis, M., et al. (2020). Rising incidence of early-onset colorectal cancer—A call to action. *Nature Reviews. Clinical Oncology*. https://doi.org/10.1038/s41571-020-00445-1.

Alexander, D. D., Weed, D. L., Cushing, C. A., & Lowe, K. A. (2011). Meta-analysis of prospective studies of red meat consumption and colorectal cancer. *European Journal of Cancer Prevention, 20*(4), 293–307. https://doi.org/10.1097/CEJ.0b013e328345f985.

Alicandro, G., Tavani, A., & La Vecchia, C. (2017). Coffee and cancer risk: A summary overview. *European Journal of Cancer Prevention, 26*(5), 424–432. https://doi.org/10.1097/cej.0000000000000341.

Alperet, D. J., Rebello, S. A., Khoo, E. Y., Tay, Z., Seah, S. S., Tai, B. C., et al. (2020). The effect of coffee consumption on insulin sensitivity and other biological risk factors for type 2 diabetes: A randomized placebo-controlled trial. *The American Journal of Clinical Nutrition, 111*(2), 448–458. https://doi.org/10.1093/ajcn/nqz306.

American Cancer Society. (2014). *Steps for increasing colorectal cancer screening rates: A manual for community health centers*. Retrieved from 2014 http://nccrt.org/wp-content/uploads/0305.60-Colorectal-Cancer-Manual_FULFILL.pdf.

Arab, L. (2010). Epidemiologic evidence on coffee and cancer. *Nutrition and Cancer, 62*(3), 271–283. https://doi.org/10.1080/01635580903407122.

Attene-Ramos, M. S., Wagner, E. D., Gaskins, H. R., & Plewa, M. J. (2007). Hydrogen sulfide induces direct radical-associated DNA damage. *Molecular Cancer Research, 5*(5), 455–459. https://doi.org/10.1158/1541-7786.Mcr-06-0439.

Aune, D., Chan, D. S., Lau, R., Vieira, R., Greenwood, D. C., Kampman, E., et al. (2011). Dietary fibre, whole grains, and risk of colorectal cancer: Systematic review and dose-response meta-analysis of prospective studies. *BMJ, 343*, d6617. https://doi.org/10.1136/bmj.d6617.

Bae, S., Ulrich, C. M., Neuhouser, M. L., Malysheva, O., Bailey, L. B., Xiao, L., et al. (2014). Plasma choline metabolites and colorectal cancer risk in the Women's Health Initiative Observational Study. *Cancer Research, 74*(24), 7442–7452. https://doi.org/10.1158/0008-5472.Can-14-1835.

Bian, X., Wu, W., Yang, L., Lv, L., Wang, Q., Li, Y., et al. (2019). Administration of Akkermansia muciniphila ameliorates dextran sulfate sodium-induced ulcerative colitis in mice. *Frontiers in Microbiology, 10*, 2259. https://doi.org/10.3389/fmicb.2019.02259.

Bouvard, V., Loomis, D., Guyton, K. Z., Grosse, Y., Ghissassi, F. E., Benbrahim-Tallaa, L., et al. (2015). Carcinogenicity of consumption of red and processed meat. *The Lancet Oncology, 16*(16), 1599–1600. https://doi.org/10.1016/s1470-2045(15)00444-1.

Bultman, S. J. (2014). Molecular pathways: Gene-environment interactions regulating dietary fiber induction of proliferation and apoptosis via butyrate for cancer prevention. *Clinical Cancer Research, 20*(4), 799–803. https://doi.org/10.1158/1078-0432.Ccr-13-2483.

Butt, M. S., & Sultan, M. T. (2011). Coffee and its consumption: Benefits and risks. *Critical Reviews in Food Science and Nutrition, 51*(4), 363–373. https://doi.org/10.1080/10408390903586412.

Cai, W. J., Wang, M. J., Ju, L. H., Wang, C., & Zhu, Y. C. (2010). Hydrogen sulfide induces human colon cancer cell proliferation: Role of Akt, ERK and p21. *Cell Biology International, 34*(6), 565–572. https://doi.org/10.1042/cbi20090368.

Cao, H., Xu, M., Dong, W., Deng, B., Wang, S., Zhang, Y., et al. (2017). Secondary bile acid-induced dysbiosis promotes intestinal carcinogenesis. *International Journal of Cancer, 140*(11), 2545–2556. https://doi.org/10.1002/ijc.30643.

Carr, P. R., Alwers, E., Bienert, S., Weberpals, J., Kloor, M., Brenner, H., et al. (2018). Lifestyle factors and risk of sporadic colorectal cancer by microsatellite instability status: A systematic review and meta-analyses. *Annals of Oncology, 29*(4), 825–834. https://doi.org/10.1093/annonc/mdy059.

Centers for Disease Control and Prevention. (2020a). *Adult obesity facts*. Retrieved from https://www.cdc.gov/obesity/data/adult.html.

Centers for Disease Control and Prevention. (2020b). *Overweight & obesity statistics*. Retrieved from https://www.niddk.nih.gov/health-information/health-statistics/overweight-obesity.

Chan, C. W. H., Law, B. M. H., Waye, M. M. Y., Chan, J. Y. W., So, W. K. W., & Chow, K. M. (2019). Trimethylamine-N-oxide as one hypothetical link for the relationship between intestinal microbiota and cancer—Where we are and where shall we go? *Journal of Cancer, 10*(23), 5874–5882. https://doi.org/10.7150/jca.31737.

Chen, F. W., Sundaram, V., Chew, T. A., & Ladabaum, U. (2017). Advanced-stage colorectal cancer in persons younger than 50 years not associated with longer duration of symptoms or time to diagnosis. *Clinical Gastroenterology and Hepatology, 15*(5), 728–737. e723. https://doi.org/10.1016/j.cgh.2016.10.038.

Chong, C. W., Wong, L. C., Teh, C. S. J., Ismail, N. H., Chan, P. Q., Lim, C. S., et al. (2020). Coffee consumption revealed sex differences in host endogenous metabolism and gut microbiota in healthy adults. *Journal of Food Biochemistry, 44*(12), e13535. https://doi.org/10.1111/jfbc.13535.

Chung, L., Thiele Orberg, E., Geis, A. L., Chan, J. L., Fu, K., DeStefano Shields, C. E., et al. (2018). Bacteroides fragilis toxin coordinates a pro-carcinogenic inflammatory cascade via targeting of colonic epithelial cells. *Cell Host & Microbe, 23*(2), 203–214.e205. https://doi.org/10.1016/j.chom.2018.01.007.

Cornelis, M. C., Byrne, E. M., Esko, T., Nalls, M. A., Ganna, A., Paynter, N., et al. (2015). Genome-wide meta-analysis identifies six novel loci associated with habitual coffee consumption. *Molecular Psychiatry, 20*(5), 647–656. https://doi.org/10.1038/mp.2014.107.

Craciun, S., & Balskus, E. P. (2012). Microbial conversion of choline to trimethylamine requires a glycyl radical enzyme. *Proceedings of the National Academy of Sciences of the United States of America, 109*(52), 21307–21312. https://doi.org/10.1073/pnas.1215689109.

Cross, A. J., & Gunter, M. J. (2018). Coffee and colorectal cancer: Grounds for prevention? *Gastroenterology, 154*(4), 790–792. https://doi.org/10.1053/j.gastro.2018.02.006.

Cross, A. J., Moore, S. C., Boca, S., Huang, W. Y., Xiong, X., Stolzenberg-Solomon, R., et al. (2014). A prospective study of serum metabolites and colorectal cancer risk. *Cancer, 120*(19), 3049–3057. https://doi.org/10.1002/cncr.28799.

Dao, M. C., Everard, A., Aron-Wisnewsky, J., Sokolovska, N., Prifti, E., Verger, E. O., et al. (2016). Akkermansia muciniphila and improved metabolic health during a dietary intervention in obesity: Relationship with gut microbiome richness and ecology. *Gut, 65*(3), 426–436. https://doi.org/10.1136/gutjnl-2014-308778.

De Filippis, F., Pasolli, E., Tett, A., Tarallo, S., Naccarati, A., De Angelis, M., et al. (2019). Distinct genetic and functional traits of human intestinal Prevotella copri strains are associated with different habitual diets. *Cell Host & Microbe, 25*(3), 444–453.e443. https://doi.org/10.1016/j.chom.2019.01.004.

De Smet, I., Van Hoorde, L., Vande Woestyne, M., Christiaens, H., & Verstraete, W. (1995). Significance of bile salt hydrolytic activities of lactobacilli. *The Journal of Applied Bacteriology, 79*(3), 292–301. https://doi.org/10.1111/j.1365-2672.1995.tb03140.x.

Depommier, C., Everard, A., Druart, C., Plovier, H., Van Hul, M., Vieira-Silva, S., et al. (2019). Supplementation with Akkermansia muciniphila in overweight and obese human volunteers: A proof-of-concept exploratory study. *Nature Medicine, 25*(7), 1096–1103. https://doi.org/10.1038/s41591-019-0495-2.

Ding, T., & Schloss, P. D. (2014). Dynamics and associations of microbial community types across the human body. *Nature, 509*(7500), 357–360. https://doi.org/10.1038/nature13178.

Donohoe, D. R., Holley, D., Collins, L. B., Montgomery, S. A., Whitmore, A. C., Hillhouse, A., et al. (2014). A gnotobiotic mouse model demonstrates that dietary fiber protects against colorectal tumorigenesis in a microbiota- and butyrate-dependent manner. *Cancer Discovery, 4*(12), 1387–1397. https://doi.org/10.1158/2159-8290.Cd-14-0501.

Edwards, B. K., Ward, E., Kohler, B. A., Eheman, C., Zauber, A. G., Anderson, R. N., et al. (2010). Annual report to the nation on the status of cancer, 1975–2006, featuring colorectal cancer trends and impact of interventions (risk factors, screening, and treatment) to reduce future rates. *Cancer*, *116*(3), 544–573. https://doi.org/10.1002/cncr.24760.

Everard, A., Belzer, C., Geurts, L., Ouwerkerk, J. P., Druart, C., Bindels, L. B., et al. (2013). Cross-talk between Akkermansia muciniphila and intestinal epithelium controls diet-induced obesity. *Proceedings of the National Academy of Sciences of the United States of America*, *110*(22), 9066–9071. https://doi.org/10.1073/pnas.1219451110.

Falony, G., Vieira-Silva, S., & Raes, J. (2015). Microbiology meets big data: The case of gut microbiota-derived trimethylamine. *Annual Review of Microbiology*, *69*, 305–321. https://doi.org/10.1146/annurev-micro-091014-104422.

Farhana, L., Nangia-Makker, P., Arbit, E., Shango, K., Sarkar, S., Mahmud, H., et al. (2016). Bile acid: A potential inducer of colon cancer stem cells. *Stem Cell Research & Therapy*, 7(1), 181. https://doi.org/10.1186/s13287-016-0439-4.

Feng, H. Y., & Chen, Y. C. (2016). Role of bile acids in carcinogenesis of pancreatic cancer: An old topic with new perspective. *World Journal of Gastroenterology*, *22*(33), 7463–7477. https://doi.org/10.3748/wjg.v22.i33.7463.

Fong, W., Li, Q., & Yu, J. (2020). Gut microbiota modulation: A novel strategy for prevention and treatment of colorectal cancer. *Oncogene*, *39*(26), 4925–4943. https://doi.org/10.1038/s41388-020-1341-1.

Fung, K. Y., Cosgrove, L., Lockett, T., Head, R., & Topping, D. L. (2012). A review of the potential mechanisms for the lowering of colorectal oncogenesis by butyrate. *The British Journal of Nutrition*, *108*(5), 820–831. https://doi.org/10.1017/s0007114512001948.

Gadaleta, R. M., Garcia-Irigoyen, O., & Moschetta, A. (2017). Bile acids and colon cancer: Is FXR the solution of the conundrum? *Molecular Aspects of Medicine*, *56*, 66–74. https://doi.org/10.1016/j.mam.2017.04.002.

Galeone, C., Turati, F., La Vecchia, C., & Tavani, A. (2010). Coffee consumption and risk of colorectal cancer: A meta-analysis of case-control studies. *Cancer Causes & Control*, *21*(11), 1949–1959. https://doi.org/10.1007/s10552-010-9623-5.

Garrett, W. S. (2015). Cancer and the microbiota. *Science*, *348*(6230), 80–86. https://doi.org/10.1126/science.aaa4972.

Garrett, W. S. (2019). The gut microbiota and colon cancer. *Science*, *364*(6446), 1133–1135. https://doi.org/10.1126/science.aaw2367.

Giovannucci, E. (1998). Meta-analysis of coffee consumption and risk of colorectal cancer. *American Journal of Epidemiology*, *147*(11), 1043–1052.

Gopalakrishnan, V., Spencer, C. N., Nezi, L., Reuben, A., Andrews, M. C., Karpinets, T. V., et al. (2018). Gut microbiome modulates response to anti-PD-1 immunotherapy in melanoma patients. *Science*, *359*(6371), 97–103. https://doi.org/10.1126/science.aan4236.

Gopal-Srivastava, R., & Hylemon, P. B. (1988). Purification and characterization of bile salt hydrolase from Clostridium perfringens. *Journal of Lipid Research*, *29*(8), 1079–1085.

Guercio, B. J., Sato, K., Niedzwiecki, D., Ye, X., Saltz, L. B., Mayer, R. J., et al. (2015). Coffee intake, recurrence, and mortality in stage III colon cancer: Results from CALGB 89803 (Alliance). *Journal of Clinical Oncology*, *33*(31), 3598–3607. https://doi.org/10.1200/jco.2015.61.5062.

Hamer, H. M., Jonkers, D., Venema, K., Vanhoutvin, S., Troost, F. J., & Brummer, R. J. (2008). Review article: The role of butyrate on colonic function. *Alimentary Pharmacology & Therapeutics*, 27(2), 104–119. https://doi.org/10.1111/j.1365-2036.2007.03562.x.

Hamilton, J. P., Xie, G., Raufman, J. P., Hogan, S., Griffin, T. L., Packard, C. A., et al. (2007). Human cecal bile acids: Concentration and spectrum. *American Journal of Physiology. Gastrointestinal and Liver Physiology*, *293*(1), G256–G263. https://doi.org/10.1152/ajpgi.00027.2007.

Han, M. A., Zeraatkar, D., Guyatt, G. H., Vernooij, R. W. M., El Dib, R., Zhang, Y., et al. (2019). Reduction of red and processed meat intake and cancer mortality and incidence: A systematic review and meta-analysis of cohort studies. *Annals of Internal Medicine*, *171*(10), 711–720. https://doi.org/10.7326/m19-0699.

Hang, S., Paik, D., Yao, L., Kim, E., Jamma, T., Lu, J., et al. (2019). Bile acid metabolites control T_H17 and T_{reg} cell differentiation. *Nature*, *576*(7785), 143–148. https://doi.org/10.1038/s41586-019-1785-z.

Hansen, L., Skeie, G., Landberg, R., Lund, E., Palmqvist, R., Johansson, I., et al. (2012). Intake of dietary fiber, especially from cereal foods, is associated with lower incidence of colon cancer in the HELGA cohort. *International Journal of Cancer*, *131*(2), 469–478. https://doi.org/10.1002/ijc.26381.

He, X., Wu, K., Zhang, X., Nishihara, R., Cao, Y., Fuchs, C. S., et al. (2019). Dietary intake of fiber, whole grains and risk of colorectal cancer: An updated analysis according to food sources, tumor location and molecular subtypes in two large US cohorts. *International Journal of Cancer*, *145*(11), 3040–3051. https://doi.org/10.1002/ijc.32382.

Hidaka, A., Harrison, T. A., Cao, Y., Sakoda, L. C., Barfield, R., Giannakis, M., et al. (2020). Intake of dietary fruit, vegetables, and fiber and risk of colorectal cancer according to molecular subtypes: A pooled analysis of 9 studies. *Cancer Research*, *80*(20), 4578–4590. https://doi.org/10.1158/0008-5472.Can-20-0168.

Higdon, J. V., & Frei, B. (2006). Coffee and health: A review of recent human research. *Critical Reviews in Food Science and Nutrition*, *46*(2), 101–123. https://doi.org/10.1080/10408390500400009.

Hildebrandt, M. A., Hoffmann, C., Sherrill-Mix, S. A., Keilbaugh, S. A., Hamady, M., Chen, Y. Y., et al. (2009). High-fat diet determines the composition of the murine gut microbiome independently of obesity. *Gastroenterology*, *137*(5), 1716–1724. e1711-1712. https://doi.org/10.1053/j.gastro.2009.08.042.

Hofmann, A. F. (1999). The continuing importance of bile acids in liver and intestinal disease. *Archives of Internal Medicine*, *159*(22), 2647–2658. https://doi.org/10.1001/archinte.159.22.2647.

Hofseth, L. J., Hebert, J. R., Chanda, A., Chen, H., Love, B. L., Pena, M. M., et al. (2020). Early-onset colorectal cancer: Initial clues and current views. *Nature Reviews. Gastroenterology & Hepatology*, *17*(6), 352–364. https://doi.org/10.1038/s41575-019-0253-4.

Hu, Y., Ding, M., Yuan, C., Wu, K., Smith-Warner, S. A., Hu, F. B., et al. (2018). Association between coffee intake after diagnosis of colorectal cancer and reduced mortality. *Gastroenterology*, *154*(4), 916–926.e919. https://doi.org/10.1053/j.gastro.2017.11.010.

Hullings, A. G., Sinha, R., Liao, L. M., Freedman, N. D., Graubard, B. I., & Loftfield, E. (2020). Whole grain and dietary fiber intake and risk of colorectal cancer in the NIH-AARP Diet and Health Study cohort. *The American Journal of Clinical Nutrition*, *112*(3), 603–612. https://doi.org/10.1093/ajcn/nqaa161.

Ijssennagger, N., Belzer, C., Hooiveld, G. J., Dekker, J., van Mil, S. W., Müller, M., et al. (2015). Gut microbiota facilitates dietary heme-induced epithelial hyperproliferation by opening the mucus barrier in colon. *Proceedings of the National Academy of Sciences of the United States of America*, *112*(32), 10038–10043. https://doi.org/10.1073/pnas.1507645112.

Ijssennagger, N., van der Meer, R., & van Mil, S. W. C. (2016). Sulfide as a mucus barrier-breaker in inflammatory bowel disease? *Trends in Molecular Medicine*, *22*(3), 190–199. https://doi.org/10.1016/j.molmed.2016.01.002.

Islami, F., Goding Sauer, A., Miller, K. D., Siegel, R. L., Fedewa, S. A., Jacobs, E. J., et al. (2018). Proportion and number of cancer cases and deaths attributable to potentially modifiable risk factors in the United States. *CA: A Cancer Journal for Clinicians*, *68*(1), 31–54. https://doi.org/10.3322/caac.21440.

Jaquet, M., Rochat, I., Moulin, J., Cavin, C., & Bibiloni, R. (2009). Impact of coffee consumption on the gut microbiota: A human volunteer study. *International Journal of Food Microbiology*, *130*(2), 117–121. https://doi.org/10.1016/j.ijfoodmicro.2009.01.011.

Jarocki, P., & Targonski, Z. (2013). Genetic diversity of bile salt hydrolases among human intestinal bifidobacteria. *Current Microbiology*, *67*(3), 286–292. https://doi.org/10.1007/s00284-013-0362-1.

Je, Y., Liu, W., & Giovannucci, E. (2009). Coffee consumption and risk of colorectal cancer: A systematic review and meta-analysis of prospective cohort studies. *International Journal of Cancer*, *124*(7), 1662–1668. https://doi.org/10.1002/ijc.24124.

Jeyakumar, A., Dissabandara, L., & Gopalan, V. (2017). A critical overview on the biological and molecular features of red and processed meat in colorectal carcinogenesis. *Journal of Gastroenterology*, *52*(4), 407–418. https://doi.org/10.1007/s00535-016-1294-x.

Jia, W., Xie, G., & Jia, W. (2018). Bile acid-microbiota crosstalk in gastrointestinal inflammation and carcinogenesis. *Nature Reviews. Gastroenterology & Hepatology*, *15*(2), 111–128. https://doi.org/10.1038/nrgastro.2017.119.

Jones, B. V., Begley, M., Hill, C., Gahan, C. G., & Marchesi, J. R. (2008). Functional and comparative metagenomic analysis of bile salt hydrolase activity in the human gut microbiome. *Proceedings of the National Academy of Sciences of the United States of America*, *105*(36), 13580–13585. https://doi.org/10.1073/pnas.0804437105.

Kamangar, F., Nasrollahzadeh, D., Safiri, S., Sepanlou, S. G., Fitzmaurice, C., Ikuta, K. S., et al. (2020). The global, regional, and national burden of oesophageal cancer and its attributable risk factors in 195 countries and territories, 1990–2017: A systematic analysis for the Global Burden of Disease Study 2017. *The Lancet Gastroenterology & Hepatology*, *5*(6), 582–597. https://doi.org/10.1016/S2468-1253(20)30007-8.

Kashino, I., Akter, S., Mizoue, T., Sawada, N., Kotemori, A., Matsuo, K., et al. (2018). Coffee drinking and colorectal cancer and its subsites: A pooled analysis of 8 cohort studies in Japan. *International Journal of Cancer*, *143*(2), 307–316. https://doi.org/10.1002/ijc.31320.

Keijzers, G. B., De Galan, B. E., Tack, C. J., & Smits, P. (2002). Caffeine can decrease insulin sensitivity in humans. *Diabetes Care*, *25*(2), 364–369. https://doi.org/10.2337/diacare.25.2.364.

Kneuertz, P. J., Chang, G. J., Hu, C. Y., Rodriguez-Bigas, M. A., Eng, C., Vilar, E., et al. (2015). Overtreatment of young adults with colon cancer: More intense treatments with unmatched survival gains. *JAMA Surgery*, *150*(5), 402–409. https://doi.org/10.1001/jamasurg.2014.3572.

Koeth, R. A., Wang, Z., Levison, B. S., Buffa, J. A., Org, E., Sheehy, B. T., et al. (2013). Intestinal microbiota metabolism of L-carnitine, a nutrient in red meat, promotes atherosclerosis. *Nature Medicine*, *19*(5), 576–585. https://doi.org/10.1038/nm.3145.

Kolb, H., Kempf, K., & Martin, S. (2020). Health effects of coffee: Mechanism unraveled? *Nutrients*, *12*(6), 1842. https://doi.org/10.3390/nu12061842.

Kühn, T., Stepien, M., López-Nogueroles, M., Damms-Machado, A., Sookthai, D., Johnson, T., et al. (2020). Prediagnostic plasma bile acid levels and colon cancer risk: A prospective study. *Journal of the National Cancer Institute*, *112*(5), 516–524. https://doi.org/10.1093/jnci/djz166.

Li, G., Ma, D., Zhang, Y., Zheng, W., & Wang, P. (2013). Coffee consumption and risk of colorectal cancer: A meta-analysis of observational studies. *Public Health Nutrition*, *16*(2), 346–357. https://doi.org/10.1017/s1368980012002601.

Liu, R., Hong, J., Xu, X., Feng, Q., Zhang, D., Gu, Y., et al. (2017). Gut microbiome and serum metabolome alterations in obesity and after weight-loss intervention. *Nature Medicine*, *23*(7), 859–868. https://doi.org/10.1038/nm.4358.

Liu, X., Liu, H., Yuan, C., Zhang, Y., Wang, W., Hu, S., et al. (2017). Preoperative serum TMAO level is a new prognostic marker for colorectal cancer. *Biomarkers in Medicine*, *11*(5), 443–447. https://doi.org/10.2217/bmm-2016-0262.

Loeb, L. A. (2016). Human cancers express a mutator phenotype: Hypothesis, origin, and consequences. *Cancer Research*, *76*(8), 2057–2059. https://doi.org/10.1158/0008-5472.Can-16-0794.

Long, S. L., Gahan, C. G. M., & Joyce, S. A. (2017). Interactions between gut bacteria and bile in health and disease. *Molecular Aspects of Medicine*, *56*, 54–65. https://doi.org/10.1016/j.mam.2017.06.002.

Long, X., Wong, C. C., Tong, L., Chu, E. S. H., Ho Szeto, C., Go, M. Y. Y., et al. (2019). Peptostreptococcus anaerobius promotes colorectal carcinogenesis and modulates tumour immunity. *Nature Microbiology*, *4*(12), 2319–2330. https://doi.org/10.1038/s41564-019-0541-3.

Louis, P., Hold, G. L., & Flint, H. J. (2014). The gut microbiota, bacterial metabolites and colorectal cancer. *Nature Reviews. Microbiology*, *12*(10), 661–672. https://doi.org/10.1038/nrmicro3344.

Ludwig, I. A., Clifford, M. N., Lean, M. E., Ashihara, H., & Crozier, A. (2014). Coffee: Biochemistry and potential impact on health. *Food & Function*, *5*(8), 1695–1717. https://doi.org/10.1039/c4fo00042k.

Ma, C., Han, M., Heinrich, B., Fu, Q., Zhang, Q., Sandhu, M., et al. (2018). Gut microbiome-mediated bile acid metabolism regulates liver cancer via NKT cells. *Science*, *360*(6391), eaan5931. https://doi.org/10.1126/science.aan5931.

Magee, E. A., Richardson, C. J., Hughes, R., & Cummings, J. H. (2000). Contribution of dietary protein to sulfide production in the large intestine: An in vitro and a controlled feeding study in humans. *The American Journal of Clinical Nutrition*, *72*(6), 1488–1494. https://doi.org/10.1093/ajcn/72.6.1488.

Maldonado, R. F., Sá-Correia, I., & Valvano, M. A. (2016). Lipopolysaccharide modification in Gram-negative bacteria during chronic infection. *FEMS Microbiology Reviews*, *40*(4), 480–493. https://doi.org/10.1093/femsre/fuw007.

Marchesi, J. R., Adams, D. H., Fava, F., Hermes, G. D., Hirschfield, G. M., Hold, G., et al. (2016). The gut microbiota and host health: A new clinical frontier. *Gut*, *65*(2), 330–339. https://doi.org/10.1136/gutjnl-2015-309990.

Martin, G., Kolida, S., Marchesi, J. R., Want, E., Sidaway, J. E., & Swann, J. R. (2018). In vitro modeling of bile acid processing by the human fecal microbiota. *Frontiers in Microbiology*, *9*, 1153. https://doi.org/10.3389/fmicb.2018.01153.

Masuda, N. (1981). Deconjugation of bile salts by Bacteroids and Clostridium. *Microbiology and Immunology*, *25*(1), 1–11. https://doi.org/10.1111/j.1348-0421.1981.tb00001.x.

Matson, V., Fessler, J., Bao, R., Chongsuwat, T., Zha, Y., Alegre, M. L., et al. (2018). The commensal microbiome is associated with anti-PD-1 efficacy in metastatic melanoma patients. *Science*, *359*(6371), 104–108. https://doi.org/10.1126/science.aao3290.

Mehta, R. S., Nishihara, R., Cao, Y., Song, M., Mima, K., Qian, Z. R., et al. (2017). Association of dietary patterns with risk of colorectal cancer subtypes classified by Fusobacterium nucleatum in tumor tissue. *JAMA Oncology*, *3*(7), 921–927. https://doi.org/10.1001/jamaoncol.2016.6374.

Micek, A., Gniadek, A., Kawalec, P., & Brzostek, T. (2019). Coffee consumption and colorectal cancer risk: A dose-response meta-analysis on prospective cohort studies. *International Journal of Food Sciences and Nutrition*, *70*(8), 986–1006. https://doi.org/10.1080/09637486.2019.1591352.

Miller, K. D., Siegel, R. L., Lin, C. C., Mariotto, A. B., Kramer, J. L., Rowland, J. H., et al. (2016). Cancer treatment and survivorship statistics, 2016. *CA: A Cancer Journal for Clinicians*, *66*(4), 271–289. https://doi.org/10.3322/caac.21349.

Mima, K., Nishihara, R., Qian, Z. R., Cao, Y., Sukawa, Y., Nowak, J. A., et al. (2016). Fusobacterium nucleatum in colorectal carcinoma tissue and patient prognosis. *Gut*, *65*(12), 1973–1980. https://doi.org/10.1136/gutjnl-2015-310101.

Morales-Oyarvide, V., Yuan, C., Babic, A., Zhang, S., Niedzwiecki, D., Brand-Miller, J. C., et al. (2019). Dietary insulin load and cancer recurrence and survival in patients with stage III colon cancer: Findings from CALGB 89803 (Alliance). *Journal of the National Cancer Institute*, *111*(2), 170–179. https://doi.org/10.1093/jnci/djy098.

Murphy, N., Jenab, M., & Gunter, M. J. (2018). Adiposity and gastrointestinal cancers: Epidemiology, mechanisms and future directions. *Nature Reviews. Gastroenterology & Hepatology*, *15*(11), 659–670. https://doi.org/10.1038/s41575-018-0038-1.

Murphy, N., Norat, T., Ferrari, P., Jenab, M., Bueno-de-Mesquita, B., Skeie, G., et al. (2012). Dietary fibre intake and risks of cancers of the colon and rectum in the European prospective investigation into cancer and nutrition (EPIC). *PLoS One*, 7(6), e39361. https://doi.org/10.1371/journal.pone.0039361.

Myers, E. A., Feingold, D. L., Forde, K. A., Arnell, T., Jang, J. H., & Whelan, R. L. (2013). Colorectal cancer in patients under 50 years of age: A retrospective analysis of two institutions' experience. *World Journal of Gastroenterology*, *19*(34), 5651–5657. https://doi.org/10.3748/wjg.v19.i34.5651.

Nguyen, L. H., Ma, W., Wang, D. D., Cao, Y., Mallick, H., Gerbaba, T. K., et al. (2020). Association between sulfur-metabolizing bacterial communities in stool and risk of distal colorectal cancer in men. *Gastroenterology*, *158*(5), 1313–1325. https://doi.org/10.1053/j.gastro.2019.12.029.

Oh, H., Kim, H., Lee, D. H., Lee, A., Giovannucci, E. L., Kang, S. S., et al. (2019). Different dietary fibre sources and risks of colorectal cancer and adenoma: A dose-response meta-analysis of prospective studies. *The British Journal of Nutrition*, *122*(6), 605–615. https://doi.org/10.1017/s0007114519001454.

Oh, H. K., Lee, J. Y., Lim, S. J., Kim, M. J., Kim, G. B., Kim, J. H., et al. (2008). Molecular cloning and characterization of a bile salt hydrolase from Lactobacillus acidophilus PF01. *Journal of Microbiology and Biotechnology*, *18*(3), 449–456.

Ohnaka, K., Ikeda, M., Maki, T., Okada, T., Shimazoe, T., Adachi, M., et al. (2012). Effects of 16-week consumption of caffeinated and decaffeinated instant coffee on glucose metabolism in a randomized controlled trial. *Journal of Nutrition and Metabolism*, *2012*, 207426. https://doi.org/10.1155/2012/207426.

Papathanasopoulos, A., & Camilleri, M. (2010). Dietary fiber supplements: Effects in obesity and metabolic syndrome and relationship to gastrointestinal functions. *Gastroenterology*, *138*(1), 65–72.e61-62. https://doi.org/10.1053/j.gastro.2009.11.045.

Perera, V., Gross, A. S., & McLachlan, A. J. (2012). Influence of environmental and genetic factors on CYP1A2 activity in individuals of South Asian and European ancestry. *Clinical Pharmacology and Therapeutics*, *92*(4), 511–519. https://doi.org/10.1038/clpt.2012.139.

Poole, R., Kennedy, O. J., Roderick, P., Fallowfield, J. A., Hayes, P. C., & Parkes, J. (2017). Coffee consumption and health: Umbrella review of meta-analyses of multiple health outcomes. *BMJ*, *359*, j5024. https://doi.org/10.1136/bmj.j5024.

Portincasa, P., & Calamita, G. (2012). Water channel proteins in bile formation and flow in health and disease: When immiscible becomes miscible. *Molecular Aspects of Medicine*, *33*(5–6), 651–664. https://doi.org/10.1016/j.mam.2012.03.010.

Quagliani, D., & Felt-Gunderson, P. (2017). Closing America's fiber intake gap: Communication strategies from a food and fiber summit. *American Journal of Lifestyle Medicine*, *11*(1), 80–85. https://doi.org/10.1177/1559827615588079.

Ramasamy, S., Singh, S., Taniere, P., Langman, M. J., & Eggo, M. C. (2006). Sulfide-detoxifying enzymes in the human colon are decreased in cancer and upregulated in differentiation. *American Journal of Physiology. Gastrointestinal and Liver Physiology*, *291*(2), G288–G296. https://doi.org/10.1152/ajpgi.00324.2005.

Reynolds, A., Mann, J., Cummings, J., Winter, N., Mete, E., & Te Morenga, L. (2019). Carbohydrate quality and human health: A series of systematic reviews and meta-analyses. *Lancet*, *393*(10170), 434–445. https://doi.org/10.1016/s0140-6736(18)31809-9.

Ridlon, J. M., Harris, S. C., Bhowmik, S., Kang, D. J., & Hylemon, P. B. (2016). Consequences of bile salt biotransformations by intestinal bacteria. *Gut Microbes*, 7(1), 22–39. https://doi.org/10.1080/19490976.2015.1127483.

Ridlon, J. M., Kang, D. J., & Hylemon, P. B. (2006). Bile salt biotransformations by human intestinal bacteria. *Journal of Lipid Research*, *47*(2), 241–259. https://doi.org/10.1194/jlr.R500013-JLR200.

Robertson, D., Wade, D., Workman, R., Woosley, R. L., & Oates, J. A. (1981). Tolerance to the humoral and hemodynamic effects of caffeine in man. *The Journal of Clinical Investigation*, *67*(4), 1111–1117. https://doi.org/10.1172/jci110124.

Roediger, W. E., Duncan, A., Kapaniris, O., & Millard, S. (1993). Reducing sulfur compounds of the colon impair colonocyte nutrition: Implications for ulcerative colitis. *Gastroenterology*, *104*(3), 802–809. https://doi.org/10.1016/0016-5085(93)91016-b.

Romano, K. A., Vivas, E. I., Amador-Noguez, D., & Rey, F. E. (2015). Intestinal microbiota composition modulates choline bioavailability from diet and accumulation of the proatherogenic metabolite trimethylamine-N-oxide. *mBio*, *6*(2), e02481. https://doi.org/10.1128/mBio.02481-14.

Rooks, M. G., & Garrett, W. S. (2016). Gut microbiota, metabolites and host immunity. *Nature Reviews. Immunology*, *16*(6), 341–352. https://doi.org/10.1038/nri.2016.42.

Rothschild, D., Weissbrod, O., Barkan, E., Kurilshikov, A., Korem, T., Zeevi, D., et al. (2018). Environment dominates over host genetics in shaping human gut microbiota. *Nature*, *555*(7695), 210–215. https://doi.org/10.1038/nature25973.

Rubinstein, M. R., Wang, X., Liu, W., Hao, Y., Cai, G., & Han, Y. W. (2013). Fusobacterium nucleatum promotes colorectal carcinogenesis by modulating E-cadherin/β-catenin signaling via its FadA adhesin. *Cell Host & Microbe*, *14*(2), 195–206. https://doi.org/10.1016/j.chom.2013.07.012.

Sartini, M., Bragazzi, N. L., Spagnolo, A. M., Schinca, E., Ottria, G., Dupont, C., et al. (2019). Coffee consumption and risk of colorectal cancer: A systematic review and meta-analysis of prospective studies. *Nutrients*, *11*(3), 694. https://doi.org/10.3390/nu11030694.

Saus, E., Iraola-Guzmán, S., Willis, J. R., Brunet-Vega, A., & Gabaldón, T. (2019). Microbiome and colorectal cancer: Roles in carcinogenesis and clinical potential. *Molecular Aspects of Medicine*, *69*, 93–106. https://doi.org/10.1016/j.mam.2019.05.001.

Scanlan, P. D., Shanahan, F., & Marchesi, J. R. (2009). Culture-independent analysis of desulfovibrios in the human distal colon of healthy, colorectal cancer and polypectomized individuals. *FEMS Microbiology Ecology*, *69*(2), 213–221. https://doi.org/10.1111/j.1574-6941.2009.00709.x.

Schmit, S. L., Rennert, H. S., Rennert, G., & Gruber, S. B. (2016). Coffee consumption and the risk of colorectal cancer. *Cancer Epidemiology, Biomarkers & Prevention*, *25*(4), 634–639. https://doi.org/10.1158/1055-9965.Epi-15-0924.

Schroeder, B. O., Birchenough, G. M. H., Ståhlman, M., Arike, L., Johansson, M. E. V., Hansson, G. C., et al. (2018). Bifidobacteria or fiber protects against diet-induced microbiota-mediated colonic mucus deterioration. *Cell Host Microbe*, *23*(1), 27–40. e27. https://doi.org/10.1016/j.chom.2017.11.004.

Schwingshackl, L., Schwedhelm, C., Hoffmann, G., Knüppel, S., Laure Preterre, A., Iqbal, K., et al. (2018). Food groups and risk of colorectal cancer. *International Journal of Cancer*, *142*(9), 1748–1758. https://doi.org/10.1002/ijc.31198.

Scientific Advisory Committee on Nutrition. (2015). *Carbohydrates and health*. Retrieved from https://assets.publishing.service.gov.uk/government/uploads/system/uploads/attachment_data/file/445503/SACN_Carbohydrates_and_Health.pdf.

Scott, A. J., Alexander, J. L., Merrifield, C. A., Cunningham, D., Jobin, C., Brown, R., et al. (2019). International Cancer Microbiome Consortium consensus statement on the role of the human microbiome in carcinogenesis. *Gut*, *68*(9), 1624–1632. https://doi.org/10.1136/gutjnl-2019-318556.

Siegel, R. L., Miller, K. D., Fuchs, H. E., & Jemal, A. (2021). Cancer statistics, 2021. *CA: A Cancer Journal for Clinicians*, *71*(1), 7–33. https://doi.org/10.3322/caac.21654.

Siegel, R. L., Miller, K. D., Goding Sauer, A., Fedewa, S. A., Butterly, L. F., Anderson, J. C., et al. (2020). Colorectal cancer statistics, 2020. *CA: A Cancer Journal for Clinicians*, *70*(3), 145–164. https://doi.org/10.3322/caac.21601.

Sivan, A., Corrales, L., Hubert, N., Williams, J. B., Aquino-Michaels, K., Earley, Z. M., et al. (2015). Commensal Bifidobacterium promotes antitumor immunity and facilitates anti-PD-L1 efficacy. *Science*, *350*(6264), 1084–1089. https://doi.org/10.1126/science.aac4255.

Slavin, J. (2013). Fiber and prebiotics: Mechanisms and health benefits. *Nutrients*, *5*(4), 1417–1435. https://doi.org/10.3390/nu5041417.

Smith, R. A., Andrews, K. S., Brooks, D., Fedewa, S. A., Manassaram-Baptiste, D., Saslow, D., et al. (2019). Cancer screening in the United States, 2019: A review of current American Cancer Society guidelines and current issues in cancer screening. *CA: A Cancer Journal for Clinicians*, *69*(3), 184–210. https://doi.org/10.3322/caac.21557.

Smith, P. M., Howitt, M. R., Panikov, N., Michaud, M., Gallini, C. A., Bohlooly, Y. M., et al. (2013). The microbial metabolites, short-chain fatty acids, regulate colonic T_{reg} cell homeostasis. *Science*, *341*(6145), 569–573. https://doi.org/10.1126/science.1241165.

Song, M., & Chan, A. T. (2018). The potential role of exercise and nutrition in harnessing the immune system to improve colorectal cancer survival. *Gastroenterology*, *155*(3), 596–600. https://doi.org/10.1053/j.gastro.2018.07.038.

Song, M., Chan, A. T., & Sun, J. (2019). Influence of the gut microbiome, diet, and environment on risk of colorectal cancer. *Gastroenterology*, *158*(2), 322–340. https://doi.org/10.1053/j.gastro.2019.06.048.

Song, M., Garrett, W. S., & Chan, A. T. (2015). Nutrients, foods, and colorectal cancer prevention. *Gastroenterology*, *148*(6), 1244–1260.e1216. https://doi.org/10.1053/j.gastro.2014.12.035.

Song, M., & Giovannucci, E. (2016). Preventable incidence and mortality of carcinoma associated with lifestyle factors among white adults in the United States. *JAMA Oncology*, *2*(9), 1154–1161. https://doi.org/10.1001/jamaoncol.2016.0843.

Song, J. J., Tian, W. J., Kwok, L. Y., Wang, Y. L., Shang, Y. N., Menghe, B., et al. (2017). Effects of microencapsulated Lactobacillus plantarum LIP-1 on the gut microbiota of hyperlipidaemic rats. *The British Journal of Nutrition*, *118*(7), 481–492. https://doi.org/10.1017/s0007114517002380.

Song, M., Wu, K., Meyerhardt, J. A., Ogino, S., Wang, M., Fuchs, C. S., et al. (2018). Fiber intake and survival after colorectal cancer diagnosis. *JAMA Oncology*, *4*(1), 71–79. https://doi.org/10.1001/jamaoncol.2017.3684.

Song, M., Wu, K., Meyerhardt, J. A., Yilmaz, O., Wang, M., Ogino, S., et al. (2018). Low-carbohydrate diet score and macronutrient intake in relation to survival after colorectal cancer diagnosis. *JNCI Cancer Spectrum*, *2*(4), pky077. https://doi.org/10.1093/jncics/pky077.

Sonnenburg, E. D., Smits, S. A., Tikhonov, M., Higginbottom, S. K., Wingreen, N. S., & Sonnenburg, J. L. (2016). Diet-induced extinctions in the gut microbiota compound over generations. *Nature*, *529*(7585), 212–215. https://doi.org/10.1038/nature16504.

Stellwag, E. J., & Hylemon, P. B. (1976). Purification and characterization of bile salt hydrolase from *Bacteroides fragilis subsp. fragilis*. *Biochimica et Biophysica Acta*, *452*(1), 165–176. https://doi.org/10.1016/0005-2744(76)90068-1.

Tai, N., Wong, F. S., & Wen, L. (2015). The role of gut microbiota in the development of type 1, type 2 diabetes mellitus and obesity. *Reviews in Endocrine & Metabolic Disorders*, *16*(1), 55–65. https://doi.org/10.1007/s11154-015-9309-0.

Tanaka, H., Hashiba, H., Kok, J., & Mierau, I. (2000). Bile salt hydrolase of Bifidobacterium longum-biochemical and genetic characterization. *Applied and Environmental Microbiology*, *66*(6), 2502–2512. https://doi.org/10.1128/aem.66.6.2502-2512.2000.

Ternes, D., Karta, J., Tsenkova, M., Wilmes, P., Haan, S., & Letellier, E. (2020). Microbiome in colorectal cancer: How to get from meta-omics to mechanism? *Trends in Microbiology*, *28*(5), 401–423. https://doi.org/10.1016/j.tim.2020.01.001.

Tian, C., Wang, W., Hong, Z., & Zhang, X. (2013). Coffee consumption and risk of colorectal cancer: A dose-response analysis of observational studies. *Cancer Causes & Control*, *24*(6), 1265–1268. https://doi.org/10.1007/s10552-013-0200-6.

Tilg, H., Adolph, T. E., Gerner, R. R., & Moschen, A. R. (2018). The intestinal microbiota in colorectal cancer. *Cancer Cell*, *33*(6), 954–964. https://doi.org/10.1016/j.ccell.2018.03.004.

Tomasetti, C., & Vogelstein, B. (2015). Cancer etiology. Variation in cancer risk among tissues can be explained by the number of stem cell divisions. *Science*, *347*(6217), 78–81. https://doi.org/10.1126/science.1260825.

Tremaroli, V., & Bäckhed, F. (2012). Functional interactions between the gut microbiota and host metabolism. *Nature*, *489*(7415), 242–249. https://doi.org/10.1038/nature11552.

Um, C. Y., McCullough, M. L., Guinter, M. A., Campbell, P. T., Jacobs, E. J., & Gapstur, S. M. (2020). Coffee consumption and risk of colorectal cancer in the Cancer Prevention Study-II Nutrition cohort. *Cancer Epidemiology*, *67*, 101730. https://doi.org/10.1016/j.canep.2020.101730.

US Department of Health and Human Services, & US Department of Agriculture. (2015). *2015–2020 Dietary guidelines for Americans* (8th ed.). Retrieved from https://health.gov/our-work/food-nutrition/previous-dietary-guidelines/2015.

US Preventive Services Task Force. (2020). *Draft recommendation statement: Colorectal cancer screening*. Retrieved from 2020 https://uspreventiveservicestaskforce.org/uspstf/draft-recommendation/colorectal-cancer-screening3#fullrecommendationstart.

Van Blarigan, E. L., Fuchs, C. S., Niedzwiecki, D., Zhang, S., Saltz, L. B., Mayer, R. J., et al. (2018). Association of Survival With Adherence to the American Cancer Society nutrition and physical activity guidelines for cancer survivors after colon cancer diagnosis: The CALGB 89803/Alliance trial. *JAMA Oncology*, *4*(6), 783–790. https://doi.org/10.1001/jamaoncol.2018.0126.

van Dam, R. M., Hu, F. B., & Willett, W. C. (2020). Coffee, caffeine, and health. *The New England Journal of Medicine*, *383*(4), 369–378. https://doi.org/10.1056/NEJMra1816604.

van Erpecum, K. J., Wang, D. Q., Moschetta, A., Ferri, D., Svelto, M., Portincasa, P., et al. (2006). Gallbladder histopathology during murine gallstone formation: Relation to motility and concentrating function. *Journal of Lipid Research*, *47*(1), 32–41. https://doi.org/10.1194/jlr.M500180-JLR200.

van Meer, S., Leufkens, A. M., Bueno-de-Mesquita, H. B., van Duijnhoven, F. J., van Oijen, M. G., & Siersema, P. D. (2013). Role of dietary factors in survival and mortality in colorectal cancer: A systematic review. *Nutrition Reviews*, *71*(9), 631–641. https://doi.org/10.1111/nure.12042.

van Zutphen, M., Kampman, E., Giovannucci, E. L., & van Duijnhoven, F. J. B. (2017). Lifestyle after colorectal cancer diagnosis in relation to survival and recurrence: A review of the literature. *Current Colorectal Cancer Reports*, *13*(5), 370–401. https://doi.org/10.1007/s11888-017-0386-1.

Vitaglione, P., Fogliano, V., & Pellegrini, N. (2012). Coffee, colon function and colorectal cancer. *Food & Function*, *3*(9), 916–922. https://doi.org/10.1039/c2fo30037k.

Vital, M., Rud, T., Rath, S., Pieper, D. H., & Schluter, D. (2019). Diversity of bacteria exhibiting bile acid-inducible 7alpha-dehydroxylation genes in the human gut. *Computational and Structural Biotechnology Journal*, *17*, 1016–1019. https://doi.org/10.1016/j.csbj.2019.07.012.

Wang, J., & Jia, H. (2016). Metagenome-wide association studies: Fine-mining the microbiome. *Nature Reviews. Microbiology*, *14*(8), 508–522. https://doi.org/10.1038/nrmicro.2016.83.

Wang, L., Tang, L., Feng, Y., Zhao, S., Han, M., Zhang, C., et al. (2020). A purified membrane protein from Akkermansia muciniphila or the pasteurised bacterium blunts colitis associated tumourigenesis by modulation of CD8(+) T cells in mice. *Gut, 69*(11), 1988–1997. https://doi.org/10.1136/gutjnl-2019-320105.

WCRF/AICR. (2018). *Continuous update project report: Diet, nutrition, physical activity and colorectal cancer.*

Wedick, N. M., Brennan, A. M., Sun, Q., Hu, F. B., Mantzoros, C. S., & van Dam, R. M. (2011). Effects of caffeinated and decaffeinated coffee on biological risk factors for type 2 diabetes: A randomized controlled trial. *Nutrition Journal, 10*, 93. https://doi.org/10.1186/1475-2891-10-93.

Wells, J. E., & Hylemon, P. B. (2000). Identification and characterization of a bile acid 7alpha-dehydroxylation operon in Clostridium sp. strain TO-931, a highly active 7alpha-dehydroxylating strain isolated from human feces. *Applied and Environmental Microbiology, 66*(3), 1107–1113. https://doi.org/10.1128/aem.66.3.1107-1113.2000.

Wijaya, A., Hermann, A., Abriouel, H., Specht, I., Yousif, N. M., Holzapfel, W. H., et al. (2004). Cloning of the bile salt hydrolase (bsh) gene from *Enterococcus faecium* FAIR-E 345 and chromosomal location of bsh genes in food enterococci. *Journal of Food Protection, 67*(12), 2772–2778. https://doi.org/10.4315/0362-028x-67.12.2772.

Wirbel, J., Pyl, P. T., Kartal, E., Zych, K., Kashani, A., Milanese, A., et al. (2019). Meta-analysis of fecal metagenomes reveals global microbial signatures that are specific for colorectal cancer. *Nature Medicine, 25*(4), 679–689. https://doi.org/10.1038/s41591-019-0406-6.

Wong, S. H., & Yu, J. (2019). Gut microbiota in colorectal cancer: Mechanisms of action and clinical applications. *Nature Reviews. Gastroenterology & Hepatology, 16*(11), 690–704. https://doi.org/10.1038/s41575-019-0209-8.

World Cancer Research Fund/American Institute for Cancer Research. (2018). *Continuous update project: Diet, nutrition, physical activiry, and colorectal cancer.* Retrieved from https://www.wcrf.org/sites/default/files/Colorectal-cancer-report.pdf.

Wu, S., Powers, S., Zhu, W., & Hannun, Y. A. (2016). Substantial contribution of extrinsic risk factors to cancer development. *Nature, 529*(7584), 43–47. https://doi.org/10.1038/nature16166.

Xu, P., Hong, F., Wang, J., Cong, Y., Dai, S., Wang, S., et al. (2017). Microbiome remodeling via the montmorillonite adsorption-excretion axis prevents obesity-related metabolic disorders. *eBioMedicine, 16*, 251–261. https://doi.org/10.1016/j.ebiom.2017.01.019.

Yachida, S., Mizutani, S., Shiroma, H., Shiba, S., Nakajima, T., Sakamoto, T., et al. (2019). Metagenomic and metabolomic analyses reveal distinct stage-specific phenotypes of the gut microbiota in colorectal cancer. *Nature Medicine, 25*(6), 968–976. https://doi.org/10.1038/s41591-019-0458-7.

Yamada, H., Kawado, M., Aoyama, N., Hashimoto, S., Suzuki, K., Wakai, K., et al. (2014). Coffee consumption and risk of colorectal cancer: The Japan Collaborative Cohort Study. *Journal of Epidemiology, 24*(5), 370–378. https://doi.org/10.2188/jea.je20130168.

Yang, Y., Weng, W., Peng, J., Hong, L., Yang, L., Toiyama, Y., et al. (2017). Fusobacterium nucleatum increases proliferation of colorectal cancer cells and tumor development in mice by activating toll-like receptor 4 signaling to nuclear factor-κB, and up-regulating expression of MicroRNA-21. *Gastroenterology, 152*(4), 851–866.e824. https://doi.org/10.1053/j.gastro.2016.11.018.

Yassour, M., Lim, M. Y., Yun, H. S., Tickle, T. L., Sung, J., Song, Y. M., et al. (2016). Sub-clinical detection of gut microbial biomarkers of obesity and type 2 diabetes. *Genome Medicine, 8*(1), 17. https://doi.org/10.1186/s13073-016-0271-6.

Yazici, C., Wolf, P. G., Carroll, T. P., Mutlu, E., Xicola, R. M., Llor, X., et al. (2015). 511 Bilophila wadsworthia is more abundant in the colonic microbiome of colorectal cancer cases compared to healthy controls. *Gastroenterology, 148*(4), S-100.

Yazici, C., Wolf, P. G., Kim, H., Cross, T. L., Vermillion, K., Carroll, T., et al. (2017). Race-dependent association of sulfidogenic bacteria with colorectal cancer. *Gut*, *66*(11), 1983–1994. https://doi.org/10.1136/gutjnl-2016-313321.

Ye, P., Xi, Y., Huang, Z., & Xu, P. (2020). Linking obesity with colorectal cancer: Epidemiology and mechanistic insights. *Cancers (Basel)*, *12*(6), 1408. https://doi.org/10.3390/cancers12061408.

York, A. (2018). Gut microbiota sways response to cancer immunotherapy. *Nature Reviews Microbiology*, *16*(3), 121. https://doi.org/10.1038/nrmicro.2018.12.

Yoshimoto, S., Loo, T. M., Atarashi, K., Kanda, H., Sato, S., Oyadomari, S., et al. (2013). Obesity-induced gut microbial metabolite promotes liver cancer through senescence secretome. *Nature*, *499*(7456), 97–101. https://doi.org/10.1038/nature12347.

Yu, X., Bao, Z., Zou, J., & Dong, J. (2011). Coffee consumption and risk of cancers: A meta-analysis of cohort studies. *BMC Cancer*, *11*, 96. https://doi.org/10.1186/1471-2407-11-96.

Zeng, H., Claycombe, K. J., & Reindl, K. M. (2015). Butyrate and deoxycholic acid play common and distinct roles in HCT116 human colon cell proliferation. *The Journal of Nutritional Biochemistry*, *26*(10), 1022–1028. https://doi.org/10.1016/j.jnutbio.2015.04.007.

Zeng, H., & Combs, G. F., Jr. (2008). Selenium as an anticancer nutrient: Roles in cell proliferation and tumor cell invasion. *The Journal of Nutritional Biochemistry*, *19*(1), 1–7. https://doi.org/10.1016/j.jnutbio.2007.02.005.

Zeng, H., Umar, S., Rust, B., Lazarova, D., & Bordonaro, M. (2019). Secondary bile acids and short chain fatty acids in the colon: A focus on colonic microbiome, cell proliferation, inflammation, and cancer. *International Journal of Molecular Sciences*, *20*(5), 1214. https://doi.org/10.3390/ijms20051214.

Zhang, X., Albanes, D., Beeson, W. L., van den Brandt, P. A., Buring, J. E., Flood, A., et al. (2010). Risk of colon cancer and coffee, tea, and sugar-sweetened soft drink intake: Pooled analysis of prospective cohort studies. *Journal of the National Cancer Institute*, *102*(11), 771–783. https://doi.org/10.1093/jnci/djq107.

CHAPTER FIVE

Immunotherapy in colorectal cancer

Parul Agarwal[a], Dung T. Le[a,*], and Patrick M. Boland[b]
[a]Sidney Kimmel Cancer Center, Johns Hopkins University, Baltimore, MD, United States
[b]Rutgers Cancer Institute of New Jersey, New Brunswick, NJ, United States
*Corresponding author: e-mail address: dle@jhmi.edu

Contents

Abstract

The immune tumor microenvironment (TME) of colorectal cancer (CRC) is a crucial contributor to disease biology, making it an important target for therapeutic intervention. The diversity of immune cell populations within various subsets of CRC has led to the discovery that immune characterization of the TME has both prognostic and predictive value for patients. The convergence of improved molecular and cellular characterization of CRC along with the widespread use of immunotherapy in solid tumors has led to a revolution in the approach to clinical care. Monoclonal antibodies (mAbs) which target key immune checkpoints, such as programmed death-1 (PD-1) and cytotoxic T-lymphocyte antigen 4 (CTLA-4), have demonstrated remarkable clinical activity in microsatellite instability-high (MSI-H) CRCs and are now used in routine

Advances in Cancer Research, Volume 151
ISSN 0065-230X
https://doi.org/10.1016/bs.acr.2021.03.002

practice. The observation that MSI-H cancers are highly infiltrated with immune cells and carry a high neoantigen load led to the successful targeting of these cancers with immunotherapy. More recently, the Food and Drug Administration (FDA) approved a PD-1 inhibitor for microsatellite stable (MSS) cancers with high tumor mutation burden. However, the anti-tumor activity of immunotherapy is rare in the majority of CRC. While immune cell characterization does provide prognostic value in these patients, these observations have not yet led to therapeutic interventions. By delineating factors that predict efficacy, resistance, and therapeutic targets, ongoing research will inform the development of effective combination strategies for the vast majority of MSS CRC and immunotherapy-resistant MSI-H cancers.

Abbreviations

AJCC	American Joint Committee on Cancer
APC	adenomatous polyposis coli
APCs	antigen presenting cells
APM	antigen presenting machinery
CD	cluster of differentiation
CDT	cytolethal distending toxin
CEA	carcinoembryonic antigen
CIMP	CpG island hypermethylation phenotype
CIN	chromosomal instability
CMS	consensus molecular subtype
COX	cyclooxygenase
CR	complete response
CRC	colorectal cancer
CT	center of tumor
CTA	cancer/testis antigens
ctDNA	circulating tumor DNA
CTL	cytotoxic T-lymphocyte
CTLA-4	cytotoxic T-lymphocyte antigen 4
DC	dendritic cell
DCR	disease control rate
DFS	disease free survival
dMMR	deficient mismatch repair
DNA	deoxyribonucleic acid
DOR	duration of response
ED	exonuclease domain
EGFR	epidermal growth factor receptor
FDA	food and drug administration
FOLFIRI	5-FU, leucovorin, irinotecan
FOLFOX	5-FU, leucovorin, oxaliplatin
FOLFOXIRI	5-FU, leucovorin, oxaliplatin, irinotecan
FSP	frame shift peptide
FU	fluorouracil
G-CSF	granulocyte stimulating factor
GTP	guanosine 5′-Triphosphate
HLA	human leukocyte antigen

HPV	human papillomavirus
HR	hazard ratio
IA	investigator-assessed
ICD	immunogenic cell death
ICI	immune checkpoint inhibitor
IDO	indolamine 2′3′-dioxygenase
IFN	interferon
IHC	immunohistochemistry
IL	interleukin
IM	invasive margin
indels	insertion/deletions
IRF	interferon regulatory factor
IS	immunoscore
JAK	Janus kinase
LAG	lymphocyte-activation gene
LOH	loss of heterozygosity
M1	type I macrophage
M2	type II macrophage
mAbs	monoclonal antibodies
mCRC	metastatic colorectal cancer
MDSC	myeloid derived suppressor cell
MHC	major histocompatibility complex
MMR	mismatch repair
mPR	major pathologic response
MSI	microsatellite instability
MSI-H	microsatellite instability-high
MSI-L	microsatellite instability-low
MSS	microsatellite stable
NF	nuclear factor
NK	natural killer
ORF	open reading frame
ORR	overall response rate
OS	overall survival
pCR	pathologic complete response
PCR	polymerase chain reaction
PD	programmed death
PD-L1	programmed death-ligand 1
PFS	progression free survival
PGE	prostaglandin E
pMMR	mismatch repair proficient
PR	partial response
SBRT	stereotactic body radiotherapy
SITC	Society of Immunotherapy of Cancer
SNV	single-nucleotide variants
TAA	tumor associated antigen
TAM	tumor associated macrophage
TAP	transporter associated with antigen processing

TCGA	The Cancer Genome Atlas
TCR	T-cell receptor
TGF-β	transforming growth factor beta
Th	T-helper
TIL	tumor infiltrating lymphocyte
TKI	tyrosine kinase inhibitor
TLR	toll-like receptor
TLS	tertiary lymphoid structures
TMB	tumor mutational burden
TMB-H	tumor mutational burden-high
TMB-L	tumor mutational burden-low
TME	tumor microenvironment
TNF	tumor necrosis factor
TNM	tumor, node, metastasis
Tregs	T-regulatory
TSA	tumor-specific antigen
TTR	time to recurrence
UICC	Union for International Cancer Control

1. Immunity in colorectal cancer

CRC remains a leading cause of cancer-related mortality and morbidity worldwide. In the United States, approximately 53,200 individuals will die from this disease in 2020, making it the second most common cause of cancer death among adults. Patients with loco-regional tumor involvement have a 5-year survival rate of 71%, and those with distant disease have a 5-year survival rate of approximately 13%. Classically, the tumor (T), node (N), and metastasis (M) staging criteria as outlined by the American Joint Committee on Cancer and Union for International Cancer Control (AJCC/UICC) has been utilized as the major prognostication tool (AJCC Cancer Staging Manual, 2018). However, disparate clinical outcomes are observed for patients in the same stage even after accounting for stage-specific pathologic risk features (Nagtegaal, Quirke, & Schmoll, 2011). A deeper understanding of the heterogeneous molecular pathways, gene expression profiles, and immune responses implicated in CRC pathogenesis have provided insights into additional elements which influence prognosis and may predict response to treatment. This discussion will focus on the immune landscape of the CRC TME.

1.1 Immune subsets/prognosis

The loss of inhibitory signals and activation of growth signals leads to the proliferation of malignant cells and subsequent tumor formation (Vogelstein et al., 1988). In addition to tumor cells, the TME consists of endothelial cells, fibroblasts and immune cells. Interactions between these malignant and non-malignant cells creates a complex and dynamic TME (Hanahan & Weinberg, 2011). A more comprehensive understanding of the role of these non-malignant cells in cancer cell recruitment, activation, and persistence in the TME has broadened the scope of anti-tumor targets (Hanahan & Coussens, 2012). In particular, characterization of the immune cells of both the adaptive and innate immune systems has provided key insights about the regulation of tumor growth and development of metastases. The innate immune system is comprised of macrophages, neutrophils, dendritic cells (DCs), and myeloid derived suppressor cells (MDSCs). Major components of the adaptive immune system include B-cells and cluster of differentiation (CD)8^+ and $CD4^+$ T-lymphocytes (Goldszmid, Dzutsev, & Trinchieri, 2014). Natural Killer (NK) cells and γδ T-cells interface in both the innate and adaptive immune systems. Immune cells in the TME are modulated by an intricate network of chemokine and cytokine signaling (Dranoff, 2004).

Key functions of adaptive immune cells in the tumor immune response are briefly outlined here. $CD8^+$ cytotoxic T-lymphocytes (CTLs) are a critical component of the adaptive anti-tumor response, capable of directly killing tumor cells (Fridman, Pagès, Sautès-Fridman, & Galon, 2012; Galli et al., 2020; Knutson & Disis, 2005; Sakaguchi, 2000). $CD4^+$ T-cells have unique cytokine repertoires and are classified into various subtypes, including T-helper (Th)1, Th2, Th17, and T-regulatory (Tregs) cells. Th1 cells generate interferon (IFN)-ɣ, interleukin (IL)-2, IL-12, and tumor necrosis factor (TNF) α cytokines. This subset favors cellular immunity by activating CTLs, NK cells, and antigen presenting cells (APCs) (Knutson & Disis, 2005). In contrast, Th2 cells produce IL-4, IL-5, IL-6, IL-10, and IL-13, suppress Th1 cytokine production, and mediate the anti-inflammatory immune response (Chtanova & Mackay, 2001; Dong & Flavell, 2001). Th17 cells produce IL-17 which induces the production of IL-6, granulocyte stimulating factor (G-CSF) and further modulates the inflammatory response (Dong, 2006). Finally, suppressive T-regs act by inhibiting the maturation of APCs, consuming IL-2 and generating immunosuppressive cytokines and metabolites (Misra, Bayry, Lacroix-Desmazes, Kazatchkine, & Kaveri, 2004; Onishi, Fehervari, Yamaguchi, & Sakaguchi, 2008;

Sakaguchi, 2000). Ultimately, whether cells of the adaptive immune system trigger a pro-tumor or anti-tumor response depends largely on context.

There is extensive interaction between the innate and adaptive immune systems, linked by several immune cell subtypes. NK cells produce pro-inflammatory and immunosuppressive cytokines, growth factors, and chemokines. They ensure self-tolerance, adapt to the surrounding environment, and display antigen-specific immunologic memory (Vivier et al., 2011). The myeloid compartment of the innate immune system also has both anti-tumor and pro-tumor properties. Tumor associated macrophages (TAMs) are categorized as Type I macrophages (M1) which demonstrate a classical activation (Th1)/pro-inflammatory phenotype or Type II macrophages (M2) which demonstrate an alternative activation (Th2)/immunosuppressive phenotype (Dandekar, Kingaonkar, & Dhabekar, 2011). DCs are a type of APC that prime CTLs, and secrete type 1 IFN (Banchereau & Steinman, 1998). On the other hand, MDSCs are generally immunosuppressive and express low levels of Major Histocompatibility Complex (MHC) class II and CD80, and induce stromal cell proliferation and extracellular matrix deposition (Bronte et al., 2000; Movahedi et al., 2008).

Importantly, overall prognosis is related to the presence or absence of tumor infiltrating lymphocytes (TILs) within specific areas of the TME (Diederichsen, Hjelmborg, Christensen, Zeuthen, & Fenger, 2003; Naito et al., 1998). In CRC, increased infiltration of $CD8^+$ T-cells, Th1 cells, M1 macrophages, DCs, and NK cells is generally associated with improved outcomes, while infiltration of M2 macrophages, MDSCs, and Th17 cells is associated with worse outcomes (Barbera-Guillem et al., 2000; Becht et al., 2016; Bindea et al., 2013; Boissière-Michot et al., 2014; Bruni, Angell, & Galon, 2020; Chen & Chen, 2014; Edin et al., 2012; Herrera et al., 2013; Liu et al., 2011; Salama et al., 2009; Väyrynen et al., 2014). The type, density, location, and functional orientation of immune cells in the TME is defined as the immune contexture (Fridman et al., 2012; Galon et al., 2006; Galon, Fridman, & Pagès, 2007). The immune contexture can be broadly classified in one of three categories: (1) immune-activated (2) immune-inflamed and (3) immune-ignorant.

Immune-activated tumors are characterized by diffuse infiltration of activated DCs, M1 macrophages, Th1 cells and CTLs. There is increased formation of tertiary lymphoid structures (TLS), upregulation of Human Leukocyte Antigen (HLA) class I/II, and amplified antigen processing and presentation. Robust activation of immune evasion pathways by immune-activated tumor cells results in increased expression of immune checkpoints as a mechanism of immune escape (Llosa et al., 2015). Immune checkpoints like PD-1 and CTLA-4 may be exploited as potential targets for treatment with anti-tumor immunotherapies. Immune-inflamed tumors are characterized by upregulation of TGF-β and complement leading to enriched infiltration of MDSCs and Th17 cells. There is increased angiogenesis, infiltration of stromal cells and a dense extracellular matrix resulting in an immunosuppressive phenotype (Becht et al., 2016). Immune-ignorant tumors are characterized by low class I MHC expression with decreased infiltration of lymphocytes and myeloid cells and diminished expression of genes implicated antigen processing and in T-cell chemotaxis (Becht et al., 2016; Bindea et al., 2013). These immune phenotypes have been described in the context of the consensus molecular subtypes of CRC, which will be discussed in detail later. Evaluation of the immune TME in CRC has highlighted the complex signaling pathways involved at the genetic, molecular, and cellular level which ultimately generate a dynamic tumor immune phenotype.

1.2 Immunoscore®

Prior work has shown that intra-tumoral immune infiltration is heterogeneous and that the immune reaction at the center of the tumor (CT) and invasive margin (IM) holds prognostic value beyond the AJCC/UICC TNM staging system (Galon et al., 2006; Mlecnik et al., 2011). The Immunoscore® (IS) was established in an effort to clinically translate the immune contexture into a reproducible, prognostic and potentially predictive biomarker for CRC (Angell & Galon, 2013; Hermitte, 2016). This score is calculated by quantifying two lymphocyte populations ($CD3^+$ lymphocytes and $CD8^+$ lymphocytes) at the CT and IM (Pagès et al., 2009). On the basis of these calculations, an IS ranging from I0 to I4 is obtained. Tumors with a score of I0 have low densities of lymphocytes in the CT and IM and tumors with a score of I4 have high densities of both cell types in the CT and IM (Galon et al., 2012).

The prognostic value of the IS has been validated across several studies, including a large international study in patients with early stage CRC sponsored by the Society of Immunotherapy of Cancer (SITC). In patients with stage I-III CRC, the density of $CD3^+$ and $CD8^+$ T-cells at the CT and IM was calculated and the mean of these four densities was translated into an IS. Risk of recurrence was categorized by IS (high, intermediate, low)

Table 1 Immunoscore® and risk of recurrence.

Immunoscore®	Mean percentile $CD3^+$ and $CD8^+$ T cells at CT and IM	Risk of recurrence
I0–1	0%–25%	High
I2	>25%–70%	Intermediate
I3–4	70%–100%	Low

as described in Table 1 (Marliot et al., 2020; Pagès et al., 2018). In patients with a high, intermediate, or low IS, recurrence rate at 3 years was 5%, 17%, and 26%, respectively. Those with a high IS demonstrated a longer 5-year overall survival (OS) when compared to a low IS (82% vs 62%, $P = 0.0004$). In a multivariate analysis, the IS was associated with time to recurrence (TTR) independent of age, sex, T-stage, N-stage and microsatellite instability (MSI) status ($P < 0.0001$) (Pagès et al., 2018). In the phase III NCCTG N0147 (Alliance) trial, patients with stage III CRC treated with folinic acid, fluorouracil, and oxaliplatin (FOLFOX) were grouped as high risk (T4 or N2) or low risk (T1–3, N1) disease and IS was calculated. Among both high risk and low risk patients, a high (I2–I4) vs a low (I0–I1) IS was associated with a statistically significant improvement in 3-year disease free survival (DFS); these results persisted after excluding mismatch repair deficient (dMMR) patients (Sinicrope et al., 2018).

The IS may hold potential for utilization as a predictive biomarker to guide adjuvant chemotherapy allocation in stage III CRC. One of the most relevant data sets comes from a subgroup analysis of patients with stage III CRCs from the SITC Immunoscore consortium. Regardless of pathologic risk status, treatment with adjuvant chemotherapy compared to no chemotherapy was associated with increased 5-year TTR in stage III CRC (32% vs 51%, $P < 0.0001$). Further analysis of TTR demonstrated that IS predicted benefit from adjuvant therapy. While there was no significant difference in TTR within the I0 and I0–1 groups (low IS), those with a score of I2 or higher saw a significant improvement in TTR and OS when adjuvant chemotherapy was employed (Mlecnik et al., 2020). In the current paradigm, patients with stage III or high-risk stage II CRC are typically recommended to receive adjuvant chemotherapy; though requiring further validation, these findings suggest that the IS has capacity to identify a population of CRC patients that derive no benefit from adjuvant chemotherapy.

Further assessment of the IS in advanced CRC and other tumor types may provide an opportunity to deepen understanding of the evolving and heterogeneous immune landscape, which is fundamental to elucidate potential mechanisms of response and resistance to treatments. As the first immune-based scoring system for cancer with prognostic and predictive potential, the IS has validated the significance of the tumor immune milieu in CRC pathogenesis and treatment. Earlier this year, the WHO classification of digestive system tumors introduced immune response as an essential and desirable diagnostic criterion for CRC, further bolstering this concept.

1.3 Neoantigens in cancer/mutational burden

The ability of the immune system to recognize and target aberrant proteins on cancer cells with subsequent tumor destruction was identified as early as the 1940s (Gross, 1943). During the process of tumorigenesis, cancers acquire thousands of somatic mutations. Acquired driver and passenger mutations in the cancer genome alter amino acid coding sequences and result in nonsynonymous mutations through single-nucleotide variants (SNV), mutational frameshifts, splice variants, or gene fusions (Hinrichs & Rosenberg, 2014; Leach, Krummel, & Allison, 1996). These mutations lead to the expression of abnormal proteins in cancer cells which subsequently undergo processing and are formed into peptides. Peptides are presented on tumor cells via MHC molecules and are recognized by T-cells as non-self proteins (Coulie et al., 1995).

There are three broad classifications of tumor antigens that interact with T-cells. Neoantigens are tumor-specific antigens (TSAs) that are generated through nonsynonymous mutations and fusions or, in tumors that are virally mediated, by viral open reading frames (ORFs) (Feng, Shuda, Chang, & Moore, 2008). Tumor-associated antigens (TAAs) have low-level expression in normal tissue, but have a greater expression in malignant tissues (Ward, Gubin, & Schreiber, 2016). Finally, cancer/testis antigens (CTAs) are expressed by tumor tissue and by reproductive tissue, but have limited expression on other normal tissues (Simpson, Caballero, Jungbluth, Chen, & Old, 2005). In order to kill cancer cells, $CD8^+$ effector T-cells must recognize tumor antigens displayed by MHC class I molecules. Given that TSAs are unique to malignant cells and not affected by T-cell tolerance, they are attractive targets for anti-tumor treatments.

In a patient with melanoma treated with autologous tumor-reactive TILs who achieved a complete response (CR), persistent neoantigen-specific clonal T-cells were identified, highlighting the likely importance of a potent neoantigen-specific immune response (Zhou, Dudley, Rosenberg, & Robbins, 2005). Cancers with a higher mutational burden express a greater number of neoantigens, which may be predictive of response to treatment with certain therapies. However, immunogenic neoantigens have been described in cancers with lower mutational burden, underscoring additional key qualities of neoantigens that may be needed in order to elicit immune responses (Le et al., 2015; Turajlic et al., 2017). The burden and fraction of clonal neoantigens (as opposed to sub-clonal neoantigens) correlates with response to immune checkpoint inhibitors (McGranahan et al., 2016). The degree to which a neoantigen demonstrates sequence homology to known antigens should also be considered, and certain mutations may result in more potent neoantigens (Turajlic et al., 2017). Additional features which affect whether an antigen is high quality include overall neoantigen expression, likelihood of loss, and ability to bind to multiple HLAs (McGranahan & Swanton, 2019).

The optimal method of directly measuring the number of neoantigens and T-cell responses has not been defined; however, tumor mutational burden (TMB) has been utilized as a surrogate marker of neoantigen formation and response to immunotherapy. TMB is a measure of the total amount of somatic coding mutations in a tumor and is most often calculated from targeted next generation sequence panels (Chalmers et al., 2017). Studies across several tumor types have demonstrated that improved response to immune checkpoint inhibitors may be correlated with a high TMB (Rizvi et al., 2015; Snyder et al., 2014; Van Allen et al., 2015; Yarchoan, Hopkins, & Jaffee, 2017). Earlier this year, the FDA approved pembrolizumab for the treatment of adult and pediatric patients with refractory unresectable or metastatic TMB-high (TMB-H) solid tumors, defined by $\geq$10 mutations/megabase (muts/Mb). This approval was based on a prospectively planned retrospective analysis of 102 patients with TMB-H tumors; however, patients with CRC were not included. When treated with pembrolizumab, those with TMB-H tumors had an overall response rate (ORR) of 29% while those in the non-TMB-H group had an ORR of only 6%. Median duration of response (DOR) was not reached, with 50% of patients having a response duration $\geq$24 months (Marabelle et al., 2020).

1.4 Immune checkpoints/immune checkpoint inhibitors in colon cancer

Maintaining the homeostasis of the immune response is essential for host survival. As such, the immune response is regulated by co-stimulatory and inhibitory signals, broadly referred to as "immune checkpoints." Activated T-cells express co-inhibitory receptors such as PD-1, CTLA-4, and lymphocyte-activation gene 3 (LAG-3). The magnitude and quality of the effector T-cell response is regulated in part by these immune checkpoints (Pardoll, 2012). Malignant cells can dysregulate expression of immune checkpoint proteins as one mechanism of immune escape (Schreiber, Old, & Smyth, 2011). In an effort to overcome this resistance, antibodies that block immune checkpoints and target lymphocyte receptors or their ligands have been developed. These immunomodulatory mAbs, known as immune checkpoint inhibitors (ICIs), have resulted in robust anti-tumor responses in many previously recalcitrant malignancies. The most widely studied immune checkpoints to date are CTLA-4 and PD-1.

CTLA-4 is a co-inhibitory receptor primarily expressed on the T-cell surface and regulates the early stages of T-cell activation by counteracting CD28 (Schwartz, 1992). Following antigen recognition, CD28 functions to augment T-cell receptor (TCR) signaling and activate T-cells. CTLA-4 outcompetes CD28 and inhibits T-cells directly, downregulates helper T-cell activity and upregulates regulatory T-cell immunosuppressive activity. The role of CTLA-4 in modulating T-cell activation was originally demonstrated in *Ctla-4* knockout mice, and resulted in a lethal immune hyper-activation phenotype (Tivol et al., 1995; Waterhouse et al., 1995). Initially, there was hesitation about targeting CTLA-4 due to concerns of immune-related toxicity, but further evaluation revealed that a therapeutic window could be achieved with partial CTLA-4 blockade (Leach et al., 1996). Based on this pivotal preclinical work, ipilimumab and tremelimumab, two fully humanized CTLA-4 antibodies, began clinical testing over 20 years ago in melanoma (Pardoll, 2012). In a phase III trial including patients with treatment refractory melanoma, a survival benefit was observed in patients treated with ipilimumab (Hodi et al., 2010). These results led to FDA approval of ipilimumab for treatment refractory metastatic melanoma in 2010.

PD-1 primarily functions in limiting the activity of T-cells in peripheral tissues and is induced when T-cells are activated at the time of an inflammatory response (Ishida, Agata, Shibahara, & Honjo, 1992). PD-1 is more widely expressed than CTLA-4 and is found on a variety of cell types including $CD4^+$ T-cells, $CD8^+$ T-cells, B cells, macrophages, and DCs (Yamazaki et al., 2002). Programmed death-ligand 1 (PD-L1) and PD-L2 are important mediators of the PD-1 pathway, and constitutive expression of PD-1 and PD-L1 leads to immune evasion and subsequent tumor proliferation in a variety of malignancies (Dong et al., 2002; Konishi et al., 2004).

The interaction between PD-1 and PD-L1/L2 hinders the proliferation and function of T effector cells, inhibits apoptosis of tumor cells, and enhances the development of immunosuppressive Tregs (Shindo et al., 2015; Terawaki et al., 2011). Blockade of the PD-1 pathway at various points can enhance anti-tumor activity in the TME by restoring effector T-cell activity and downstream immune signaling. Pembrolizumab is a humanized IgG4 kappa mAb that targets the PD-1 membrane protein on the surface of T-cells. In a phase I trial in patients with previously treated advanced melanoma refractory to ipilimumab, approximately 25% of patients demonstrated an objective response when treated with pembrolizumab, leading to FDA approval of this agent in 2013 for previously treated melanoma (Robert et al., 2014).

In the years following, several additional ICIs targeting various immune checkpoints have been developed, are utilized in clinical practice, and are under clinical investigation including nivolumab (anti-PD-1), durvalumab (anti-PD-L1), avelumab (anti-PD-L1), and atezolizumab (anti-PD-L1). Few direct comparisons exist among the various ICIs, but as the use of these agents expands, more data regarding efficacy and toxicity profiles is emerging (Duan et al., 2020).

2. Colorectal cancer genetics and subtyping

It is widely believed that CRCs evolve from colonic epithelial cells to adenomas to carcinomas through a series of genetic changes. Fearon and Vogelstein initially described this process, with multiple later studies adding depth of understanding and complexity (Vogelstein et al., 1988). Particularly within sporadic MSS colorectal cancers, mutational inactivation of the Adenomatous Polyposis Coli (*APC*) gene is the most frequent mutation seen; this loss of negative feedback allows for the unfettered activation of β-catenin, with its nuclear translocation inducing constitutive activation of the Wnt signaling pathway (Markowitz & Bertagnolli, 2009). There are at least three overarching pathways by which CRCs develop: a chromosomal instability (CIN) pathway, a mutator phenotype/mismatch repair pathway or a hypermethylation phenotype (CpG island hypermethylation phenotype, CIMP) pathway (Markowitz & Bertagnolli, 2009). CIN tumors are characterized by alterations in large portions of chromosomes resulting in aneuploidy, sub-chromosomal genomic amplifications and loss of heterozygosity (Pino & Chung, 2010). $CIMP^+$ tumors demonstrate aberrant promoter methylation of tumor suppressor genes and often, hypermethylation of the mismatch repair gene *MLH1* leading to MSI; CIMP$^+$ tumors account for the majority of sporadic tumors with MSI (Weisenberger et al., 2006).

The genetic and epigenetic context of CRCs likely plays a significant role in response or lack thereof to immunotherapeutic approaches. Recently, increased attention has also been drawn to the local milieu in which a tumor forms, with the tumor microbiome of rising interest. Considering the most common clinically utilized molecular categorizations, MSI holds the greatest relevance with respect to the use of immunotherapy. Another subtyping system, the Consensus Molecular Subtype (CMS) criteria, has primarily been employed for research purposes. Some recent efforts have utilized this classification for prospective enrollment in rational interventional trials. Herein we will review current subtyping schemas in CRC and potential implications for immune targeting strategies.

2.1 Microsatellite instability (MSI)

It has been estimated that approximately 15% of all CRCs are microsatellite unstable (MSI-H), while the remaining 85% are microsatellite stable (MSS) (Boland & Goel, 2010). The mismatch repair (MMR) proteins are responsible for correction of single nucleotide base mispairings, small insertions, and small deletions occurring via routine deoxyribonucleic acid (DNA) replication (Chung & Rustgi, 1995). There are multiple MMR family members, with four key genes most directly associated with genomic instability: *MLH1*, *MSH2*, *MSH6*, and *PMS2*. On the other hand, MSI refers to a molecular phenotype witnessed in the absence of functional MMR proteins. Inactivation of key MMR proteins can occur in the setting of a deleterious germline MMR gene mutation (Lynch Syndrome), via hypermethylation induced *MLH1* promoter silencing (sporadic), or less commonly via somatic bi-allelic MMR gene mutations (Boland & Goel, 2010; Haraldsdottir et al., 2014). Whatever the mechanism, dysfunction of key MMR proteins results in greatly increased rates of slipped-strand mismatches of short repeating bi- or tri-nucleotides (microsatellites), which is detected as MSI on a polymerase chain reaction (PCR) based assay.

Historically, nomenclature has referred to these tumors in one of two ways, based upon the test performed to evaluate the presence of functioning mismatch repair proteins: immunohistochemistry (IHC) or PCR. The IHC based assay involves staining of the four key MMR proteins. Absence of protein detection categorizes the tumor as being MMR deficient (dMMR), while the presence of all proteins signifies that a tumor is MMR proficient (pMMR). The MSI-assay classically includes evaluation of five selected microsatellite loci. If $\geq$2 microsatellites are unstable, the tumor is deemed MSI-H. Tumors with no unstable microsatellites are deemed MSS. Those that exhibit instability in 1/5 markers are referred to as MSI-low (MSI-L); the majority of MSI-L tumors have no MMR defect and thus, are treated similarly to MSS tumors. For clinical purposes, dMMR and MSI-H refer to the same tumor types, while pMMR and MSS are also generally overlapping, with a small percentage of tumors displaying discrepancies between IHC and PCR assays. MMR testing is routine in CRC and done as reflex testing in many situations. However, through novel approaches, current next-generation sequencing technologies are capable of detecting MSI with excellent concordance (97%), 95% sensitivity and 98% specificity (Trabucco et al., 2019). This manner of testing may represent the future standard both across cancers and within CRC, though the turn-around time of the assay will remain a critical factor driving practice in many situations.

The deficient mismatch repair proteins within MSI-H cancers makes them unique, as these tumors exhibit very high mutational rates. This was described very clearly through the cancer genome atlas (TCGA), where up to 16% of CRCs were hypermutated, the majority being MSI-H tumors (Guinney et al., 2015). The mechanism of genomic instability results in greatly increased rates of insertion/deletions (indels), specifically, which can induce translational frameshift, producing a robust source of tumor neoantigens. MSI-H CRCs have long been recognized to have additional features which differentiate them from MSS tumors. They most commonly originate in the right side of the colon, are poorly differentiated, and are mucinous (Gafa et al., 2000). CIMP^{+} MSI-H tumors more frequently occur in older patients, women and smokers. Most importantly, MSI-H tumors often possess a robust lymphocytic infiltrate, with a substantial portion being activated T-cells, in contrast to most CRCs (Schwitalle et al., 2008). Despite the increased frequency of unfavorable prognostic features (such as poor differentiation), MSI-H CRCs carry a better prognosis, particularly in early stage disease (Popat, Hubner, & Houlston, 2005). The reasons for this are not fully elucidated, though anti-tumor immunity may be instrumental in preventing the dissemination of these tumors. Perhaps reflecting this protective effect, MSI is detected in CRC in approximately 10%–20% of early stage cancers, but just 3%–4% of Stage IV colon cancers (Koopman et al., 2009).

The etiology of anti-tumor immunity in MSI-H CRC is thought to be due to the markedly increased mutational rate and the resultant accumulation of large numbers of frameshift mutations (Schwitalle, Linnebacher, Ripberger, Gebert, & von Knebel Doeberitz, 2004). MSI was shown to be an early event in the development of adenomas within Lynch Syndrome-affected individuals (Yurgelun et al., 2012). Elegant work further showed that TILs isolated from patients with MSI-H CRCs recognize MSI frame shift peptides (FSPs); specifically, these T-cells demonstrate cytotoxic activity against other MSI-H CRC cells, with HLA-specificity, suggesting HLA class I-mediated antigen processing and presentation. While these responses were also seen among individuals with germline MMR mutations (Lynch Syndrome) not affected by CRC, FSP-specific T-cell responses were largely absent in those with MSS CRCs (Schwitalle et al., 2008).

2.2 DNA polymerase exonuclease domain (POLE/POLD1) mutant tumors

A small subset of MSS colon cancers are genomically distinct by virtue of a somatic DNA polymerase exonuclease domain (ED) mutation: either POLE or POLD1. Pathogenic mutations of this kind negatively affect the proofreading function of these polymerases, with evidence of predisposition to multiple cancers, including CRC (Mur et al., 2020). These defects result in the so-called "ultramutator" phenotype, also described in the TCGA dataset (Guinney et al., 2015). In colorectal cancer, POLE mutations are the primary alteration seen, though this is a rare finding, perhaps with a frequency near 1% of all CRCs (Domingo et al., 2016; Hino et al., 2019). Available data suggests that patients affected with these POLE-mutated tumors are on average younger and more likely to be male. Similar to MSI-H tumors, the prevalence appears to be higher in early stage disease. Very small case numbers limit firm additional conclusions, but in one report, POLE mutated CRCs possessed significantly increased densities of $CD8^+$, $CD45RO^+$, and $CD8^+$ $CD45RO^+$ cells, as compared to both POLE wildtype and POLE non-ED mutated tumors (Mo et al., 2020).

2.3 Consensus molecular subtypes (CMSs)

Multiple different subtyping systems have been proposed for CRC. In 2015, a multi-national group reported out their collaborative efforts to classify CRCs based on a gene-expression based system, forming four consensus molecular subtypes (CMSs). The four subtypes were based upon data from 18 data sets, including data from 4151 patients, largely drawn from primary tumors and early stage CRCs (Guinney et al., 2015). The four subtypes with frequency were CMS1 (MSI immune, 14%), CMS2 (canonical, 37%), CMS3 (metabolic, 13%), and CMS4 (mesenchymal, 23%).

Significant biologic differences exist between the groups. CMS1 tumors are characterized by immune infiltration and activation, with high mutational counts. Not surprisingly, there is considerable overlap with MSI and CIMP$^+$ CRCs, as well as *BRAF* mutated tumors. Of relevance, PD-1 activation is enriched in this subgroup. CMS2 tumors are characterized by high somatic copy number alterations (SCNAs), as well as activation of Wnt and MYC signaling. CMS3 tumors have prominent metabolic derangements and an overrepresentation of *KRAS* mutations. Finally, CMS4 tumors are characterized by TGF-β activation, with marked upregulation of stromal/mesenchymal pathways, as well as angiogenic signaling. CMS4 CRCs have the worst prognosis and are more commonly diagnosed at stage III or IV.

Looking at this slightly differently, a pan-cancer immunogenomic analysis of the TCGA identified six immune subtypes (C1–C6), characterized by differences in macrophage and lymphocyte signatures, Th1:Th2 cell ratio, neoantigen load, aneuploidy, expression of immunomodulatory genes and prognosis. The majority of CRCs fell into the C1 (*wound healing*) subtype, described by elevated expression of angiogenic genes, a high proliferation rate, and a Th2 cell bias (Thorsson et al., 2018). A subsequent analysis, also utilizing TCGA data, similarly suggested C1 to be the predominant phenotype (77%) consistently across all but CMS1 tumors (Soldevilla et al., 2019). Seventeen percent of CRCs aligned with the C2 (IFN-*γ dominant*) category, characterized by the highest M1/M2 macrophage polarization, the highest CD8$^+$ T-cell densities, and the greatest TCR diversity. Perhaps not surprisingly, the C2 subtype was the major phenotype within CMS1 tumors (53%) and was under-represented among CMS2–4 tumors (8%–13%). The C3 *inflammatory* subtype was characterized by a high Th1:Th2 ratio and the highest Th17 gene expression, with minor enrichment among CMS3 and CMS4 tumors (6% and 7% respectively). Finally, C4 *lymphocyte depleted* subtype was predominantly seen in CMS3 (4%), and the few C6 *TGF-β dominant* cases all belonged to the CMS4 subgroup (2.3%). No CRC samples belonged to the C5 *immunologically quiet* subtype.

This subtyping data may prove useful in later analysis of clinical studies to understand why certain tumors respond or not to therapy. Another hope is that these subtypes, along with TMB, TIL density or MSI status, could be used to select patients for rationale therapeutic strategies. Recent efforts have demonstrated both the feasibility of developing a validated clinical CMS assay and enrolling a select CMS-defined population to rationale therapeutics (NCT034365630) (Morris et al., 2021).

3. The tumor microbiome

The study of microbes as pathogenic forces in cancer has a long history, and multiple virally mediated cancers are well known, with human papillomavirus (HPV)-associated cervical cancer being a prime example. In this case, cancer is induced directly by microbial DNA via insertional mutagenesis (Munger et al., 2004). In many other malignancies, the role of microbes as pathogenic vs commensal organisms is somewhat less clear. The last decade has seen renewed interest in study of the human microbiome, particularly as it pertains to cancer, in part fueled by the increasing power of genomic sequencing platforms. While the human genome is largely static and contains approximately 20,000 genes, the human microbiome may change significantly over time, with an estimated 10–20 million non-redundant genes (Lloyd-Price et al., 2017). However, it has become clear that the tumor microbiota is distinctly different from that within the adjacent healthy mucosa; for instance, Fusobacterium is enriched in CRCs, with relative depletion of other bacterial phyla, such as Bacteroidetes (Kostic et al., 2012). Evaluation of differing tumor types demonstrates distinct microbial compositions, primarily intracellular in location and found within both the neoplastic cells and the associated immune cells (Nejman et al., 2020). This highlights the impact the microbiome may have on tumor evolution.

Microbes can interact in multiple ways with non-diseased human organ systems, as well as nascent malignancies. There is strong animal evidence that the microbiome is capable of exerting genotoxic effects, for instance, via the enumeration of reactive oxygen species (Kipanyula et al., 2013). Additionally, specific mutational signatures found in a subset of CRCs can be generated in human organoids to which genotoxic bacteria have been added (Pleguezuelos-Manzano et al., 2020). Certain bacterial toxins may also provoke carcinogenesis. *Escherichia coli* and *Campylobacter jejuni* are among the bacteria which produce cytolethal distending toxin (CDT), which can induce double-strand DNA breaks. Within inoculated $APC^{MIN/+}$ mouse models, CDT promotes the development of CRCs, a process which is suppressed by either antibiotic driven eradication of campylobacter jejuni or deleterious mutation of the CDT gene (He et al., 2019).

Bacterial metabolism of host dietary nutrients, bile salts, and xenobiotics may also play an important role in linking diet to cancer risk. For example, intestinal bacteria can affect the fermentation of fiber to key short chain fatty acids, such as butyrate, a compound which possesses significant anti-neoplastic properties (O'Keefe et al., 2015). In turn, the presence and concentration of these metabolites may serve to shape the local immune response (Campbell et al., 2020).

In some instances, the resident bacteria have capacity to induce a chronic inflammatory state, a condition well linked to carcinogenesis in CRC and which may exert a detrimental immunologic impact (Wu et al., 2009). *Fusobacterium nucleatum* can induce activation of the nuclear factor-κB (NF-κB) pathway. Further, *fusobacterium nucleatum* has been described to recruit tumor infiltrating myeloid cells, and detection in tumors is inversely associated with T-cell density (Kostic et al., 2013; Mima et al., 2015). Clearly, therapeutic anti-microbial interventions would seem of interest, particularly as these pertain to the immunologic targeting of tumors. However, broad-spectrum antibiotic use prior to ICIs is consistently linked to lack of clinical response in multiple tumor types, suggesting a complex, dynamic relationship between the tumor microbiome and therapeutics (Derosa et al., 2018). Additional work in mouse models, utilizing inoculation of selected bacterial populations, has suggested that various species can enhance or inhibit effect of PD-1 inhibition, pointing to the potential utility of more sophisticated targeted strategies in the clinical setting (Routy et al., 2018).

4. Immunotherapy in dMMR/MSI-H tumors

4.1 Clinical data with ICIs

Treatment with ICIs initially demonstrated limited clinical activity in non-selected CRC populations. In a multicenter phase I dose escalation trial of an anti-PD-L1 agent (BMS-9365559) in previously treated cancers, no responses were seen in patients with CRC (Brahmer et al., 2012). Similarly, in a phase II trial of tremelimumab (anti-CTLA-4) in a heavily pretreated population of metastatic CRC (mCRC), 44 of 45 treated patients did not receive a second dose of tremelimumab. However, one patient demonstrated a partial response (PR) with a median duration of response (DOR) of 6 months, indicating a signal for further study (Chung et al., 2010). The clinical activity of ICIs in dMMR mCRC was first seen in a phase I dose escalation study of nivolumab (anti-PD-1) in treatment refractory solid tumors. One patient with dMMR metastatic CRC demonstrated a PR at

8 weeks and in a follow-up analysis, a durable complete response (CR) lasting greater than 3 years was seen (Brahmer et al., 2010; Lipson et al., 2013). As knowledge of the immunogenic environment of dMMR CRC grew and evidence of anti-tumor responses to immune checkpoint inhibition was seen in a variety of malignancies, several studies were initiated to explore the therapeutic potential of immune checkpoint inhibition in specific subsets of CRC.

4.1.1 Treatment refractory setting

In a phase II multi-center trial, KEYNOTE-016, patients with previously treated dMMR metastatic CRC ($n=10$), pMMR metastatic CRC ($n=18$), and dMMR non-CRC ($n=7$) received pembrolizumab (anti-PD-1) every 2 weeks. Rates of immune-related ORR and 20-week immune-related progression free survival (PFS) were as follows: dMMR CRC 40% and 78%, pMMR CRC 0% and 11%, and dMMR non-CRC 71% and 67%, respectively. Moreover, whole exome sequencing demonstrated a significantly higher somatic mutational burden in dMMR vs pMMR tumors (1782 vs 73, respectively $P=0.002$) (Le et al., 2015). An updated publication of KEYNOTE-016 included 86 patients with 12 types of dMMR metastatic cancers, of which nearly half had dMMR mCRC. A similar ORR was observed in patients with dMMR mCRC (52%) and dMMR non-CRC (54%), with an impressive complete radiographic response rate of 21%. Notably, the presence of Lynch syndrome (seen in 48% of patients) did not influence outcomes (Le et al., 2017). On the basis of these findings, pembrolizumab received breakthrough therapy designation for the treatment of patients with MSI-H CRC.

KEYNOTE-164 was a single arm phase II study evaluating pembrolizumab in patients with previously treated, dMMR mCRC. Cohort A ($n=61$) included patients treated with at least two prior lines of standard therapy and Cohort B ($n=63$) included patients treated with at least one prior line of therapy. ORR in both cohorts was 33%, and at a median follow-up of 31 months (cohort A) and 24 months (cohort B) the median DOR was not reached in either cohort (Le et al., 2019). KEYNOTE-158 evaluated patients with metastatic dMMR non-CRC ($n=233$) who had received at least one prior therapy. At a median follow-up of 13.4 months, ORR was 34.3% and median PFS was 4.1 months (Marabelle et al., 2020). In a pooled analysis of KEYNOTE-016, KEYNOTE-164, KEYNOTE-158, KEYNOTE-012, and KEYNOTE-28, a total of 149 patients with heavily pre-treated dMMR cancers were evaluated, including 90 patients with dMMR metastatic CRC. ORR was 39.6% and

DOR of greater than 6 months was seen in 78% of patients. These results led to the FDA approval of pembrolizumab for dMMR/MSI-H mCRC that has progressed following chemotherapy and dMMR/MSI-H metastatic solid tumors refractory to prior treatment, the first FDA approval of a therapy for a tissue-agnostic indication (Marcus, Lemery, Keegan, & Pazdur, 2019).

CheckMate 142 was a multi-cohort phase II study, which treated patients with dMMR mCRC with nivolumab, as monotherapy or in combination with ipilimumab, and included cohorts in both the treatment refractory and frontline setting. In the previously treated cohort with nivolumab monotherapy, 31% (23 of 74) of patients achieved an investigator-assessed objective response rate (IA-ORR), and 69% had disease control for at least 12 weeks. Response to nivolumab was independent of PD-L1 status, history of Lynch syndrome, and *BRAF*/*KRAS* status (Overman et al., 2017). In one combination immunotherapy cohort of CheckMate 142, patients with previously treated dMMR/MSI-H mCRC ($n = 119$), received nivolumab 3 mg/kg and ipilimumab 1 mg/kg every 3 weeks (four doses) followed by nivolumab 3 mg/kg every 2 weeks. At a median follow-up of 13.4 months, ORR was 55% and disease control rate (DCR) was 80% (Andre et al., 2018). Results from CheckMate 142 led to the FDA approval of nivolumab with or without ipilimumab in previously treated dMMR mCRC. Based upon this data, it appears there is likely greater activity of dual ICI therapy in MSI-H CRC, consistent with the superiority seen in some other cancers (e.g., melanoma). However, it is important to note that these cohorts were non-randomized, making direct comparison potentially misleading; definitive data remains lacking.

4.1.2 Treatment naïve setting

With the evidence of sustained and robust responses in refractory MSI-H CRC, it was hypothesized that use of ICIs as first-line treatment in this subset would improve outcomes. KEYNOTE-177 was a phase III trial in which patients with treatment naïve metastatic dMMR CRC ($n = 307$) were randomized in a 1:1 ratio to pembrolizumab vs investigator's choice of doublet chemotherapy ± bevacizumab or cetuximab. The chemotherapy arm consisted of 5-fluorouracil (5-FU) and leucovorin plus either oxaliplatin (mFOLFOX6) or irinotecan (FOLFIRI). The primary endpoints were PFS and OS. At a median follow-up of 32 months, pembrolizumab demonstrated a statistically significant improvement in PFS when compared to chemotherapy (16.5 months vs 8.2 months, respectively, ($P < 0.00002$)). Moreover, radiographic responses were seen more frequently with PD-1 blockade when compared to chemotherapy (44% vs 33%) (Andre et al., 2020). Consistent with prior data, responses to pembrolizumab were highly durable, with the median DOR not yet reached (range 2.3+ to 41.4+ months). Overall survival is not yet mature for evaluation, and there has been a high rate of crossover from the chemotherapy to the immunotherapy group, a feature that may affect this outcome. Regardless, interim results from KEYNOTE-177 led to the FDA approval of pembrolizumab for front-line treatment of unresectable or metastatic dMMR/MSI-H CRC in June 2020.

In a treatment naïve cohort of CheckMate 142, patients received nivolumab 3 mg/kg every 3 weeks and low dose ipilimumab 1 mg/kg every 6 weeks with a primary endpoint of IA-ORR. At a median follow-up of 29 months, ORR was 69% and median DOR, PFS, and OS were not reached. Rates of PFS and OS at 24 months were 74% and 79%, respectively. At the time of interim analysis, 19 patients had discontinued study treatment without further therapy and no new safety signals were identified (Lenz et al., 2020). These encouraging data have led to additional phase III studies evaluating 1st line ICI use in this patient population, both with use of the strategy of combined ICIs and with PD-L1 added to chemotherapy (NCT04008030, NCT02997228).

4.1.3 Neoadjuvant/adjuvant setting

Preclinical work has suggested that the neoadjuvant administration of immune checkpoint blockade may be more effective than later stage applications (Liu et al., 2016). Furthermore, preliminary data in melanoma with combination ipilimumab and nivolumab demonstrated the potential improved efficacy of immunotherapy when initiated in primary intact, early stage disease (Blank et al., 2018). As MSI-H CRC is identified more commonly in early stage CRC, there is rationale to explore intensified investigation in this setting.

In the exploratory NICHE study (NCT030264140), patients with stage I-III dMMR CRC ($n=20$) and pMMR CRC ($n=15$) received a single dose of neoadjuvant ipilimumab on day 1 and two doses of nivolumab on days 1 and 15. The primary endpoint was safety and feasibility. Secondary endpoints included histopathological response rate and change in T-cell infiltration. Remarkably, a pathologic response was observed in 100% of dMMR tumors. Nineteen patients had a major pathologic response (mPR), defined as $\leq$10% residual viable tumor in the surgical specimen.

Of those 19 patients, 12 patients had a pathologic complete response (pCR). One patient with 18% residual viable tumor was defined as having a partial response (Chalabi et al., 2020).

Two ongoing large phase III trials are evaluating the addition of either PD-1 blockade (pembrolizumab, NCT02912559) or PD-L1 blockade (avelumab, NCT03827044) to standard of care adjuvant chemotherapy in stage III dMMR/MSI-H colon cancers. Circulating tumor DNA (ctDNA) as a marker for recurrence of disease and response to ICIs in the adjuvant setting is also under evaluation in both phase I and phase II studies (NCT0383359, NCT03803553). Combinatorial approaches with chemotherapy and radiotherapy, oncolytic viruses, vaccines, and novel ICIs represent strategies that may be employed to improve responses to ICIs. Select trial results, as well as ongoing and/or planned trials utilizing ICIs in dMMR/MSI-H CRC are outlined in Tables 2 and 3.

4.2 Biomarkers of response

While it is clear that dMMR/MSI-H is a predictive biomarker for response to treatment with ICIs, a proportion of patients with dMMR/MSI-H tumors do not benefit from treatment with these agents. As the use of ICIs has rapidly multiplied, improved predictors of response to these agents remain an area of continued focus and great need.

Both TMB and PD-L1 expression have shown promise as potential markers of response to ICIs in several malignancies (Cristescu et al., 2018; Garon et al., 2015; Panda et al., 2017; Samstein et al., 2019). However, this is not consistently seen in CRC, either within MSI-H or MSS tumors. In the updated analysis of KEYNOTE-016, 12 patients demonstrated primary refractory disease to treatment with pembrolizumab. In an analysis including three of these refractory patients, there was no significant difference between the average number of nonsynonymous mutations between the responders and the non-responders (1644 vs 1413, respectively, $P=0.67$) (Le et al., 2017). In contrast, in a series of 22 patients with dMMR CRC, mutational burden had the strongest association with ORR and PFS by multivariate and univariate analysis (Schrock et al., 2019).

While the immune contexture influences prognosis, it has not been consistently utilized as predictive marker for response to ICIs. In a small series, densities of $CD3^+$ and $CD8^+$ T-cell densities at the CT and IM were found to be higher in dMMR CRCs that had benefited from PD-1 blockade, though there was considerable variability in this small cohort (Chakrabarti

Table 2 Selected trial results in dMMR/MSI-H CRC.

Name; NCT number	Phase	Treatment	Setting	Primary Endpoints	ORR	PFS	OS
KEYNOTE-016; NCT01876511	II	Pembrolizumab	Treatment refractory metastatic cancers – Cohort A: dMMR CRC – Cohort B: pMMR CRC – Cohort C: dMMR non-CRC	Immune related (IR) ORR and IR- PFS	Cohort A: 40% Cohort B: 0% Cohort C: 71%	IR-PFS at 20 weeks: Cohort A: 78% Cohort B: 11% Cohort C: 67%	Median OS not reached in cohorts A and C; median OS of 5 months in cohort B
KEYNOTE-016; NCT01876511 Update	II	Pembrolizumab	Treatment refractory metastatic dMMR cancers – Cohort A: CRC – Cohort B: non-CRC	ORR and PFS	Cohort A: 52% Cohort B: 54%	Median PFS not reached in either cohort	Median OS not reached in either cohort
KEYNOTE-164; NCT02460198	II	Pembrolizumab	Treatment refractory dMMR/MSI-H mCRC – Cohort A: ≥2 prior lines of therapy – Cohort B: ≥1 prior line of therapy	Blinded independent central review (BICR)-ORR	Cohort A: 33% Cohort B: 33%	Estimated PFS Cohort A: 12 month: 34%, 24 month: 31% Cohort B: 12 month: 41%, 24 month: 37%	Estimated OS Cohort A: 12 month: 72%, 24 month: 55% Cohort B: 12 month: 76%, 24 month: 63%
CheckMate 142; NCT02060188	II	Nivolumab	Treatment refractory dMMR/MSI-H mCRC (85% ≥ 2 lines prior therapy)	Investigator assessed (IA)-ORR	IA-ORR: 31%	At 12 months, IA-PFS of 50%	At 12 months, IA-OS of 73%

Continued

Table 2 Selected trial results in dMMR/MSI-H CRC.—cont'd

Name; NCT number	Phase	Treatment	Setting	Primary Endpoints	ORR	PFS	OS
	II	Nivolumab and ipilimumab → nivolumab	Treatment refractory dMMR/MSI-H mCRC (76% ≥ 2 lines prior therapy)—Nivolumab 3 mg/kg and ipilimumab 1 mg/kg × 4 doses → nivolumab 3 mg/kg every 2 weeks	IA-ORR	IA-ORR: 55%	At 12 months, PFS rate of 71%, median PFS not reached	At 12 months, OS rate of 85%, median OS not reached
	II	Nivolumab and Ipilimumab	Treatment naïve dMMR/MSI-H mCRC—Nivolumab 3 mg/kg every two weeks and ipilimumab 1 mg/kg every 6 weeks	IA-ORR	At a median follow-up of 29 months, IA-ORR: 69%	At 24 months, PFS rate of 74%, median PFS not reached	At 24 months, OS rate of 79%, median OS not reached
KEYNOTE-177 NCT02563002	III	Pembrolizumab vs investigator's choice chemotherapy	Treatment naive dMMR/MSI-H mCRC—Experimental Arm A: pembrolizumab—Active Comparator Arm B: mFOLFOX6 or FOLFIRI ±bevacizumab or cetuximab	PFS, OS	Arm A: 43.8% Arm B: 33.1%	At a median follow-up of 32.4 months, median PFS: -Arm A: 16.5 months-Arm B: 8.2 months	OS data not yet mature
NICHE; NCT03026140	II	Nivolumab and ipilimumab	Resectable (Stage I, II, III) dMMR and pMMR CRC—Nivolumab 3 mg/kg (day 1 and day 15) and ipilimumab 1 m/kg (day 1) → surgery	Safety & feasibility	Pathologic response rate: dMMR: 100% pMMR: 27%	not reported	not reported

Table 3 Selected ongoing trials in dMMR/MSI-H CRC.

Name; NCT number	Phase	Design and cohort	Primary endpoint
Metastatic			
CheckMate 8HW; NCT04008030	III	dMMR/MSI-H mCRC irrespective of prior treatment status	PFS assessed by BICR
		– Experimental Arm A: Nivolumab	
		– Experimental Arm B: Nivolumab + Ipilimumab	
		– Active Comparator Arm C: Investigator's choice chemotherapy	
COMMIT SWOG1610; NCT02997228	III	Treatment naive dMMR/MSI-H mCRC	PFS
		– Active Comparator Arm A: mFOLFOX6 + Bevacizumab	
		– Experimental Arm B: Atezolizumab	
		– Experimental Arm C: mFOLFOX6 + Bevacizumab + Atezolizumab	
SAMCO; NCT03186326	II	dMMR/MSI-H mCRC with progression after standard 1st line chemotherapy ± targeted therapy	PFS
		– Active Comparator: Arm A: Standard Chemotherapy (FOLOFX or FOLFIRI ± targeted therapy)	
		– Experimental Arm B: Avelumab 10 mg/kg every 2 weeks	
NCT04001101	II	Previously treated metastatic dMMR/MSI-H solid tumors, including CRC	Out of field ORR
		– Active Comparator Arm: Radiation and Pembrolizumab	
		– Placebo Arm: Pembrolizumab	
NCT03607890	II	Metastatic dMMR/MSI-H cancers resistant to prior PD-(L)1 inhibitor	ORR
		– Experimental Cohort 1 (co-administered): Nivolumab (480 mg) and relatlimab (160 mg)	
		– Experimental Cohort 2 (sequential administration): Nivolumab (480 mg) and relatlimab (960 mg)	

Continued

Table 3 Selected ongoing trials in dMMR/MSI-H CRC.—cont'd

Name; NCT number	Phase	Design and cohort	Primary endpoint
ATAPEMBRO; NCT04014530	I/II	dMMR, pMMR mCRC or dMMR Endometrial Carcinoma (EC)	Treatment-emergent adverse event, MTD of Ataluren, ORR
		– Experimental Phase I: dMMR or pMMR mCRC, dMMR EC. Dose escalation Ataluren (premature stop codon suppressor) to determine maximally tolerated dose (MTD) and Pembrolizumab every 3 weeks	
		– Experimental Phase II: dMMR mCRC or EC. Ataluren at MTD and Pembrolizumab every 3 weeks	
		– Experimental Phase II: pMMR mCRC. Ataluren at MTD and Pembrolizumab Q3W	
Garnet; NCT02715284	I	Previously treated advanced solid tumors $\geq$ Stage IIIB, including dMMR/MSI-H CRC	ORR, several cohort specific primary outcomes
		– Experimental Part 2B, Cohort F: non-EC dMMR/MSI-H & POLE-Mutant cancers. Dostarlimab (anti-pd-1 antibody) 500 mg every 3 weeks $\times$ 4 cycles $\rightarrow$ 1000 mg every 6 weeks	
NCT03436563	I/II	Previously treated dMMR/MSI-H CRC, CMS4 CRC, dMMR/MSI-H non-CRC, MSS CRC	ORR, clearance of ctDNA
		– Experimental Arm: M7824 (anti-PD-L1/TGFbetaRII fusion protein)	
QUILT-3.055; NCT032228667	II	Previously treated dMMR/MSI tumors with prior treatment with PD-1/PD-L1	ORR
		– Intervention: N-803 (IL-15 super agonist) and ICI (anti-PD-1 or PD-L1)	
NCT04118933	II	Previously treated metastatic dMMR/MSI-H CRC	ORR
		– Intervention: JS001 (anti-PD-1 antibody)	
NCT02983578	II	Advanced and Refractory Pancreatic, NSCLC, and dMMR CRC	adverse events, correlatives
		– Intervention: Danvatirsen (STAT3 inhibitor) and durvalumab	

NCT03104439	II	Previously treated CRC and pancreatic adenocarcinoma, including MSI-H CRC	DCR
		– Experimental: Nivolumab, ipilimumab, and radiation therapy	
Adjuvant			
ATOMIC Alliance A021502; NCT02912559	III	Resected stage III dMMR/MSI-H CRC	DFS
		– Experimental Arm: mFOLFOX6 and atezolizumab	
		– Active Comparator Arm II: mFOLFOX6	
NCT03832569	I/Pilot	MSI-H/dMMR solid tumors following adjuvant therapy with positive ctDNA	Proportion of patients with ctDNA clearance
		– Intervention: Pembrolizumab	
NCT03803552	III	Stage III CRC following adjuvant chemotherapy	DFS, clearance rate of ctDNA
		– Experimental: ctDNA positive MSI-H CRC treated with nivolumab	
POLEM; NCT03827044	III	Stage III dMMR or POLE exonuclease domain mutant CRC	3 year DFS
		– Experimental Arm: Avelumab for 6 months + Fluoropyrimidine-based adjuvant therapy (± oxaliplatin)	
		– Comparator Arm: Fluoropyrimidine-based adjuvant therapy (± oxaliplatin)	
Neoadjuvant			
NCT04082572	II	MSI-H/dMMR locally advanced solid cancers, including CRC	Safety of neoadjuvant pembrolizumab, rate of pathologic complete response (pCR)
		– Experimental: Treatment Pembrolizumab	

Continued

Table 3 Selected ongoing trials in dMMR/MSI-H CRC.—cont'd

Name; NCT number	Phase	Design and cohort	Primary endpoint
NCT04165772	II	Stage II or Stage III dMMR/MSI-H rectal adenocarcinoma.	pCR or complete clinical response at 12 months
		– Experimental: TSR-042 (anti-PD-1 antibody) upto 6 months → standard chemoradiation and total mesorectal excision (TME) → TSR-042	
NCT03926338	I/II	Resectable dMMR/MSI-H CRC	rate of treatment-related surgery delay, rate of grade 3 or higher treatment related adverse events
		– Experimental Arm A: Toripalimab (anti-PD-1 antibody) and celecoxib → colectomy	
		– Experimental Arm B: Toripalimab → colectomy	
NCT04304209	II/III	Clinical stage cT3-4N0M0 or cTxN+M0 dMMR/MSI-H rectal cancer	pCR
		– Experimental cohort A: Neoadjuvant Sintilimab (anti-PD-1 antibody) → curative surgery or watch and wait → Sintilimab ± Capeox according to pathologic response	
		– Experimental cohort B, Arm 1: Neoadjuvant sintilimab, capeox and concurrent radiotherapy, → curative surgery or watch and wait → adjuvant Capeox	
		– Experimental cohort B, Arm 2: 4 cycles neoadjuvant concurrent chemoradiation therapy with capeox, → curative surgery or watch and wait → adjuvant Capeox chemotherapy	
EA2201; NCT not assigned	II	dMMR/MSI-H T3-4Nx or TxN+ rectal cancer	pCR
		– Intervention: Nivolumab and ipilimumab q28 days x 2 cycles → short course radiation → 2 additional cycles nivolumab and ipilimumab	

et al., 2019). In an upcoming phase II trial, patients with pMMR metastatic CRC and high immune infiltrates will receive PD-1 blockade and chemotherapy (NCT04262687). Immune infiltration will be prospectively evaluated, providing additional insight into the predictive value of immune infiltrate and response to ICI. Both germline (Lynch) and sporadic dMMR/MSI-H tumors show robust immune infiltration, and no significant difference in response has been seen between these two groups when treated with anti-PD-1 therapy (Le et al., 2017).

The role of KRAS status as a predictor of response to immunotherapy also requires further characterization. In a subgroup analysis from KEYNOTE-177, the relative benefit of PD-1 blockade was most pronounced in RAS wildtype tumors (HR 0.44) as opposed to KRAS/NRAS mutated tumors (HR 1.19). The significance of this finding is difficult to interpret as RAS mutation status was unavailable in 30% of the cohort, and the overall number of patients with KRAS/NRAS mutations in this study was small (Andre et al., 2020). In contrast, in the CheckMate 142 and KEYNOTE-164 studies, ORR and DCR was independent of KRAS/NRAS mutated disease. The tumor immune microenvironment of KRAS mutated CRC has been previously characterized as poorly immunogenic, indicating that the impact of KRAS status in response to ICIs is likely modulated by additional factors.

5. Immunotherapy in microsatellite stable (MSS) colorectal cancers

5.1 Immune checkpoint inhibitors (PD-1 ± CTLA-4)

Multiple studies have looked at the use of ICIs in advanced MSS CRC. One arm of the aforementioned multi-cohort KEYNOTE-028 trial enrolled 23 patients with PD-L1 positive advanced CRC and treated them with pembrolizumab (O'Neil et al., 2017). Ninety-six percent of the patients had MSS tumors. The only clinical response or durable control witnessed was in the lone patient with an MSI-H CRC. While two additional patients experienced slight decreases in tumor size, the effect was not prolonged, such that the median PFS was 1.8 months and median survival 5.3 months. This data is consistent with Keynote-016, where no responses were observed among the patients with MSS tumors (Le et al., 2015). These data also highlight the lack of effect for this class of drugs among unselected patients with advanced MSS CRCs. As previously mentioned, CTLA-4 inhibition as monotherapy similarly failed to produce any signal of meaningful activity (Chung et al., 2010).

On the basis that combination therapy with dual CTLA-4 and PD-1 blockade is superior to either PD-1 or CTLA-4 monotherapy in some tumor types, investigation of this combination has been carried out in MSS tumors (Larkin et al., 2019). One unpublished early readout of Checkmate 142 described the efficacy of two dose combinations of ipilimumab and nivolumab in MSS CRC cohorts (Overman et al., 2016). Neither dosing combination proved compelling, though one patient achieved a response. The combination of PD-L1 and CTLA-4 inhibition was later explored via use of durvalumab and tremelimumab in the CCTG CO.26 study (Chen et al., 2020). In this randomized phase II study, patients received either combination ICI therapy or best supportive care. Interestingly, the study met its primary endpoint, improving survival per the pre-planned analysis (HR = 0.72, P = 0.07). While disease control was low, it appeared higher in the combined ICI arm (23% vs 7%); however, responses were uncommon overall and the median PFS was no different. The combination has yet to be further explored. Despite the 'positive' result, activity does not appear generalizable, and thus, the clinical significance is dubious.

In MSS CRC, TMB has gained some interest as a potential biomarker of benefit for ICI therapy. Just as it is recognized that dMMR/MSI-H tumors were characterized by very high mutational rates, there is recognition that a small portion of non-MSI-H tumors possessed significantly elevated rates of mutations, even beyond POLE mutant CRCs. In one analysis, 2.9% of MSS CRCs were categorized as TMB-H, as defined by a TMB of at least 11.7 muts/Mb, inclusive of synonymous and non-synonymous mutations. Approximately 21% of these cases had a detectable POLE mutation, without any POLD1 mutations detected (Fabrizio et al., 2018). Distinct from other MSS TMB-H tumors, POLE mutated tumors have a markedly increased mutational burden of greater than 100 muts/Mb, with a mean of 200–220 muts/Mb (Campbell et al., 2017; Hu et al., 2020). From initial case reports, POLE mutant cancers, including CRCs responded exceptionally well to PD-1 therapy (Mehnert et al., 2016; Silberman et al., 2019; Wang, Gong, Tu, Lee, & Fakih, 2018). A recent multi-national study enrolled 16 patients with pMMR POLE mutated cancers, including seven CRCs, and treated them with nivolumab monotherapy. Six (38%) of the patients achieved a response, including 5/7 (71%) with CRC. Ten (62%) achieved disease control (stable disease or response); the benefit was restricted to patients with mutations established to be pathogenic or of unknown significance. All five patients with non-pathogenic mutations exhibited progressive disease (Benoit, ESMO 2020).

Non-POLE mutated TMB-H MSS CRCs are rare and represent a heterogeneous population, with TMB being a continuum rather than there being a clear threshold. In this context, there appears to be a rare subset of patients with TMB-H MSS CRCs with sufficient neoantigen burden to derive benefit from PD-1 monotherapy. Optimal identification methods remain unknown. The TAPUR study enrolled patients with mCRC and a TMB of ≥9 (93% via Foundation One) and treated them with pembrolizumab (Meiri et al., 2020). Three (11%) responses were seen, with seven of the 27 (26%) patients benefiting from treatment for at least 6 months. However, of the three patients that received 1 year of treatment or greater, two had TMBs of 41 and 54.3 muts/Mb, with the third having a TMB of 17 muts/Mb. This contrasts to the majority of enrolled patients having TMBs in the range of 10–11 muts/Mb, and raises the possibility that these were, in fact, MSI-H patients, otherwise misclassified. Accuracy in the classification of dMMR/MSI status, as it pertains to ICI benefit, has been previously cited as a small but significant issue (Cohen et al., 2019). Of importance when considering TMB, the multi-cohort KN-158 study, which evaluated use of pembrolizumab and retrospectively assessed TMB status and correlation to response, did not enroll any patients with CRC. Thus, while pembrolizumab is now approved for TMB-H tumors (TMB ≥ 10 by Foundation One CDx), supportive data in the setting of a CRC diagnosis is very limited. Thus, there is muted enthusiasm for the treatment of non-POLE mutated, non-MSI-H CRCs with approved ICIs.

While metastatic studies have been underwhelming in MSS CRC, recent interest has increased in utilizing the neoadjuvant space for therapeutic evaluation in early stage CRC. Responses to monotherapy or combination therapy remain infrequent, but are detected in MSS CRC. The NICOLE study utilized two doses of neoadjuvant nivolumab monotherapy (NCT04123925). From early results, of the 18 pMMR/MSS tumors included, two demonstrated a mPR, one of these being a pCR. Compared to control cohorts, tumors that were exposed to nivolumab exhibited greater intra-tumoral CTL densities within and a lower proportion of CTLs at the tumor: stromal interface as opposed to within the tumor (Avallone et al., 2020). As mentioned earlier, in the similarly designed NICHE study, preoperative nivolumab and ipilimumab induced pathologic response in 4/15 (27%) pMMR colon cancers, with achievement of two pCR and one mPR (Chalabi et al., 2020). Baseline PD-1+ CTL density was the sole biomarker which predicted response in the early stage pMMR CRCs. Neoadjuvant studies may help to greatly enhance our understanding of the TME interplay with both standard therapies and immunotherapeutics. A potential shortfall of this approach is that the TME of primary tumors and metastatic sites can differ quite significantly. The extent to which these treatments alter the TME in the context of metastatic disease may vary widely and thus could limit effective translation to the advanced disease setting. Table 4 presents results from select studies in MSS CRC.

5.2 Hypothesized mechanisms of resistance

Patients treated with immunotherapeutic agents generally fall into one of three categories: (1) those in whom response is achieved and ongoing (responders), (2) those in whom the tumor never responds (innate resistance), and (3) those in whom response is achieved, followed by progression (acquired resistance). Uniform definitions of resistance are of critical importance in reaching optimal understanding and eventually overcoming these differing forms of resistance. The Immunotherapy Resistance Taskforce, formed by the SITC, recently made recommendations surrounding optimal parameters for defining primary (innate) vs secondary (acquired resistance) in the clinical setting, to serve as a framework for future trial planning and interpretation (Kluger et al., 2020).

At this time, the primary characteristics that determine immunoresponsiveness are incompletely understood. Inefficient T-cell generation, inadequate T-cell effector function, and impaired formation of T-cell memory have been proposed as three broad categories responsible for failure of these drugs (Jenkins, Barbie, & Flaherty, 2018). While a paucity of preexisting immune response may be a primary driver of innate resistance, other factors that drive resistance may otherwise see considerable overlap between innate and acquired resistance.

Inefficient T-cell generation would seem to be one of the major factors driving innate resistance to ICIs. Such a state may be due to tumor intrinsic or tumor extrinsic factors. The lesser presence of key antigen presenting cells, including DCs, has been described in CRC. The density of DCs in primary tumors is below that seen within normal colonic mucosa, and DCs are virtually absent in metastases (Schwaab, Weiss, Schned, & Barth, 2001). Suppressive cytokines (such as CCL2, CXCL1, CXCL5 and VEGF) may hamper DC maturation and anti-tumor function, thus preventing the formation of an adaptive anti-tumor immune response (Michielsen et al., 2011). Tumor-derived prostaglandin E2 (PGE_2) has been linked both to DC presence and quality as well as $CD8^+$ CTL survival and function

Table 4 Selected Trial Results in pMMR/MSS CRC.

Name; NCT number	Phase	Treatment	Setting	Primary Endpoints	ORR	PFS	OS
KEYNOTE-028; NCT02054806	I	Pembrolizumab	Treatment refractory PD-L1 tumors—mCRC Cohort ($n=23$)	ORR	All patients: 4% MSS tumors: 0%	1.8 months	5.3 months
NCT00313794	II	Tremelimumab	Treatment refractory mCRC ($n=49$)	ORR	2%	2.3 months	4.8 months
CheckMate 142; NCT02060188	II	Nivolumab + ipilimumab	Treatment refractory pMMR/MSS mCRC cohorts: Cohort 1: Nivo 1 mg/kg + Ipi 3 mg/kg → Nivo 3 mg/kg ($n=10$) Cohort 2: Nivo 3 mg/kg + Ipi 1 mg/kg → Nivo 3 mg/kg ($n=10$)	ORR, Safety	Cohort 1: 10% Cohort 2: 0%	Cohort 1: 2.3 months Cohort 2: 1.3 months	Cohort 1: 11.5 months Cohort 2: 3.7 months
CCTG CO.26	II	Arm 1: Durvalumab + tremelimumab Arm 2: Best supportive care	Refractory mCRC ($n=180$)	OS	Arm 1: 2% Arm 2: 0%	Arm 1: 1.8 months Arm 2: 1.9 months	Arm 1: 6.6 months Arm 2: 4.1 months (HR 0.72)
TAPUR; NCT02693535	II	Pembrolizumab	Refractory advanced cancer Cohort: TMB-H mCRC ($n=27$)	DCR	11%	9.3 weeks	51.9 months
COTEZO Imblaze370; NCT02060188	III	Arm A: Atzeolizumab + cobimetinib Arm B: Atezolizumab Arm C: Regorafenib	Treatment refractory mCRC ($n=363$)	OS	Arm A: 3% Arm B: 2% Arm C: 2%	Arm A: 1.9 months Arm B: 1.9 months Arm C: 2 months	Arm A: 8.9 months Arm B: 7.1 months Arm C: 8.5 months

(Chen et al., 2015; Zelenay et al., 2015). A possible avenue to reverse this defect is the use of cyclooxygenase (COX) or PGE_2 receptor 4 (EP4) inhibition; both potentiate PD-1 inhibition in preclinical models (Lu et al., 2020; Zelenay et al., 2015).

While the role of neoantigen quantity and quality has been discussed, the processing and presentation of neoantigens merits further attention. Tumor genetic and epigenetic modifications which alter the landscape of tumor neoantigens are hypothesized drivers of resistance at multiple levels. In CRC, mutations in antigen presenting machinery (APM) genes, *TAP1*, *TAP2*, and beta-2-microglobulin (*B2M*), are seen frequently, particularly so in MSI-H tumors (Kloor et al., 2005). Epigenetic mechanisms, such as hypermethylation-induced silencing are known to additionally downregulate the transporter associated with antigen processing (TAP) proteins 1 and 2 (Snahnicanova et al., 2020). The loss of both *TAP1* and *TAP2* has been linked to loss of MHC I antigen expression. Downregulation of MHC Class I expression has been linked to poor prognosis in CRC (Watson et al., 2006). On the other hand, an increased density of CTLs is seen in *TAP1*, *TAP2* and MHC I positive CRC cases, suggesting that the absent function of key APM components may prevent TIL infiltration (Kasajima et al., 2010). Looking one step further, the knockout of *B2M* in a lung cancer mouse model confers resistance to PD-1 blockade in an otherwise sensitive tumor (Gettinger et al., 2017). Still, the situation remains complicated, particularly in MSI-H tumors. While the loss of *B2M* has been described in both innate and acquired resistance, other reports have documented that a sizeable percentage of MSI-H tumors (24%), still obtain benefit despite the presence of this frequent alteration (Gurjao et al., 2019; Middha et al., 2019; Zaretsky et al., 2016).

Clearly, neoantigen burden and antigen presenting machinery are not the sole factors responsible for determining response to ICI. A recent meta-analysis of PD-1/L1 inhibitor monotherapy studies suggested that TMB accounts for 55% of the difference in response across tumor types, leaving a large amount of inter-tumor variability without explanation (Yarchoan et al., 2017). The decreased responsiveness seen in MSS CRCs, compared to other tumor types with similar spectra of mutational burden, suggests the presence of additional inhibitory factors. One intrinsic genomic contributor is Wnt/β-catenin signaling, a broadly activated pathway in CRC which has been strongly linked to immune exclusion across cancers (Luke, Bao, Sweis, Spranger, & Gajewski, 2019). Canonical Wnt/β-catenin signaling is initiated by the binding of a key family protein to cell surface receptors, resulting in translocation of β-catenin and transcriptional activation. A TCGA analysis demonstrated a link between activated signaling and an absence of T-cell infiltration (Grasso et al., 2018). Components of the pathway may also influence the infiltration of DCs (Pacella et al., 2018). Of therapeutic relevance, in multiple pre-clinical models the inhibition of signaling augments dendritic and $CD8^+$ T-cell infiltration and increases tumor clearance (Pacella et al., 2018; Wang et al., 2020).

Studies have described deficiencies in signaling pathways in PD-1 resistant tumors, such as JAK/STAT or IFN-γ, which can hamper T-cell activation and engagement. In one analysis of four patients with melanoma with acquired resistance (response, followed by progression months later), tumor biopsies at progression demonstrated mutations in interferon-receptor associated Janus kinase 1 (JAK1) or Janus kinase 2 (JAK2) genes concurrent with loss of heterozygosity (LOH) of the wild-type allele in two of the patients (Zaretsky et al., 2016). In vitro, these mutations confer a lack of response to IFN-γ, along with repressed PD-L1 and MHC class I expression. JAK1/2 mutations have also been identified in the setting of primary resistance to PD-1 therapy, including in dMMR colon cancer (Shin et al., 2017). In the same vein, melanomas with genomic defects in the IFN-γ signaling pathway demonstrate resistance to CTLA-4 therapy (Gao et al., 2016).

Tumor extrinsic factors may also contribute significantly to the efficacy of immunotherapeutic strategies, including the presence of inhibitory cytokines (e.g., IL-6, IL-10, TGF-β) or inhibitory cell populations (e.g., T-regs, M2-macrophages, MDSCs). The latter may both exhibit direct inhibition, but also propagate the production of suppressive cytokines in the tumor microenvironment. Clinical data specific to CRC and differential immunotherapeutic effect is limited. However, in gastric cancer and lung cancer, the frequency of PD-1 + CTLs relative to PD-1 + Tregs predicts the efficacy of PD-1 blockade, with PD-1 inhibition augmenting PD-1+ Treg cell-mediated suppressive function. While overall macrophage density is associated with improved outcome, high M2 macrophage density confers a worse prognosis (Yang et al., 2020). The PGE_2 receptor 4 (EP4), has been identified as a master regulator of immunosuppressive myeloid cells, promoting M2 macrophage polarization. Again, EP4 antagonism appears to be a potent immunosensitizer (Lu et al., 2020).

An additional source of resistance, which may be either tumor extrinsic or intrinsic, is the baseline expression or inducible upregulation of additional immune checkpoints. MSI-H colon cancers have been characterized to upregulate multiple immune checkpoints, including PD-1, PD-L1,

CTLA-4, LAG-3, and indolamine 2′3′-dioxygenase (IDO-1) (Llosa et al., 2015). Clinical data for PD-1 and CTLA-4 has been discussed and is currently being explored, but other checkpoints also are of interest. In a murine model of CRC (MC-38), PD-L1 blockade induces $CD4^+$ and $CD8^+$ T-cell subsets, which frequently express LAG-3 and PD-1 and upregulated the NKG2A receptor. Dual LAG-3 and PD-1 inhibition enhanced survival in this model, highlighting the potential importance of this and other immune checkpoints (Beyrend et al., 2019). In clinical settings, early data has demonstrated that the strategy of LAG-3 inhibition may hold promise, as 33% of patients in a small study of PD-1 refractory melanoma benefited from combined LAG-3 and PD-1 inhibition (Atkinson et al., 2020).

5.3 Immunotherapeutic combinatorial approaches

5.3.1 Combinations with chemotherapy

As ICIs have entered the clinical arena, interest in how these might combine with chemotherapy, both for good and for bad, has arisen. While one school of thought has long supposed that chemotherapy, due to its immunosuppressive influence, would negatively affect the impact of ICIs, other preclinical data has suggested a net positive impact with certain combinations. The anti-metabolites 5-FU and gemcitabine are described to eliminate the suppressive myeloid population with selectivity (Le et al., 2009; Vincent et al., 2010). Oxaliplatin has also been reported to suppress this population of cells; in contrast, in other studies, irinotecan prevented MDSC apoptosis, rather than invoking expansion of MDSCs (Kanterman et al., 2014; Kim & Kim, 2019).

Immunogenic cell death (ICD) refers to a distinct form of cell death that facilitates T-cell responses against unearthed TSAs. ICD is classically defined as: (1) occurring in immunocompetent hosts, though largely absent in immunocompromised animals, and (2) producing a situation where the susceptible treated cancer cells are capable of functioning as a vaccine in another syngeneic model, which is then challenged with untreated cells of the same origin (Garg et al., 2017). To this end, oxaliplatin with or without TAS-102 has been suggested to induce ICD in preclinical models (Limagne et al., 2019; Tesniere et al., 2010). 5-FU, alone or as part of a combination, also increases $CD8^+$ T-cells at tumor sites (Taniura et al., 2020).

In multiple clinical trials, increased densities of intratumoral T-cells are detected following the use of standard chemotherapeutics. EORTC 40983 compared patients who were randomized to receive 3 months of preoperative FOLFOX to those who underwent hepatic metastatectomies alone. Patients who received FOLFOX had higher $CD3^+$ lymphocytes (1.85 ×) and $CD117^+$ mast cells (4.8 ×) within the hepatic tumors, as well as lower macrophage counts within the normal hepatic tissue and at the tumor-liver interface (Tanis et al., 2015). Consistent with this, an analysis of the OSLO-COMET study demonstrated that a short time between neo-adjuvant chemotherapy and hepatic resection was associated with increased T-cell density within the tumor, while the choice of regimen (oxaliplatin vs irinotecan-based) and achievement of response were not associated with density (Dagenborg et al., 2020). The full biologic impact of chemotherapy associated changes in the microenvironment (pro- vs anti- tumor) and the degree to which an anti-tumor immune response is invoked remains incompletely understood.

To date, several trials adding ICIs to chemotherapy have been reported. However, there is not a clear sign of improved outcomes. FOLFOX combined with either pembrolizumab or a combination of atezolizumab and bevacizumab yielded similar response rates to that expected with FOLFOX alone: 53% and 44%, respectively (Bendell et al., 2015; Shahda et al., 2017). Despite this, there remains interest in evaluating combinations with chemotherapy (with and without angiogenic therapies), particularly in the metastatic setting. The MEDETREME study is combining durvalumab, tremelimumab and 3 months of FOLFOX in patients with RAS-mutated MSS tumors. Phase Ib study results noted an acceptable safety profile, with five complete responses (25%) and an estimated 12-month PFS rate of 50% (Fumet JD ASCO 2020). These results are encouraging, though extremely preliminary. Similar studies are ongoing with PD-1/L1 inhibition and triplet chemotherapy. The AtezoTRIBE study will add atezolizumab to FOLFOXIRI and bevacizumab, while the NIVACOR study will evaluate nivolumab with FOLFOXIRI + bevacizumab in RAS/BRAF mutated patients with metastatic disease (NCT03721653, NCT04072198).

5.3.2 Combinations with anti-VEGF/R therapy

Tumor hypoxia is closely linked to angiogenesis, as well as recruitment and activation of immunosuppressive cellular populations, namely Tregs and MDSCs (Deng et al., 2019; Facciabene et al., 2011). On the other hand, suppression of vascular endothelial growth factor (VEGF)-receptor (VEGFR) axis signaling through use of either a tyrosine kinase inhibitor (TKI)-sunitinib or a mAb reduces Treg populations in syngeneic models. Interestingly, patients with metastatic CRC have higher percentages of peripheral blood Tregs ($CD25^+$ $Foxp3^+$ $CD127^{lo}$ in $CD4^+$ cells) than healthy volunteers and see a significant decrease in levels when bevacizumab is added to chemotherapy (Terme et al., 2013). In a syngeneic CRC model, the multikinase TKI, regorafenib, demonstrated potent reduction in tumor vascularization, inhibited tumor growth, and markedly reduced macrophage infiltration. When compared to a murine VEGFR2 mAb (DC101), the TKI specifically and significantly reduced quantities of $TIE2^+$ TAMs, a strongly proangiogenic, inflammatory cell subset, which is induced with use of the mAb (Abou-Elkacem et al., 2013). Combinations of PD-1 inhibitors with VEGFR-inhibiting multikinase TKIs have demonstrated synergistic tumor-suppressive activity in murine CRC models (Wang et al., 2020).

PD-1/L1 and VEGF/VEGFR inhibitor combinations have seen approval in several disease types, though it is unknown whether the activity represents synergy or is simply an additive effect (Finn et al., 2020; Rini et al., 2019). Regardless, given the pre-clinical data and standard use of VEGF/VEGFR targeting therapies in metastatic CRC, the combination of ICI and angiogenic signaling blockade has garnered significant clinical attention. The addition of atezolizumab to a combination of bevacizumab and capecitabine has been evaluated in two studies, MODUL and BACCI. In MODUL, when atezolizumab was added to maintenance chemotherapy in the first line, it conferred no significant PFS advantage, though there were small numerical advantages for ORR, DCR, and DOR (Grothey et al., 2018). In the refractory setting, the BACCI trial demonstrated that the addition of atezolizumab to bevacizumab and capecitabine provided a small PFS benefit (HR 0.725, $P=0.051$), with a trend toward higher response rate (8.54% vs 4.35%, $P=0.5$) (Mettu NB, ESMO 2019). Still, in both trials, any positive effect was rather limited, and without biomarkers or other means to improve patient selection, broad use of these interventions would not yield reliable clinically significant benefit.

Preliminary clinical data has also supported a possible role for VEGFR-targeting TKIs in combination with ICIs. The REGONIVO (EPOC1603) study treated a cohort of 25 patients with mCRC with the combination of regorafenib and nivolumab, 24 of whom were pMMR or MSS. Among the pMMR patients, the ORR was 33.3% with eight responses, with the majority ongoing at the time of reporting. Interestingly, patients with lung metastases appeared more likely to respond than those with liver metastases (Fukuoka et al., 2020). Since this report, a few series have been reported which have failed to replicate these results (Li et al., 2020; Wang et al., 2020). Additional confirmatory studies are underway (NCT04126733, NCT03657641). Similarly, lenvantinib, a VEGFR multikinase TKI, is being evaluated in combination with PD-1 blockade in multiple tumor types (Taylor et al., 2020). Preliminary data from this combination also suggests clinical activity, with 22% (7/32) of patients with mCRC achieving an objective response (Lwin et al., 2020).

5.3.3 Combinations with RAS pathway inhibition

The RAS-RAF-MAPK signaling pathway is well established to play a central role in the transformation of CRCs, as well as driving cellular proliferation, survival and progression. Gene expression analysis suggest that anti-tumor Th1/cytotoxic immunity is suppressed in KRAS mutated colon cancer, with the effect modulated by CMS subtype; the subset of KRAS mutated CMS2 tumors possesses lower expression of Th1 associated genes, as well as the lowest infiltration of CTLs, neutrophils, and Th1 cells (Lal et al., 2018). Additional research has suggested that KRAS suppresses interferon regulatory factor 2 (IRF2), which ultimately results in CXCR2-driven MDSC migration into the tumor microenvironment (Liao et al., 2019). Preclinical investigation showed that downstream inhibition of this signaling pathway overcame PD-1 resistance in syngeneic mouse models (Liu et al., 2015), sparking considerable interest in moving the combination of the PD-1/L1 and RAS-MAPK pathway inhibition to the clinic. Early data with the MEK inhibitor cobimetinib, in combination with atezolizumab, was promising; however, the subsequent randomized phase III study, IMblaze370, failed to demonstrate benefit (Eng et al., 2019). Only 3% of patients achieved responses with the combination, at the cost of significant toxicity. Despite this, multiple investigations of MEK inhibition remain active, with the focus shifting to combination with chemotherapy or the use dual checkpoint inhibition.

Agents targeting KRAS G12C have recently emerged with encouraging preliminary activity, representing a significant cancer breakthrough. These drugs bind to the mutant cysteine in G12C, which sits directly adjacent to the P2 pocket of the inactive GDP-bound RAS isoform. This interaction prevents GTP binding and transition of KRAS to the active state; in turn this strongly blocks downstream signaling, demonstrated by almost complete pERK inhibition, in a neoplastic-cell specific fashion (Canon et al., 2019). In syngeneic models, KRAS G12C inhibition produced cures in some portion of CT-26 $KRAS^{G12C}$ murine tumors treated with AMG 510, an outcome dependent upon T-cells. When PD-1 inhibition was added to AMG 510, 9/10 (90%) mice achieved lasting complete responses. Immunophenotyping demonstrated a significant increase in CTLs within tumors after AMG 510 monotherapy or with the combination, but not with PD-1 inhibition alone (Canon et al., 2019). Interestingly, this effect was not seen with use of a MEK inhibitor in the same model. Clinical trials are currently evaluating the KRAS G12C and PD-1 inhibitor combination in various tumor types (NCT04185883, NCT03785249).

Within the same pathway, therapies which block signaling at the level of the epidermal growth factor receptor (EGFR), particularly mAbs, have been postulated to have beneficial immunologic impact on the TME. Anti-EGFR mAbs induce a tumor-specific CTL response in mouse models (Garrido et al., 2011). Use of these drugs is linked to more robust $CD8^+$ tumor infiltration in CRC liver metastases over that seen with chemotherapy alone or with anti-VEGF therapy (Van den Eynde et al., 2018). This has prompted several investigations of ICIs with anti-EGFR monotherapy (NCT02713373, NCT04561336, NCT03442569) or as a cocktail with anti-EGFR mAbs and chemotherapy (NCT03174405, NCT03608046, NCT04513951). To date, it is unclear to what degree, if any, the addition of PD-1 or CTLA-4 blockade will improve outcomes in these populations.

5.3.4 Combinations with radiation

Radiation is utilized as standard of care in many patients with rectal cancer, and at times for metastatic disease, as an alternative to resection. Dose and fractionation can differ depending upon the scenario and may have varying immunologic impact. Chemoradiation has been shown in increase $CD8^+$ tumor infiltration in rectal cancers, without increasing Treg density (Shinto et al., 2014). Further, the induction of systemic tumor response beyond the radiation field, the abscopal effect, has long been thought to have an immunologic basis (Demaria et al., 2004). On the other hand, conventional doses of radiation therapy are sufficient to induce cell death among mature T-cells (Heylmann, Rodel, Kindler, & Kaina, 2014). High dose radiation, more akin to stereotactic body radiotherapy (SBRT), induces vigorous $CD8^+$ T-cell tumor infiltration and suppression of MDSCs (Filatenkov et al., 2015). This response is reliant on APCs and, thus, the radiation exposure is likened to serving as an anti-tumor vaccine. PD-L1 inhibition is synergistic with radiation in these pre-clinical models, though dose and schedule seem to matter, highlighting the balance of immune suppression and stimulation conferred by radiation (Deng et al., 2014; Kosinsky et al., 2018).

These data have sparked multiple trials evaluating radiation and immunotherapy combinations. Pembrolizumab and SBRT have been utilized in the pre-operative oligometastatic setting, demonstrating an increase in TILs both within target lesions and within non-radiated lesions (Deming et al., 2020). The clinical impact of this approach remains uncertain, as elucidated by a separate small phase II study which combined SBRT (8 Gy × 3 fractions) with the PD-1 inhibitor, AMP-224, and demonstrated no objective responses (0/17) (Duffy et al., 2016). More recently, investigators combined nivolumab and ipilimumab with SBRT (8 Gy × 3 fractions) in MSS mCRC and responses were seen in four (10%) patients including one complete response, with several patients continuing on study therapy (Parikh et al., 2020). Additional investigation is underway to fully evaluate the activity of this combination (NCT04575922). In the meantime, two important early stage rectal cancer studies are ongoing: NRG-GI002 is evaluating the addition of pembrolizumab to pre-operative chemoradiation, while FR-2 utilizes the PD-L1 inhibitor, durvalumab, following pre-operative chemoradiation (NCT02921256, NCT0310204). The results and long-term follow-up are awaited.

5.3.5 Combinations with bi-specific antibodies

Bispecific antibodies have been the subject of intense investigation, though clinical success has only been established in hematologic malignancies to this point. There are multiple variations in construction of bispecific Ab, with the overarching intent being to link the binding of tumor antigen to an immune effector cell, stimulating anti-tumor immune engagement. Naturally, there is interest in examining the value of these therapies on their own and also in combination with ICIs. CEA-TCB (RG7802; RO6958688; cibisatamab) is an IgG1 bispecific heterodimeric antibody which binds with one arm to the CD3ε chain within the T-cell receptor complex on T-cells and with the other two arms to the tumor cell expressed protein, carcinoembryonic antigen (CEA). Pre-clinical models demonstrated that the simultaneous binding of CEA-TCB to tumor and T-cells leads to the formation of immunologic synapses, T-cell activation, secretion of cytotoxic granules and tumor cell lysis (Bacac et al., 2016). In phase I studies, CEA-TCB was studied both as monotherapy and in combination with the PD-L1 inhibitor, atezolizumab. Activity in preliminary reports appeared greater with the combination, producing two (18%) responses among MSS patients and nine (82%) with disease control (response or stable disease) (Tabernero et al., 2017). Safety concerns seem to have hampered the development, though additional study is ongoing now with the addition of an anti-CD20 mAb, obinutuzumab, as pre-treatment (NCT03866239).

5.3.6 Future directions

Given that one of the major obstacles in MSS colorectal cancer seems to be the lack of a robust innate anti-tumor immune response, treatments meant to directly stimulate an immune response, such as vaccines, seem a natural choice for investigation. To date, vaccine strategies have largely failed in cancer, and multiple agents have not made it past initial testing in CRC, despite pre-clinical support for efficacy (Mocellin, Rossi, Lise, & Nitti, 2004). In the era of ICIs, there is renewed interest to reexamine vaccines alone or combined with ICIs. In similar hope of stimulating an in vivo response, toll-like receptor (TLR) agonists and oncolytic viruses are under active investigation. Finally, multiple investigational studies targeting additional immune checkpoints are in progress or planned, whether intended to provoke additional stimulatory signals (e.g., OX40, CD40, CD137, ICOS, GITR) or to block inhibitory signals (e.g., TIM3, LAG3, TIGIT, VISTA, CD73). A summary of select ongoing studies is presented in Table 5.

Table 5 Selected Ongoing Trials in pMMR/MSS CRC with ICIs.

Name; NCT number	Phase	Design and cohort	Primary endpoint
Chemotherapy ± VEGF inhibition			
METIMMOX; NCT03388190	II	Randomized Phase II in 1st line MSS mCRC 5-FU + LV + Oxaliplatin (FLOT) ± nivolumab	PFS
AtezoTRIBE; NCT03721653	II	Randomized Phase II in 1st line mCRC: FOLFOXIRI + bevacizumab ± atezolizumab	PFS
NIVACOR; NCT04072198	II	1st line RAS mutant or BRAFmut mCRC: FOLFOXIRI + bevacizumab + nivolumab	ORR
MEDITREME NCT03202758	Ib/II	1st line KRAS mutant mCRC: FOLFOX + durvalumab + tremelimumab	Safety PFS
POCHI NCT04262687	II	pMMR/MSS mCRC with high immune score CAPOX + bevacizumab + pembrolizumab	ORR
MAYA NCT03832621	II	Refractory MSS MGMT silenced mCRC Temozolomide + nivolumab + ipilimumab	8 month PFS

Table 5 Selected Ongoing Trials in pMMR/MSS CRC with ICIs.—cont'd

Name; NCT number	Phase	Design and cohort	Primary endpoint
ARETHUSA NCT03519412	II	Multi-cohort: MGMT silenced pMMR refractory mCRC Temozolomide → pembrolizumab (if TMB $\geq$20)	ORR
VEGF/VEGFR inhibition			
NCT04126733	II	Refractory pMMR/MSS mCRC: Regorafenib + nivolumab	ORR
NCT03657641	I/II	Refractory mCRC Regorafenib + pembrolizumab	Toxicity, PFS, OS
REGOMUNE; NCT03475953	II	Refractory pMMR mCRC Regorafenib + avelumab	ORR
NCT04362839	II	Refractory pMMR/MSS mCRC Regorafenib + nivolumab + ipilimumab	Recommended Dose Level
EGFR/RAS/MAPK inhibition			
NCT03442569	II	RAS wt MSS refractory mCRC Panitumumab + nivolumab + ipilimumab	ORR
AVETUXIRI NCT03608046	II	RAS wt refractory MSS mCRC Cetuximab + irinotecan + avelumab	ORR
NCT03475004	II	Refractory mCRC Binimetinib + bevacizumab + pembrolizumab	ORR
KEYNOTE-651; NCT03374254	I	pMMR CRC Cohort A (refractory)—binimetinib + pembrolizumab Cohort C (1st line)—FOLFOX + binimetinib + pembrolizumab Cohort E (2nd line)—FOLFIRI + binimetinib + pembrolizumab	Safety ORR
NCT03271047	I/II	RAS mutant MSS mCRC Binimetinib + nivolumab ± ipilimumab	Safety ORR
NCT03377361	I/II	RAS mutant MSS mCRC Trametinib + nivolumab ± ipilimumab	Safety ORR
NCT04044430	I/II	BRAF V600E mutant MSS mCRC Binimetinib + encorafenib + nivolumab	ORR Safety

Continued

Table 5 Selected Ongoing Trials in pMMR/MSS CRC with ICIs.—cont'd

Name; NCT number	Phase	Design and cohort	Primary endpoint
NCT04017650	I/II	BRAF V600E mutant MSS mCRC Encorafenib + cetuximab + nivolumab	ORR Safety
CodeBreak101 NCT04185883	Ib	KRAS G12C mutant mCRC AMG 510 + anti-PD-1	Safety
Radiation			
NCT04503694	II	Locally advanced rectal cancer Short course RT (5Gy × 5 fx) + regorafenib + nivolumab	pCR rate
NCT02837263	I	pMMR/MSS mCRC (oligometastatic) SBRT + pembrolizumab	1 year recurrence rate
NCT04575922	II	Refractory MSS mCRC RT + nivolumab + ipilimumab	DCR

Other targets		**Target**		
NCT03866239	I	CEA-TCB	Refractory MSS mCRC Obinutuzumab + Cibisatamab + Atezolizumab	Safety ORR
NCT02834052	I/II	TLR3	Refractory pMMR mCRC Poly-ICLC + pembrolizumab	Safety ORR
NCT04119830	II	TLR3	Refractory pMMR/MSS mCRC Rintatolimod + pembrolizumab	ORR
NCT03435640	I/II	TLR7/8 IL-2	mCRC NKTR-262 + NKTR-214 ± nivolumab	Safety ORR
NCT03507699	I	TLR-9	Refractory mCRC CMP-001 + nivolumab + ipilimumab ± SBRT	Safety
NCT04166383	II	Anti-angiogenic	Refractory MSS mCRC VB-111 + nivolumab	Safety ORR
STOPTRAFFIC-1; NCT04599140	I/II	CXCR1 CXCR2	Refractory RAS mutated MSS mCRC SX-682 + nivolumab	Safety ORR
NCT03711058	I/II	AKT	Solid tumors and MSS mCRC Copanlisib and nivolumab	Safety ORR
NCT03983954	I	5T4 TAA	Solid tumors, including mCRC Naptumomab + durvalumab	Safety ORR

Table 5 Selected Ongoing Trials in pMMR/MSS CRC with ICIs.—cont'd

Other targets		Target		
NCT03576963	I/II	DNMT	Refractory CIMP+ MSS mCRC Guadecitabine + nivolumab	Safety
EORTC1615 NCT04031872	I/II	TGF-β	mCRC with TGF-β signature LY3200882 (galunisertib) + capecitabine	Safety ORR
NCT04117087	II	KRAS	KRAS mutated resected CRC or pancreatic cancer KRAS peptide vaccine + poly-ICLC + nivolumab + ipilimumab	Safety T-cell response

6. Conclusions

The activity of monoclonal antibodies targeting PD-1 and CTLA-4 in MSI-H colorectal cancers has markedly changed the prognosis and treatment landscape for these tumors. The benefits of ICIs or other immunotherapeutic strategies have yet to be realized in MSS CRCs and can be further built upon in the MSI-H subset. Most likely, combinatorial strategies which engage anti-tumor DCs and CTLs, while damping MDSC and additional inhibitory signals will be necessary in these more resistant tumors. Improved understanding of TME, both from pre-clinical work and clinical-translational work, will be critical to employing new agents and strategies in an optimal manner, such that full benefit may be gained.

References

Abou-Elkacem, L., Arns, S., Brix, G., Gremse, F., Zopf, D., Kiessling, F., et al. (2013). Regorafenib inhibits growth, angiogenesis, and metastasis in a highly aggressive, orthotopic colon cancer model. *Molecular Cancer Therapeutics*, *12*(7), 1322–1331. https://doi.org/10.1158/1535-7163.MCT-12-1162.

AJCC. (2018). In S. Edge, M. B. Amin, F. Greene, D. R. Byrd, R. K. Brookland, M. K. Washington, & L. R. Meyer (Eds.), *AJCC Cancer Staging Manual* (8th ed.). Chicago: Springer International Publishing.

Andre, T., Lonardi, S., Wong, M., Lenz, H.-J., Gelsomino, F., Aglietta, M., et al. (2018). Nivolumab + ipilimumab combination in patients with DNA mismatch repair-deficient/microsatellite instability-high (dMMR/MSI-H) metastatic colorectal cancer (mCRC): First report of the full cohort from CheckMate-142. *Journal of Clinical Oncology*, *36*(4_Suppl), 553. https://doi.org/10.1200/JCO.2018.36.4_suppl.553.

Andre, T., Shiu, K. K., Kim, T. W., Jensen, B. V., Jensen, L. H., Punt, C., et al. (2020). Pembrolizumab in microsatellite-instability-high advanced colorectal cancer. *The New England Journal of Medicine*, *383*(23), 2207–2218. https://doi.org/10.1056/NEJMoa2017699.

Angell, H., & Galon, J. (2013). From the immune contexture to the Immunoscore: The role of prognostic and predictive immune markers in cancer. *Current Opinion in Immunology*, *25*(2), 261–267. https://doi.org/10.1016/j.coi.2013.03.004.

Atkinson, V., Khattak, A., Haydon, A., Eastgate, M., Roy, A., Prithviraj, P., et al. (2020). Eftilagimod alpha, a soluble lymphocyte activation gene-3 (LAG-3) protein plus pembrolizumab in patients with metastatic melanoma. *Journal for Immunotherapy of Cancer*, *8*(2), e001681. https://doi.org/10.1136/jitc-2020-001681.

Avallone, A., De Stefano, A., Pace, U., Catteau, A., Di Gennaro, E., Tatangelo, F., et al. (2020). Neoadjuvant nivolumab in early stage colorectal cancer. *Annals of Oncology*, *31*(Suppl_4), S449.

Bacac, M., Fauti, T., Sam, J., Colombetti, S., Weinzierl, T., Ouaret, D., et al. (2016). A novel carcinoembryonic antigen T-cell bispecific antibody (CEA TCB) for the treatment of solid tumors. *Clinical Cancer Research*, *22*(13), 3286–3297. https://doi.org/10.1158/1078-0432.CCR-15-1696.

Banchereau, J., & Steinman, R. M. (1998). Dendritic cells and the control of immunity. *Nature*, *392*(6673), 245–252. https://doi.org/10.1038/32588.

Barbera-Guillem, E., Nelson, M. B., Barr, B., Nyhus, J. K., May, K. F., Jr., Feng, L., et al. (2000). B lymphocyte pathology in human colorectal cancer. Experimental and clinical therapeutic effects of partial B cell depletion. *Cancer Immunology, Immunotherapy*, *48*(10), 541–549. https://doi.org/10.1007/pl00006672.

Becht, E., de Reyniès, A., Giraldo, N. A., Pilati, C., Buttard, B., Lacroix, L., et al. (2016). Immune and stromal classification of colorectal cancer is associated with molecular subtypes and relevant for precision immunotherapy. *Clinical Cancer Research*, *22*(16), 4057–4066. https://doi.org/10.1158/1078-0432.Ccr-15-2879.

Bendell, J. C., Powderly, J. D., Lieu, C. H., Echkardt, S. G., Hurwitz, H., Hochster, H. S., et al. (2015). Safety and efficacy of MPDL3280A (anti-PDL1) in combination with bevacizumab (bev) and/or FOLFOX in patients (pts) with metastatic colorectal cancer (mCRC). *Journal of Clinical Oncology*, *33*(3_Suppl), 704.

Beyrend, G., van der Gracht, E., Yilmaz, A., van Duikeren, S., Camps, M., Höllt, T., et al. (2019). PD-L1 blockade engages tumor-infiltrating lymphocytes to co-express targetable activating and inhibitory receptors. *Journal for Immunotherapy of Cancer*, 7(1), 217. https://doi.org/10.1186/s40425-019-0700-3.

Bindea, G., Mlecnik, B., Tosolini, M., Kirilovsky, A., Waldner, M., Obenauf, A. C., et al. (2013). Spatiotemporal dynamics of intratumoral immune cells reveal the immune landscape in human cancer. *Immunity*, *39*(4), 782–795. https://doi.org/10.1016/j.immuni.2013.10.003.

Blank, C. U., Rozeman, E. A., Fanchi, L. F., Sikorska, K., van de Wiel, B., Kvistborg, P., et al. (2018). Neoadjuvant versus adjuvant ipilimumab plus nivolumab in macroscopic stage III melanoma. *Nature Medicine*, *24*(11), 1655–1661. https://doi.org/10.1038/s41591-018-0198-0.

Boissière-Michot, F., Lazennec, G., Frugier, H., Jarlier, M., Roca, L., Duffour, J., et al. (2014). Characterization of an adaptive immune response in microsatellite-instable colorectal cancer. *OncoImmunology*, *3*, e29256. https://doi.org/10.4161/onci.29256.

Boland, C. R., & Goel, A. (2010). Microsatellite instability in colorectal cancer. *Gastroenterology*, *138*(6). https://doi.org/10.1053/j.gastro.2009.12.064. 2073-2087.e2073.

Brahmer, J. R., Drake, C. G., Wollner, I., Powderly, J. D., Picus, J., Sharfinan, W. H., et al. (2010). Phase I study of single-agent anti-programmed death-1 (MDX-1106) in refractory solid tumors: Safety, clinical activity, pharmacodynamics, and immunologic correlates. *Journal of Clinical Oncology*, *28*(19), 3167–3175. https://doi.org/10.1200/jco.2009.26.7609.

Brahmer, J. R., Tykodi, S. S., Chow, L. Q. M., Hwu, W.-J., Topalian, S. L., Hwu, P., et al. (2012). Safety and activity of anti–PD-L1 antibody in patients with advanced cancer. *New England Journal of Medicine*, *366*(26), 2455–2465. https://doi.org/10.1056/NEJMoa1200694.

Bronte, V., Apolloni, E., Cabrelle, A., Ronca, R., Serafini, P., Zamboni, P., et al. (2000). Identification of a CD11b(+)/Gr-1(+)/CD31(+) myeloid progenitor capable of activating or suppressing CD8(+) T cells. *Blood*, *96*(12), 3838–3846.

Bruni, D., Angell, H. K., & Galon, J. (2020). The immune contexture and immunoscore in cancer prognosis and therapeutic efficacy. *Nature Reviews Cancer*, *20*(11), 662–680. https://doi.org/10.1038/s41568-020-0285-7.

Campbell, B. B., Light, N., Fabrizio, D., Zatzman, M., Fuligni, F., de Borja, R., et al. (2017). Comprehensive analysis of hypermutation in human cancer. *Cell*, *171*(5), 1042–1056. e1010. https://doi.org/10.1016/j.cell.2017.09.048.

Campbell, C., McKenney, P. T., Konstantinovsky, D., Isaeva, O. I., Schizas, M., Verter, J., et al. (2020). Bacterial metabolism of bile acids promotes generation of peripheral regulatory T cells. *Nature*, *581*(7809), 475–479. https://doi.org/10.1038/s41586-020-2193-0.

Canon, J., Rex, K., Saiki, A. Y., Mohr, C., Cooke, K., Bagal, D., et al. (2019). The clinical KRAS(G12C) inhibitor AMG 510 drives anti-tumour immunity. *Nature*, *575*(7781), 217–223. https://doi.org/10.1038/s41586-019-1694-1.

Chakrabarti, S., Huebner, L. J., Finnes, H. D., Muranyi, A., Clements, J., Singh, S., et al. (2019). Intratumoral CD3 + and CD8 + T-cell densities in patients with DNA mismatch repair–deficient metastatic colorectal Cancer receiving programmed cell Death-1 blockade. *JCO Precision Oncology*, *3*, 1–7. https://doi.org/10.1200/po.19.00055.

Chalabi, M., Fanchi, L. F., Dijkstra, K. K., Van den Berg, J. G., Aalbers, A. G., Sikorska, K., et al. (2020). Neoadjuvant immunotherapy leads to pathological responses in MMR-proficient and MMR-deficient early-stage colon cancers. *Nature Medicine*, *26*(4), 566–576. https://doi.org/10.1038/s41591-020-0805-8.

Chalmers, Z. R., Connelly, C. F., Fabrizio, D., Gay, L., Ali, S. M., Ennis, R., et al. (2017). Analysis of 100,000 human cancer genomes reveals the landscape of tumor mutational burden. *Genome Medicine*, *9*(1), 34. https://doi.org/10.1186/s13073-017-0424-2.

Chen, J., & Chen, Z. (2014). The effect of immune microenvironment on the progression and prognosis of colorectal cancer. *Medical Oncology*, *31*(8), 82. https://doi.org/10.1007/s12032-014-0082-9.

Chen, E. X., Jonker, D. J., Loree, J. M., Kennecke, H. F., Berry, S. R., Couture, F., et al. (2020). Effect of combined immune checkpoint inhibition vs best supportive care alone in patients with advanced colorectal cancer: The Canadian Cancer Trials Group CO.26 Study. *JAMA Oncology*, *6*(6), 831–838. https://doi.org/10.1001/jamaoncol.2020.0910.

Chen, J. H., Perry, C. J., Tsui, Y. C., Staron, M. M., Parish, I. A., Dominguez, C. X., et al. (2015). Prostaglandin E2 and programmed cell death 1 signaling coordinately impair CTL function and survival during chronic viral infection. *Nature Medicine*, *21*(4), 327–334. https://doi.org/10.1038/nm.3831.

Chtanova, T., & Mackay, C. R. (2001). T cell effector subsets: Extending the Th1/Th2 paradigm. *Advances in Immunology*, *78*, 233–266. https://doi.org/10.1016/s0065-2776(01)78005-4.

Chung, K. Y., Gore, I., Fong, L., Venook, A., Beck, S. B., Dorazio, P., et al. (2010). Phase II study of the anti-cytotoxic T-lymphocyte-associated antigen 4 monoclonal antibody, tremelimumab, in patients with refractory metastatic colorectal cancer. *Journal of Clinical Oncology*, *28*(21), 3485–3490. https://doi.org/10.1200/JCO.2010.28.3994.

Chung, D. C., & Rustgi, A. K. (1995). DNA mismatch repair and cancer. *Gastroenterology*, *109*(5), 1685–1699. https://doi.org/10.1016/0016-5085(95)90660-6.

Cohen, R., Hain, E., Buhard, O., Guilloux, A., Bardier, A., Kaci, R., et al. (2019). Association of primary resistance to immune checkpoint inhibitors in metastatic colorectal cancer with misdiagnosis of microsatellite instability or mismatch repair deficiency status. *JAMA Oncology*, *5*(4), 551–555. https://doi.org/10.1001/jamaoncol.2018.4942.

Coulie, P. G., Lehmann, F., Lethé, B., Herman, J., Lurquin, C., Andrawiss, M., et al. (1995). A mutated intron sequence codes for an antigenic peptide recognized by cytolytic T lymphocytes on a human melanoma. *Proceedings of the National Academy of Sciences of the United States of America*, *92*(17), 7976–7980. https://doi.org/10.1073/pnas.92.17.7976.

Cristescu, R., Mogg, R., Ayers, M., Albright, A., Murphy, E., Yearley, J., et al. (2018). Pan-tumor genomic biomarkers for PD-1 checkpoint blockade-based immunotherapy. *Science*, *362*(6411). https://doi.org/10.1126/science.aar3593.

Dagenborg, V. J., Marshall, S. E., Yaqub, S., Grzyb, K., Boye, K., Lund-Iversen, M., et al. (2020). Neoadjuvant chemotherapy is associated with a transient increase of intratumoral T-cell density in microsatellite stable colorectal liver metastases. *Cancer Biology & Therapy*, *21*(5), 432–440. https://doi.org/10.1080/15384047.2020.1721252.

Dandekar, R. C., Kingaonkar, A. V., & Dhabekar, G. S. (2011). Role of macrophages in malignancy. *Annals of Maxillofacial Surgery*, *1*(2), 150–154. https://doi.org/10.4103/2231-0746.92782.

Demaria, S., Ng, B., Devitt, M. L., Babb, J. S., Kawashima, N., Liebes, L., et al. (2004). Ionizing radiation inhibition of distant untreated tumors (abscopal effect) is immune mediated. *International Journal of Radiation Oncology, Biology, Physics*, *58*(3), 862–870. https://doi.org/10.1016/j.ijrobp.2003.09.012.

Deming, D. A., Emmerich, P., Turk, A. A., Lubner, S. J., Uboha, N. V., LoConte, N. K., et al. (2020). Pembrolizumab (Pem) in combination with stereotactic body radiotherapy (SBRT) for resectable liver oligometastatic MSS/MMR proficient colorectal cancer (CRC). *Journal of Clinical Oncology*, *38*, 4046 (suppl; abstr 4046).

Deng, J., Li, J., Sarde, A., Lines, J. L., Lee, Y. C., Qian, D. C., et al. (2019). Hypoxia-induced VISTA promotes the suppressive function of myeloid-derived suppressor cells in the tumor microenvironment. *Cancer Immunology Research*, 7(7), 1079–1090. https://doi.org/10.1158/2326-6066.CIR-18-0507.

Deng, L., Liang, H., Burnette, B., Beckett, M., Darga, T., Weichselbaum, R. R., et al. (2014). Irradiation and anti-PD-L1 treatment synergistically promote antitumor immunity in mice. *The Journal of Clinical Investigation*, *124*(2), 687–695. https://doi.org/10.1172/JCI67313.

Derosa, L., Hellmann, M. D., Spaziano, M., Halpenny, D., Fidelle, M., Rizvi, H., et al. (2018). Negative association of antibiotics on clinical activity of immune checkpoint inhibitors in patients with advanced renal cell and non-small-cell lung cancer. *Annals of Oncology*, *29*(6), 1437–1444. https://doi.org/10.1093/annonc/mdy103.

Diederichsen, A. C., Hjelmborg, J., Christensen, P. B., Zeuthen, J., & Fenger, C. (2003). Prognostic value of the CD4+/CD8+ ratio of tumour infiltrating lymphocytes in colorectal cancer and HLA-DR expression on tumour cells. *Cancer Immunology, Immunotherapy*, *52*(7), 423–428. https://doi.org/10.1007/s00262-003-0388-5.

Domingo, E., Freeman-Mills, L., Rayner, E., Glaire, M., Briggs, S., Vermeulen, L., et al. (2016). Somatic POLE proofreading domain mutation, immune response, and prognosis in colorectal cancer: A retrospective, pooled biomarker study. *The Lancet Gastroenterology & Hepatology*, *1*(3), 207–216. https://doi.org/10.1016/S2468-1253(16)30014-0.

Dong, C. (2006). Diversification of T-helper-cell lineages: Finding the family root of IL-17-producing cells. *Nature Reviews Immunology*, *6*(4), 329–334. https://doi.org/10.1038/nri1807.

Dong, C., & Flavell, R. A. (2001). Th1 and Th2 cells. *Current Opinion in Hematology*, *8*(1), 47–51.

Dong, H., Strome, S. E., Salomao, D. R., Tamura, H., Hirano, F., Flies, D. B., et al. (2002). Tumor-associated B7-H1 promotes T-cell apoptosis: A potential mechanism of immune evasion. *Nature Medicine*, *8*(8), 793–800. https://doi.org/10.1038/nm730.

Dranoff, G. (2004). Cytokines in cancer pathogenesis and cancer therapy. *Nature Reviews Cancer*, *4*(1), 11–22. https://doi.org/10.1038/nrc1252.

Duan, J., Cui, L., Zhao, X., Bai, H., Cai, S., Wang, G., et al. (2020). Use of immunotherapy with programmed cell death 1 vs programmed cell death ligand 1 inhibitors in patients with cancer: A systematic review and meta-analysis. *JAMA Oncology*, *6*(3), 375–384. https://doi.org/10.1001/jamaoncol.2019.5367.

Duffy, A. G., Makarova-Rusher, O. V., Pratt, D., Kleiner, D. E., Fioravanti, S., Walker, M., et al. (2016). A pilot study of AMP-224, a PD-L2 Fc fusion protein, in combination with stereotactic body radiation therapy (SBRT) in patients with metastatic colorectal cancer. *Journal of Clinical Oncology*, *34*, 560 (suppl 4S; abstr 560).

Edin, S., Wikberg, M. L., Dahlin, A. M., Rutegård, J., Öberg, Å., Oldenborg, P. A., et al. (2012). The distribution of macrophages with a M1 or M2 phenotype in relation to prognosis and the molecular characteristics of colorectal cancer. *PLoS One*, 7(10). https://doi.org/10.1371/journal.pone.0047045, e47045.

Eng, C., Kim, T. W., Bendell, J., Argiles, G., Tebbutt, N. C., Di Bartolomeo, M., et al. (2019). Atezolizumab with or without cobimetinib versus regorafenib in previously treated metastatic colorectal cancer (IMblaze370): A multicentre, open-label, phase 3, randomised, controlled trial. *The Lancet Oncology*, *20*(6), 849–861. https://doi.org/10.1016/S1470-2045(19)30027-0.

Fabrizio, D. A., George, T. J., Jr., Dunne, R. F., Frampton, G., Sun, J., Gowen, K., et al. (2018). Beyond microsatellite testing: Assessment of tumor mutational burden identifies subsets of colorectal cancer who may respond to immune checkpoint inhibition. *Journal of Gastrointestinal Oncology*, *9*(4), 610–617. https://doi.org/10.21037/jgo.2018.05.06.

Facciabene, A., Peng, X., Hagemann, I. S., Balint, K., Barchetti, A., Wang, L. P., et al. (2011). Tumour hypoxia promotes tolerance and angiogenesis via CCL28 and T(reg) cells. *Nature*, *475*(7355), 226–230. https://doi.org/10.1038/nature10169.

Feng, H., Shuda, M., Chang, Y., & Moore, P. S. (2008). Clonal integration of a polyomavirus in human Merkel cell carcinoma. *Science*, *319*(5866), 1096–1100. https://doi.org/10.1126/science.1152586.

Filatenkov, A., Baker, J., Mueller, A. M., Kenkel, J., Ahn, G. O., Dutt, S., et al. (2015). Ablative tumor radiation can change the tumor immune cell microenvironment to induce durable complete remissions. *Clinical Cancer Research*, *21*(16), 3727–3739. https://doi.org/10.1158/1078-0432.CCR-14-2824.

Finn, R. S., Qin, S., Ikeda, M., Galle, P. R., Ducreux, M., Kim, T. Y., et al. (2020). Atezolizumab plus bevacizumab in unresectable hepatocellular carcinoma. *The New England Journal of Medicine*, *382*(20), 1894–1905. https://doi.org/10.1056/NEJMoa1915745.

Fridman, W. H., Pagès, F., Sautès-Fridman, C., & Galon, J. (2012). The immune contexture in human tumours: Impact on clinical outcome. *Nature Reviews. Cancer*, *12*(4), 298–306. https://doi.org/10.1038/nrc3245.

Fukuoka, S., Hara, H., Takahashi, N., Kojima, T., Kawazoe, A., Asayama, M., et al. (2020). Regorafenib plus nivolumab in patients with advanced gastric or colorectal cancer: an open-label, dose-escalation, and dose-expansion phase Ib trial (REGONIVO, EPOC1603). *Journal of Clinical Oncology*, *38*(18), 2053–2061. https://doi.org/10.1200/jco.19.03296.

Gafa, R., Maestri, I., Matteuzzi, M., Santini, A., Ferretti, S., Cavazzini, L., et al. (2000). Sporadic colorectal adenocarcinomas with high-frequency microsatellite instability. *Cancer*, *89*(10), 2025–2037.

Galli, F., Aguilera, J. V., Palermo, B., Markovic, S. N., Nisticò, P., & Signore, A. (2020). Relevance of immune cell and tumor microenvironment imaging in the new era of immunotherapy. *Journal of Experimental & Clinical Cancer Research*, *39*(1), 89. https://doi.org/10.1186/s13046-020-01586-y.

Galon, J., Costes, A., Sanchez-Cabo, F., Kirilovsky, A., Mlecnik, B., Lagorce-Pagès, C., et al. (2006). Type, density, and location of immune cells within human colorectal tumors predict clinical outcome. *Science*, *313*(5795), 1960–1964. https://doi.org/10.1126/science.1129139.

Galon, J., Fridman, W.-H., & Pagès, F. (2007). The adaptive immunologic microenvironment in colorectal cancer: A novel perspective. *Cancer Research*, *67*(5), 1883–1886. https://doi.org/10.1158/0008-5472.Can-06-4806.

Galon, J., Pagès, F., Marincola, F. M., Angell, H. K., Thurin, M., Lugli, A., et al. (2012). Cancer classification using the immunoscore: A worldwide task force. *Journal of Translational Medicine*, *10*, 205. https://doi.org/10.1186/1479-5876-10-205.

Gao, J., Shi, L. Z., Zhao, H., Chen, J., Xiong, L., He, Q., et al. (2016). Loss of IFN-gamma pathway genes in tumor cells as a mechanism of resistance to anti-CTLA-4 therapy. *Cell*, *167*(2), 397–404. e399. https://doi.org/10.1016/j.cell.2016.08.069.

Garg, A. D., More, S., Rufo, N., Mece, O., Sassano, M. L., Agostinis, P., et al. (2017). Trial watch: Immunogenic cell death induction by anticancer chemotherapeutics. *Oncoimmunology*, *6*(12), e1386829. https://doi.org/10.1080/2162402X.2017.1386829.

Garon, E. B., Rizvi, N. A., Hui, R., Leighl, N., Balmanoukian, A. S., Eder, J. P., et al. (2015). Pembrolizumab for the treatment of non–small-cell lung cancer. *New England Journal of Medicine*, *372*(21), 2018–2028. https://doi.org/10.1056/NEJMoa1501824.

Garrido, G., Rabasa, A., Sanchez, B., Lopez, M. V., Blanco, R., Lopez, A., et al. (2011). Induction of immunogenic apoptosis by blockade of epidermal growth factor receptor activation with a specific antibody. *Journal of Immunology*, *187*(10), 4954–4966. https://doi.org/10.4049/jimmunol.1003477.

Gettinger, S., Choi, J., Hastings, K., Truini, A., Datar, I., Sowell, R., et al. (2017). Impaired HLA class I antigen processing and presentation as a mechanism of acquired resistance to immune checkpoint inhibitors in lung cancer. *Cancer Discovery*, 7(12), 1420–1435. https://doi.org/10.1158/2159-8290.CD-17-0593.

Goldszmid, R. S., Dzutsev, A., & Trinchieri, G. (2014). Host immune response to infection and cancer: Unexpected commonalities. *Cell Host & Microbe*, *15*(3), 295–305. https://doi.org/10.1016/j.chom.2014.02.003.

Grasso, C. S., Giannakis, M., Wells, D. K., Hamada, T., Mu, X. J., Quist, M., et al. (2018). Genetic mechanisms of immune evasion in colorectal cancer. *Cancer Discovery*, *8*(6), 730–749. https://doi.org/10.1158/2159-8290.CD-17-1327.

Gross, L. (1943). Intradermal immunization of C3H mice against a sarcoma that originated in an animal of the same line. *Cancer Research*, *3*(5), 326–333.

Grothey, A., Tabernero, J., Arnold, D., De Gramont, A., Ducreux, M. P., O'Dwyer, P. J., et al. (2018). Fluoropyrimidine (FP) + bevacizumab (BEV) + atezolizumab vs FP/BEV in BRAFwt metastatic colorectal cancer (mCRC): Findings from Cohort 2 of MODUL—A multicentre, randomized trial of biomarker-driven maintenance treatment following first-line induction therapy. *Annals of Oncology*, *29*(8), viii714–5.

Guinney, J., Dienstmann, R., Wang, X., de Reynies, A., Schlicker, A., Soneson, C., et al. (2015). The consensus molecular subtypes of colorectal cancer. *Nature Medicine*, *21*(11), 1350–1356. https://doi.org/10.1038/nm.3967.

Gurjao, C., Liu, D., Hofree, M., AlDubayan, S. H., Wakiro, I., Su, M. J., et al. (2019). Intrinsic resistance to immune checkpoint blockade in a mismatch repair-deficient colorectal cancer. *Cancer Immunology Research*, 7(8), 1230–1236. https://doi.org/10.1158/2326-6066.CIR-18-0683.

Hanahan, D., & Coussens, L. M. (2012). Accessories to the crime: Functions of cells recruited to the tumor microenvironment. *Cancer Cell*, *21*(3), 309–322. https://doi.org/10.1016/j.ccr.2012.02.022.

Hanahan, D., & Weinberg, R. A. (2011). Hallmarks of cancer: The next generation. *Cell*, *144*(5), 646–674. https://doi.org/10.1016/j.cell.2011.02.013.

Haraldsdottir, S., Hampel, H., Tomsic, J., Frankel, W. L., Pearlman, R., de la Chapelle, A., et al. (2014). Colon and endometrial cancers with mismatch repair deficiency can arise from somatic, rather than germline, mutations. *Gastroenterology, 147*(6), 1308–1316. e1301 https://doi.org/10.1053/j.gastro.2014.08.041.

He, Z., Gharaibeh, R. Z., Newsome, R. C., Pope, J. L., Dougherty, M. W., Tomkovich, S., et al. (2019). Campylobacter jejuni promotes colorectal tumorigenesis through the action of cytolethal distending toxin. *Gut, 68*(2), 289–300. https://doi.org/10.1136/gutjnl-2018-317200.

Hermitte, F. (2016). Biomarkers immune monitoring technology primer: Immunoscore® colon. *Journal for Immunotherapy of Cancer, 4*(1), 57. https://doi.org/10.1186/s40425-016-0161-x.

Herrera, M., Herrera, A., Domínguez, G., Silva, J., García, V., García, J. M., et al. (2013). Cancer-associated fibroblast and M2 macrophage markers together predict outcome in colorectal cancer patients. *Cancer Science, 104*(4), 437–444. https://doi.org/10.1111/cas.12096.

Heylmann, D., Rodel, F., Kindler, T., & Kaina, B. (2014). Radiation sensitivity of human and murine peripheral blood lymphocytes, stem and progenitor cells. *Biochimica et Biophysica Acta, 1846*(1), 121–129. https://doi.org/10.1016/j.bbcan.2014.04.009.

Hino, H., Shiomi, A., Kusuhara, M., Kagawa, H., Yamakawa, Y., Hatakeyama, K., et al. (2019). Clinicopathological and mutational analyses of colorectal cancer with mutations in the POLE gene. *Cancer Medicine, 8*(10), 4587–4597. https://doi.org/10.1002/cam4.2344.

Hinrichs, C. S., & Rosenberg, S. A. (2014). Exploiting the curative potential of adoptive T-cell therapy for cancer. *Immunological Reviews, 257*(1), 56–71. https://doi.org/10.1111/imr.12132.

Hodi, F. S., O'Day, S. J., McDermott, D. F., Weber, R. W., Sosman, J. A., Haanen, J. B., et al. (2010). Improved survival with ipilimumab in patients with metastatic melanoma. *The New England Journal of Medicine, 363*(8), 711–723. https://doi.org/10.1056/NEJMoa1003466.

Hu, H., Cai, W., Wu, D., Hu, W., Dong Wang, L., Mao, J., et al. (2020). Ultra-mutated colorectal cancer patients with POLE driver mutations exhibit distinct clinical patterns. *Cancer Medicine, 10*, 135–142. https://doi.org/10.1002/cam4.3579.

Ishida, Y., Agata, Y., Shibahara, K., & Honjo, T. (1992). Induced expression of PD-1, a novel member of the immunoglobulin gene superfamily, upon programmed cell death. *The EMBO Journal, 11*(11), 3887–3895.

Jenkins, R. W., Barbie, D. A., & Flaherty, K. T. (2018). Mechanisms of resistance to immune checkpoint inhibitors. *British Journal of Cancer, 118*(1), 9–16. https://doi.org/10.1038/bjc.2017.434.

Kanterman, J., Sade-Feldman, M., Biton, M., Ish-Shalom, E., Lasry, A., Goldshtein, A., et al. (2014). Adverse immunoregulatory effects of 5FU and CPT11 chemotherapy on myeloid-derived suppressor cells and colorectal cancer outcomes. *Cancer Research, 74*(21), 6022–6035. https://doi.org/10.1158/0008-5472.CAN-14-0657.

Kasajima, A., Sers, C., Sasano, H., Johrens, K., Stenzinger, A., Noske, A., et al. (2010). Down-regulation of the antigen processing machinery is linked to a loss of inflammatory response in colorectal cancer. *Human Pathology, 41*(12), 1758–1769. https://doi.org/10.1016/j.humpath.2010.05.014.

Kim, N. R., & Kim, Y. J. (2019). Oxaliplatin regulates myeloid-derived suppressor cell-mediated immunosuppression via downregulation of nuclear factor-kappaB signaling. *Cancer Medicine, 8*(1), 276–288. https://doi.org/10.1002/cam4.1878.

Kipanyula, M. J., Seke Etet, P. F., Vecchio, L., Farahna, M., Nukenine, E. N., & Nwabo Kamdje, A. H. (2013). Signaling pathways bridging microbial-triggered inflammation and cancer. *Cellular Signalling, 25*(2), 403–416. https://doi.org/10.1016/j.cellsig.2012.10.014.

Kloor, M., Becker, C., Benner, A., Woerner, S. M., Gebert, J., Ferrone, S., et al. (2005). Immunoselective pressure and human leukocyte antigen class I antigen machinery defects in microsatellite unstable colorectal cancers. *Cancer Research, 65*(14), 6418–6424. https://doi.org/10.1158/0008-5472.CAN-05-0044.

Kluger, H. M., Tawbi, H. A., Ascierto, M. L., Bowden, M., Callahan, M. K., Cha, E., et al. (2020). Defining tumor resistance to PD-1 pathway blockade: Recommendations from the first meeting of the SITC immunotherapy resistance taskforce. *Journal for Immunotherapy of Cancer, 8*(1). https://doi.org/10.1136/jitc-2019-000398.

Knutson, K. L., & Disis, M. L. (2005). Tumor antigen-specific T helper cells in cancer immunity and immunotherapy. *Cancer Immunology, Immunotherapy, 54*(8), 721–728. https://doi.org/10.1007/s00262-004-0653-2.

Konishi, J., Yamazaki, K., Azuma, M., Kinoshita, I., Dosaka-Akita, H., & Nishimura, M. (2004). B7-H1 expression on non-small cell lung cancer cells and its relationship with tumor-infiltrating lymphocytes and their PD-1 expression. *Clinical Cancer Research, 10*(15), 5094–5100. https://doi.org/10.1158/1078-0432.Ccr-04-0428.

Koopman, M., Kortman, G. A., Mekenkamp, L., Ligtenberg, M. J., Hoogerbrugge, N., Antonini, N. F., et al. (2009). Deficient mismatch repair system in patients with sporadic advanced colorectal cancer. *British Journal of Cancer, 100*(2), 266–273. https://doi.org/10.1038/sj.bjc.6604867.

Kosinsky, Y., Dovedi, S. J., Peskov, K., Voronova, V., Chu, L., Tomkinson, H., et al. (2018). Radiation and PD-(L)1 treatment combinations: Immune response and dose optimization via a predictive systems model. *Journal for Immunotherapy of Cancer, 6*(1), 17. https://doi.org/10.1186/s40425-018-0327-9.

Kostic, A. D., Chun, E., Robertson, L., Glickman, J. N., Gallini, C. A., Michaud, M., et al. (2013). Fusobacterium nucleatum potentiates intestinal tumorigenesis and modulates the tumor-immune microenvironment. *Cell Host & Microbe, 14*(2), 207–215. https://doi.org/10.1016/j.chom.2013.07.007.

Kostic, A. D., Gevers, D., Pedamallu, C. S., Michaud, M., Duke, F., Earl, A. M., et al. (2012). Genomic analysis identifies association of Fusobacterium with colorectal carcinoma. *Genome Research, 22*(2), 292–298. https://doi.org/10.1101/gr.126573.111.

Lal, N., White, B. S., Goussous, G., Pickles, O., Mason, M. J., Beggs, A. D., et al. (2018). KRAS mutation and consensus molecular subtypes 2 and 3 are independently associated with reduced immune infiltration and reactivity in colorectal cancer. *Clinical Cancer Research, 24*(1), 224–233. https://doi.org/10.1158/1078-0432.CCR-17-1090.

Larkin, J., Chiarion-Sileni, V., Gonzalez, R., Grob, J. J., Rutkowski, P., Lao, C. D., et al. (2019). Five-year survival with combined nivolumab and Ipilimumab in advanced melanoma. *The New England Journal of Medicine, 381*(16), 1535–1546. https://doi.org/10.1056/NEJMoa1910836.

Le, D. T., Durham, J. N., Smith, K. N., Wang, H., Bartlett, B. R., Aulakh, L. K., et al. (2017). Mismatch repair deficiency predicts response of solid tumors to PD-1 blockade. *Science, 357*(6349), 409–413. https://doi.org/10.1126/science.aan6733.

Le, H. K., Graham, L., Cha, E., Morales, J. K., Manjili, M. H., & Bear, H. D. (2009). Gemcitabine directly inhibits myeloid derived suppressor cells in BALB/c mice bearing 4T1 mammary carcinoma and augments expansion of T cells from tumor-bearing mice. *International Immunopharmacology, 9*(7–8), 900–909. https://doi.org/10.1016/j.intimp.2009.03.015.

Le, D. T., Kim, T. W., Van Cutsem, E., Geva, R., Jäger, D., Hara, H., et al. (2019). Phase II open-label study of pembrolizumab in treatment-refractory, microsatellite instability–high/mismatch repair–deficient metastatic colorectal cancer: KEYNOTE-164. *Journal of Clinical Oncology, 38*(1), 11–19. https://doi.org/10.1200/JCO.19.02107.

Le, D. T., Uram, J. N., Wang, H., Bartlett, B. R., Kemberling, H., Eyring, A. D., et al. (2015). PD-1 blockade in tumors with mismatch-repair deficiency. *New England Journal of Medicine, 372*(26), 2509–2520. https://doi.org/10.1056/NEJMoa1500596.

Leach, D. R., Krummel, M. F., & Allison, J. P. (1996). Enhancement of antitumor immunity by CTLA-4 blockade. *Science, 271*(5256), 1734–1736. https://doi.org/10.1126/science.271.5256.1734.

Lenz, H.-J., Lonardi, S., Zagonel, V., Cutsem, E. V., Limon, M. L., Wong, M., et al. (2020). Nivolumab (NIVO) + low-dose ipilimumab (IPI) as first-line (1L) therapy in microsatellite instability-high/mismatch repair-deficient (MSI-H/dMMR) metastatic colorectal cancer (mCRC): Two-year clinical update. *Journal of Clinical Oncology, 38*(15_suppl), 4040. https://doi.org/10.1200/JCO.2020.38.15_suppl.4040.

Li, J., Cong, L., Liu, J., Peng, L., Wang, J., Feng, A., et al. (2020). The efficacy and safety of regorafenib in combination with anti-PD-1 antibody in refractory microsatellite stable metastatic colorectal cancer: A retrospective study. *Frontiers in Oncology, 10*, 594125. https://doi.org/10.3389/fonc.2020.594125.

Liao, W., Overman, M. J., Boutin, A. T., Shang, X., Zhao, D., Dey, P., et al. (2019). KRAS-IRF2 Axis drives immune suppression and immune therapy resistance in colorectal cancer. *Cancer Cell, 35*(4), 559–572.e557. https://doi.org/10.1016/j.ccell.2019.02.008.

Limagne, E., Thibaudin, M., Nuttin, L., Spill, A., Derangere, V., Fumet, J. D., et al. (2019). Trifluridine/tipiracil plus oxaliplatin improves PD-1 blockade in colorectal cancer by inducing immunogenic cell death and depleting macrophages. *Cancer Immunology Research*, 7(12), 1958–1969. https://doi.org/10.1158/2326-6066.CIR-19-0228.

Lipson, E. J., Sharfman, W. H., Drake, C. G., Wollner, I., Taube, J. M., Anders, R. A., et al. (2013). Durable cancer regression off-treatment and effective reinduction therapy with an anti-PD-1 antibody. *Clinical Cancer Research, 19*(2), 462–468. https://doi.org/10.1158/1078-0432.Ccr-12-2625.

Liu, J., Blake, S. J., Yong, M. C. R., Harjunpää, H., Ngiow, S. F., Takeda, K., et al. (2016). Improved efficacy of neoadjuvant compared to adjuvant immunotherapy to eradicate metastatic disease. *Cancer Discovery, 6*(12), 1382–1399. https://doi.org/10.1158/2159-8290.Cd-16-0577.

Liu, J., Duan, Y., Cheng, X., Chen, X., Xie, W., Long, H., et al. (2011). IL-17 is associated with poor prognosis and promotes angiogenesis via stimulating VEGF production of cancer cells in colorectal carcinoma. *Biochemical and Biophysical Research Communications, 407*(2), 348–354. https://doi.org/10.1016/j.bbrc.2011.03.021.

Liu, L., Mayes, P. A., Eastman, S., Shi, H., Yadavilli, S., Zhang, T., et al. (2015). The BRAF and MEK inhibitors Dabrafenib and Trametinib: Effects on immune function and in combination with immunomodulatory antibodies targeting PD-1, PD-L1, and CTLA-4. *Clinical Cancer Research, 21*(7), 1639–1651. https://doi.org/10.1158/1078-0432.CCR-14-2339.

Llosa, N. J., Cruise, M., Tam, A., Wicks, E. C., Hechenbleikner, E. M., Taube, J. M., et al. (2015). The vigorous immune microenvironment of microsatellite instable colon cancer is balanced by multiple counter-inhibitory checkpoints. *Cancer Discovery, 5*(1), 43–51. https://doi.org/10.1158/2159-8290.Cd-14-0863.

Lloyd-Price, J., Mahurkar, A., Rahnavard, G., Crabtree, J., Orvis, J., Hall, A. B., et al. (2017). Strains, functions and dynamics in the expanded human microbiome project. *Nature, 550*(7674), 61–66. https://doi.org/10.1038/nature23889.

Lu, W., Yu, W., He, J., Liu, W., Yang, J., Lin, X., et al. (2020). Reprogramming immunosuppressive myeloid cells facilitates immunotherapy for colorectal cancer. *EMBO Molecular Medicine*, e12798. https://doi.org/10.15252/emmm.202012798.

Luke, J. J., Bao, R., Sweis, R. F., Spranger, S., & Gajewski, T. F. (2019). WNT/beta-catenin pathway activation correlates with immune exclusion across human cancers. *Clinical Cancer Research, 25*(10), 3074–3083. https://doi.org/10.1158/1078-0432.CCR-18-1942.

Lwin, Z., Gomez-Roca, C., Saada-Bouzid, E., Yanez, E., Longo Munoz, F., Im, S. A., et al. (2020). LEAP-005: Phase II study of lenvatinib (len) plus pembrolizumab (pembro) in patients (pts) with previously treated advanced solid tumours. *Annals of Oncology, 31*-(suppl_4), S1170.

Marabelle, A., Fakih, M., Lopez, J., Shah, M., Shapira-Frommer, R., Nakagawa, K., et al. (2020). Association of tumour mutational burden with outcomes in patients with advanced solid tumours treated with pembrolizumab: Prospective biomarker analysis of the multicohort, open-label, phase 2 KEYNOTE-158 study. *The Lancet Oncology*, *21*(10), 1353–1365. https://doi.org/10.1016/s1470-2045(20)30445-9.

Marabelle, A., Le, D. T., Ascierto, P. A., Di Giacomo, A. M., De Jesus-Acosta, A., Delord, J. P., et al. (2020). Efficacy of pembrolizumab in patients with noncolorectal high microsatellite instability/mismatch repair-deficient cancer: Results from the phase II KEYNOTE-158 study. *Journal of Clinical Oncology*, *38*(1), 1–10. https://doi.org/10.1200/jco.19.02105.

Marcus, L., Lemery, S. J., Keegan, P., & Pazdur, R. (2019). FDA approval summary: Pembrolizumab for the treatment of microsatellite instability-high solid Tumors. *Clinical Cancer Research*, *25*(13), 3753–3758. https://doi.org/10.1158/1078-0432.Ccr-18-4070.

Markowitz, S. D., & Bertagnolli, M. M. (2009). Molecular origins of cancer: Molecular basis of colorectal cancer. *The New England Journal of Medicine*, *361*(25), 2449–2460. https://doi.org/10.1056/NEJMra0804588.

Marliot, F., Chen, X., Kirilovsky, A., Sbarrato, T., El Sissy, C., Batista, L., et al. (2020). Analytical validation of the immunoscore and its associated prognostic value in patients with colon cancer. *Journal for Immunotherapy of Cancer*, *8*(1), e000272. https://doi.org/10.1136/jitc-2019-000272.

McGranahan, N., Furness, A. J. S., Rosenthal, R., Ramskov, S., Lyngaa, R., Saini, S. K., et al. (2016). Clonal neoantigens elicit T cell immunoreactivity and sensitivity to immune checkpoint blockade. *Science*, *351*(6280), 1463–1469. https://doi.org/10.1126/science.aaf1490.

McGranahan, N., & Swanton, C. (2019). Neoantigen quality, not quantity. *Science Translational Medicine*, *11*(506), eaax7918. https://doi.org/10.1126/scitranslmed.aax7918.

Mehnert, J. M., Panda, A., Zhong, H., Hirshfield, K., Damare, S., Lane, K., et al. (2016). Immune activation and response to pembrolizumab in POLE-mutant endometrial cancer. *The Journal of Clinical Investigation*, *126*(6), 2334–2340. https://doi.org/10.1172/JCI84940.

Meiri, E., Garrett-Mayer, E., Halabi, S., Mangat, P. K., Shrestha, S., Ahn, E. R., et al. (2020). Pembrolizumab (P) in patients (Pts) with colorectal cancer (CRC) with high tumor mutational burden (HTMB): Results from the Targeted Agent and Profiling Utilization Registry (TAPUR) Study. *Journal of Clinical Oncology*, *38*(4_suppl), 133 (February 01, 2020).

Michielsen, A. J., Hogan, A. E., Marry, J., Tosetto, M., Cox, F., Hyland, J. M., et al. (2011). Tumour tissue microenvironment can inhibit dendritic cell maturation in colorectal cancer. *PLoS One*, *6*(11), e27944. https://doi.org/10.1371/journal.pone.0027944.

Middha, S., Yaeger, R., Shia, J., Stadler, Z. K., King, S., Guercio, S., et al. (2019). Majority of B2M-mutant and -deficient colorectal carcinomas achieve clinical benefit from immune checkpoint inhibitor therapy and are microsatellite instability-high. *JCO Precision Oncology*, *3*. https://doi.org/10.1200/PO.18.00321.

Mima, K., Sukawa, Y., Nishihara, R., Qian, Z. R., Yamauchi, M., Inamura, K., et al. (2015). Fusobacterium nucleatum and T cells in colorectal carcinoma. *JAMA Oncology*, *1*(5), 653–661. https://doi.org/10.1001/jamaoncol.2015.1377.

Misra, N., Bayry, J., Lacroix-Desmazes, S., Kazatchkine, M. D., & Kaveri, S. V. (2004). Cutting edge: Human CD4+ CD25+ T cells restrain the maturation and antigen-presenting function of dendritic cells. *The Journal of Immunology*, *172*(8), 4676–4680. https://doi.org/10.4049/jimmunol.172.8.4676.

Mlecnik, B., Bifulco, C., Bindea, G., Marliot, F., Lugli, A., Lee, J. J., et al. (2020). Multicenter International Society for Immunotherapy of Cancer Study of the consensus immunoscore for the prediction of survival and response to chemotherapy in stage III colon cancer. *Journal of Clinical Oncology, 38*(31), 3638–3651. https://doi.org/10.1200/jco.19.03205.

Mlecnik, B., Tosolini, M., Kirilovsky, A., Berger, A., Bindea, G., Meatchi, T., et al. (2011). Histopathologic-based prognostic factors of colorectal cancers are associated with the state of the local immune reaction. *Journal of Clinical Oncology, 29*(6), 610–618. https://doi.org/10.1200/jco.2010.30.5425.

Mo, S., Ma, X., Li, Y., Zhang, L., Hou, T., Han-Zhang, H., et al. (2020). Somatic POLE exonuclease domain mutations elicit enhanced intratumoral immune responses in stage II colorectal cancer. *Journal for Immunotherapy of Cancer, 8*(2). https://doi.org/10.1136/jitc-2020-000881.

Mocellin, S., Rossi, C. R., Lise, M., & Nitti, D. (2004). Colorectal cancer vaccines: Principles, results, and perspectives. *Gastroenterology, 127*(6), 1821–1837. https://doi.org/10.1053/j.gastro.2004.07.012.

Morris, J. S., Luthra, R., Liu, Y., Duose, D. Y., Lee, W., Reddy, N. G., et al. (2021). Development and validation of a gene signature classifier for consensus molecular subtyping of colorectal carcinoma in a CLIA-certified setting. *Clinical Cancer Research, 27*(1), 120–130. https://doi.org/10.1158/1078-0432.CCR-20-2403.

Movahedi, K., Guilliams, M., Van den Bossche, J., Van den Bergh, R., Gysemans, C., Beschin, A., et al. (2008). Identification of discrete tumor-induced myeloid-derived suppressor cell subpopulations with distinct T cell–suppressive activity. *Blood, 111*(8), 4233–4244. https://doi.org/10.1182/blood-2007-07-099226.

Munger, K., Baldwin, A., Edwards, K. M., Hayakawa, H., Nguyen, C. L., Owens, M., et al. (2004). Mechanisms of human papillomavirus-induced oncogenesis. *Journal of Virology, 78*(21), 11451–11460. https://doi.org/10.1128/JVI.78.21.11451-11460.2004.

Mur, P., Garcia-Mulero, S., Del Valle, J., Magraner-Pardo, L., Vidal, A., Pineda, M., et al. (2020). Role of POLE and POLD1 in familial cancer. *Genetics in Medicine, 22*, 2089–2100. https://doi.org/10.1038/s41436-020-0922-2.

Nagtegaal, I. D., Quirke, P., & Schmoll, H. J. (2011). Has the new TNM classification for colorectal cancer improved care? *Nature Reviews. Clinical Oncology, 9*(2), 119–123. https://doi.org/10.1038/nrclinonc.2011.157.

Naito, Y., Saito, K., Shiiba, K., Ohuchi, A., Saigenji, K., Nagura, H., et al. (1998). CD8+T cells infiltrated within cancer cell nests as a prognostic factor in human colorectal cancer. *Cancer Research, 58*(16), 3491–3494.

Nejman, D., Livyatan, I., Fuks, G., Gavert, N., Zwang, Y., Geller, L. T., et al. (2020). The human tumor microbiome is composed of tumor type-specific intracellular bacteria. *Science, 368*(6494), 973–980. https://doi.org/10.1126/science.aay9189.

O'Keefe, S. J., Li, J. V., Lahti, L., Ou, J., Carbonero, F., Mohammed, K., et al. (2015). Fat, fibre and cancer risk in African Americans and rural Africans. *Nature Communications, 6*, 6342. https://doi.org/10.1038/ncomms7342.

O'Neil, B. H., Wallmark, J. M., Lorente, D., Elez, E., Raimbourg, J., Gomez-Roca, C., et al. (2017). Safety and antitumor activity of the anti-PD-1 antibody pembrolizumab in patients with advanced colorectal carcinoma. *PLoS One, 12*(12), e0189848. https://doi.org/10.1371/journal.pone.0189848.

Onishi, Y., Fehervari, Z., Yamaguchi, T., & Sakaguchi, S. (2008). Foxp3+ natural regulatory T cells preferentially form aggregates on dendritic cells in vitro and actively inhibit their maturation. *Proceedings of the National Academy of Sciences of the United States of America, 105*(29), 10113–10118. https://doi.org/10.1073/pnas.0711106105.

Overman, M. J., Kopetz, S., McDermott, R. S., Leach, J., Lonardi, S., Lenz, H.-J., et al. (2016). Nivolumab ± ipilimumab in treatment (tx) of patients (pts) with metastatic colorectal cancer (mCRC) with and without high microsatellite instability (MSI-H): CheckMate-142 interim results. *Journal of Clinical Oncology, 34*(15_suppl), 3501. https://doi.org/10.1200/JCO.2016.34.15_suppl.3501.

Overman, M. J., McDermott, R., Leach, J. L., Lonardi, S., Lenz, H. J., Morse, M. A., et al. (2017). Nivolumab in patients with metastatic DNA mismatch repair-deficient or microsatellite instability-high colorectal cancer (CheckMate 142): An open-label, multicentre, phase 2 study. *The Lancet Oncology, 18*(9), 1182–1191. https://doi.org/10.1016/s1470-2045(17)30422-9.

Pacella, I., Cammarata, I., Focaccetti, C., Miacci, S., Gulino, A., Tripodo, C., et al. (2018). Wnt3a neutralization enhances T-cell responses through indirect mechanisms and restrains tumor growth. *Cancer Immunology Research, 6*(8), 953–964. https://doi.org/10.1158/2326-6066.CIR-17-0713.

Pagès, F., Kirilovsky, A., Mlecnik, B., Asslaber, M., Tosolini, M., Bindea, G., et al. (2009). In situ cytotoxic and memory T cells predict outcome in patients with early-stage colorectal Cancer. *Journal of Clinical Oncology, 27*(35), 5944–5951. https://doi.org/10.1200/jco.2008.19.6147.

Pagès, F., Mlecnik, B., Marliot, F., Bindea, G., Ou, F.-S., Bifulco, C., et al. (2018). International validation of the consensus immunoscore for the classification of colon cancer: A prognostic and accuracy study. *The Lancet, 391*(10135), 2128–2139. https://doi.org/10.1016/S0140-6736(18)30789-X.

Panda, A., Betigeri, A., Subramanian, K., Ross, J. S., Pavlick, D. C., Ali, S., et al. (2017). Identifying a clinically applicable mutational burden threshold as a potential biomarker of response to immune checkpoint therapy in solid tumors. *JCO Precision Oncology, 1*, 1–13. https://doi.org/10.1200/po.17.00146.

Pardoll, D. M. (2012). The blockade of immune checkpoints in cancer immunotherapy. *Nature Reviews. Cancer, 12*(4), 252–264. https://doi.org/10.1038/nrc3239.

Parikh, A. R., Clark, J. W., Wo, J. Y.-L., Yeap, B. Y., Allen, J. N., Blazkowsky, L. S., et al. (2020). A phase II study of ipilimumab and nivolumab with radiation in microsatellite stable (MSS) metastatic colorectal adenocarcinoma (mCRC). *Journal of Clinical Oncology, 37*, 3514. 2019 (suppl; abstr 3514).

Pino, M. S., & Chung, D. C. (2010). The chromosomal instability pathway in colon cancer. *Gastroenterology, 138*(6), 2059–2072. https://doi.org/10.1053/j.gastro.2009.12.065.

Pleguezuelos-Manzano, C., Puschhof, J., Rosendahl Huber, A., van Hoeck, A., Wood, H. M., Nomburg, J., et al. (2020). Mutational signature in colorectal cancer caused by genotoxic pks(+) E. coli. *Nature, 580*(7802), 269–273. https://doi.org/10.1038/s41586-020-2080-8.

Popat, S., Hubner, R., & Houlston, R. S. (2005). Systematic review of microsatellite instability and colorectal cancer prognosis. *Journal of Clinical Oncology, 23*(3), 609–618. https://doi.org/10.1200/JCO.2005.01.086.

Rini, B. I., Plimack, E. R., Stus, V., Gafanov, R., Hawkins, R., Nosov, D., et al. (2019). Pembrolizumab plus Axitinib versus Sunitinib for advanced renal-cell carcinoma. *The New England Journal of Medicine, 380*(12), 1116–1127. https://doi.org/10.1056/NEJMoa1816714.

Rizvi, N. A., Hellmann, M. D., Snyder, A., Kvistborg, P., Makarov, V., Havel, J. J., et al. (2015). Mutational landscape determines sensitivity to PD-1 blockade in non–small cell lung cancer. *Science, 348*(6230), 124–128. https://doi.org/10.1126/science.aaa1348.

Robert, C., Ribas, A., Wolchok, J. D., Hodi, F. S., Hamid, O., Kefford, R., et al. (2014). Anti-programmed-death-receptor-1 treatment with pembrolizumab in ipilimumab-refractory advanced melanoma: A randomised dose-comparison cohort of a phase 1 trial. *The Lancet, 384*(9948), 1109–1117. https://doi.org/10.1016/S0140-6736(14)60958-2.

Routy, B., Le Chatelier, E., Derosa, L., Duong, C. P. M., Alou, M. T., Daillere, R., et al. (2018). Gut microbiome influences efficacy of PD-1-based immunotherapy against epithelial tumors. *Science, 359*(6371), 91–97. https://doi.org/10.1126/science.aan3706.

Sakaguchi, S. (2000). Regulatory T cells: Key controllers of immunologic self-tolerance. *Cell, 101*(5), 455–458. https://doi.org/10.1016/S0092-8674(00)80856-9.

Salama, P., Phillips, M., Grieu, F., Morris, M., Zeps, N., Joseph, D., et al. (2009). Tumor-infiltrating FOXP3+ T regulatory cells show strong prognostic significance in colorectal cancer. *Journal of Clinical Oncology, 27*(2), 186–192. https://doi.org/10.1200/jco.2008.18.7229.

Samstein, R. M., Lee, C.-H., Shoushtari, A. N., Hellmann, M. D., Shen, R., Janjigian, Y. Y., et al. (2019). Tumor mutational load predicts survival after immunotherapy across multiple cancer types. *Nature Genetics, 51*(2), 202–206. https://doi.org/10.1038/s41588-018-0312-8.

Schreiber, R. D., Old, L. J., & Smyth, M. J. (2011). Cancer immunoediting: Integrating immunity's roles in cancer suppression and promotion. *Science, 331*(6024), 1565–1570. https://doi.org/10.1126/science.1203486.

Schrock, A. B., Ouyang, C., Sandhu, J., Sokol, E., Jin, D., Ross, J. S., et al. (2019). Tumor mutational burden is predictive of response to immune checkpoint inhibitors in MSI-high metastatic colorectal cancer. *Annals of Oncology, 30*(7), 1096–1103. https://doi.org/10.1093/annonc/mdz134.

Schwaab, T., Weiss, J. E., Schned, A. R., & Barth, R. J., Jr. (2001). Dendritic cell infiltration in colon cancer. *Journal of Immunotherapy: Official Journal of the Society for Biological Therapy, 24*(2), 130–137.

Schwartz, R. H. (1992). Costimulation of T lymphocytes: The role of CD28, CTLA-4, and B7/BB1 in interleukin-2 production and immunotherapy. *Cell, 71*(7), 1065–1068. https://doi.org/10.1016/s0092-8674(05)80055-8.

Schwitalle, Y., Kloor, M., Eiermann, S., Linnebacher, M., Kienle, P., Knaebel, H. P., et al. (2008). Immune response against frameshift-induced neopeptides in HNPCC patients and healthy HNPCC mutation carriers. *Gastroenterology, 134*(4), 988–997. https://doi.org/10.1053/j.gastro.2008.01.015.

Schwitalle, Y., Linnebacher, M., Ripberger, E., Gebert, J., & von Knebel Doeberitz, M. (2004). Immunogenic peptides generated by frameshift mutations in DNA mismatch repair-deficient cancer cells. *Cancer Immunity, 4*, 14.

Shahda, S., Noonan, A. M., Bekaii-Saab, T. S., O'Neill, B. H., Sehdev, A., Lalib, W., et al. (2017). A phase II study of pembrolizumab in combination with mFOLFOX6 for patients with advanced colorectal cancer. *Journal of Clinical Oncology, 35*(15_suppl), 3541 (May 20, 2017).

Shin, D. S., Zaretsky, J. M., Escuin-Ordinas, H., Garcia-Diaz, A., Hu-Lieskovan, S., Kalbasi, A., et al. (2017). Primary resistance to PD-1 blockade mediated by JAK1/2 mutations. *Cancer Discovery*, 7(2), 188–201. https://doi.org/10.1158/2159-8290.CD-16-1223.

Shindo, Y., Yoshimura, K., Kuramasu, A., Watanabe, Y., Ito, H., Kondo, T., et al. (2015). Combination immunotherapy with 4-1BB activation and PD-1 blockade enhances antitumor efficacy in a mouse model of subcutaneous tumor. *Anticancer Research, 35*(1), 129–136.

Shinto, E., Hase, K., Hashiguchi, Y., Sekizawa, A., Ueno, H., Shikina, A., et al. (2014). CD8 + and FOXP3+ tumor-infiltrating T cells before and after chemoradiotherapy for rectal cancer. *Annals of Surgical Oncology, 21*(Suppl 3), S414–S421. https://doi.org/10.1245/s10434-014-3584-y.

Silberman, R., Steiner, D. F., Lo, A. A., Gomez, A., Zehnder, J. L., Chu, G., et al. (2019). Complete and prolonged response to immune checkpoint blockade in POLE-mutated colorectal cancer. *JCO Precision Oncology, 3*, 1–5. https://doi.org/10.1200/Po.18.00214.

Simpson, A. J., Caballero, O. L., Jungbluth, A., Chen, Y. T., & Old, L. J. (2005). Cancer/testis antigens, gametogenesis and cancer. *Nature Reviews. Cancer, 5*(8), 615–625. https://doi.org/10.1038/nrc1669.

Sinicrope, F. A., Shi, Q., Hermitte, F., Heying, E. N., Benson, A. B., Gill, S., et al. (2018). Immunoscore to provide prognostic information in low- (T1-3N1) and high-risk (T4 or N2) subsets of stage III colon carcinoma patients treated with adjuvant FOLFOX in a phase III trial (NCCTG N0147; Alliance). *Journal of Clinical Oncology*, *36*(4_suppl), 614. https://doi.org/10.1200/JCO.2018.36.4_suppl.614.

Snahnicanova, Z., Kasubova, I., Kalman, M., Grendar, M., Mikolajcik, P., Gabonova, E., et al. (2020). Genetic and epigenetic analysis of the beta-2-microglobulin gene in microsatellite instable colorectal cancer. *Clinical and Experimental Medicine*, *20*(1), 87–95. https://doi.org/10.1007/s10238-019-00601-7.

Snyder, A., Makarov, V., Merghoub, T., Yuan, J., Zaretsky, J. M., Desrichard, A., et al. (2014). Genetic basis for clinical response to CTLA-4 blockade in melanoma. *New England Journal of Medicine*, *371*(23), 2189–2199. https://doi.org/10.1056/NEJMoa1406498.

Soldevilla, B., Carretero-Puche, C., Gomez-Lopez, G., Al-Shahrour, F., Riesco, M. C., Gil-Calderon, B., et al. (2019). The correlation between immune subtypes and consensus molecular subtypes in colorectal cancer identifies novel tumour microenvironment profiles, with prognostic and therapeutic implications. *European Journal of Cancer*, *123*, 118–129. https://doi.org/10.1016/j.ejca.2019.09.008.

Tabernero, J., Melero, M., Ros, W., Argiles, G., Marabelle, A., Rodriguez-Ruiz, M. E., et al. (2017). Phase Ia and Ib studies of the novel carcinoembryonic antigen (CEA) T-cell bispecific (CEA CD3 TCB) antibody as a single agent and in combination with atezolizumab: Preliminary efficacy and safety in patients with metastatic colorectal cancer (mCRC). *Journal of Clinical Oncology*, *35*(15_suppl), 3002 (May 20, 2017).

Tanis, E., Julie, C., Emile, J. F., Mauer, M., Nordlinger, B., Aust, D., et al. (2015). Prognostic impact of immune response in resectable colorectal liver metastases treated by surgery alone or surgery with perioperative FOLFOX in the randomised EORTC study 40983. *European Journal of Cancer*, *51*(17), 2708–2717. https://doi.org/10.1016/j.ejca.2015.08.014.

Taniura, T., Iida, Y., Kotani, H., Ishitobi, K., Tajima, Y., & Harada, M. (2020). Immunogenic chemotherapy in two mouse colon cancer models. *Cancer Science*, *111*(10), 3527–3539. https://doi.org/10.1111/cas.14624.

Taylor, M. H., Schmidt, E. V., Dutcus, C., Pinheiro, E. M., Yasuhiro, F., Lubiniecki, G., et al. (2020). The LEAP program: Lenvatinib plus pembrolizumab for the treatment of advanced solid tumors. *Future Oncology*, *17*, 637–648. https://doi.org/10.2217/fon-2020-0937.

Terawaki, S., Chikuma, S., Shibayama, S., Hayashi, T., Yoshida, T., Okazaki, T., et al. (2011). IFN-α directly promotes programmed cell death-1 transcription and limits the duration of T cell-mediated immunity. *Journal of Immunology*, *186*(5), 2772–2779. https://doi.org/10.4049/jimmunol.1003208.

Terme, M., Pernot, S., Marcheteau, E., Sandoval, F., Benhamouda, N., Colussi, O., et al. (2013). VEGFA-VEGFR pathway blockade inhibits tumor-induced regulatory T-cell proliferation in colorectal cancer. *Cancer Research*, *73*(2), 539–549. https://doi.org/10.1158/0008-5472.CAN-12-2325.

Tesniere, A., Schlemmer, F., Boige, V., Kepp, O., Martins, I., Ghiringhelli, F., et al. (2010). Immunogenic death of colon cancer cells treated with oxaliplatin. *Oncogene*, *29*(4), 482–491. https://doi.org/10.1038/onc.2009.356.

Thorsson, V., Gibbs, D. L., Brown, S. D., Wolf, D., Bortone, D. S., Ou Yang, T. H., et al. (2018). The immune landscape of cancer. *Immunity*, *48*(4), 812–830.e814. https://doi.org/10.1016/j.immuni.2018.03.023.

Tivol, E. A., Borriello, F., Schweitzer, A. N., Lynch, W. P., Bluestone, J. A., & Sharpe, A. H. (1995). Loss of CTLA-4 leads to massive lymphoproliferation and fatal multiorgan tissue destruction, revealing a critical negative regulatory role of CTLA-4. *Immunity*, *3*(5), 541–547. https://doi.org/10.1016/1074-7613(95)90125-6.

Trabucco, S. E., Gowen, K., Maund, S. L., Sanford, E., Fabrizio, D. A., Hall, M. J., et al. (2019). A novel next-generation sequencing approach to detecting microsatellite instability and pan-tumor characterization of 1000 microsatellite instability-high cases in 67,000 patient samples. *The Journal of Molecular Diagnostics*, *21*(6), 1053–1066. https://doi.org/10.1016/j.jmoldx.2019.06.011.

Turajlic, S., Litchfield, K., Xu, H., Rosenthal, R., McGranahan, N., Reading, J. L., et al. (2017). Insertion-and-deletion-derived tumour-specific neoantigens and the immunogenic phenotype: A pan-cancer analysis. *The Lancet Oncology*, *18*(8), 1009–1021. https://doi.org/10.1016/S1470-2045(17)30516-8.

Van Allen, E. M., Miao, D., Schilling, B., Shukla, S. A., Blank, C., Zimmer, L., et al. (2015). Genomic correlates of response to CTLA-4 blockade in metastatic melanoma. *Science*, *350*(6257), 207–211. https://doi.org/10.1126/science.aad0095.

Van den Eynde, M., Mlecnik, B., Bindea, G., Fredriksen, T., Church, S. E., Lafontaine, L., et al. (2018). The link between the multiverse of immune microenvironments in metastases and the survival of colorectal cancer patients. *Cancer Cell*, *34*(6), 1012–1026. e1013. https://doi.org/10.1016/j.ccell.2018.11.003.

Väyrynen, J. P., Sajanti, S. A., Klintrup, K., Mäkelä, J., Herzig, K. H., Karttunen, T. J., et al. (2014). Characteristics and significance of colorectal cancer associated lymphoid reaction. *International Journal of Cancer*, *134*(9), 2126–2135. https://doi.org/10.1002/ijc.28533.

Vincent, J., Mignot, G., Chalmin, F., Ladoire, S., Bruchard, M., Chevriaux, A., et al. (2010). 5-fluorouracil selectively kills tumor-associated myeloid-derived suppressor cells resulting in enhanced T cell-dependent antitumor immunity. *Cancer Research*, *70*(8), 3052–3061. https://doi.org/10.1158/0008-5472.CAN-09-3690.

Vivier, E., Raulet, D. H., Moretta, A., Caligiuri, M. A., Zitvogel, L., Lanier, L. L., et al. (2011). Innate or adaptive immunity? The example of natural killer cells. *Science*, *331*(6013), 44–49. https://doi.org/10.1126/science.1198687.

Vogelstein, B., Fearon, E. R., Hamilton, S. R., Kern, S. E., Preisinger, A. C., Leppert, M., et al. (1988). Genetic alterations during colorectal-tumor development. *The New England Journal of Medicine*, *319*(9), 525–532. https://doi.org/10.1056/NEJM198809013190901.

Wang, C., Chevalier, D., Saluja, J., Sandhu, J., Lau, C., & Fakih, M. (2020). Regorafenib and Nivolumab or Pembrolizumab combination and circulating tumor DNA response assessment in refractory microsatellite stable colorectal cancer. *The Oncologist*, *25*(8), e1188–e1194. https://doi.org/10.1634/theoncologist.2020-0161.

Wang, C., Gong, J., Tu, T. Y., Lee, P. P., & Fakih, M. (2018). Immune profiling of microsatellite instability-high and polymerase epsilon (POLE)-mutated metastatic colorectal tumors identifies predictors of response to anti-PD-1 therapy. *Journal of Gastrointestinal Oncology*, *9*(3), 404–415. https://doi.org/10.21037/jgo.2018.01.09.

Wang, Y., Wei, B., Gao, J., Cai, X., Xu, L., Zhong, H., et al. (2020). Combination of fruquintinib and anti-PD-1 for the treatment of colorectal cancer. *Journal of Immunology*, *205*(10), 2905–2915. https://doi.org/10.4049/jimmunol.2000463.

Wang, C., Yan, J., Yin, P., Gui, L., Ji, L., Ma, B., et al. (2020). Beta-catenin inhibition shapes tumor immunity and synergizes with immunotherapy in colorectal cancer. *Oncoimmunology*, *9*(1), 1809947. https://doi.org/10.1080/2162402X.2020.1809947.

Ward, J. P., Gubin, M. M., & Schreiber, R. D. (2016). The role of neoantigens in naturally occurring and therapeutically induced immune responses to cancer. *Advances in Immunology*, *130*, 25–74. https://doi.org/10.1016/bs.ai.2016.01.001.

Waterhouse, P., Penninger, J. M., Timms, E., Wakeham, A., Shahinian, A., Lee, K. P., et al. (1995). Lymphoproliferative disorders with early lethality in mice deficient in Ctla-4. *Science*, *270*(5238), 985–988. https://doi.org/10.1126/science.270.5238.985.

Watson, N. F., Ramage, J. M., Madjd, Z., Spendlove, I., Ellis, I. O., Scholefield, J. H., et al. (2006). Immunosurveillance is active in colorectal cancer as downregulation but not complete loss of MHC class I expression correlates with a poor prognosis. *International Journal of Cancer*, *118*(1), 6–10. https://doi.org/10.1002/ijc.21303.

Weisenberger, D. J., Siegmund, K. D., Campan, M., Young, J., Long, T. I., Faasse, M. A., et al. (2006). CpG island methylator phenotype underlies sporadic microsatellite instability and is tightly associated with BRAF mutation in colorectal cancer. *Nature Genetics*, *38*(7), 787–793. https://doi.org/10.1038/ng1834.

Wu, S., Rhee, K. J., Albesiano, E., Rabizadeh, S., Wu, X., Yen, H. R., et al. (2009). A human colonic commensal promotes colon tumorigenesis via activation of T helper type 17 T cell responses. *Nature Medicine*, *15*(9), 1016–1022. https://doi.org/10.1038/nm.2015.

Yamazaki, T., Akiba, H., Iwai, H., Matsuda, H., Aoki, M., Tanno, Y., et al. (2002). Expression of programmed death 1 ligands by murine T cells and APC. *Journal of Immunology*, *169*(10), 5538–5545. https://doi.org/10.4049/jimmunol.169.10.5538.

Yang, Z., Zhang, M., Peng, R., Liu, J., Wang, F., Li, Y., et al. (2020). The prognostic and clinicopathological value of tumor-associated macrophages in patients with colorectal cancer: A systematic review and meta-analysis. *International Journal of Colorectal Disease*, *35*(9), 1651–1661. https://doi.org/10.1007/s00384-020-03686-9.

Yarchoan, M., Hopkins, A., & Jaffee, E. M. (2017). Tumor mutational burden and response rate to PD-1 inhibition. *New England Journal of Medicine*, *377*(25), 2500–2501. https://doi.org/10.1056/NEJMc1713444.

Yurgelun, M. B., Goel, A., Hornick, J. L., Sen, A., Turgeon, D. K., Ruffin, M. T. T., et al. (2012). Microsatellite instability and DNA mismatch repair protein deficiency in Lynch syndrome colorectal polyps. *Cancer Prevention Research (Philadelphia, Pa.)*, *5*(4), 574–582. https://doi.org/10.1158/1940-6207.CAPR-11-0519.

Zaretsky, J. M., Garcia-Diaz, A., Shin, D. S., Escuin-Ordinas, H., Hugo, W., Hu-Lieskovan, S., et al. (2016). Mutations associated with acquired resistance to PD-1 blockade in melanoma. *The New England Journal of Medicine*, *375*(9), 819–829. https://doi.org/10.1056/NEJMoa1604958.

Zelenay, S., van der Veen, A. G., Bottcher, J. P., Snelgrove, K. J., Rogers, N., Acton, S. E., et al. (2015). Cyclooxygenase-dependent tumor growth through evasion of immunity. *Cell*, *162*(6), 1257–1270. https://doi.org/10.1016/j.cell.2015.08.015.

Zhou, J., Dudley, M. E., Rosenberg, S. A., & Robbins, P. F. (2005). Persistence of multiple tumor-specific T-cell clones is associated with complete tumor regression in a melanoma patient receiving adoptive cell transfer therapy. *Journal of Immunotherapy (Hagerstown, MD)*, *28*(1), 53–62. https://doi.org/10.1097/00002371-200501000-00007.

CHAPTER SIX

Racial and ethnic disparities in colorectal cancer incidence and mortality

John M. Carethers*
Division of Gastroenterology and Hepatology, Department of Internal Medicine, and Department of Human Genetics and Rogel Cancer Center, University of Michigan, Ann Arbor, MI, United States
*Corresponding author: e-mail address: jcarethe@umich.edu

Contents

Abstract

The occurrence of colorectal cancer (CRC) shows a large disparity among recognized races and ethnicities in the U.S., with Black Americans demonstrating the highest incidence and mortality from this disease. Contributors for the observed CRC disparity appear to be multifactorial and consequential that may be initiated by structured societal issues (e.g., low socioeconomic status and lack of adequate health insurance) that facilitate abnormal environmental factors (through use of tobacco and alcohol, and poor diet composition that modifies one's metabolism, microbiome and local immune microenvironment) and trigger cancer-specific immune and genetic changes (e.g., localized inflammation and somatic driver gene mutations). Mitigating the disparity by prevention through CRC screening has been demonstrated; this has not been

Advances in Cancer Research, Volume 151
ISSN 0065-230X
https://doi.org/10.1016/bs.acr.2021.02.007

adequately shown once CRC has developed. Acquiring additional knowledge into the science behind the observed disparity will inform approaches towards abating both the incidence and mortality of CRC between U.S. racial and ethnic groups.

Abbreviations

5-FU	5-fluorouracil
CI	confidence interval
CIMP	CpG island methylator phenotype
CORI	Clinical Outcomes Research Initiative
COVID-19	coronavirus disease 2019
CRC	colorectal cancer
EMAST	elevated microsatellite alterations at selected tetranucleotide repeats
FAP	familial adenomatous polyposis
FIT	fecal immunochemical test
HRT	hormone replacement therapy
IL6	interleukin 6
MMR	DNA mismatch repair
MSI-H	microsatellite instability-high
mt-sDNA	multi-target stool DNA test
NSAID	non-steroidal anti-inflammatory drug
PD-L1	programmed death ligand-1
PLCO	Prostate, Lung, Colorectal, and Ovarian Cancer Screening Trial
SEER	Surveillance, Epidemiology, and End Results
SSA	sessile serrated adenoma

1. Epidemiology of CRC disparities in racial/ethnic populations

Colorectal cancer (CRC) will affect 149,500 persons and is expected to kill 52,980 in the U.S. during 2021, despite it being a largely preventable disease with an achievable high utilization of screening in the population (Carethers, 2015, 2020; Siegel, Miller, Fuchs, & Jemal, 2021). The distribution of CRC is not even across U.S. subpopulations; there is marked difference in CRC incidence, cancer stage, and cancer mortality by race and ethnicity. In particular, Black Americans show the highest incidence, and have the highest mortality among major U.S. racial and ethnic groups (Ashktorab, Brim, Kupfer, & Carethers, 2017). Data from the Surveillance, Epidemiology, and End Results (SEER) program reveal that Black American's overall incidence for CRC is 41.9 per 100,000, as compared to that of White Americans of 37.0 per 100,000 (Table 1). After Black Americans, Native Americans have the second highest CRC incidence rate

Table 1 SEER U.S. age-adjusted colorectal cancer incidence rates, 2000–2017, and age-adjusted colorectal cancer mortality rates, 2000–2018.

	All	White	Black	Asian/PI	Am Indian	Hispanic
Incidence						
Overall	37.3	37.0	41.9	31.7	39.3	33.5
Male	42.4	41.8	49.4	37.2	38.0	38.4
Female	32.9	32.8	36.5	27.3	39.9	29.6
Early-onset (20–44 years)		6.7	7.9	6.3		
Deaths						
Overall	13.1	12.9	16.8	8.9	14.0	10.8
Male	15.8	15.5	21.3	10.8	16.4	14.0
Female	10.9	10.8	13.8	7.3	12.0	8.3

Data shown are per 100,000 population (adjusted to the 2000 U.S. standard population), and obtained from seer.cancer.gov. Data for early-onset colorectal cancer incidence rates per 100,000 are derived from SEER in Ashktorab H, et al, *Dig Dis Sci* 2016;61:3026–3030.

at 39.3 per 100,000. The CRC incidence disparity for Black Americans is observed in patients under the age of 45 years, indicating that factors that allow for that disparity begin earlier than the typical CRC screening age of 50 years. Data for CRC mortality follows the pattern of the incidence rates, with Black Americans showing the highest mortality among racial/ethnic groups at 16.8 per 100,000 and Native Americans second at 14.0 per 100,000, as compared to White Americans at 12.9 per 100,000. Incidence and mortality rates for males are higher than females across all subpopulations (Table 1).

While the overall trends for CRC incidence have fallen since the 1980s for not so clear reasons initially, further decrease has been accentuated by the implementation of national screening efforts in the U.S. (Ashktorab, Brim et al., 2017; Carethers, 2015) The only exception to this is the rising rate of CRC among persons under the age of 50 years, a population that does not routinely undergo CRC screening (Ashktorab, Vilmenay et al., 2016; Carethers, 2016a; Hussan et al., 2020) (see also chapter "Early age onset colorectal cancer" by Patel et al.). The overall trend for decreased incidence and mortality is spread among racial/ethnic populations. For instance, since 2000, CRC incidence rates for Black Americans dropped from 60 per 100,000 to 41.9 per 100,000 by 2017, and rates for White Americans dropped from 56 per 100,000 to 37.0 per 100,000 (Fig. 1A). CRC incidence

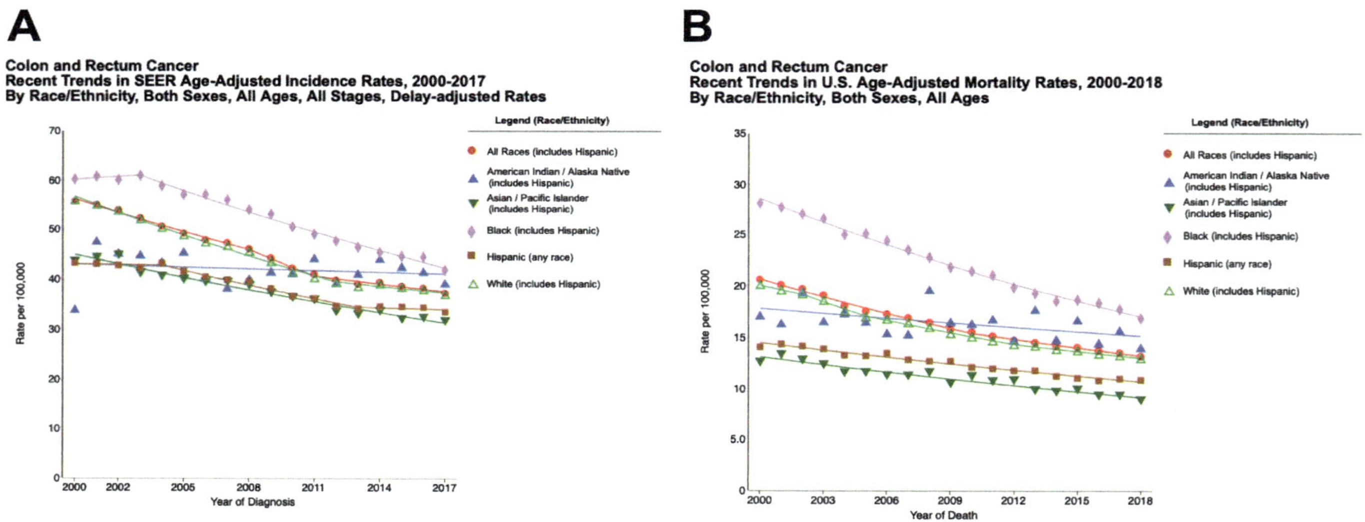

Fig. 1 Trends in SEER age-adjusted colorectal cancer incidence and mortality per 100,000 (adjusted to the 2000 U.S. standard population), by race/ethnicity. (A) Trends for SEER age-adjusted colorectal cancer incidence, 2000–2017. (B) Trends for SEER age-adjusted colorectal cancer mortality, 2000–2018. *Data obtained from seer.cancer.gov.*

rates for Native Americans "appear" stable over the 2000–2017 time period; this is likely because CRCs among Native Americans were undercounted in the past, perhaps due to shorter lifespans from comorbid conditions such as diabetes mellitus, precluding the development and tallying of CRC among older populations who are more likely to develop CRC. Improved medical care among Native Americans over time have provided for longer lifespans and increased observation of the development of CRCs. The rate of CRC incidence among Native Americans is approximating the CRC incidence among Black Americans, and this may be due to more gains in cancer prevention applied in the Black American population as compared to the Native American population. As with CRC incidence, CRC mortality rates have dropped amongst all racial/ethnic groups (Fig. 1B). And as with incidence, CRC mortality rates for Black and Native Americans are above the average mortality rates for all races (Fig. 1B). It should be noted that Asian Americans and Hispanic Americans demonstrate lower than average incidence and mortality rates for CRC. However, there are likely subpopulations among both Hispanics and Asian Americans that showcase high incidence and/or mortality due to both populations being heterogeneous in their genomic origins (Zavala et al., 2021).

Data for disease stage presentation, which is most robust for Black and White Americans, also show a disparity, affecting survival (DeSantis, Miller, Sauer, Jemal, & Siegel, 2019; Siegel et al., 2021). For instance, at disease presentation, 37% of Black Americans present with localized CRC, where the disease is easily curable by surgical or endoscopic removal, compared to 38% of White Americans, and only 32% of Black Americans present with regional disease as compared to 36% of White Americans, where the disease is generally curable by a combination of surgery and chemotherapy. However, 26% of Black Americans present with metastatic CRC as compared to only 22% of White Americans, where the disease is often fatal (Siegel et al., 2021). Overall 5-year CRC survival rates for Black Americans have been consistently 6–12% below that of White Americans, with Black American overall survival rates at 61% and White American survival rates at 67% in 2017 (Fig. 2A). While the disparity in survival is very small for localized disease between Black and White Americans (87% vs 90%) (Fig. 2B), the disparity is slightly wider with regional disease (71% vs 75%, respectively) (Fig. 2C) and wider for metastatic disease (12% vs 17%, respectively) (Fig. 2D).

CRC is the third most common cancer diagnosed and second deadliest cancer among all Americans (DeSantis et al., 2019; Siegel et al., 2021). The Black:White incidence ratio for CRC is 1.13, indicating that for every 100

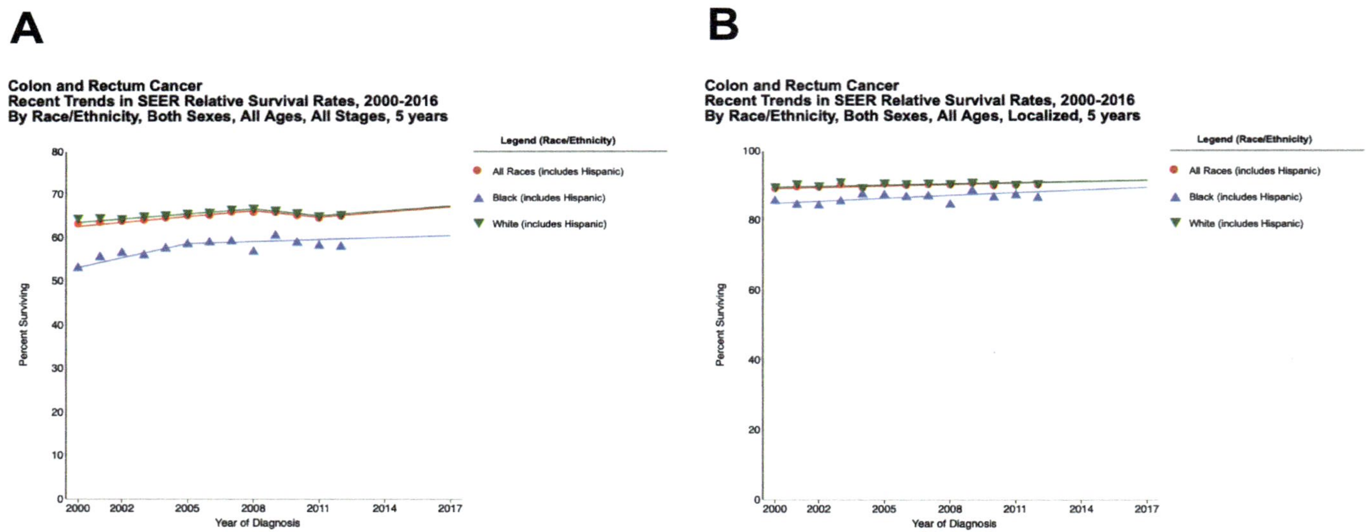

Fig. 2 Trends in SEER relative survival rates per 100,000 (adjusted to the 2000 U.S. standard population), 2000–2016, by race/ethnicity. (A) Trends for SEER relative survival rates for all stages of colorectal cancer. (B) Trends for SEER relative survival rates for localized stage of colorectal cancer. (C) Trends for SEER relative survival rates for regional stage of colorectal cancer. (D) Trends for SEER relative survival rates for distant stage of colorectal cancer. *Data obtained from seer.cancer.gov.*

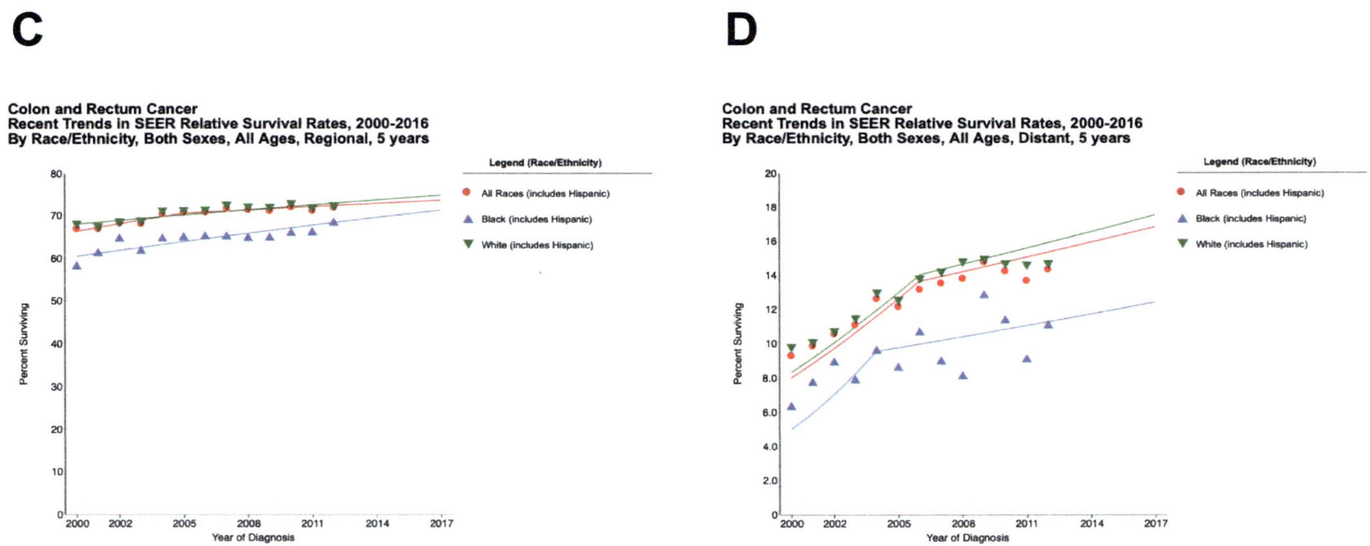
C
Colon and Rectum Cancer
Recent Trends in SEER Relative Survival Rates, 2000-2016
By Race/Ethnicity, Both Sexes, All Ages, Regional, 5 years
Percent Surviving
Year of Diagnosis
Legend (Race/Ethnicity)
All Races (includes Hispanic)
Black (includes Hispanic)
White (includes Hispanic)
D
Colon and Rectum Cancer
Recent Trends in SEER Relative Survival Rates, 2000-2016
By Race/Ethnicity, Both Sexes, All Ages, Distant, 5 years
Percent Surviving
Year of Diagnosis
Legend (Race/Ethnicity)
All Races (includes Hispanic)
Black (includes Hispanic)
White (includes Hispanic)

Fig. 2—Cont'd

CRCs in White Americans, there are 113 CRCs in Black Americans (Ashktorab, Brim et al., 2017; Carethers, 2018). The Black:White mortality ratio for CRC is 1.32, again meaning for every 100 CRC deaths in White Americans, there are 132 CRC deaths among Black Americans (Ashktorab, Brim et al., 2017; Carethers, 2018). These ratios indicate a disparity for CRC incidence (1.13), but also a growth in the degree of disparity once the disease is diagnosed for mortality (1.32) among Black Americans. Additionally, Black Americans present 3–7 years earlier in median age for CRC diagnosis than that for White Americans, with the median age of 63 years for Black men and 64 years for Black women, compared to 66 years for White men and 70 years for White women (Carethers, 2015, 2018; Siegel et al., 2021). In essence, the CRC incidence curve for Black Americans is shifted to earlier ages, with the proportion of CRCs under the age of 50 years for Black Americans twice the proportion of CRCs under the age of 50 years for White Americans (Carethers, 2018). Prior analysis showed that if CRC screening for African Americans were to begin at age 45 years that the proportion of CRCs under the age of 45 would equal the proportion of CRCs under the age of 50 years for White Americans (Carethers, 2005, 2018). Furthermore, there are anatomical site differences, with Black Americans consistently showing more right-sided colon cancers than other racial/ethnic groups; this is highly relevant as the detection of and survival from right-sided colon cancers are lower than the left side of the colon (Brenner, Chang-Claude, Seiler, Rickert, & Hoffmeister, 2011; Carethers, 2005, 2011, 2015; Carethers et al., 2014; Nishihara et al., 2013).

The multitude of observations over several decades make no doubt of a disparity for CRC that decreases survival among Black Americans as compared to any other racial/ethnic group in the U.S. There are likely multifactorial drivers that lead to the creation of the disparity that are explored below. These include the mirroring of CRC patterns with earlier and right-sided location and development of clinically significant precursor adenomas, the contribution of socioeconomic inequities that might drive the biology within the colon creating the disparity and the consequences of lifelong induced biological differences on the behavior of CRC, and effects of missed screening in populations, enhancing the disparity.

2. Root causes for CRC risk disparity in racial/ethnic populations

Epidemiological studies link several factors that put a person at risk for CRC. About 65% of the risk comes from environmental factors, whereas

35% of the risk comes from genetic factors (Lichtenstein, Holm, Verkasalo, et al., 2000). Environmental factors include those contributions that indirectly or directly influence the colonic epithelium, such as downstream effects of ingesting a diet rich in red meat, high fat or high calories, acquisition of obesity and excess body mass index, ingesting a low fiber diet, use of tobacco products, and possessing low serum levels for calcium or vitamin D (Carethers & Jung, 2015; Grady & Carethers, 2008; O'Keefe, Ou, Aufreiter, et al., 2009; Satia et al., 2005) (see also chapter "Personalized nutrition for colorectal cancer: The interplay of diet, lifestyle, and the gut microbiota in colorectal cancer prevention and survivorship" by Wang and Song). Environmental factors may have varying degrees of effect among populations and are potentially modifiable for persons (Fig. 3). A second modifiable risk factor is CRC screening utilization, which may mitigate much of the environmental effects if performed population-wide (Fig. 3) (Grubbs et al., 2013). High screening utilization has been shown to eliminate the CRC incidence and nearly erase the mortality disparity among Black Americans (Grubbs et al., 2013).

The genetic or heritable susceptibility component for CRC risk is a non-modifiable factor that is largely related to a person's family history of cancer, the age of onset of those cancers in relatives, and the person's racial/ethnic ancestry (Fig. 3). Heritable susceptibility factors generally include significant family history of cancer; persons in this category are removed from the average population screening pool and placed in a more focused surveillance pool to assess colonic neoplasia at earlier and more frequent intervals. Persons may also be assessed through genetic counseling and appropriate genetic testing for high and low risk alleles for CRC predisposition syndromes (Carethers & Stoffel, 2015; Rex et al., 2017; Stoffel & Carethers, 2020) (Fig. 3). At present, Black and Hispanic Americans with

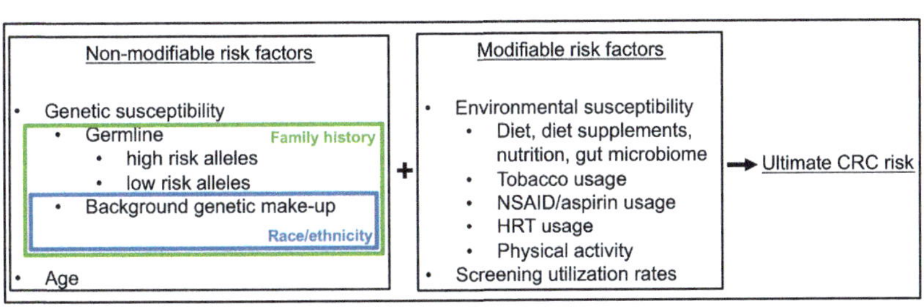

Fig. 3 Non-modifiable and modifiable patient risk factors associated with the development of colorectal cancer. Green box includes elements that make up family history, and the blue box contains elements that make up race/ethnicity. NSAID, non-steroidal anti-inflammatory drug; HRT, hormone replacement therapy; CRC, colorectal cancer.

a family history have the lowest likelihood of participation in screening (Perencevich, Ojha, Steyerberg, & Syngal, 2013). Furthermore, Black Americans are less likely to know their parental history of cancer than White Americans (Kupfer, McCaffrey, & Kim, 2006), and family members are less likely to tell relatives about the finding of colonic polyps (Murff, Peterson, Fowke, et al., 2008). This lack of information, lack of knowledge or lack of transmittal of information from Black Americans may falsely put one in the sporadic average-risk screening category rather than in the category with a positive family history where screening starts earlier. Reported heritable susceptibility factors often do not contemplate a contribution from racial/ethnic ancestry. Age is another non-modifiable powerful predictor for CRC risk, with exponential rise of CRCs after the age of 50 years (Perencevich et al., 2013). Historically, nearly 95% of all CRCs happened after the age of 50 years leading to the original recommendations that screening start for those at average risk at this age (Winawer, Fletcher, Miller, et al., 1997); recent trends suggest that this proportion has shrunk with the occurrence of more early-onset CRCs (Hussan et al., 2020; Peterse et al., 2018; Siegel et al., 2021). Specifically, for Black Americans who have the highest rates for CRC in the U.S., the proportion of CRCs over the age of 50 years was 89% as compared to 95% for the general population, meaning more cancer in this population were not screened for because of the higher proportion under the age of 50 years (Carethers, 2005, 2018). The observation that Black Americans present with CRC at younger ages than White Americans has several implications. First, there is less time from age 50 years to the average age of onset for CRC in the majority population, shortening the potential window period for screening and intervention. Second, earlier onset of CRC may have implications for other family members and their risk, and the approach to their screening. Third, younger persons have a higher likelihood of possessing a heritable syndrome that would shift a person from screening to intense surveillance. And fourth, since most CRCs originate from adenomas, implies that adenomas may be initiated earlier in Black Americans (Carethers, 2015, 2016a; Carethers & Stoffel, 2015; Stoffel & Carethers, 2020). Race/ethnicity as a heritable factor was not included in CRC screening recommendations in the same way as family history or age until 2017, despite the multitude of studies showing a disparity in incidence and mortality from CRC among Black Americans (Rex et al., 2017) (Fig. 3).

Defining drivers for risk that are disparate between populations is often difficult unless all diverse populations are studied relatively equally with acquired data to gain insights for potential interventions. Most racial/ethnic groups in the U.S. have not been studied or had the same level of participation in clinical trials as the majority population. However, there are a number of potential contributing factors that may generate the incidence and mortality disparity from CRC, particularly for Black Americans. These include: (a) lower level of education, and lower socioeconomic status, (b) lower level of insurance coverage, less access to medical care particularly preventive approaches, (c) ingesting diets based on socioeconomic ability that are more conducive towards CRC development, (d) lower levels of physical activity and higher rates of obesity, (e) higher rates of tobacco and alcohol usage, (f) lower utilization of screening modalities for CRC, (g) lower utilization of aspirin/NSAIDs and hormone replacement therapy that are associated with reduced risk for CRC, and (h) generational mistrust of the U.S. health system (Carethers & Doubeni, 2020; DeSantis et al., 2019; Green, Peters-Lewis, Percac-Lima, et al., 2008; Kim, Pe'rez-Stable, Wong, et al., 2008; Lumpkins et al., 2013; Tammana & Laiyemo, 2014). The majority of these contributing factors inferred to explain the CRC disparity for Black Americans are in theory correctable. This CRC disparity has been reported for decades (Carethers, 1999; DeSantis et al., 2019; Siegel et al., 2021), suggesting that the relevant environmental, social and behavior factors are difficult to modify (either at the personal or societal level). Interestingly, there are societal, environmental and behavior factors that should confer similar CRC risk between Black Americans and Hispanic Americans, yet the CRC incidence and mortality are lower for Hispanic Americans as compared to Black Americans (Table 1). This observation suggests additional factors involving biology and/or genetic background could be contributors, particularly in the right microenvironment of the colon (Carethers, 2015).

The coronavirus disease 2019 (COVID-19) pandemic has sharpened the focus on root causes for disparities in racial/ethnic populations, since this disease, like CRC, disproportionately affects these populations (Carethers, 2021a,b; Newman, Winn, & Carethers, 2021). At least a proportion of the origin of the disparity for some racial/ethnic populations lie in systemic structural disadvantages for these populations that ultimately place them at higher risk for CRC (Fig. 4). Generational disadvantages in the U.S. overall initiates a cascade of consequential lifelong effects that increase risk for CRC, commencing with lower socioeconomic status of families and individuals that lessen the ability to obtain higher education, and determines what neighborhoods one lives in. This in turn limits the type of employment

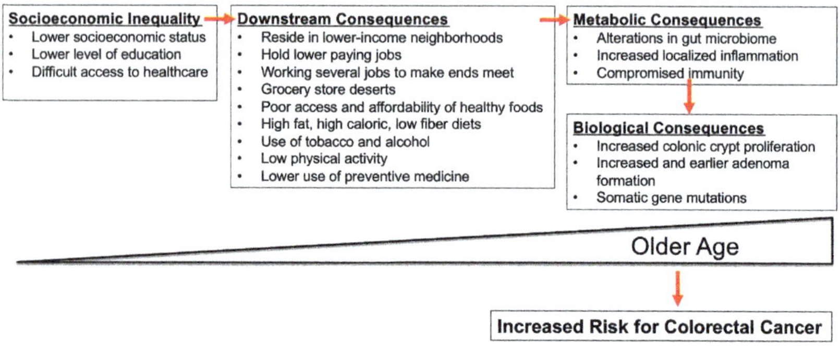

Fig. 4 Connections and consequences initiated by socioeconomic disparities for ultimate risk and development of colorectal cancer.

one can hold, and more likely that their neighborhood environment may be devoid of grocery stores but with ample liquor or party stores, with easy accessibility to tobacco and alcohol instead of fresh fruit, vegetables, and protein sources. The neighborhood environment is more likely to lack open park spaces for physical activity, and access to medical care may be limited by the type and density of medical practitioners. Over time, with altered diet and relative physical inactivity and less than optimal preventive care, one's gut microbiome is altered, and it along with obesity increases localized tissue inflammation and compromises healthy immunity. These in turn have direct biological consequences on colonic crypt proliferation rates, increasing the possibility of biological and environmental induced somatic gene mutations that lead to colonic neoplasia (Fig. 4). It may be possible to intervene at any point of this process to reduce CRC risk; this includes lifestyle changes which remain challenging to implement to have large impact (Carethers, 2015) and screening, which if implemented with high levels of utilization, may fully mitigate much of the disparity (Grubbs et al., 2013).

3. Biological and genetic contributions to CRC risk in racial/ethnic populations

The adenoma-to-carcinoma histologic and genetic progression of CRC formation has been elucidated over the past 30 years (Fig. 5) (Cancer Genome Atlas Network, 2012; Carethers & Jung, 2015; Grady & Carethers, 2008). Monogenetic heritable conditions such as Familial Adenomatous Polyposis (FAP), caused by a germline mutation in the *APC* gene, and Lynch syndrome, caused by a germline mutation in one of the DNA mismatch repair (MMR) genes (*MSH2*, *MLH1*, *MSH6*, *PMS2*, *EPCAM*)

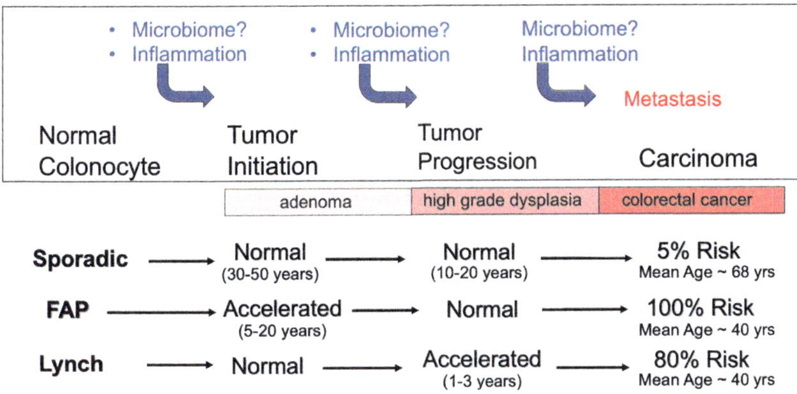

Fig. 5 Adenoma-to-carcinoma sequence in the human colon for sporadic, familial adenomatous polyposis (FAP) and Lynch colorectal cancers, with potential environmental modifiers and ultimate population risk of cancer. Note that sporadic colorectal cancer has a dwell time of up to 5 decades for tumor initiation and up to 2 additional decades for tumor progression; tumor initiation is greatly accelerated in FAP while tumor progression is greatly accelerated in Lynch syndrome. Environmental influences that modify the local microbiome and inflammatory microenvironment may modify the rate of progression of the adenoma-to-carcinoma sequence at any step.

speed up portions of the adenoma-to-carcinoma process compared to that of sporadic CRCs as a consequence of the germline mutations and subsequent effects on cellular growth pathways in the transformed colonocyte (Fig. 5) (Carethers & Stoffel, 2015; Stoffel & Carethers, 2020). What is not as clearly elucidated as the genetics are the environmental effects on colonocyte epithelial behavior and the adenoma-to-carcinoma sequence, which may be heavily influenced by contributions from the gut microbiome and localized inflammatory pathways (Fig. 5) (Brennan & Garrett, 2016; Carethers, Koi, & Tseng-Rogenski, 2015; Dejea et al., 2014, 2018; Koi, Okita, & Carethers, 2018; Lee et al., 2012; Munakata et al., 2019; Okita et al., 2020; Tseng-Rogenski et al., 2012; Tseng-Rogenski, Hamaya, Choi, & Carethers, 2015). It is these environmental influences that are more likely to drive the contribution for CRC incidence and mortality disparities.

3.1 Anatomical site of CRCs and frequency and site of precursor adenomas

The intracolonic location of CRCs and precursor adenomas have implications for population and subpopulation screening. Non-invasive screening tests such as fecal immunochemical test (FIT) and multi-target stool DNA test (mt-sDNA) may help identify CRCs (and much less so adenomas) anywhere in the colon, whereas flexible sigmoidoscopy is limited to the terminal 60 cm of the colorectum (Carethers, 2020; Rex et al., 2017). Moreover, proximal cancers and adenomas (orad to the colonic splenic flexure) may have a different genetic makeup, such as CpG island methylator phenotype (CIMP), microsatellite instability-high (MSI-H) and show different biological behavior (such as flat superficial spreading morphology, rapid growth, serrated pattern) and grow to larger sizes without colonic obstruction in the liquid stool containing proximal colon (Carethers, 2011). The majority of data for anatomical CRC site in regards to race/ethnicity is for Black Americans.

Black Americans with CRC have a 7 to 15% higher frequency prevalence of proximal distribution compared to White Americans in multiple studies (Carethers et al., 2014; Popat, Hubner, & Houlston, 2005; Schroy, Coe, Chen, O'Brien, & Heeren, 2013; Shavers, 2007; Xicola, Gagnon, Clark, et al., 2014). This remains true among population-based studies in which MSI-H CRCs, which are predominantly in the proximal colon, are removed from analyses, keeping an ~10% increase for microsatellite stable (MSS) proximal cancers for Black Americans over White Americans (Carethers et al., 2014; Xicola et al., 2014). This increased prevalence of proximal cancers implies a need to perform colonoscopy rather than flexible sigmoidoscopy to be the more effective endoscopic screening modality; however, colonoscopy is not equally effective in mortality reduction for proximal and distal CRCs (Baxter et al., 2009; Brenner et al., 2011; Nishihara et al., 2013). Colonoscopy produces a 67–84% reduction for left-sided cancers and 1–56% reduction for right-sided cancers (Baxter et al., 2009; Brenner et al., 2011; Nishihara et al., 2013). Given this, even if colonoscopy was fully employed as the screening modality in Black Americans, there might be a disparity remaining due to the excess of right-sided CRC deaths in Black Americans and the reduced sensitivity of colonoscopy for the right colon. Still, colonoscopy remains the gold standard and best test to detect, visualize and remove proximal lesions.

Adenomas, the precursor to most CRCs, mirror the anatomical location differences observed for CRCs in Black Americans. Advanced adenomas (>9 mm in diameter, villous histology or high-grade dysplasia) are direct precursors to CRC and pose the highest risk to patients. Data from the Clinical Outcomes Research Initiative (CORI) revealed that Black Americans had 7.7% advanced adenomas as compared to 6.2% White Americans among 141,413 average-risk persons (Lieberman, Holub, Eisen, Kraemer, & Morris, 2005). Among 327,785 average risk persons in CORI, advanced adenomas were more frequent in Black Americans at every 5-year age interval over the age of 50 years compared to White Americans (age 50–54 years, 7.1% vs 6.2%; age 55–59 years, 8.5% vs 7.4%; age 60–64 years, 11.5% vs 8.6%; age 65–69 years, 12.0% vs 9.7%) (Lieberman et al., 2014). Proximal adenomas were also more prevalent in Black Americans compared to White Americans, with an odds ratio of 1.15 (95%CI 1.03–1.29) in CORI and 1.26 (95% CI 1.04–1.54) from a Kaiser Permanente study (Corley et al., 2013; Lieberman et al., 2014). Indeed, 53.3% of all advanced polyps were proximal among Black Americans compared to 50.6% for White Americans (Lieberman et al., 2014). The Kaiser Permanente study also demonstrated a higher odds ratio for advanced adenomas for Native Americans (1.48, 95%CI 0.85–2.56) as well as for proximal advanced adenomas (1.55, 95%CI 0.79–3.02) compared to White Americans (Corley et al., 2013).

3.2 Sessile serrated lesions

In addition to advanced adenomas, sessile serrated adenomas (SSAs) are high risk for CRC advancement. SSAs are the likely precursor to the flat or laterally spreading cancers identified in the proximal colon. They are often blend in with the colonic mucosa background and have a mucous cap and are difficult to detect without enhanced endoscopic methods. SSAs are much more likely to contain cancer than their typical adenomatous counterparts. These lesions genetically often demonstrate CIMP and may be MSI-H due to methylation of the promotor of the DNA MMR gene *MLH1* (Carethers, 2011; Leggett & Whitehall, 2010). Additionally, SSAs and sessile serrated cancers may contain mutations in *BRAF* but lack mutations in *KRAS* (Carethers, 2011; Leggett & Whitehall, 2010).

One potential assumption is that the high risk SSAs might be more common in certain racial/ethnic populations, particularly for Black Americans, as a way to explain both a right-sided prevalence and higher mortality from CRC. However, there is scant data in this regard for this hypothesis. One study examining colonoscopic findings among 1681 Black and 1172 White Americans identified 0.2% and 0.3% of dysplastic serrated lesions, respectively (Schroy et al., 2013). Among a cohort of 12,085 patients with colonoscopies of which 83% were Black Americans, 252 patients (2.5%) had a sessile serrated lesion that tended to be distal (Ashktorab et al., 2019). Molecular characterization revealed 55.6% of SSAs with $BRAF^{V600E}$ mutation without significant CIMP or MSI-H, perhaps due to the distal location in this study (Ashktorab et al., 2019). Within a Black American population, there is an association of SSAs with the presence of endometrial polyps (Ashktorab et al., 2020).

3.3 Genomic instability: MSI-H and EMAST

CRC is ultimately a genetic disease; a culmination of environmental influences combined with innate genetic damage in a colonic stem cell that alters normally regulated growth pathways to slowly acquire neoplastic characteristics and transform into CRC (Carethers & Jung, 2015; Grady & Carethers, 2008). Environmental factors may affect each stage of pathogenesis as well as metastasis (Fig. 5).

Among sporadic CRCs, inactivation of the DNA MMR gene *MLH1* is the epigenetic cause for the majority of hypermutated tumors (containing a high tumor mutational burden) and generation of microsatellite instability-high (MSI-H), a biomarker demonstrating frameshift mutation of mono- and di-nucleotide microsatellite sequences as a consequence of loss of functional DNA MMR (Cancer Genome Atlas Network, 2012; Carethers, 2016b, 2017; Carethers & Jung, 2015; Grady & Carethers, 2008). MSI-H occurs in ~15% of all CRCs, with cancers more often located in the proximal colon with subepithelial lymphoid aggregates and increased intraepithelial lymphocytes as a response to immunologically-activated truncated protein neoantigens derived from frameshifted coding sequences (Carethers et al., 2014; Carethers & Jung, 2015; Koi & Carethers, 2017; Lee et al., 2012; Raeker & Carethers, 2020). Patients with hypermutated tumors have longer survival as compared to same-staged patients with non-hypermutated tumors in the absence of any therapy, and further demonstrate improvement with immune checkpoint therapy treatment for which these tumors are more sensitive (André et al., 2020; Carethers, 2016b, 2017; Carethers & Jung, 2015). Patients with hypermutated tumors have less response to standardized 5-fluorouracil-containing (5-FU) chemotherapy, as functional DNA MMR is one mechanism to execute the cytotoxicity of 5-FU (Carethers et al., 1999, 2004; Hamaya et al., 2015; Iwaizumi, Tseng-Rogenski, & Carethers, 2011; Suzuki et al., 2016; Tajima, Hess, Cabrera, Kolodner, & Carethers, 2004; Tajima, Iwaizumi, Tseng-Rogenski, Cabrera, & Carethers, 2011). A population-based study demonstrated a difference in proportion of MSI-H/hypermutated tumors among White (14.1%) and Black Americans (6.6%) (Carethers et al., 2014). A lower frequency of MSI-H/hypermutated CRCs among Black Americans was confirmed in a meta-analysis of the few studies that contained this population in their cohorts, with an odds ratio of 0.78 for CRCs to be MSI-H compared to White American CRCs; however due to very few studies, the result did not reach statistical significance

(Ashktorab, Ahuja et al., 2016). The lower apparent prevalence of MSI-H/hypermutated CRCs among Black Americans might contribute to a disparity in survival as a group, and may further limit the use of immune checkpoint inhibitors that could prolong survival. The lower MSI-H/hypermutated prevalence infers that a higher proportion of Black Americans may respond to 5-FU treatment.

Another mechanism to inactivate DNA MMR involves inflammatory pathways (Carethers et al., 2015; Koi, Tseng-Rogenski, & Carethers, 2018; Tseng-Rogenski et al., 2012, 2015, 2020), a consequence of the local environment. Interleukin 6 (IL6), which also triggers release of reactive oxygen species, signals down its JAK/STAT3 kinase pathway to shift the DNA MMR gene MSH3 from its normal place in the nucleus to the cytosol, where it cannot repair DNA (Tseng-Rogenski et al., 2012, 2015, 2020). MSH3 heterodimerizes with MSH2 in the nucleus to form MutSβ and repair dinucleotide or longer microsatellites (Carethers, 2016b, 2017; Carethers et al., 2015; Koi, Tseng-Rogenski et al., 2018; Munakata et al., 2019; Raeker & Carethers, 2020; Tseng-Rogenski et al., 2012, 2015, 2020), and when MSH3 is shifted to the cytosol (without MSH2), long-repeat frameshift mutations occur and accumulate in the nucleus. These inflammation-associated microsatellite alterations, also known as EMAST (elevated microsatellite alterations at selected tetranucleotide repeats) are not detected on regular MSI-H panels as they only include mono- and di-nucleotide markers and no tetra-nucleotide markers (Raeker & Carethers, 2020). CRCs have intraepithelial and intra-tumoral inflammation, showing an intimate relationship with the epithelium that likely effects IL6 levels locally to trigger EMAST (Lee et al., 2010, 2012; Tseng-Rogenski et al., 2015). EMAST can be identified in 50% of all CRCs, and patients whose CRCs demonstrate EMAST tend to be staged as advanced cancers and show shorter survival, unlike patients with MSI-H CRCs (Garcia et al., 2012; Koi et al., 2016). Persons with EMAST CRCs are responsive to 5-FU (Hamaya et al., 2015). The only analysis of EMAST by race/ethnicity is among Black Americans for rectal cancer. From a population-based cohort for which 33% of rectal cancers showed EMAST, Black Americans had 49% tumors showing EMAST, and White Americans had 26% tumors showing EMAST (Devaraj et al., 2010). This data suggests that inflammation is more common with tumors from Black Americans, and in the case of EMAST, may contribute towards

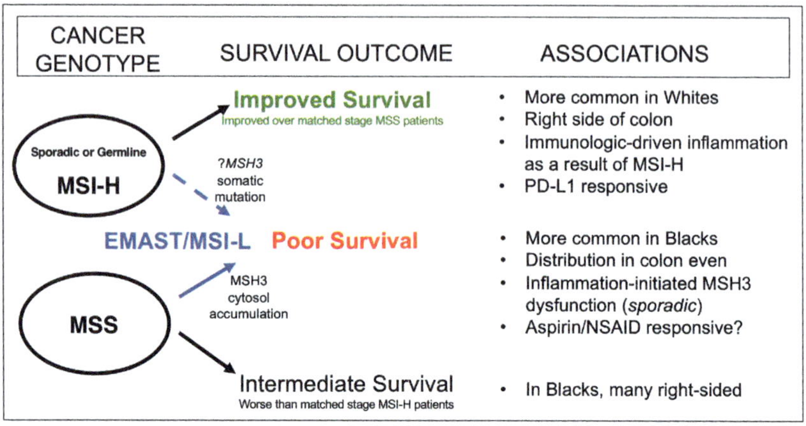

Fig. 6 Summary of patient associations and outcomes from colorectal cancer genomic instability genotypes. EMAST, elevated microsatellite alterations at selected tetranucleotide repeats; MSI-H, microsatellite instability-high; MSS, microsatellite stable; PD-L1, programmed death ligand-1; NSAID, non-steroidal anti-inflammatory drug.

the disparity with lower survival. For instance, aspirin/NSAIDs had an adverse effect on the survival of White Americans >65 years of age with CRC as compared to Black Americans, which may be due to higher levels of nascent inflammation in Black Americans (Koi et al., 2020).

A summation of MSI-H and EMAST genotypes on survival and associations between Black and White Americans with CRC is shown in Fig. 6.

3.4 Tumor immune environment

The immune environment can be the product of genomic instability, as in the case of MSI-H CRCs where frameshifted generated neoantigens induce a lymphocytic inflammation, and can cause genomic instability, as in the case of EMAST CRCs where IL6 triggers inactivation of DNA MMR (Carethers, 2017). Immune cells play key roles in the pathogenesis of CRC and patient survival (Koi & Carethers, 2017). High density of lymphocytes at the center and at the invasive margin of the CRC can predict improved patient survival compared to low density, irrespective of tumor stage, and was better at survival prediction than stage (Galon et al., 2006). These lymphocytes are mainly $CD8^+$ and granzyme B^+ cytotoxic T cells, as well as $CD45RO^+$ memory T cells (Mlecnik et al., 2016). For MSI-H/hypermutated CRCs, the expressed frameshifted peptides that are recognized as neoantigens by the immune system induce memory T cell differentiation as well as inducing the expression of PD-1 receptors on tumor epithelial cells (Mlecnik et al., 2016). The overriding factor for survival prediction, however, irrespective of being an MSI-H or MSS tumor, seems to the $CD8^+$/$CD45RO^+$ T cell immune response (Mlecnik et al., 2016). In a population-based study that analyzed over 500 CRCs, MSI-H CRCs demonstrated higher $CD8^+$ T cell counts than MSS counterparts; Black American MSI-H and MSS CRCs lacked the high counts of $CD8^+$ T cells as compared to White American MSI-H and MSS CRCs (Carethers et al., 2014). Additionally, Black American MSS CRCs showed significantly less intra-tumoral, intraepithelial, and invasive boarder granzyme B^+ T cells than White American MSS CRCs (Basa et al., 2016). Genetic expression analysis verified lower *GZMB* expression (encoding granzyme B) in Black American CRCs compared to White Americans using two separate cohorts, as well as lower expression of *PDL1* (encoding the immune checkpoint ligand PD-L1) and higher numbers of exhausted $CD8^+$ T cells (Peredes et al., 2020). These findings suggest that there may be defective immune defense mechanisms against CRCs in African Americans that could contribute to the disparity.

3.5 Diet-induced and microbiome effects on colonic environment

Specific gut microbiota can affect adenoma growth and CRC progression and metastasis (Bullman et al., 2017; Dejea et al., 2014, 2018). However, data is scant as it relates to any racial/ethnic differences. Sulfidogenic bacteria produce hydrogen sulfide that triggers pro-inflammatory pathways and hyperproliferation, and has been shown to be more abundant from non-involved biopsies among Black American CRC patients compared to White American CRC patients (Yazici et al., 2017). Other pro-inflammatory bacteria such as *Enterobacter* species and *Fusobacterium nucleatum* were significantly demonstrated more prevalent among Black Americans compared to White Americans undergoing screening colonoscopy, and demonstrated decreased microbial diversity (Farhana et al., 2018). While a consistent diet directs a persistently stable microbiota make-up, the gut microbiota can be dramatically modified by changes in diet (see also Chapter 3). Mucosal biopsies obtained after 2-week diet swaps between native Africans (on an African diet) and Black Americans (on a Western diet) demonstrate for Black Americans a marked lowering of colonocyte proliferation, reduced intraepithelial lymphocyte inflammation, increase volume of availability of the short chain fatty acid butyrate (a normal fuel molecule for colonocytes made by gut microbes) and decrease in the secondary bile acid deoxycholic acid, thought to be pro-carcinogenic agents, compared to baseline biopsies (O'Keefe et al., 2015). The opposite occurred in native Africans after ingesting the Western diet after 2 weeks, demonstrating a doubling of colonocyte proliferation and increase in deoxycholic acid (O'Keefe et al., 2015). Thus, even short-term diet changes can have a profound effect on modifying the microbiome, colonocyte proliferation and luminal contents of the colon. A lifelong Westernized diet might be extrapolated as increased risk for colonic neoplasia development for Black Americans, as compared to native Africans. This observation also suggests that Black Americans might be able to reduce CRC risk with diet manipulation. This would likely require lifestyle changes or a prolonged period of time to obtain risk reduction.

3.6 Germline and somatic genetics

The heritable adenomatous polyposis syndromes (e.g., Lynch syndrome, FAP, *MYH*-associated polyposis, polymerase proofreading associated polyposis, Familial Colorectal Cancer Type X) appear to exist in multiple racial/ethnic populations, but there is no known specific predilection for Black Americans or any other racial/ethnic group (Carethers, 1999; Carethers & Stoffel, 2015). Among Black American families with Lynch syndrome, the proportion of DNA MMR mutations obtained favored mutation of *MLH1* over other DNA MMR genes and represented two-thirds of cases, with a cumulative cancer risk similar to Lynch families of European descent (Guindalini et al., 2015). There were multiple novel *MLH1* mutations observed in the Black American Lynch families that were not seen in national mutation databases, suggesting genetic diversity in the mutational spectrum of *MLH1* (Guindalini et al., 2015). Similarly, the hamartomatous polyposis syndromes (e.g., *PTEN* hamartoma syndrome, juvenile polyposis, Peutz-Jeghers syndrome) have appeared in multiple racial/ethnic groups (Carethers, 1999; Stoffel & Carethers, 2020).

Non-inherited sporadic CRCs genetically develop through the accumulation of somatic (non-germline) mutations of critical cell regulatory genes to inactivate their growth regulatory abilities, as well as activate oncogenes that drive incessant proliferation signals (Cancer Genome Atlas Network, 2012; Carethers & Jung, 2015; Grady & Carethers, 2008). Each individual's CRC has a unique mutation profile, but there are overall common mutation patterns based on the type of genomic instability present in the CRC. For instance, hypermutated CRCs have hundreds to thousands of accumulated mutations per megabase of DNA in their cancer genome, which are typically the result of promotor methylation of *MLH1* (causing MSI-H) or less frequently mutation of *POLE* (Cancer Genome Atlas Network, 2012). MSI-H/hypermutated CRCs show a spectrum of accumulated somatic mutations that result from frameshifts in genes with coding microsatellites, such as *ACVR2*, *TGFBR2*, *MSH6*, *MSH3*, etc., along with *BRAF* mutations (Cancer Genome Atlas Network, 2012; Carethers & Jung, 2015). Non-hypermutated CRCs show a small range of 1 to 8 accumulated somatic driver mutations, and commonly demonstrate mutations in *APC*, *TP53*, *KRAS*, *TTN*, and PIK3CA (Cancer Genome Atlas Network, 2012). The Cancer Genome Atlas could not specifically determine any racial/ethnic differences in somatic gene mutations largely due to the paucity of diverse patient CRCs examined (Cancer Genome Atlas Network, 2012). In a separate cohort comparing 103 Black and 129 White American CRCs, three mutated genes were found exclusively among Black Americans: *EPHA6* (mutated frequency 5.83%), *FLCN* (mutated frequency of 2.91%), and *HTR1F* (mutated frequency of 2.91%) (Guda et al., 2015). Careful examination of *EPHA6* showed missense and splice site mutations, and *FLCN* showed frameshift insertions and nonsense mutations—all consistent with being deleterious mutations (Guda et al., 2015). These exclusive mutations within Black American CRCs raise the possibility that *EPHA6* and *FLCN* may be unique driver mutations in this population group; it remains to be determined if these mutated genes contribute in any way towards a disparity in outcome. Finally, Black American CRCs tend to have a higher prevalence of *KRAS* mutation over White and Asian Americans, which may contribute to poor outcomes for the observed disparity (Staudacher et al., 2017; Yoon et al., 2015).

4. Screening utilization contributions to CRC risk in racial/ethnic populations, and mitigation strategies

Screening for CRC is highly cost-effective to reduce the prevalence of CRC in the general population (Carethers, 2015; Rex et al., 2017). Depending on the type of test utilized to screen, screening may find cancers and precursor adenomas, and locate CRCs at earlier stages compared to unscreened patients (Carethers, 2015, 2020; Rex et al., 2017). Colonoscopy and FIT are considered first-tiered screening tests, with fecal DNA tests, computed tomography colonoscopy, and flexible sigmoidoscopy considered second-tiered tests (Carethers, 2020; Rex et al., 2017). Colonoscopy is the ultimate test to be completed irrespective of the initial screening positive test due to its ability to completely visualize the colon, and is diagnostic and therapeutic. Because CRCs generally begin as adenoma precursors, there is evidence that removal of adenomas (via polypectomy) that may result from screening reduces the incidence and mortality from CRC, and the protective effect from CRC mortality is durable (Nishihara et al., 2013; Shaukat, Mongin, Geisser, et al., 2013; Zauber, Winawer, O'Brien, et al., 2012). Thus, the more the average population-at-risk is screened, the lower the incidence and prevalence of CRC should follow. National U.S. cancer-related and gastrointestinal organizations have suggested and push screening targets to be 80% of the population-at-risk; current achievements for CRC screening in the U.S. are in the mid-60% range, a steady improvement over the past 2 decades (Carethers & Doubeni, 2020; DeSantis et al., 2019). However, uneven utilization across U.S. subpopulations may contribute towards a CRC disparity. Among Black Americans, screening by any test was utilized in 62% of the eligible population > 50 years of age as compared to 65% among White Americans (Carethers & Doubeni, 2020; DeSantis et al., 2019). Additionally, non-colonoscopic CRC screening tests that are positive require a follow-up colonoscopy with potential biopsy and/or removal of lesions. Black Americans within the PLCO cancer screening trial were less likely than White Americans to have a follow-up colonoscopy within 1 year of an abnormal screening test (Laiyemo, Doubeni, Pinsky, et al., 2010). This reduced lack of follow-up colonoscopy was even apparent among Black Americans who presented with rectal bleeding, a situation that is not screening but evaluative for "red flag" symptoms for possible CRC presence (Shields, Stoffel, Chung, et al., 2014). A study estimated that differences in screening accounted for 42% of the disparity for CRC incidence and 19% of the mortality between Black and White Americans (Lansdorp-Vogelaar et al., 2012).

There are several likely barriers for differences in screening utilization. These range from fatalism and fear of diagnosis, scheduling and implementation of screening, to lack of provider recommendation for screening (Carethers, 2015; Carethers & Doubeni, 2020; Green et al., 2008; Tammana & Laiyemo, 2014). Efforts at patient and physician education to increase screening utilization has not improved overall rates (Carethers & Doubeni, 2020; Kupfer, Carr, & Carethers, 2015). A successful approach to improve colonoscopy utilization for populations that are underutilizing it has been the implementation of professional or community-based peer navigators (Carethers & Doubeni, 2020; Chen et al., 2008; Grubbs et al., 2013; Jandorf et al., 2013; Kupfer et al., 2015; Naylor, Ward, & Polite, 2012). Navigation programs have been shown to substantially improve colonoscopy completion rates (Chen et al., 2008; Grubbs et al., 2013; Jandorf et al., 2013; Naylor et al., 2012). The Delaware Cancer Consortium performed 10,000 patient navigations for colonoscopic CRC screening among Black and White Americans over a 9-year period (Grubbs et al., 2013). There was a dramatic change in stage at diagnosis for Black Americans during the 9-year period, with distant CRC disease dropping from 23% to 7%, regional disease dropping from 56% to 33%, and local disease rising from 15% to 50% where CRC is curable (Grubbs et al., 2013). Furthermore, over the 9-year study, the overall number of patients navigated rose for both Black Americans (47.8% to 73.5%) and White Americans (58.0–74.7%) and the CRC incidence dropped for Black Americans (from 68/100,000 to 48/100,000) and White Americans (from 60/100,000 to 48/100,000), totally eliminating the CRC incidence disparity (Grubbs et al., 2013). In a similar vein, the CRC mortality, generally a lagging measure to incidence, dropped for Black Americans (from 31.27/100,000 to 18.35/100,000) and for White Americans (from 19.45/100,000 to 16.94/100,000), nearly eliminating the mortality disparity (Grubbs et al., 2013). This remarkable study shows the power of navigation to raise CRC screening for all subpopulations, and essentially proves that increasing the rate of CRC screening for the whole population can eliminate the disparities in CRC incidence and mortality.

Even though a well-orchestrated navigation colonoscopic screening program can potentially eliminate CRC disparities, there are still some additional separate challenges for equity. For instance, the quality of colonoscopy and the quality of the endoscopist is important for adequate screening; among Medicare enrollees, Black Americans more often received colonoscopies from providers with lower polyp detection rates than White Americans, and were 31% more likely to be diagnosed with interval cancer (Fedewa, Flanders, Ward, et al., 2017). The COVID-19 pandemic has caused delay of in-person CRC screening, which even with just a 6-month delay, will lead to increases in diagnosed CRCs over the next decade

(Carethers, Sengupta, Blakey, Ribas, & D'Souza, 2020). The COVID-19-induced screening delay is predicted to be much worse in already disadvantaged populations, some of which may only receive screening because of navigation (Carethers et al., 2020). A shift to at-home non-invasive CRC screening tests during the pandemic seems appropriate with the reduction in in-person screening; aspects of this remain challenging particular for disadvantaged populations because of (a) less likely to receive follow-up colonoscopy for a positive non-invasive test, (b) adequate ability to pay when the colonoscopy is billed as a diagnostic versus screening test, and individual fallout from COVID-19-induced economic difficulties, and (c) loss of navigation to complete the initial and follow-up tests (Carethers, 2021a; Carethers et al., 2020; Newman et al., 2021). Some enhancement strategies for participation with colonoscopy and non-invasive screening are listed in Table 2.

Table 2 Enhancement strategies for increased participation via navigation for colorectal cancer screening.

Colonoscopy	Non-invasive Screening
Navigation personnel's racial/ethnic background similar to patient's background; use native language	Telemedicine (phone or video) instructions and advice from virtual assistants
Multifaceted points of communication and execution: (a) provides general education about the procedure, including its importance in reducing cancer risk, (b) ensure prep is picked up and/or delivered, (c) instructions and coaching on prep utilization and completion, (d) arrange transportation to and from colonoscopy site, (e) arrange observer post-procedure with follow-up contact within hours post procedure	Post-navigation follow-up after test evaluation for transmission of results and next steps
Mitigates screening costs through insurance and other means for underinsured patients	Move to colonoscopy navigation if non-invasive test is positive
With healthcare provider, communicates results of colonoscopy and any pathology, and next steps	Persistent community education presence on importance of colorectal cancer screening for racial/ethnic groups

5. Future directions

There are a number of biological, genetic, immunologic, and screening utilization differences observed for Black Americans with CRC (where there is the most data for any U.S. racial/ethnic population) compared to White Americans (Table 3). There is growing data that Native Americans have higher proportions of precursor adenomas as well as increased CRC incidence and mortality rates (Corley et al., 2013; Siegel et al., 2021). Socioeconomic inequalities are likely a great contributor as a root cause with subsequent alterations in biology with the consequence of increasing CRC risk. These factors are difficult to change at the societal and often at the person level. The great equalizer appears to be prevention, with evidence that navigated colonoscopic screening can eliminate the CRC incidence and mortality disparity in a disadvantaged subpopulation as well as further reduce it for the majority population (Grubbs et al., 2013). While an aggressive screening strategy does not change the biology of adenoma or CRC formation, it certainly interrupts the process and is beneficial to the entire screened population.

Table 3 Observed biological, genetic, immunologic and screening difference risks regarding Black American colorectal cancer patients and their tumors.

Biological differences for Black American CRC risk as compared to White CRC risk	References
Increased number of adenomas >9 mm	Corley et al. (2013), Lieberman et al. (2005, 2014)
Increased proximal number of adenomas >9 mm	Corley et al. (2013), Lieberman et al. (2014)
Earlier onset of sporadic CRC	DeSantis et al. (2019), Siegel et al. (2021)
Increased proximal CRCs	Carethers et al. (2014)
Increase sulfidogenic bacteria in colon	Yazici et al. (2017)
Increased pro-inflammatory *Fusobacterium* and *Enterobacter* species in colon	Farhana et al. (2018)

Continued

Table 3 Observed biological, genetic, immunologic and screening difference risks regarding Black American colorectal cancer patients and their tumors.—cont'd

Biological differences for Black American CRC risk as compared to White CRC risk	References
Genetic differences for Black American CRC risk	
Decreased frequency of MSI-H CRCs	Ashktorab, Ahuja et al. (2016), Carethers et al. (2014)
Increased frequency of inflammation-associated microsatellite alterations/EMAST	Devaraj et al. (2010)
Unique somatic *FLCN*, *EPHA6*, and *HTR1F* mutation	Guda et al. (2015)
Increased frequency of *KRAS* mutation	Staudacher et al. (2017), Yoon et al. (2015)
Immunologic differences for Black American CRC risk	
Decreased high numbers of $CD8^+$ T lymphocytes within CRC	Carethers et al. (2014)
Decreased numbers of granzyme B^+ T lymphocytes withing CRC	Basa et al. (2016), Peredes et al. (2020)
Screening and surveillance differences for Black American CRC risk	
Lower frequency of population CRC screening uptake	DeSantis et al. (2019)
Lower frequency of colonoscopy screening uptake	DeSantis et al. (2019)
Lower frequency of follow-up after positive non-invasive CRC screening test	Laiyemo et al. (2010), Lansdorp-Vogelaar et al. (2012), Shields et al. (2014)

Even if the U.S. achieves 80% eligible screening utilization, there will still be a smaller number of individuals presenting with CRC, and there will likely continually be disparities. A multi-society group that included the National Cancer Institute outlined some suggestions for areas to be studied to further eliminate cancer disparities (Polite et al., 2017). These and other considerations (Kupfer et al., 2015) for reducing CRC disparities at the research, clinical and educational fronts are listed in Table 4.

Table 4 Some research, clinical and educational approaches to reduce colorectal cancer disparities.

Research	Clinical	Education
Better assess biological and environmental determinants of CRC incidence	Utilize patient navigation more broadly to increase colonoscopy screening rates for CRC	Physician education regarding physician biases for recommendation for CRC screening
Determine biological, environmental, and system-level determinants of post-diagnosis CRC care	Lower financial and insurance barriers to non-invasive CRC screening utilization, with concomitant colonoscopy utilization for screened-positive patients	Patient education to address barriers such as fear and mistrust for screening
Re-design prevention, drug and treatment clinical trials to acknowledge and address CRC disparities	Assess population CRC incidence by age and race/ethnicity every 5–10 years to potentially allow 95% eligible to be screened	Advance community engagement strategies throughout the cancer care continuum

Acknowledgments

Supported by the United States Public Health Service (R01 CA206010) and the Rogel Cancer Center of the University of Michigan. The funders had no role in study design, data collection and analysis, decision to publish, or preparation of the manuscript.

Disclosure of potential conflicts of interest

No potential conflicts of interest are disclosed.

References

André, T., Shiu, K. K., Kim, T. W., Jensen, B. V., Jensen, L. H., Punt, C., et al. (2020). Pembrolizumab in microsatellite-instability-high advanced colorectal cancer. *The New England Journal of Medicine*, *383*, 2207–2218.

Ashktorab, H., Vilmenay, K., Brim, H., Laiyemo, A. O., Kibreab, A., & Nouraie, M. (2016). Colorectal cancer in young African Americans: Is it time to revisit guidelines and prevention? *Digestive Diseases and Sciences*, *61*, 3026–3030.

Ashktorab, H., Ahuja, S., Kannan, L., Llor, X., Ellis, N., Xicola, R. M., et al. (2016). A meta-analysis of MSI frequency and race in colorectal cancer. *Oncotarget*, *7*, 34546–34557.

Ashktorab, H., Brim, H., Kupfer, S. S., & Carethers, J. M. (2017). Racial disparity in gastrointestinal cancer risk. *Gastroenterology*, *153*, 910–923.

Ashktorab, H., Delker, D., Kanth, P., Goel, A., Carethers, J. M., & Brim, H. (2019). Molecular characterization of sessile serrated adenoma/polyps from a large African American cohort. *Gastroenterology*, *157*, 572–574.

Ashktorab, H., Sherif, Z., Tarjoman, T., Azam, S., Lee, E., Shokrani, B., et al. (2020). Elevated risk for sessile serrated polyps in African Americans with endometrial polyps. *Digestive Diseases and Sciences*, *65*, 2686–2690.

Basa, R. C. B., Davies, V., Li, X., Murali, B., Shah, J., Yang, B., et al. (2016). Decreased anti-tumor cytotoxic immunity among colon cancers from African Americans. *PLoS One*, *11*, e0156660.

Baxter, N. N., Goldwasser, M. A., Paszat, L. F., Saskin, R., Urbach, D. R., & Rabeneck, L. (2009). Association of colonoscopy and death from colorectal cancer. *Annals of Internal Medicine*, *150*, 1–8.

Brennan, C. A., & Garrett, W. S. (2016). Gut microbiota, inflammation, and colorectal cancer. *Annual Review of Microbiology*, *70*, 395–411.

Brenner, H., Chang-Claude, J., Seiler, C. M., Rickert, A., & Hoffmeister, M. (2011). Protection from colorectal cancer after colonoscopy: A population-based, case-control study. *Annals of Internal Medicine*, *154*, 22–30.

Bullman, S., Pedamallu, C. S., Sicinska, E., Clancy, T. E., Zhang, X., Cai, D., et al. (2017). Analysis of *Fusobacterium* persistence and antibiotic response in colorectal cancer. *Science*, *358*, 1443–1448.

Cancer Genome Atlas Network. (2012). Comprehensive molecular characterization of human colon and rectal cancer. *Nature*, *487*, 330–337.

Carethers, J. M. (1999). Racial and ethnic factors in the genetic pathogenesis of colorectal cancer. *Journal of the Association for Academic Minority Physicians*, *10*, 59–67.

Carethers, J. M. (2005). Should African Americans be screened for colorectal cancer earlier? *Nature Clinical Practice. Gastroenterology & Hepatology*, *2*, 352–353.

Carethers, J. M. (2011). One colon lumen but two organs. *Gastroenterology*, *141*, 411–412.

Carethers, J. M. (2015). Screening for colorectal cancer in African Americans: Determinants and rationale for an earlier age to commence screening. *Digestive Diseases and Sciences*, *60*, 711–721.

Carethers, J. M. (2016a). The increasing incidence of colorectal cancers diagnosed in subjects under age 50 among races: Cracking the conundrum. *Digestive Diseases and Sciences*, *61*, 2767–2769.

Carethers, J. M. (2016b). Hereditary, sporadic and metastatic colorectal cancers are commonly driven by specific spectrums of defective DNA mismatch repair components. *Transactions of the American Clinical and Climatological Association*, *127*, 81–97.

Carethers, J. M. (2017). Microsatellite instability pathway and EMAST in colorectal cancer. *Current Colorectal Cancer Reports*, *13*, 73–80.

Carethers, J. M. (2018). Clinical and genetic factors to inform reducing colorectal cancer disparities in African Americans. *Frontiers in Oncology*, *8*, 531.

Carethers, J. M. (2020). Fecal DNA testing for colorectal cancer screening. *Annual Review of Medicine*, *71*, 59–69.

Carethers, J. M. (2021a). Insights into disparities observed with COVID-19. *Journal of Internal Medicine*, *289*, 11. https://doi.org/10.1111/joim.13199. Online ahead of print.

Carethers, J. M. (2021b). Rectifying COVID-19 disparities with treatment and vaccination. *JCI Insight*, *6*, e147800.

Carethers, J. M., & Doubeni, C. A. (2020). Causes of socioeconomic disparities in colorectal cancer and intervention framework and strategies. *Gastroenterology*, *158*, 354–367.

Carethers, J. M., & Jung, B. H. (2015). Genetics and genetic biomarkers in sporadic colorectal cancer. *Gastroenterology*, *149*, 1177–1190.

Carethers, J. M., & Stoffel, E. M. (2015). Lynch syndrome and Lynch syndrome mimics: The growing complex landscape of hereditary colon cancer. *World Journal of Gastroenterology*, *21*, 9253–9261.

Carethers, J. M., Chauhan, D. P., Fink, D., Nebel, S., Bresalier, R. S., Howell, S. B., et al. (1999). Mismatch repair proficiency and *in vitro* response to 5-fluorouracil. *Gastroenterology*, *117*, 123–131.

Carethers, J. M., Smith, E. J., Behling, C. A., Nguyen, L., Tajima, A., Doctolero, R. T., et al. (2004). Use of 5-fluorouracil and survival in patients with microsatellite unstable colorectal cancer. *Gastroenterology*, *126*, 394–401.

Carethers, J. M., Murali, B., Yang, B., Doctolero, R. T., Tajima, A., Basa, R., et al. (2014). Influence of race on microsatellite instability and $CD8^+$ T cell infiltration in colon cancer. *PLoS One*, *9*, e100461.

Carethers, J. M., Koi, M., & Tseng-Rogenski, S. (2015). EMAST is a form of microsatellite instability that is initiated by inflammation and modulates colorectal cancer progression. *Genes*, *6*, 185–205.

Carethers, J. M., Sengupta, R., Blakey, R., Ribas, A., & D'Souza, G. (2020). Disparities in cancer prevention in the COVID-19 era. *Cancer Prevention Research*, *13*, 893–896.

Chen, L. A., Santos, S., Jandorf, L., Cristie, J., Castillo, A., Winkel, G., et al. (2008). A program to enhance completion of screening colonoscopy among urban minorities. *Clinical Gastroenterology and Hepatology*, *6*, 443–450.

Corley, D. A., Jensen, C. D., Marks, A. R., Zhao, W. K., de Boer, J., Levin, T. R., et al. (2013). Variation of adenoma prevalence by age, sex, race, and colon location in a large population: Implications for screening and quality programs. *Clinical Gastroenterology and Hepatology*, *11*, 172–180.

Dejea, C. M., Wick, E. C., Hechenbleikner, E. M., White, J. R., Mark Welch, J. L., Rossetti, B. J., et al. (2014). Microbiota organization is a distinct feature of proximal colorectal cancers. *Proceedings of the National Academy of Sciences of the United States of America*, *111*, 18321–18326.

Dejea, C. M., Faith, P., Craig, J. M., Boleij, A., Taddese, R., Geis, A. L., et al. (2018). Patients with familial adenomatous polyposis harbor colonic biofilms containing tumorigenic bacteria. *Science*, *359*, 592–597.

DeSantis, C. E., Miller, K. D., Sauer, A. G., Jemal, A., & Siegel, R. L. (2019). Cancer statistics for African Americans, 2019. *CA: a Cancer Journal for Clinicians*, *69*, 211–233.

Devaraj, B., Lee, A., Cabrera, B. L., Miyai, K., Luo, L., Ramamoorthy, S., et al. (2010). Relationship of EMAST and microsatellite instability among patients with rectal cancer. *Journal of Gastrointestinal Surgery*, *14*, 1521–1528.

Farhana, L., Antaki, F., Murshed, F., Mahmud, H., Judd, S. L., Nangia-Makker, P., et al. (2018). Gut microbiome profiling and colorectal cancer in African Americans and Caucasian Americans. *World Journal of Gastrointestinal Pathophysiology*, *9*, 47–58.

Fedewa, S. A., Flanders, W. D., Ward, K. C., et al. (2017). Racial and ethnic disparities in interval colorectal cancer incidence: A population-based cohort study. *Annals of Internal Medicine*, *166*, 857–866.

Galon, J., Costes, A., Sanchez-Cabo, F., Kirilovsky, A., Mlecnik, B., Lagorce-Pagès, C., et al. (2006). Type, density, and location of immune cells within human colorectal tumors predict clinical outcome. *Science*, *313*, 1960–1964.

Garcia, M., Choi, C., Kim, H. R., Daoud, Y., Toiyama, Y., Takahashi, M., et al. (2012). Association between recurrent metastasis from stage II and III primary colorectal tumors and moderate microsatellite instability. *Gastroenterology*, *143*, 48–50.

Grady, W. M., & Carethers, J. M. (2008). Genomic and epigenetic instability in colorectal cancer pathogenesis. *Gastroenterology*, *135*, 1079–1099.

Green, A. R., Peters-Lewis, A., Percac-Lima, S., et al. (2008). Barriers to screening colonoscopy for low-income Latino and white patients in an urban community health center. *Journal of General Internal Medicine, 23*, 834–840.

Grubbs, S. S., Polite, B. N., Carney, J., Jr., Bowser, W., Rogers, J., Katurakes, N., et al. (2013). Eliminating racial disparities in colorectal cancer in the real world: It took a village. *Journal of Clinical Oncology, 31*, 1928–1930.

Guda, K., Veigl, M. L., Varadan, V., Nosrati, A., Ravi, L., Lutterbaugh, J., et al. (2015). Novel recurrently mutated genes in African American colon cancers. *Proceedings of the National Academy of Sciences of the United States of America, 112*, 1149–1154.

Guindalini, R. S., Win, A. K., Gulden, C., Lindor, N. M., Newcomb, P. A., Haile, R. W., et al. (2015). Mutation spectrum and risk of colorectal cancer in African American families with Lynch syndrome. *Gastroenterology, 149*, 1446–1453.

Hamaya, Y., Guarinos, C., Tseng-Rogenski, S. S., Iwaizumi, M., Das, R., Jover, R., et al. (2015). Efficacy of 5-fluorouracil adjuvant therapy for patients with EMAST-positive stage II/III colorectal cancers. *PLoS One, 10*, e0127591.

Hussan, H., Patel, A., Le Roux, M., Cruz-Monserrate, Z., Porter, K., Clinton, S. K., et al. (2020). Rising incidence of colorectal cancer in young adults corresponds with increasing surgical resections in obese patients. *Clinical and Translational Gastroenterology, 11*, e00160.

Iwaizumi, M., Tseng-Rogenski, S., & Carethers, J. M. (2011). DNA mismatch repair proficiency executing 5-fluorouracil cytotoxicity in colorectal cancer cells. *Cancer Biology & Therapy, 12*, 756–764.

Jandorf, L., Cooperman, J. L., Stossel, L. M., Itzkowitz, S., Thompson, H. S., Villagra, C., et al. (2013). Implementation of culturally targeted patient navigation system for screening colonoscopy in a direct referral system. *Health Education Research, 28*, 803–815.

Kim, S. E., Pe'rez-Stable, E. J., Wong, S., et al. (2008). Association between cancer risk perception and screening behavior among diverse women. *Archives of Internal Medicine, 168*, 728–734.

Koi, M., & Carethers, J. M. (2017). The colorectal cancer immune microenvironment and approach to immunotherapies. *Future Oncology, 13*, 1633–1647.

Koi, M., Garcia, M., Choi, C., Kim, H. R., Koike, J., Hemmi, H., et al. (2016). Microsatellite alterations with allelic loss at 9p24.2 signify less-aggressive colorectal cancer metastasis. *Gastroenterology, 150*, 944–955.

Koi, M., Okita, Y., & Carethers, J. M. (2018). *Fusobacterium nucleatum* infection in colorectal cancer: Linking inflammation, DNA mismatch repair and genetic and epigenetic alterations. *Journal of the Anus, Rectum and Colon, 2*, 37–46.

Koi, M., Tseng-Rogenski, S. S., & Carethers, J. M. (2018). Inflammation-associated microsatellite alterations: Mechanisms and significance in the prognosis of patients with colorectal cancer. *World Journal of Gastrointestinal Oncology, 10*, 1–14.

Koi, M., Okita, Y., Takeda, K., Koeppe, E., Stoffel, E. M., Galanko, J. A., et al. (2020). Co-morbid risk factors and NSAID use among White and Black Americans that predicts overall survival from diagnosed colon cancer. *PLoS One, 15*, e0239676.

Kupfer, S. S., McCaffrey, S., & Kim, K. E. (2006). Racial and gender disparities in hereditary colorectal cancer risk assessment: The role of family history. *Journal of Cancer Education, 21*, S32–S36.

Kupfer, S. S., Carr, R. M., & Carethers, J. M. (2015). Reducing colorectal cancer risk among African Americans. *Gastroenterology, 149*, 1302–1304.

Laiyemo, A. O., Doubeni, C., Pinsky, P. F., et al. (2010). Race and colorectal cancer disparities: Health-care utilization vs different cancer susceptibilities. *Journal of the National Cancer Institute, 102*, 538–546.

Lansdorp-Vogelaar, I., Kuntz, K. M., Knudsen, A. B., van Ballegooijen, M., Zauber, A. G., & Jemal, A. (2012). Contribution of screening and survival differences to racial disparities in colorectal cancer rates. *Cancer Epidemiology, Biomarkers & Prevention, 21*, 728–736.

Lee, S. Y., Chung, H., Devaraj, B., Iwaizumi, M., Han, H. S., Hwang, D. Y., et al. (2010). Microsatellite alterations at selected tetranucleotide repeats are associated with morphologies of colorectal neoplasias. *Gastroenterology, 139*, 1519–1525.

Lee, S.-Y., Miyai, K., Han, H. S., Hwang, D.-Y., Seong, M. K., Chung, H., et al. (2012). Microsatellite instability, EMAST, and morphology associations with T cell infiltration in colorectal neoplasia. *Digestive Diseases and Sciences, 57*, 72–78.

Leggett, B., & Whitehall, V. (2010). Role of the serrated pathway in colorectal cancer pathogenesis. *Gastroenterology, 138*, 2088–2100.

Lichtenstein, P., Holm, N. V., Verkasalo, P. K., et al. (2000). Environmental and heritable factors in the causation of cancer—Analyses of cohorts of twins from Sweden, Denmark, and Finland. *The New England Journal of Medicine, 343*, 78–85.

Lieberman, D. A., Holub, J., Eisen, G., Kraemer, D., & Morris, C. D. (2005). Prevalence of polyps greater than 9 mm in a consortium of diverse clinical practice settings in the United States. *Clinical Gastroenterology and Hepatology, 3*, 798–805.

Lieberman, D. A., Williams, J. L., Holub, J. L., Morris, C. D., Logan, J. R., Eisen, G. M., et al. (2014). Race, ethnicity, and sex affect risk for polyps >9 mm in average-risk individuals. *Gastroenterology, 147*, 351–358.

Lumpkins, C., Cupertino, P., Young, K., Daley, C., Yeh, H., & Greiner, K. (2013). Racial/ethnic variations in colorectal cancer screening self-efficacy, fatalism and risk perception in a safety-net clinic population: Implications for tailored interventions. *Journal of Community Medicine & Health Education, 3*, 1–4.

Mlecnik, B., Bindea, G., Angell, H. K., Maby, P., Angelova, M., Tougeron, D., et al. (2016). Integrative analyses of colorectal cancer show immunoscore is a stronger predictor of patient survival than microsatellite instability. *Immunity, 44*, 698–711.

Munakata, K., Koi, M., Kitajima, T., Tseng-Rogenski, S. S., Uemura, M., Matsuno, H., et al. (2019). Inflammation-associated microsatellite alterations caused by MSH3 dysfunction are prevalent in ulcerative colitis and increase with neoplastic advancement. *Clinical and Translational Gastroenterology, 10*, e00105.

Murff, H. J., Peterson, N. B., Fowke, J. H., et al. (2008). Colonoscopy screening in African Americans and Whites with affected first-degree relatives. *Archives of Internal Medicine, 168*, 625–631.

Naylor, K., Ward, J., & Polite, B. N. (2012). Interventions to improve care related to colorectal cancer among racial and ethnic minorities: A systematic review. *Journal of General Internal Medicine, 27*, 1033–1046.

Newman, L., Winn, R., & Carethers, J. M. (2021). Similarities in risk for COVID-19 and cancer disparities. *Clinical Cancer Research, 27*, 24–27.

Nishihara, R., Wu, K., Lochhead, P., Morikawa, T., Liao, X., Qian, Z. R., et al. (2013). Long-term colorectal-cancer incidence and mortality after lower endoscopy. *The New England Journal of Medicine, 369*, 1095–1105.

O'Keefe, S. J., Ou, J., Aufreiter, S., et al. (2009). Products of the colonic microbiota mediate the effects of diet on colon cancer risk. *The Journal of Nutrition, 139*, 2044–2048.

O'Keefe, S. J., Li, J. V., Lahti, L., Ou, J., Carbonero, F., Mohammed, K., et al. (2015). Fat, fibre and cancer risk in African Americans and rural Africans. *Nature Communications, 6*, 6342.

Okita, Y., Koi, M., Takeda, K., Ross, R., Mukherjee, B., Koeppe, E., et al. (2020). *Fusobacterium nucleatum* infection correlates with two types of microsatellite alterations in colorectal cancer and triggers DNA damage. *Gut Pathogens, 12*, 46.

Peredes, J., Zabaleta, J., Garai, J., Ji, P., Imtiaz, S., Spagnardi, M., et al. (2020). Immune-related gene expression, cellular cytotoxicity and cytokine secretion is reduced among African American colon cancer patients. *Frontiers in Oncology, 10*, 1498.

Perencevich, M., Ojha, R. P., Steyerberg, E. W., & Syngal, S. (2013). Racial and ethnic variations in the effects of family history of colorectal cancer on screening compliance. *Gastroenterology, 145*, 775–781.

Peterse, E. F. P., Meester, R. G. S., Siegel, R. L., Chen, J. C., Dwyer, A., Ahnen, D. J., et al. (2018). The impact of the rising colorectal cancer incidence in young adults on the optimal age to start screening: Microsimulation analysis I to inform the American Cancer Society colorectal cancer screening guideline. *Cancer, 124*, 2964–2973.

Polite, B. N., Adams-Campbell, L. L., Brawley, O. W., Bickell, N., Carethers, J. M., Flowers, C. R., et al. (2017). Charting the future of cancer health disparities research: A position statement from the American Association for Cancer Research, the American Cancer Society, the American Society of Clinical Oncology, and the National Cancer Institute. *Cancer Research*, 77, 4548–4555.

Popat, S., Hubner, R., & Houlston, R. S. (2005). Systematic review of microsatellite instability and colorectal cancer prognosis. *Journal of Clinical Oncology, 23*, 609–618.

Raeker, M. O., & Carethers, J. M. (2020). Immunological features with DNA microsatellite alterations in patients with colorectal cancer. *Journal of Cancer Immunology, 2*, 116–127.

Rex, D. K., Boland, C. R., Dominitz, J. A., Giardiello, F. M., Johnson, D. A., Kaltenbach, T., et al. (2017). Colorectal cancer screening: Recommendations for physicians and patients from the U.S. multi-society task force on colorectal cancer. *Gastroenterology, 153*, 307–323.

Satia, J. A., Keku, T., Galanko, J. A., Martin, C., Doctolero, R. T., Tajima, A., et al. (2005). Diet, lifestyle, and genomic instability in the North Carolina Colon Cancer Study. *Cancer Epidemiology, Biomarkers & Prevention, 14*, 429–436.

Schroy, P. C., 3rd, Coe, A., Chen, C. A., O'Brien, M. J., & Heeren, T. C. (2013). Prevalence of advanced colorectal neoplasia in white and black patients undergoing screening colonoscopy in a safety-net hospital. *Annals of Internal Medicine, 159*, 13–20.

Shaukat, A., Mongin, S. J., Geisser, M. S., et al. (2013). Long-term mortality after screening for colorectal cancer. *The New England Journal of Medicine, 369*, 1106–1114.

Shavers, V. L. (2007). Racial/ethnic variation in the anatomic subsite location of in situ and invasive cancers of the colon. *Journal of the National Medical Association, 99*, 733–748.

Shields, H. M., Stoffel, E. M., Chung, D. C., et al. (2014). Disparities in evaluation of patients with rectal bleeding 40 years and older. *Clinical Gastroenterology and Hepatology, 12*, 669–675.

Siegel, R. L., Miller, K. D., Fuchs, H. E., & Jemal, A. (2021). Cancer statistics, 2021. *CA: a Cancer Journal for Clinicians, 71*, 7–33.

Staudacher, J. J., Yazici, C., Bul, V., Zeidan, J., Khalid, A., Xia, Y., et al. (2017). Increased frequency of KRAS mutations in African Americans compared with Caucasians in sporadic colorectal cancer. *Clinical and Translational Gastroenterology, 8*, e124.

Stoffel, E. M., & Carethers, J. M. (2020). Current approaches to germline cancer genetic testing. *Annual Review of Medicine, 71*, 85–102.

Suzuki, S., Iwaizumi, M., Tseng-Rogenski, S., Hamaya, Y., Miyajima, H., Kanaoka, S., et al. (2016). Production of truncated MBD4 protein by frameshift mutation in DNA mismatch repair-deficient cells enhances 5-fluorouracil sensitivity that is independent of *hMLH1* status. *Cancer Biology & Therapy, 17*, 760–780.

Tajima, A., Hess, M. T., Cabrera, B. L., Kolodner, R. D., & Carethers, J. M. (2004). The mismatch repair complex hMutSα recognizes 5-fluoruracil-modified DNA: Implications for chemosensitivity and resistance. *Gastroenterology, 127*, 1678–1684.

Tajima, A., Iwaizumi, M., Tseng-Rogenski, S., Cabrera, B. L., & Carethers, J. M. (2011). Both hMutSα and hMutSβ complexes participate in 5-fluoruracil cytotoxicity. *PLoS One*, *6*, e28117.

Tammana, V. S., & Laiyemo, A. O. (2014). Colorectal cancer disparities: Issues, controversies and solutions. *World Journal of Gastroenterology*, *20*, 869–876.

Tseng-Rogenski, S., Chung, H., Wilk, M. B., Zhang, S., Iwaizumi, M., & Carethers, J. M. (2012). Oxidative stress induces nuclear-to-cytosol shift of hMSH3, a potential mechanism for EMAST in colorectal cancer cells. *PLoS One*, *7*, e50616.

Tseng-Rogenski, S., Hamaya, Y., Choi, D., & Carethers, J. M. (2015). Interleukin 6 alters localization of hMSH3, leading to DNA mismatch repair defects in colorectal cancer cells. *Gastroenterology*, *148*, 579–589.

Tseng-Rogenski, S. S., Munakata, K., Choi, D. Y., Martin, P. K., Mehta, S., Koi, M., et al. (2020). The human DNA MMR protein MSH3 contains nuclear localization and export signals that enable nuclear-cytosolic shuttling in response to inflammation. *Molecular and Cellular Biology*, *40*, e00029-20.

Winawer, S. J., Fletcher, R. H., Miller, L., et al. (1997). Colorectal cancer screening: Clinical guidelines and rationale. *Gastroenterology*, *112*, 594–642.

Xicola, R., Gagnon, M., Clark, J. R., et al. (2014). Excess of proximal microsatellite-stable colorectal cancer in African Americans from a multi-ethnic study. *Clinical Cancer Research*, *20*, 4962–4970.

Yazici, C., Wolf, P. G., Kim, H., Cross, T. L., Vermillion, K., Carroll, T., et al. (2017). Race dependent association of sulfidogenic bacteria with colorectal cancer. *Gut*, *66*, 1983–1994.

Yoon, H. H., Shi, Q., Alberts, S. R., Goldberg, R. M., Thibodeau, S. N., Sargent, D. J., et al. (2015). Racial differences in *BRAF/KRAS* mutation rates and survival in stage III colon cancer patients. *Journal of the National Cancer Institute*, *107*, djv186.

Zauber, A. G., Winawer, S. J., O'Brien, M. J., et al. (2012). Colonoscopic polypectomy and long-term prevention of colorectal-cancer deaths. *The New England Journal of Medicine*, *366*, 687–696.

Zavala, V., Bracci, P. M., Carethers, J. M., Carvajal-Carmona, L., Coggins, N. B., Cruz-Correa, M. R., et al. (2021). Cancer health disparities in US racial/ethnic minorities. *British Journal of Cancer*, *124*, 315–332.

CHAPTER SEVEN

Genomic and epigenomic biomarkers in colorectal cancer: From diagnosis to therapy

Juan Ruiz-Bañobre[a,b] and Ajay Goel[c,*]
[a]Medical Oncology Department, University Clinical Hospital of Santiago de Compostela, University of Santiago de Compostela (USC), CIBERONC, Santiago de Compostela, Spain
[b]Translational Medical Oncology Group (Oncomet), Health Research Institute of Santiago (IDIS), University Clinical Hospital of Santiago de Compostela, University of Santiago de Compostela (USC), CIBERONC, Santiago de Compostela, Spain
[c]Department of Molecular Diagnostics and Experimental Therapeutics, Beckman Research Institute of City of Hope Comprehensive Cancer Center, Duarte, CA, United States
*Corresponding author: e-mail address: ajgoel@coh.org

Contents

Abstract

Colorectal cancer (CRC) is the second leading cause of cancer-related deaths in the United States. Despite ongoing efforts aimed at increasing screening for CRC and early detection, and development of more effective therapeutic regimens, the overall morbidity

Advances in Cancer Research, Volume 151
ISSN 0065-230X
https://doi.org/10.1016/bs.acr.2021.02.008

and mortality from this malignancy remains a clinical challenge. Therefore, identifying and developing genomic and epigenomic biomarkers that can improve CRC diagnosis and help predict response to current therapies are of paramount importance for improving survival outcomes in CRC patients, sparing patients from toxicity associated with current regimens, and reducing the economic burden associated with these treatments. Although efforts to develop biomarkers over the past decades have achieved some success, the recent availability of high-throughput analytical tools, together with the use of machine learning algorithms, will likely hasten the development of more robust diagnostic biomarkers and improved guidance for clinical decision-making in the coming years. In this chapter, we provide a systematic and comprehensive overview on the current status of genomic and epigenomic biomarkers in CRC, and comment on their potential clinical significance in the management of patients with this fatal malignancy, including in the context of precision medicine.

1. Introduction

Colorectal cancer (CRC) is the third leading cause of cancer-related deaths in the western world. Despite ongoing efforts aimed at increasing screening of individuals and improving strategies for early detection, ~20% of patients still present with metastatic disease at diagnosis, and ~35% of patients who undergo surgery for primary disease experience relapse (Siegel et al., 2020). Over the past 3 decades, the median overall survival (OS) of patients with metastatic CRC (mCRC) has gradually increased because of the implementation of combined chemotherapy regimens as well as the introduction of targeted molecular therapies against epidermal growth factor receptor (EGFR), angiogenic factors, and, more recently, *BRAF*. In addition, the discovery that immunotherapy is effective for patients with microsatellite instability-high (MSI-H) or deficient DNA mismatch-repair (dMMR) mCRC has introduced a new therapeutic option for this subset of patients (Ruiz-Bañobre, Kandimalla, & Goel, 2019).

CRC develops through sequential accumulation of genetic and epigenetic alterations in precursor lesions (mainly adenomas and serrated polyps). These precursor lesions progressively develop a more aggressive phenotype, acquiring a range of dysplastic features and eventually progressing to adenocarcinoma, the most malignant stage (Jung, Hernández-Illán, Moreira, Balaguer, & Goel, 2020). Greater understanding of the molecular pathways involved in carcinogenesis of CRC is needed to improve the detection of therapeutically targetable vulnerabilities and to develop more effective early-diagnostic strategies in order to detect disease before it progresses to malignancy. To date, three pathways have been well established in CRC carcinogenesis: chromosome instability (CIN), microsatellite instability (MSI), and the serrated pathway. These pathways are characterized by multistep genetic and epigenetic alterations, involving several oncogenes and tumor suppressor genes, as described below.

1.1 CIN pathway

The CIN phenotype is observed in as many as 80% of sporadic CRCs, and is characterized by chromosome changes such as somatic copy-number variations, deletions, insertions, amplifications, or loss of heterozygosity (Grady & Carethers, 2008). Mutation of the *APC* gene is a key event in CRC tumors that exhibit CIN, and is considered the earliest genomic event in CRC carcinogenesis through this pathway (Powell et al., 1992). Loss of APC activity results in translocation of beta-catenin to the nucleus and subsequent activation of the WNT signaling pathway, which leads to constitutive activation of other relevant genes such as *MYC*, *CCND1*, *VEGF*, and *PPARD* (Mann et al., 1999). The WNT signaling pathway is activated in nearly all CRCs that exhibit CIN, and APC mutations have been identified in more than 80% of these cancers (Guinney et al., 2015). Later mutations that are characteristic of CRC arise in a sequential manner after mutation of *APC* as lesions progress toward adenocarcinoma. Activating mutations in *KRAS* often arise following acquisition of mutations in *APC*, and are present in 40–50% of CRCs (Santini et al., 2008). Later in the adenoma–carcinoma sequence, activating mutations in *PIK3CA* and inactivating mutations in *P53* arise in ~15% and 60% of CRCs, respectively (Baker et al., 1990; Liao et al., 2012). Loss of heterozygosity at chromosome 18q, which harbors various tumor suppressor genes from the TGF-beta pathway, such as *SMAD2* and *SMAD4*, is also found in more than 70% of CRCs (Fearon & Vogelstein, 1990).

1.2 MSI pathway

MSI is the primary mechanism for the development of colorectal tumors with a hypermutable phenotype. CRC that arises due to MSI develops more rapidly than CRC arising from CIN, having an estimated tumor development time of a few years as compared to approximately 2 decades for the CIN pathway (Aust, Sommer, & Baretton, 2012). CRCs arising due to MSI are often characterized by the presence of single-nucleotide substitutions or insertions and deletions within microsatellite repeats. Microsatellites are short tandem repeat DNA sequences spread throughout the human genome.

Because they are formed of highly repetitive sequences, they have a higher propensity for acquiring mutations than do other regions of the genome. The process of repairing errors in microsatellite repeats is tightly governed by an intact MMR system (Ruiz-Bañobre & Goel, 2019). MMR deficiency resulting from somatic or germline DNA alterations in any of the key MMR genes (*MLH1*, *MSH2*, *MSH6*, and *PMS2*), or from deletions in the *EPCAM* gene (which leads to constitutional repression of *MSH2* gene expression through promoter methylation), is the initiating event for MSI-driven carcinogenesis. The loss of *MLH1* expression due to biallelic *MLH1* promoter methylation is the key somatic event responsible for ∼75–80% of sporadic cancers with MSI. Overall, MSI is observed in ∼12–15% of sporadic CRCs and in the majority of CRCs from patients with Lynch syndrome and its variants (Muir–Torre or Turcot's syndromes) (Ruiz-Bañobre & Goel, 2019). Moreover, sporadic CRCs with MSI frequently exhibit a CpG island methylator phenotype (CIMP) and harbor *BRAF V600E* mutations, while *APC* and *P53* mutations are found in a smaller proportion of MSI-H CRCs compared with MSS CRCs (Nguyen, Goel, & Chung, 2020; Rajagopalan et al., 2002; The Cancer Genome Atlas TCGA Research Network, 2012).

1.3 Serrated pathway

The serrated pathway represents the origin of ∼15% of CRCs, and describes the progression of precursor malignant lesions that present as serrated polyps (Toyota et al., 1999). The term "serrated polyp" encompasses a heterogeneous spectrum of lesions that include, in decreasing order of prevalence, benign hyperplastic polyps (the most frequent subtype), sessile serrated adenomas (SSAs), and traditional serrated adenomas (TSAs; the less frequent subtype) (McCarthy, Serra, & Chetty, 2019). SSAs and TSAs are more advanced lesions that have a higher propensity for malignant progression. The earliest driving molecular event in the serrated pathway appears to be *BRAF V600E* mutation. This mutation is present in a large proportion of microvesicular hyperplastic polyps, which subsequently progress to SSA, then to SSA with dysplasia, and finally to CRC (Davies et al., 2002). Following the appearance of the *BRAF V600E* mutation, serrated tumors can follow two different routes and develop either into MSI CRC through acquisition of CIMP and methylation of the MLH1 promoter, or into microsatellite stable (MSS) CRCs (Guinney et al., 2015). In addition, *RNF43* mutation constitutes a subsequent important genomic event for the serrated pathway in both MSI and MSS CRCs (Yan et al., 2017).

TSAs are less prevalent than SSAs (TSAs represent 1–7% of all serrated lesions) and therefore are not characterized as well, but they often exhibit three somewhat distinct molecular patterns: *BRAF*-mutated and CIMP-high (CIMP-H), *KRAS*-mutated and CIMP-low, or *BRAF* and *KRAS* wild-type. In addition, TSAs almost never display an MSI phenotype (McCarthy et al., 2019).

Although differences and similarities in the molecular profiles for these three pathways are well defined, a deeper temporospatial understanding of these carcinogenesis pathways will help for more rational biomarker development. Conceptually, events that occur early during carcinogenesis will serve better as potential biomarkers for the early detection of precursor lesions. In contrast, molecular profiles from both early and later lesions will be useful for identifying biomarkers for prognostication, and therapeutic benefit prediction. However, as we will discuss in this chapter, the biomarker field is complex, and it is not simple to identify robust biomarkers, thus reinforcing the importance of comprehensive molecular profiling in a clearly defined clinical context (Fig. 1).

2. Diagnostic biomarkers

Recently, the incidence of CRC has been rising in adults under the age of 50, and such CRC is referred to as early-onset CRC. This increasing incidence suggests that CRC screening programs for early detection should be expanded to a younger, asymptomatic, average-risk population. Although there is a lack of consensus within the medical community regarding this issue, some medical societies are discussing implementing currently available screening strategies across a wider range of ages, including younger individuals (USPSTF. US Preventive Services Task Force, 2020). Regarding screening strategies, these can be prioritized into several tiers: the first tier could include colonoscopy, fecal immunochemical test (FIT), and fecal occult blood test (FOBT); the second tier computed tomographic colonography, FIT-DNA, and flexible sigmoidoscopy; and the third tier capsule endoscopy. Although colonoscopy is considered the gold standard for CRC screening, it has several limitations, including invasiveness, cost, and low compliance. In contrast, FOBT and FIT, which are the most commonly used non-invasive screening tests in Europe and other Western countries, lack sensitivity and specificity compared to colonoscopy, especially for precursor lesions (Quintero et al., 2012). Given the challenges associated with these commonly used screening approaches, there is a clear need

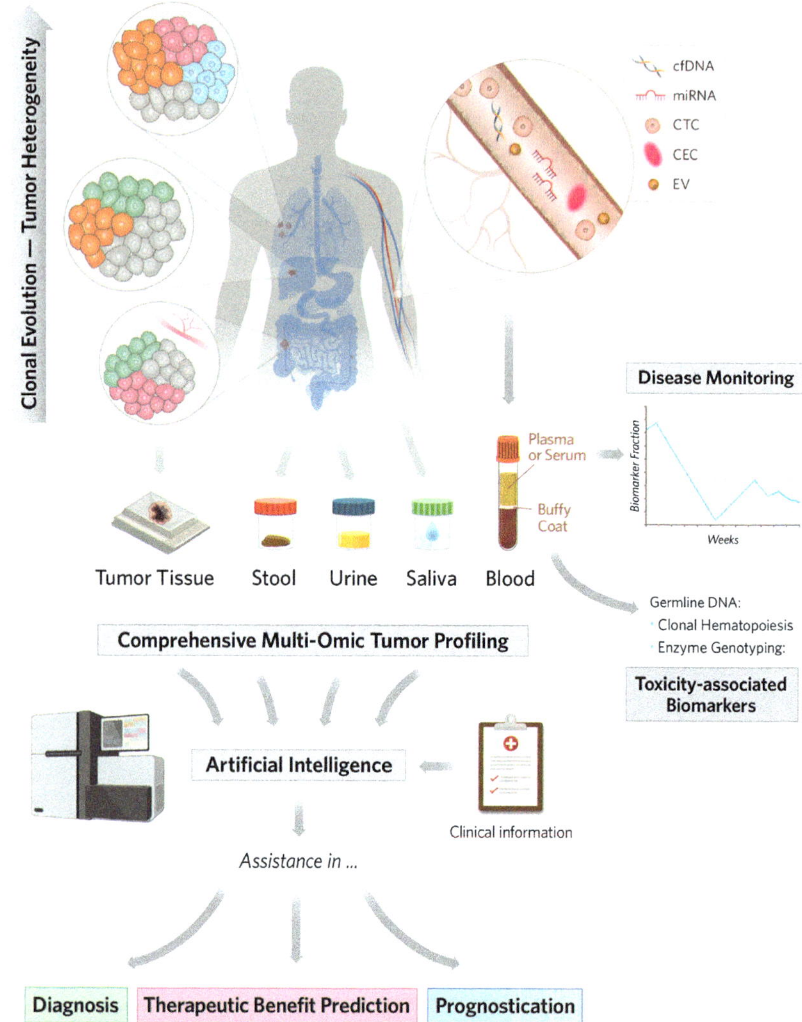

Fig. 1 Schematic and general overview of the different biospecimen sources and biomarkers involved in diagnosis, prognosis, and prediction of clinical benefit of therapies for CRC. Starting from a single ancestral clone, cancer cells undergo different routes of clonal evolution, which are responsible for temporospatial tumor heterogeneity. Although a single tumor tissue biopsy cannot capture all of the tumor variability, it is postulated that different types of liquid biopsies and biospecimens can capture this heterogeneity over time. In addition, there are many sources and types of biomarkers that can be detected in tumor tissue and other biospecimens, with blood, stool, saliva, and urine being the most promising. To date, for CRC early-diagnosis, the most frequently studied biospecimens have been stool and blood; for prognosis and clinical benefit prediction tumor tissue and blood have been most frequently studied. Moreover, because of their capacity to recapitulate tumor heterogeneity, liquid biopsies can facilitate disease monitoring and therapeutic response, especially through serial blood analyses. Identification of genotypic markers encoding drug-metabolizing enzymes is also possible by analyzing germline DNA from peripheral blood mononuclear cells. Last, implementation of novel high-throughput molecular analytic techniques, together with integration of multi-omic approaches from multiple sources with clinical and epidemiologic data using artificial intelligence, will accelerate biomarker development and assist in the clinical decision-making process. Abbreviations: CEC, circulating endothelial cells; cfDNA, cell-free DNA; CRC, colorectal cancer; CTC, circulating tumor cell; EV, extracellular vesicle; miRNA, microRNA.

for developing more robust, non-invasive strategies to diagnose premalignant and malignant colorectal lesions.

These strategies are likely to include novel biomarkers for early detection and improved, clinically useful devices and kits that leverage existing and new biomarkers. Some molecular alterations are particularly appealing for potential use as non-invasive biomarkers, especially alterations that can be detected in stool or blood. For example, epigenetic changes, either on their own or, more recently, in combination with DNA mutations and circulating cells, have emerged as front runners for the development of early detection biomarkers in colorectal neoplasia (which includes both precursor lesions and CRC). In this section we provide a summary of genomic and epigenomic diagnostic biomarkers for CRC (Table 1), highlighting well-established assays and describing some of the most promising and novel potential biomarkers in the field.

2.1 microRNAs (miRNAs)

The potential of miRNAs as biomarker candidates lies in their small size, relatively limited numbers as compared to protein-coding genes/mRNAs, and stability in a variety of biological specimens, such as tissue, blood, and stool (Bovell et al., 2012; Cortez et al., 2011; Esteller, 2011; Lu et al., 2005; Volinia et al., 2006). In addition, there is a wide variety of routine laboratory techniques for identifying and quantifying miRNAs (e.g., microarrays, quantitative reverse transcription PCR) in virtually all specimen types, which would further support application of miRNAs as biomarkers. Over the past decade, the number of studies investigating miRNAs in CRC has increased exponentially, and efforts to evaluate their potential as biomarkers have particularly increased. Despite this enthusiasm, only a few well-designed studies have thus far been conducted that include large patient cohorts, well-defined patient populations, and analysis of independent validation cohorts. None of the biomarkers evaluated in these studies have yet been adopted in clinical practice because of the lack of prospective randomized studies comparing them with current gold standard screening methods. Below, we discuss some of the most promising miRNAs for diagnostic purposes.

2.1.1 miR-92a

The presence of miR-92a has shown remarkable utility for discriminating patients with CRC from healthy controls when measured in blood or stool samples. In plasma samples, miR-92a showed good performance for

Table 1 Established and potential biomarkers in CRC diagnosis.

References	Biomarker	Specimen	Test	Accuracy	Regulatory status
miRNAs					
Ng et al. (2009) Uratani et al. (2016) Yau, Tang, Harriss, Dickins, and Polytarchou (2019)	miR-92a	Blood	NA	CRC: *AUROC: 0.89 (plasma)* *Sensitivity: 0.89* *Specificity: 0.70* A: *AUROC: 0.75 (serum)* *Sensitivity: 0.65* *Specificity: 0.79* AA: AUROC: *0.84 (serum)*	–
		Stool	NA	CRC: *AUROC: 0.80* *Sensitivity: 0.47* *Specificity: 0.91* A: *AUROC: 0.64* *Sensitivity: 0.43* *Specificity: 0.91*	–

Peng et al. (2017) Toiyama et al. (2013) Yau et al. (2019)	miR-21	Blood	NA	CRC: *AUROC: 0.85 (plasma)/0.87 (serum)* *Sensitivity: 0.69/0.75* *Specificity: 0.86/0.84* A: *AUROC: 0.80 (serum)* *Sensitivity: 0.77* *Specificity: 0.81*	–
		Stool	NA	CRC: *AUROC: 0.84* *Sensitivity: 0.60* *Specificity: 0.86* A: *AUROC: 0.77* *Sensitivity: 0.60* *Specificity: 0.83*	–
Yau et al. (2019)	2-miRNA panel: miR-92 miR-21	Stool	NA	CRC: *AUROC: 0.84* *Sensitivity: 0.54* *Specificity: 0.88*	–

Continued

Table 1 Established and potential biomarkers in CRC diagnosis.—cont'd

References	Biomarker	Specimen	Test	Accuracy	Regulatory status
Duran-Sanchon et al. (2020)	GBM model[a]: miR-421 miR-27a-3p	Stool	NA	CRC and AA: *AUROC: 0.74 (training set)/0.63 (test set)* *Sensitivity: 0.74/0.67* *Specificity: 0.63/0.60* CRC: *AUROC: 0.86/0.74* *Sensitivity: 0.96/0.96* *Specificity: 0.36/0.33* AA: *AUROC: 0.71/0.64* *Sensitivity: 0.61/0.59* *Specificity: 0.71/0.69*	–
Liu, Klein, et al. (2018) and Liu, Liu, et al. (2018)	4-miRNA panel: miR-21 miR-92a miR-29a miR-125b	Blood (serum)	NA	CRC: *AUROC: 0.95* *Sensitivity: 0.85* *Specificity: 0.99*	–
Chang et al. (2016)	2-miRNA panel: miR-223 miR-92a	Blood (plasma)	NA	CRC: *AUROC: 0.91* *Sensitivity: 0.97* *Specificity: 0.75*	–

Herreros-Villanueva et al. (2019)	6-miRNA panel: miR-19a miR-19b miR-15b miR-29a miR-335 miR-18a	Blood (plasma)	NA	CRC and AA: *AUROC: 0.92* *Sensitivity: 0.85* *Specificity: 0.90*	–
Sazanov, Kiselyova, Zakharenko, Romanov, and Zaraysky (2017)	2-miRNA panel: miR-21 miR-92a	Saliva	NA	CRC: *Sensitivity: 0.97* *Specificity: 0.91*	–
Rapado-González et al. (2019)	5-miRNA panel: miR-186-5p miR-29a-3p miR-29c-3p miR-766-3p miR-491-5p	Saliva	NA	CRC: *AUROC: 0.75* *Sensitivity: 0.72* *Specificity: 0.67*	–
DNA methylation					
Church et al. (2014)	SEPT9	Blood (plasma)	Epi proColon	CRC: *Sensitivity: 0.68* *Specificity: 0.79* AN: *Sensitivity: 0.25* *Specificity: 0.79* AA: *Sensitivity: 0.22* *Specificity: 0.79*	Registered trademark. FDA approval: alternative for CRC screening[b]

Continued

Table 1 Established and potential biomarkers in CRC diagnosis.—cont'd

References	Biomarker	Specimen	Test	Accuracy	Regulatory status
Baek et al. (2009) Carmona et al. (2013) Chen et al. (2005) Fu et al. (2018) Itzkowitz et al. (2007) Kisiel et al. (2013) Li et al. (2009) Lu et al. (2014)	VIM	Blood (plasma)	NA	CRC: *Sensitivity: 0.59* *Specificity: 0.93*	–
		Stool	ColoSure	CRC: *Sensitivity: 0.81* *Specificity: 0.95*	Trademark. Not FDA approval
Jung et al. (2020)	SFRP2	Blood (plasma and serum)	NA	CRC: *Sensitivity: 0.64–0.67* *Specificity: 0.97–1.00* A: *Sensitivity: 0.06–0.81* *Specificity: 0.73–1.00*	–
		Stool	NA	CRC: *Sensitivity: 0.56–0.94* *Specificity: 0.77–0.97* A: *Sensitivity: 0.28–0.76* *Specificity: 0.55–1.00*	–

Luo et al. (2020)	cd-score (multi-region panel)	Blood (plasma)	NA	CRC: *AUROC: 0.96* *Sensitivity: 0.88* *Specificity: 0.90*	–
Luo et al. (2020)	cg10673833 (single probe)	Blood (plasma)	NA	CRC: *AUROC: 0.90* *Sensitivity: 0.90* *Specificity: 0.87*	–
Kandimalla et al. (2020)	Multi-region panel	Blood (plasma)	EpiPanGI-Dx	CRC: *AUROC: 0.98*	–
ColonES Product Sheet (2020)	Multi-region panel	Blood (plasma)	ColonEs	CRC and A: *Sensitivity: 0.91 (A)/ 0.97 (I)/0.94 (II–IV)* *Specificity: 0.99*	–
Multi-modal					
USPSTF. US Preventive Services Task Force et al. (2020)	Multi-modal panel: KRAS mutations (x7) NDRG4 m.s. BMP3 m.s. FIT	Stool	Cologuard	CRC: *Sensitivity: 0.93* *Specificity: 0.85* AN: *Sensitivity: 0.47* *Specificity: 0.89* AA: *Sensitivity: 0.43* *Specificity: 0.89*	Registered trademark. FDA approval: CRC screening[c]

Continued

Table 1 Established and potential biomarkers in CRC diagnosis.—cont'd

References	Biomarker	Specimen	Test	Accuracy	Regulatory status
Friedland et al. (2020)	Multi-modal panel: Circulating GI endothelial cells Somatic mutations[d] SEPT9 m.s.	Blood	FirstSight	CRC: *Sensitivity: 1.00* *Specificity: 0.90* AA: *Sensitivity: 0.80*	Trademark
Cohen et al. (2018)	Multi-modal panel[e]: 8 proteins Mutations	Blood (plasma)	CancerSEEK	CRC: *AUROC: 0.85*	–
Duran-Sanchon et al. (2020)	GBM model[a]: miR-421 miR-27a-3p FIT	Stool	NA	CRC and AA: *AUROC: 0.72 (training set)/0.70 (test set)* *Sensitivity: 0.74/0.68* *Specificity: 0.63/0.64* CRC: *AUROC: 0.90/0.93* *Sensitivity: 0.96/0.97* *Specificity: 0.36/0.43* AA: *AUROC: 0.70/0.64* *Sensitivity: 0.61/0.49* *Specificity: 0.71/0.71*	–

Metabolomic					
Deng et al. (2017)	Metabolites: Succinic acid Ascorbic acid Carnitine	Urine	PolypDx	HP and A: *Sensitivity: 0.43* *Specificity: 0.91*	Trademark. Available for CLIA-certified laboratories

[a]miRNA-based predictive model included miR-421 and miR-27a-3p, along with age and sex.
[b]Alternative in average-risk individuals who refuse other screening modalities.
[c]CRC screening in average-risk population.
[d]Somatic mutations in oncogenes and tumor suppressor genes.
[e]8 proteins (cancer antigen 125, carcinoembryonic antigen, cancer antigen 19–9, proactin, hepatocyte growth factor, osteopontin, myeloperoxidase, and tissue inhibitor of metalloproteinases 1) and mutations (SNVs or INDELs) in 1933 distinct genomic positions.

Abbreviations: A, adenoma; AA, advanced adenoma; CRC, colorectal cancer; AN, advanced neoplasia; AUROC, area under the receiver operating characteristic curve; CLIA, Clinical Laboratory Improvement Amendments; FIT, fecal immunochemical test; GBM, gradient boosting machine; GI, gastrointestinal; HP, hyperplastic polyp; m.s., methylation status; NA, not available.

detecting CRC, with an area under the receiver operating characteristic curve (AUROC), sensitivity, and specificity of 0.89, 89%, and 70%, respectively (Ng et al., 2009). In the context of adenoma detection, serum miR-92a levels yielded an AUROC of 0.75, and corresponding sensitivity and specificity of 65% and 79%, respectively (Uratani et al., 2016). Recently, Yau et al. (2019) reported the results of a meta-analysis involving 17 studies evaluating the diagnostic potential of several miRNAs, specifically in stool samples. Expression of miR-92a in stool had an AUROC, sensitivity, and specificity of 0.80, 47%, and 91%, respectively, for detecting CRC (Yau et al., 2019), and of 0.64, 43%, and 91%, respectively, for detecting adenoma (Yau et al., 2019). Intriguingly, the diagnostic accuracy of miR-92a for adenoma detection was higher using blood than stool samples (Uratani et al., 2016; Yau et al., 2019).

2.1.2 miR-21

The presence of miR-21 has demonstrated high sensitivity and specificity for detecting CRC in both blood and stool samples, although results from stool samples showed greater variability. In a meta-analysis of 11 studies conducted between 2012 and 2014 and involving almost 800 patients, the overall AUROC, sensitivity, and specificity of circulating miR-21 for blood-based early detection of CRC were 0.86, 72%, and 84%, respectively (Peng et al., 2017). For the diagnosis of colorectal adenomas, miR-21 expression in serum showed an AUROC of 0.80 with a sensitivity and specificity of 76.8% and 81.1%, respectively (Toiyama et al., 2013). In a recent meta-analysis, miR-21 expression in stool samples yielded an AUROC, sensitivity, and specificity of 0.84, 60%, and 86%, respectively, for CRC (Yau et al., 2019). In the same study, the overall AUROC, sensitivity, and specificity for adenoma detection were 0.77, 60%, and 83%, respectively (Yau et al., 2019).

2.1.3 miRNA panels

Although miRNA panels have been suggested as a potentially more accurate and robust diagnostic approach than the use of a single miRNA biomarker, these approaches have not yielded consistent findings among studies. For example, in some studies, the analysis of miR-21 and miR-92a in blood samples for the non-invasive diagnosis of CRC yielded sensitivities and specificities ranging from 90% to 89%, and 90% to 96%, respectively, whereas in another cohort, the combination of these two biomarkers yielded a sensitivity of 68% and a specificity of 91% (Liu et al., 2013). In contrast, in another study, a serum-based four-miRNA panel (miR-21, miR-29a, miR-92a and miR-125b) showed an excellent performance for CRC diagnosis, with an AUROC of 0.95, sensitivity of 85%, and specificity of 99% (Liu, Klein, et al., 2018; Liu, Liu, et al., 2018). Along similar lines, a two-miRNA panel (miR-223 and miR-92a) exhibited an AUROC of 0.91, sensitivity of 97%, and specificity of 75% for detection of CRC in blood specimens from a cohort of 291 patients with CRC and 452 self-reported healthy controls (Chang et al., 2016). On the other hand, a recent meta-analysis evaluating the potential for CRC diagnosis using a combination panel of miR-21 and miR-92a detected in stool samples yielded an AUROC of 0.84, sensitivity of 54%, and specificity of 88% (Yau et al., 2019). A plasma-based six-miRNA panel (miR-19a, miR-19b, miR-15b, miR-29a, miR-335, and miR-18a), when analyzed in a cohort of 300 individuals, could accurately identify healthy controls from patients with advanced colorectal neoplasms (CRC and advanced adenoma), with an AUROC of 0.92, sensitivity of 85%, and specificity of 90% (Herreros-Villanueva et al., 2019).

Recently, Duran-Sanchon et al. (2020) proposed a new miRNA-based model to discriminate patients with advanced neoplasms (CRC and advanced adenomas) from those with non-relevant findings at colonoscopy. The authors conducted a study composed of four stages (discovery in tissue samples, technical validation in a subset of stool samples from patients of the discovery phase, clinical validation in an independent set of stool samples, and predictive modeling) and identified a panel of three miRNAs (miR-421, miR-130b-3p, and miR-27a-3p) that were significantly upregulated in fecal samples obtained from participants with advanced neoplasms. Subsequently, they developed a predictive model by combining stool expression levels of miR-421, miR27a-3p, and hemoglobin. This model could identify patients with CRC with an AUROC of 0.93, sensitivity of 97%, and specificity of 43%, as compared with an AUROC of 0.67, sensitivity of 100%, and a specificity of 31% for fetal hemoglobin concentration alone. However, the accuracy of this model for identifying patients with advanced neoplasms was significantly lower when compared with participants who had non-relevant findings at colonoscopy (AUROC of 0.62, sensitivity of 62%, specificity of 58%) or normal colonoscopy findings (AUROC of 0.59, sensitivity of 43%, specificity of 63%). One of the major limitations of this study, together with the small sample size of the validation set ($n = 189$), was the fact that it included only individuals with a positive result from a FIT, and therefore, as the authors stated, further studies are needed in FIT-naive subjects.

Although saliva is a less explored bodily fluid for CRC diagnosis, it has recently gained attention as a potential biospecimen for non-invasive liquid biopsy because it is easy and safe to collect (Malathi, Mythili, & Vasanthi, 2014). Several studies have reported the existence of various salivary molecular indicators of local and systemic disorders, including cancer (Malathi et al., 2014). In this context, Sazanov et al. recently evaluated the expression levels of miR-21 in plasma and saliva samples from CRC patients (Sazanov et al., 2017). miR-21 showed a plasma sensitivity and specificity of 65% and 85%, respectively, for discriminating between CRC patients ($n=34$) and healthy controls ($n=34$), and even greater sensitivity (97%) and specificity (91%) in saliva. The high sensitivity and specificity values for salivary miR-21 reflect the potential of this saliva-based test for CRC diagnostic purposes. Another study recently identified a set of five miRNAs (miR-186-5p, miR-29a-3p, miR-29c-3p, miR-766-3p, and miR-491-5p) that were significantly upregulated in saliva from CRC patients (Rapado-González et al., 2019). Interestingly, the combined analysis of these five miRNAs yielded an AUROC of 0.754, sensitivity of 72%, and specificity of 67% for differentiating CRC patients ($n=51$) from healthy individuals ($n=37$). Unfortunately, and in line with Sazanov et al. study (Sazanov et al., 2017), the authors did not explore the accuracy of the panel for identifying adenoma or adenoma and CRC. Furthermore, although both saliva-based studies showed promising results, they must still be evaluated in validation trials in larger retrospective and prospective cohorts.

In general, studies conducted to date on potential miRNA biomarkers share two main characteristics that limit their external validity—relatively small sample size and the lack of detailed clinical-pathological information on the CRC population evaluated in the study. These limitations must be addressed, and the miRNA biomarkers or panels evaluated in randomized prospective studies before they can be translated to the clinic.

2.2 Long noncoding RNAs (lncRNAs)

In recent years, lncRNAs have gained increasing attention as potential biomarkers for CRC. In contrast to miRNAs, the precise number of functional lncRNAs remains unclear, because of ongoing discovery of new lncRNAs. Despite promising early results as biomarkers, research on lncRNAs still focuses mostly on discovery of new lncRNAs and their functions. Studies exploring the potential of lncRNAs as biomarkers in large patient cohorts are still elusive. Although tissue-derived lncRNAs have been the primary analyte in the most studies, a few studies have also analyzed blood-derived lncRNAs to determine their potential as non-invasive biomarkers. The lncRNA HOTAIR has been analyzed in both serum and tissues and is upregulated in early stages of CRC (Svoboda et al., 2014; Zhao, Song, Zhang, Kuerban, & Wang, 2015). Along similar lines, upregulation of the lncRNA colon cancer-associated transcript 1 (CCAT1) in tumor tissue and blood also seems to be an early event during colorectal carcinogenesis (Alaiyan et al., 2013; He et al., 2014; Nissan et al., 2012; Ozawa et al., 2017). Evaluation of *HOTAIR* and *CCAT1* in a panel combining both markers revealed that upregulated expression of both markers had greater sensitivity and specificity for diagnosing CRC in plasma samples than of either lncRNA separately (Zhao et al., 2015). These two lncRNAs were also recently evaluated in a stool-based lncRNA extended panel (Gharib et al., 2021). The study population consisted of 150 participants, including a training and a validation set, and a group of 30 subjects with colon polyps. Expression levels of lncRNAs were evaluated by quantitative real-time PCR. To design a predictive panel, the investigators selected 10 significantly dysregulated lncRNAs, including *CCAT1*, *CCAT2*, *H19*, *HOTAIR*, *HULC*, *MALAT1*, *PCAT1*, *MEG3*, *PTENP1*, and *TUSC7*. The diagnostic performance of the panel in terms of distinguishing CRC as compared to healthy tissue yielded an AUROC of 0.8554 in the training set and 0.8465 in the validation set. The panel's AUROC for early-stage CRC (I–II TNM stages) was 0.7871 in the training set and 0.8121 in the validation set, and for advanced CRC (III–IV TNM stages) the AUROC was 0.9281 in the training set and 0.9236 in the validation set. The corresponding AUROC values for CRC vs colon polyp were 0.9228 (I–IV TNM stages), 0.9042 (I–II TNM stages), and 0.9362 (III–IV TNM stages). Again, in addition to the small sample size of the cohort included in the study, there is a lack of information regarding the clinical-pathological characteristics of the polyps analyzed in the study beyond the sex and age of the corresponding subjects.

2.3 Histone modifications

Although some studies have evaluated histone modifications as non-invasive diagnostic biomarkers, the data are still preliminary and further studies are required to determine the feasibility and utility of this approach. Methylation levels of histone 3 at lysine 4 are higher in CRC and in adenomas than in normal colonic mucosa, and acetylation levels of histone 3 at lysine 27 and histone 4 at lysine 12 are markedly greater in CRC compared to normal colonic mucosa (Ashktorab et al., 2009; Karczmarski et al., 2014; Nakazawa et al., 2012). In an attempt to identify non-invasive biomarkers and to take advantage of the stability of nucleosomes within the circulation, some studies have demonstrated the potential of using histone modifications in circulating nucleosomes as CRC diagnostic biomarkers. Chromatin immunoprecipitation studies revealed that reduced levels of trimethylation of histone 3 at lysine 9 and of histone 4 at lysine 20 in circulating nucleosomes were present in patients with CRC as compared with healthy control individuals (Gezer et al., 2013, 2015). Although histone modifications have a central role in cancer pathogenesis, their lack of cancer-specificity and various technical limitations associated with their use as quantitative analytes are, to date, some of the major obstacles to their use as non-invasive biomarkers.

2.4 DNA methylation

One of the first epigenetic alterations to be discovered in human cancer was the low level of DNA methylation in tumors as compared to normal tissue (Feinberg & Vogelstein, 1983). The loss of methylation is primarily due to hypomethylation of repetitive DNA sequences (e.g., LINE, SINE and Alu elements) and demethylation of coding regions and introns (Feinberg & Tycko, 2004). During carcinogenesis, the degree of hypomethylation of genomic DNA increases as a lesion progresses from a benign proliferation of cells to an invasive cancer (Fraga et al., 2004). In addition to hypomethylation of genomic DNA, hypermethylation of specific CpG islands in the promoter regions of tumor suppressor genes is also a major event in the origin of many cancers. All of these changes in DNA methylation modulate gene expression: gene silencing via hypermethylation of CpG islands in promoters, gene activation via hypomethylation of CpG-poor gene promoters, and oncogene overexpression via hypermethylation of gene bodies (Yang et al., 2014).

DNA methylation profiling provides several advantages over somatic mutation analysis for detecting cancer, including greater clinical sensitivity and dynamic range, multiple detectable methylation target regions, greater prevalence in precancerous lesions and early-stage cancers, and presence of multiple altered CpG sites within each targeted genomic region (Heyn & Esteller, 2012; Laird, 2003). Importantly, altered methylation is also seen as a field effect in colonic mucosa that predisposes normal tissue to neoplastic transformation (Luo, Yu, & Grady, 2014). In addition, the alterations in CpG methylation are relatively constant in each type of cancer, whereas there are usually no predominant somatic mutations. Despite the relatively high frequency of somatic mutations in cancers, the patterns of mutations are highly heterogeneous in individual patients, making somatic mutations less than ideal markers for early detection of cancers (Kandoth et al., 2013). Given favorable characteristics of DNA methylation markers and strong supporting evidence of their utility as biomarkers, several assays using DNA methylation biomarkers have been commercialized during the last decade, are used in current clinical practice, or have even entered clinical guidelines.

One of the most widely studied non-invasive DNA methylation biomarkers for CRC diagnosis is methylation of the *SEPT9* gene in plasma. *SEPT9* encodes septin 9, a GTP-binding protein involved in actin dynamics, cytoskeletal remodeling, vesicle trafficking, and exocytosis. Multiple studies have evaluated the diagnostic accuracy of this methylation biomarker in large cohorts of patients with CRC, and in these studies sensitivity and specificity ranged from 48% to 90% and 73% to 97%, respectively (Bergheim et al., 2018; Church et al., 2014; Fu et al., 2018; Song, Peng, et al., 2017; Song, Jia, Peng, Xiao, & Li, 2017; Tänzer et al., 2010; Wu et al., 2016). In 2016, this biomarker (commercialized as the Epi proColon test by Epigenomics) was approved by the US FDA as the first molecular blood-based assay for CRC screening. This test is not recommended for routine CRC screening, and represents an alternative only for individuals who refuse other screening modalities such as FOBT, FIT, flexible sigmoidoscopy, or colonoscopy. In one of the largest studies in which *SEPT9* plasma methylation was analyzed using Epi proColon test 2.0 methodology, this biomarker yielded an overall sensitivity of 73.7% and specificity of 97% in a large cohort (300 patients with CRC and 568 healthy control individuals) (He et al., 2018). However, findings of both this study and a 2017 meta-analysis agree that this test performed statistically significantly better in patients with advanced stage (III–IV) CRC than in those with early-stage (I–II) CRC (He et al., 2018; Song, Jia, et al., 2017). The Epi proColon test requires only a 10 mL blood sample, out of which ~3.5 mL of plasma is used to obtain free circulating DNA. Methylated *SEPT9* DNA within the plasma is amplified via PCR, along with β-actin as an internal control. Because this screening test only requires a blood draw, which can be performed in conjunction with other routine laboratory exams, this test is attractive to patients and has high compliance rates.

In an effort to further improve the diagnostic accuracy of Epi proColon test, which, as noted above, is suboptimal for early-stage CRC, other groups have attempted to combine *SEPT9* plasma methylation levels with additional biomarkers, such as FIT or other plasma-based methylated genes (for example, SHOX2 and ALX4) (Bergheim et al., 2018; Tänzer et al., 2010; Wu et al., 2016). Wu et al. evaluated a combination of *SEPT9* plasma methylation, FIT and serum carcinoembryonic antigen (CEA) levels for CRC screening that included all CRC stages, and achieved an overall sensitivity of 97.2% (Wu et al., 2016). Although the majority of the data for this test were obtained from case–control or cohort studies, *SEPT9* plasma methylation has also been analyzed in asymptomatic, intermediate-risk populations of healthy individuals to further assess its diagnostic potential as an alternative to FIT or colonoscopy (Church et al., 2014). However, when Song, Jia, et al. (2017) compared the data from this study with that from a meta-analysis of FIT in a similar population, they found lower sensitivity (68% vs 79%) and specificity (80% vs 94%) for *SEPT9* plasma methylation than for FIT. One of the major limitations of the *SEPT9* plasma methylation biomarker is its poor sensitivity (ranges from 7.9% to 38.7%) for identifying precursor lesions (adenomas) (Fu et al., 2018; He et al., 2018; Song, Peng, et al., 2017; Wu et al., 2016). *SEPT9* plasma methylation levels exhibited the highest sensitivity (83.3%) in a subgroup of patients with villous adenoma, suggesting this biomarker has potential for identifying advanced adenomas (Song, Peng, et al., 2017). However, the number of patients included in this study was small ($n = 18$), highlighting the need for the development of more ambitious approaches for detecting precancerous lesions and early-stage CRC (Song, Peng, et al., 2017).

Methylation of the *VIM* gene has also been evaluated as a non-invasive biomarker for CRC diagnosis. *VIM* encodes the intermediate filament protein vimentin, which, together with microtubules and actin microfilaments, constitutes the cytoskeleton. The diagnostic accuracy of VIM methylation for CRC appears be greater for stool samples than for blood specimens. For example, the sensitivity and specificity of *VIM* methylation were 81% and 95%, respectively, in stool samples, with similar values seen across CRC stages (Baek et al., 2009; Carmona et al., 2013; Chen et al., 2005; Fu et al., 2018; Itzkowitz et al., 2007; Kisiel et al., 2013; Li et al., 2009; Lu et al., 2014). In contrast, in plasma samples, *VIM* methylation had a sensitivity and specificity of 59% and 93%, respectively, with substantially greater sensitivity in advanced disease stages (Li et al., 2009). Given the performance of *VIM* methylation in stool samples, this biomarker has been commercialized as the ColoSure test (LabCorp) (Ned, Melillo, & Marrone, 2011), but has not yet obtained US FDA clearance or approval for use as a CRC screening test.

In contrast to *VIM* methylation, methylation of *secreted frizzled-related protein 2* (*SFRP2*) has shown more promising results for adenomas. *SFRP2* methylation has been studied for detection of precancerous lesions (both adenomas and hyperplastic polyps). The sensitivity of plasma-based *SFRP2* methylation for detecting adenomas ranged from 6.4% to 81.1%, and the specificity from 73% to 100% (Barták et al., 2017; Tang et al., 2011), while the sensitivity and specificity of stool-based *SFRP2* methylation ranged from 27.8% to 76% and from 55% to 100%, respectively (Glöckner et al., 2009; Park et al., 2017; Zhang, Zhu, Wu, Zhang, & Qi, 2014).

As high-throughput tools for molecular development continue to improve, new studies are emerging to evaluate the potential of DNA methylation biomarkers through the use of more comprehensive approaches. In 2020, Luo et al. (2020) identified CRC-specific methylation signatures by comparing CRC tissues to normal blood leukocytes. Using machine learning algorithms, the authors developed a predictive diagnostic model (cd-score) by using cell-free DNA (cfDNA) samples from a cohort of 801 patients with CRC and 1021 normal controls. In the discovery dataset, the cd-score discriminated patients with CRC ($n = 528$) from normal controls ($n = 674$) with an AUROC of 0.96, sensitivity of 87.5%, and specificity of 89.9%. This accuracy was confirmed in a validation dataset of 273 CRC patients and 347 normal controls, yielding an AUROC of 0.96, sensitivity of 87.9%, and specificity of 89.6%. With the goal of making screening methods for cancer as simple as possible, the authors prospectively investigated the efficiency of using the methylation status of CpG site cg10673833 for detecting CRC as well as precancerous lesions in high-risk populations. To identify an appropriate population, all enrolled participants ($n = 16{,}890$) were invited to take a cancer risk assessment using an established Clinical Cancer Risk Score System (Chen et al., 2019). This identified 1493 participants (ages range: 45–75 years) who were considered to be at high risk for CRC. These high-risk subjects were subsequently scheduled to undergo screening colonoscopy and were recruited into the study to undergo methylation profiling at the time of the screening. The cg10673833 methylation test identified 19 of 21 participants with CRC and 7 of 8 participants with CRC in situ (diagnosis as high-grade dysplasia), with an AUROC of 0.90, sensitivity of 89.7%, and specificity of 86.8%. The positive predictive value and negative predictive value were 0.118 (95% CI 0.101–0.138) and 0.998 (95% CI 0.993–0.999), respectively. For advanced precancerous lesions, the sensitivity was 33.3%, much higher than the positivity rate for subjects who did not have CRC or advanced precancerous lesions (12.1%).

More recently in 2020, Liu et al. (2020) reported the results of a prospective case-controlled substudy from the Circulating Cell-free Genome Atlas (CCGA; NCT02889978) and the STRIVE (NCT03085888) projects. This substudy assessed the ability of targeted methylation analysis of circulating cfDNA to detect and localize multiple cancer types across all stages at high specificity. The CCGA study was designed to determine if genome-wide cfDNA sequencing in combination with machine learning algorithms could detect and localize a large number of cancer types at sufficiently high specificity to be considered for a general population-based cancer screening program. In previous discovery work in a first CCGA substudy, whole-genome bisulfite sequencing to interrogate genome-wide methylation patterns outperformed whole-genome sequencing and targeted sequencing approaches to interrogate copy-number variants and single-nucleotide variants (SNVs)/small insertions and deletions, respectively (Liu, Klein, et al., 2018; Liu, Liu, et al., 2018; Oxnard et al., 2019). In addition, targeted sequencing with SNV-based classification was significantly confounded by clonal hematopoiesis of indeterminate potential (Swanton et al., 2018), suggesting that such a test would require concurrent sequencing of white blood cells to return accurate results. The cfDNA methylation-based classifier achieved consistently high overall specificity between the cross-validated training and independent validation sets (>99%) across all cancer types (>50) with a sensitivity of ~55%. Incorporation of clinical baseline demographic information and blood sample quality metrics to the model did not sufficiently improve sensitivity (<10%). Although this study did not evaluate the accuracy specifically across CRC stages, both the training and validation cohorts included patients with CRC (training, $n=122$ with CRC; validation, $n=53$ with CRC) among all the cancer types included (Liu et al., 2020). Because not all participants with cancer were asymptomatic, additional studies in an asymptomatic screening population are currently ongoing.

Similar efforts to identify methylation-based biomarkers are ongoing that are specifically focused on gastrointestinal (GI) cancers. One of the most promising tests is the EpiPanGI-Dx (Kandimalla et al., 2020), a cfDNA methylation fingerprint for the early detection of several GI cancers, including CRC, esophageal squamous cell and adenocarcinoma (ESCC and EAC), gastric cancer (GC), hepatocellular carcinoma (HCC), and pancreatic ductal adenocarcinoma (PDAC). Using a tissue-based genome-wide DNA methylation analysis of these GI cancers to select the most informative differentially methylated regions (DMRs), Kandimalla et al. (2020) developed a novel three distinct categories of DMR panels and sequenced 300 plasma specimens from all GI cancers, as well as age-matched healthy controls, with 40× coverage. They found (1) cancer-specific biomarker panels with AUROC values of 0.98 (CRC), 0.94 (ESCC), 0.90 (EAC), 0.90 (GC), 0.98 (HCC), and 0.85 (PDAC); (2) a pan-GI cancer biomarker panel that detected all GI cancers with an AUROC of 0.90; and (3) a multi-cancer prediction panel, EpiPanGI-Dx, with a prediction accuracy around 0.85 for most GI cancers. All three groups of DMR panels, after being trained and tested using the cfDNA cohorts, achieved excellent diagnostic accuracy as indicated by AUROC values ranging from 0.74 to 0.98, even for each of the early-stage GI cancers. This study represents the first specific GI cancer cfDNA methylation test that has potential to be applied in the clinic for the early detection of all GI cancers.

Recently, Chen et al. (2020) described a new blood-based cancer diagnostic test based on cfDNA methylation analysis, the PanSeer assay. Using publicly available microarray and whole-genome bisulfite sequencing data from The TCGA and genomic regions known to be cancer-related in the literature, as well as internal reduced representation bisulfite sequencing data from a variety of cancer tissues, the authors compiled a targeted panel of 595 genomic regions for further interrogation in plasma samples. Because tumor DNA tends to be rare in plasma, especially in patients with early-stage cancer, and because conventional methods for constructing sequencing libraries that incorporate bisulfite conversion and double-stranded ligation typically have a high DNA loss rate (Aigrain, Gu, & Quail, 2016), the authors used the Singlera library construction method. This method uses semi-targeted PCR, which requires only a single ligation event and a single PCR primer per amplicon, allowing single-molecule counting at a higher molecular recovery rate than conventional methods (Gansauge & Meyer, 2013; Zheng et al., 2014). Preliminary results of PanSeer on plasma samples from 605 asymptomatic individuals, 191 of whom were later diagnosed with stomach, esophageal, colorectal, lung, or liver cancer within 4 years of blood draw, demonstrated an accuracy of cancer diagnosis of 95%. Using a similar approach, another plasma-based assay of circulating tumor methylated DNA (ColonES, Singlera Genomics) showed impressive preliminary results, with sensitivities of 91%, 97%, and 94% for detecting adenoma, stage I CRC, and stage II–IV CRC, respectively (ColonES Product Sheet, 2020). The reported preliminary specificity of the test is 99%. Although additional studies are needed in larger population-based cohorts, DNA methylation genome-wide approaches are leading candidates for biomarker panels and seem to be the future for molecular blood-based diagnostics for CRC and many other tumor types.

2.5 Multi-modal approach

Although epigenetic alterations are among the most well-characterized CRC diagnostic biomarkers, considering that CRC evolves through acquisition of many genetic and epigenetic alterations, combinations of multiple types of molecular and cellular biomarkers have also been evaluated to see if they enable more accurate detection of colorectal polyps and CRC. Indeed, several promising combinations have been proposed to improve diagnostic performance. Among the tests that combine different biological approaches, currently, only Cologuard (Exact Sciences), a stool-based multi-target panel, is approved by the US FDA for CRC screening. Cologuard tests three genomic biomarkers (seven point mutations in *KRAS* and methylation status of *NDRG4* and *BMP3*) and includes a quantitative enzyme-linked immunosorbent assay for the presence of hemoglobin. In a study of ~10,000 intermediate-risk individuals, Imperiale et al. (2014) found that Cologuard had a significantly higher sensitivity for CRC detection compared to FIT (92.3% vs 73.8%). More importantly, the sensitivity for detecting advanced precancerous lesions (advanced adenomas or sessile serrated lesions $\geq$1 cm) was almost twofold higher for Cologuard than for FIT (42.4% vs 23.8%). However, this increased sensitivity came at the cost of lower specificity (89.8% [Cologuard] vs 96.4% [FIT]) in patients with negative colonoscopy (Imperiale et al., 2014). Given these promising results, Cologuard has been included in the United States Preventive Services Task Force and National Comprehensive Cancer Network (NCCN) guidelines as a CRC screening option at 3-year intervals, which is on equal standing with the other traditional screening options (colonoscopy, FIT, and FOBT) (USPSTF). Ongoing trials are prospectively evaluating the utility of Cologuard in other scenarios, such as longitudinal testing in an average-risk population at 3-year intervals (NCT02419716), in an average-risk population aged 45–49 (NCT03728348), and with FIT to model stool-based molecular surveillance approaches to inform health policy decisions (NCT02715141).

At the 2020 ASCO Virtual Scientific Program, Friedland et al. (2020) presented interim results of a study of a new promising diagnostic test, the FirstSight assay. FirstSight is a blood-based assay that evaluates three biomarkers: (1) circulating GI epithelial cells, (2) validated somatic oncogene and tumor suppressor mutations, and (3) methylation of *SEPT9* in cfDNA. Based on results for these three biomarkers, the assay calculates a diagnostic score (CMx Score). FirstSight was evaluated in 354 patients who had no prior history of CRC and were scheduled to undergo a colonoscopy. Eighty-six percent of the patients were asymptomatic and 14% had reported symptoms or a positive-FIT result. Prior to colonoscopy, patient blood samples were analyzed using the FirstSight assay. In this study, FirstSight achieved specificity of 90% and sensitivity of 100% for detecting CRC and sensitivity of 75.5% for detecting advanced adenomas. Overall, the test had a sensitivity of 79.7% for detecting advanced adenomas and CRC. In addition, there was a significant association between CMx score and polyp size (F value = 5.80, p-value = 0.017), but not for DNA mutation (F value = 1.29, p-value = 0.263) or methylation status (F value = 0.34, p-value = 0.560) and polyp size. Similarly, there was a significant association between CMx score and the number of polyps (F value = 23.71, p-value < 0.0001), but again not for DNA mutation (F value = 1.57, p-value = 0.210) or methylation status (F value = 1.34, p-value = 0.248) and the number of polyps. These results suggest that CMx score, which incorporates circulating epithelial cells, provided predictive information for polyp sizes and number above and beyond DNA mutation and methylation status alone.

Analysis of cancer-associated mutations in cfDNA has been also explored in conjunction with plasma protein analysis. In 2018, Cohen et al. (2018) reported on a new blood test, called CancerSEEK, for cancer diagnosis. CancerSEEK evaluates plasma levels of eight proteins (cancer antigen 125, CEA, cancer antigen 19–9, proactin, hepatocyte growth factor, osteopontin, myeloperoxidase, and tissue inhibitor of metalloproteinases) together with the presence of mutations (SNVs or INDELs) in 1933 distinct genomic sites. The presence of a mutation in an assayed gene or elevated levels of any of these proteins would classify a patient as positive. In this study, 1005 patients with non-metastatic, clinically detected cancers of the ovary, liver, stomach, pancreas, esophagus, colorectum, lung, or breast were evaluated using CancerSEEK. The median sensitivity of CancerSEEK among the eight cancer types evaluated was 70% (78% for stage III, 73% for stage II, and 43% for stage I cancers). The specificity of CancerSEEK was >99% across cancers of all stages. Moreover, the test could localize the anatomic site of origin of the cancer in a median of 83% of the patients. Given that driver gene mutations are not usually tissue-specific, the vast majority of the localization information was derived from protein markers. The accuracy of prediction for primary cancer prediction varied by tumor type, but the test was most accurate for CRC (84%).

Although it is beyond the scope of this article, many other approaches are being evaluated for diagnosing polyps and CRC in various bodily fluids. For example, PolypDx (Deng et al., 2017), is a metabolomic-based urine test for detecting adenomas. Although PolypDx was originally designed on a nuclear magnetic resonance platform, it was subsequently developed as a mass spectrometry-based urine metabolomic test to detect three metabolites: succinic acid, ascorbic acid, and carnitine. Detection of these metabolites used in combination with three clinical features (age, sex, and smoking status) yielded higher sensitivity (43%) and similar specificity (91%) to two different FITs (sensitivities of 18% and 21%, and specificities of 97% and 92%, respectively) for detecting hyperplastic polyps and adenomas (Deng et al., 2017).

2.6 Primary tumor of origin

Tissue-based markers are also being developed to guide CRC diagnosis, and there are various molecular platforms that can help define the primary tumor of origin in cases classified as cancer of unknown primary (CUP). Initial work to identify original tumor sites was performed using algorithms based on genome-wide gene expression profiles. This work achieved an accuracy of 88% for primary tumors within 56 categories and of 78% for CUP (Ojala, Kilpinen, & Kallioniemi, 2011). However, despite an accuracy rate of 78.6% for site prediction in silico, results suggested that comprehensive genome-wide profiling of gene expression by microarray analysis might not yet be suitable for clinical application in patients with CUP. Currently, 2 assays, a 92 gene RT- PCR assay and a 64 tissue-specific miRNA assay are commercially available for prediction of tumor sites of origin (Erlander et al., 2011; Meiri et al., 2012). Since comprehensive gene expression profiling can be influenced by irrelevant variants, and dimension reduction is necessary because of the large number of variables, the 92 gene RT-PCR assay might be more informative as a result of the limited number of biologically relevant genes on which it is based (Ma et al., 2006). Other methods that are based on miRNAs, which can regulate the expression of large numbers of protein-coding genes, also have achieved a high accuracy rate (Rosenfeld et al., 2008; Varadhachary et al., 2011). In this regard, recently a miRNA-based tissue signature has been proposed for the specific diagnosis of CRC with mucinous differentiation (Ruiz-Bañobre et al., 2020). Although data are still very preliminary and validation studies are needed, this signature could be especially useful for classifying CRC in those cases where a small tumor biopsy is not able to capture the whole histological nature of the tumor.

Classifiers of cancer type based on DNA methylation profiles, which can overcome the inherent limitations of working with RNA in formalin-fixed paraffin-embedded (FFPE) tissue samples, are also being developed. One such platform, the EPICUP, showed a specificity of 99.6%, a sensitivity of 97.7%, a positive predictive value of 88.6%, and a negative predictive value of 99.9% in a validation set of 7691 tumors (Moran et al., 2016). DNA methylation profiling predicted a primary cancer of origin in 188 (87%) of 216 patients with CUP. In addition, patients who received a tumor type-specific therapy consistent with their EPICUP diagnoses showed improved OS compared with those who received empiric therapy. Although this approach shows promise for achieving a real impact in daily clinical practice, the efficacy of any given assay must be demonstrated in randomized clinical trials, a scenario that has been a serious limitation in the majority of previous studies (Hainsworth et al., 2012).

3. Prognostic and predictive biomarkers

During recent decades, in parallel with the development of antineoplastic strategies, substantial efforts have been devoted to identifying biomarkers that can help in decision-making in the clinic. Although biomarkers likely have complex associations with patient outcomes, robust discrimination between prognostic and predictive biomarkers is challenging in clinical studies that aim to evaluate therapeutic benefits. To demonstrate that a biomarker is predictive of treatment benefit, biomarker status for all patients must be obtained, including patients who were treated with the agent of interest and untreated patients, preferably in the context of a randomized study. A formal statistical test of the treatment-by-biomarker interaction should be significant. To establish whether a marker is purely prognostic, it must be shown that there is significant association between the biomarker and outcome, regardless of treatment, and that treatment effects do not depend on the biomarker. Finally, a biomarker may have both predictive and prognostic implications (Ballman, 2015). This prognostic–predictive complexity is partly driven by the search for more effective therapies for patients who have a poor prognosis if treated with standard therapies (Sveen, Kopetz, & Lothe, 2020). Therefore, genetic alterations that are classically associated with CRC that has a poor prognosis are now targets of some of the most promising targeted therapies and, consequently, are also considered response prediction biomarkers for these cancers. Because the number of prognostic and/or predictive biomarkers is rising, in this article we have chosen to focus on biomarkers that are already established in clinical practice or those with a very promising future. Our discussion is structured to follow the typical clinical path, from the adjuvant to the palliative setting.

3.1 Adjuvant setting

Regarding biomarkers used in the adjuvant setting, only MSI-H/dMMR is thus far established as a useful prognostic indicator. The first reports of a favorable prognostic association of MSI in CRC were published in 1993 (Lothe et al., 1993; Thibodeau, Bren, & Schaid, 1993). This association has since been confirmed for OS and disease-free survival by meta-analyses comparing data from 1277 (Popat, 2004) and 12,782 (Guastadisegni, Colafranceschi, Ottini, & Dogliotti, 2010) patients with primary MSI-H CRCs to those with MSS CRCs. Retrospective analyses of data from randomized trials showed this association also applies to patients with stage II and III colon cancer patients who have not received adjuvant chemotherapy (Hutchins et al., 2011; Sargent et al., 2010). A corresponding association of MSI with clinicopathological features has also been reported for prognosis; patients with proximal tumors have a greater favorable prognostic association of MSI than those with distal tumors (Sinicrope et al., 2013). Various studies report stronger effects of MSI in stage II CRC (Klingbiel et al., 2015; Merok et al., 2013), similar prognostic associations across stages II and III CRC (Benatti et al., 2005; Sinicrope et al., 2011), or even a stronger effect in stage III CRC (Samowitz et al., 2001). However, no statistically significant interactions with clinicopathological features, including tumor stage, were observed in a pooled analysis of data from 7326 stage II or III colon cancers (Dienstmann et al., 2017), suggesting that the prognostic value of MSI is independent of disease stage.

The prognostic implications of MSI after adjuvant treatment might be confounded by also having a predictive value for 5-FU-based chemotherapy, thus illustrating prognostic–predictive biomarker complexity (Vilar & Gruber, 2010). Loss of MMR function could result in failure to recognize and respond to incorporation of 5-FU into tumor DNA (Jo & Carethers, 2006). However, many retrospective analyses have shown inconsistent results regarding the effects of 5-FU-based regimens in patients with

MSI-H tumors (Vilar & Gruber, 2010). Most studies that compared the effects of 5-FU-based chemotherapy to no treatment failed to show significant improvements in OS or disease-free survival (DFS) in patients with MSI-H/dMMR tumors, even though 5-FU-based chemotherapy significantly improved the outcomes of patients with MSS or MMR-proficient colon cancers or CRCs (Benatti et al., 2005; Carethers et al., 2004; Hutchins et al., 2011; Lanza et al., 2006; Sargent et al., 2010). However, only two of these studies revealed a statistically significant interaction between MSI status and the response to chemotherapy (Jover et al., 2009; Ribic et al., 2003). A meta-analysis including data from 396 patients with MSI-H CRCs revealed a substantial degree of heterogeneity with respect to the effects of 5-FU-based chemotherapy, and a lack of benefit could not be definitively confirmed (Guastadisegni et al., 2010).

MSI status is not predictive of a lack of benefit from the combination chemotherapies typically used in patients with CRC. The addition of oxaliplatin to 5-FU-based adjuvant chemotherapy regimens improves patient survival (André et al., 2004), and this effect is also seen among patients with dMMR stage II or III CRC (Cohen et al., 2020; Flejou et al., 2013; Tougeron et al., 2016). Furthermore, patients with dMMR tumors have better survival outcomes after treatment with FU, leucovorin, and oxaliplatin (FOLFOX) compared with those with MMR-proficient tumors (in one study the association was tumor location dependent) (Gavin et al., 2012; Sinicrope et al., 2013, 2014; Zaanan et al., 2011), consistent with the favorable prognostic effect of MSI. Paradoxically, metastatic MSI-H tumors are aggressive and are associated with inferior PFS and OS outcomes relative to metastatic tumors for MSS CRC (Heinemann et al., 2018; Kim et al., 2016; Tran et al., 2011; Venderbosch et al., 2014). Although other reports indicate that no prognostic associations exist and additional data are needed, the survival benefits associated with primary MSI-H CRCs seem to be lost in the metastatic setting (Jin et al., 2018; Margonis et al., 2018). On the other hand, although neoadjuvant therapy is not currently a standard therapeutic strategy in CRC, the combination of nivolumab plus ipilimumab has yielded encouraging results in CRC patients in this setting, especially among those with MSI-H/dMMR tumors. In a recent study reporting the results of a phase II NICHE trial (NCT03026140) (Chalabi et al., 2020), among 35 patients who received both nivolumab and ipilimumab and were evaluable for efficacy (two patients with pMMR CRCs were deemed ineligible post-surgery), 100% of those with dMMR CRC and 27% with pMMR CRC had pathologically relevant responses. The majority of responders (19/20 with dMMR CRC and 3/15 with pMMR CRC) had major pathologically relevant responses, including 12 complete responses. Subsequent organoid-based investigations suggested that the lower response rates of patients with pMMR CRCs reflected a lack of highly immunogenic T-cell antigens, as opposed to other tumor-intrinsic factors (Chalabi et al., 2020).

In 2012, Ebert et al. (2012) proposed the epigenetic biomarker methylation of *TFAP2E* in tumor tissue specimens as a promising negative predictive biomarker for 5-FU-based chemotherapy. The authors suggested the lack of response to 5-FU was probably mediated by DKK4, a downstream effector of *TFAP2E* that is implicated in chemoresistance to 5-FU in CRC cell lines (Xi, Formentini, Nakajima, Kornmann, & Ju, 2008; Xi, Nakajima, Schmitz, Chu, & Ju, 2006). Although the expectations were high, this study presented various important limitations for its clinical usefulness. Primarily, it interrogated a relatively small cohort of patients with CRC ($n = 220$), which was actually a combined collection of patients with non-metastatic and mCRC from four different prospective trials that were analyzed together as one large cohort. Second, only a very small subset of the entire cohort was analyzed for methylation and expression of *TFAP2E*, as well as expression of the DKK4 protein. Third, although all patients received 5-FU-based chemotherapy, their regimens differed and also included either oxaliplatin, irinotecan and/or cetuximab, or radiotherapy (chemoradiotherapy). Unfortunately, in a subsequent study the negative predictive potential of *TFAP2E* was not confirmed (Murcia et al., 2018). Although methylation of *TFAP2E* intron 3 is tumor-related, it did not correlate with loss of TFAP2E protein expression, and, more importantly, *TFAP2E* methylation did not play any role in predicting response to 5-FU-based chemotherapy in patients with CRC.

In January 2016, Dalerba et al. (2016) reported a novel approach to the problem of identifying patients with CRC who might benefit from adjuvant chemotherapy. They reasoned that the presence of a stem cell-like state could be associated with more aggressive tumors, and performed a bioinformatic search for a gene expression signature obtained from populations of stem cells and progenitor cells. By mining a large, preexisting database of CRCs, the authors identified a panel of 16 genes for which expression was inversely related to the stem cell-like state. *CDX2* was the most clinically actionable of these genes because it could be detected using immunohistochemical analysis. The investigators performed a series of validation analyses involving multiple independent datasets, which is a necessary approach for data-mining research. The first analysis confirmed an inverse relationship between *CDX2* expression and patient outcomes in which

CDX2-negative tumors (present in 6.9% of patients in the discovery dataset) were associated with significantly lower rates of 5-year DFS compared to *CDX2*-positive tumors (41% vs 74%). A validation dataset was created by immunohistochemical analysis to confirm CDX2 protein expression. Analysis of the validation dataset, in which 13% of the patients had CDX2-negative tumors, confirmed 5-year survival rates of 48% among patients with CDX2-negative tumors vs 71% among patients with CDX2-positive tumors. However, this finding was not sufficient to prove that the subgroup of patients with a worse natural history would benefit from adjuvant chemotherapy; instead they might be less responsive to treatment. To address this, the investigators focused on patients with stage II CRC and confirmed that CDX2-negative cancers were associated with significantly lower rates of survival than were CDX2-positive cancers (48%–51% vs 80%–87%). Finally, they used an expanded database to demonstrate that the benefit observed with the administration of chemotherapy in terms of disease-free survival in CDX2-negative cohorts was superior to that observed in CDX2-positive cohorts in both the stage II subgroup and the stage III subgroup. However, despite the rigorous bioinformatics analysis, the number of patients who had stage II CRC and CDX2-negative tumors was small, and so this result is not definitive. This retrospective study requires prospective confirmation with uniform interventions. In addition, the immunohistochemical analysis was performed on tissue microarrays, which facilitated rapid throughput but may have underestimated the heterogeneity of CDX2 expression throughout the tumor. Furthermore, these findings raise the question of what mechanism might be at work in silencing *CDX2*; the answer to this question could lead to the discovery of new approaches to treating the fundamental problem. Meanwhile, given these limitations, use of CDX2 has not entered clinical practice to indicate (or not) if adjuvant chemotherapy should be given to stage II CRC patients.

On the other hand, based on the cfDNA methylation analysis, Luo et al. (2020) recently developed a combined score (cp-score) for the prognostic purposes in CRC. This score also includes clinical characteristics such as age, gender, primary tumor site, and TNM stage. Using machine learning algorithms and proportional hazards regression methods, the authors conducted a variable selection on the training set ($n = 528$, events$=157$) and built a composite score on the validation set ($n = 273$, events$=77$). Uni-Cox and LASSO-Cox methods were implemented to reduce the dimensionality, and a Cox-model was constructed for prognostication using a five-marker panel, which dichotomized the patients into high-risk and low-risk groups. To characterize the discrimination potential of the composite score, TNM stage, CEA levels, primary tumor location, and the combination of all the existing biomarkers, the investigators applied time-dependent receiver operating characteristic curves. Multivariate Cox regression analysis indicated that the cp-score was highly correlated with the risk of death and was an independent prognostic factor in both the training and validation sets. As expected, TNM stage, CEA status, and primary tumor location were also prognostic factors for survival of patients with CRC. An integrative model combining the cp-score and clinical characteristics demonstrated greater prognostic accuracy in both the training (AUROC of 0.82) and validation (AUROC of 0.87) cohorts. Finally, in an attempt to develop an easier-to-use tool, the authors constructed a nomogram with a point scale of the next four variables: cp-score, CEA concentration, TNM stage, and primary tumor location.

The possibility that circulating tumor DNA (ctDNA) could be a useful prognostic marker for minimal residual disease was also been suggested for a small series of advanced CRC patients who underwent resection of liver metastases (Diehl et al., 2008). This approach has been subsequently demonstrated in localized CRC (Reinert et al., 2019; Tie et al., 2016, 2021; Tie, Cohen, Wang, Christie, et al., 2019; Tie, Cohen, Wang, Li, et al., 2019), and many ongoing clinical trials are evaluating the role of ctDNA status in adjuvant therapy decision-making and the impact on patient survival outcomes (Coakley, Garcia-Murillas, & Turner, 2019).

Finally, although it is not a genomic or epigenomic biomarker itself, Immunoscore, deserves specific attention in this section. Proposed for the first time in 2006 by Galon et al. (2006), Immunoscore was recently recognized in the current *Localized Colon Cancer ESMO Clinical Guidelines* as a useful tool for prognostication in the non-metastatic colon cancer (CC) setting. Immunoscore has been validated in a huge prospective cohort of 2681 stage I–III CC patients, and was a strong predictor for time to recurrence (TTR), OS, and DFS, independent of patient age, sex, MSI, and other relevant prognostic factors. Moreover, Immunoscore showed the highest relative contribution to the risk of all clinical parameters, including the UICC TNM staging system (Pagès et al., 2018). Based on this data, the *ESMO Clinical Guidelines* concluded that Immunoscore could help refine the prognosis for colon cancer patients with early-stage disease in conjunction with the TNM staging system (level of evidence III—prospective cohort studies, grade of recommendation C—insufficient evidence for efficacy or benefit does not outweigh the risk or the disadvantages, optional). Oppositely, the NCCN Colon Guidelines, to date, do not even mention Immunoscore, probably because of unclear evidence for its prognostic value for predicting the risk of recurrence and death in stage II or III CC separately or for its predictive role in predicting adjuvant chemotherapy benefit. In an attempt to clarify these questions specifically in stage III CC, the International Society for Immunotherapy of Cancer Immunoscore Consortium conducted a retrospective study to evaluate Immunoscore in 763 patients with TNM stage III CC from two retrospective cohorts: cohort 1 (Canada/United States) and cohort 2 (Europe/Asia) (Mlecnik et al., 2020). In this analysis, patients with a high Immunoscore also had the lowest risk of recurrence, and a significantly prolonged TTR, OS, and DFS. The association of Immunoscore with TTR was independent of major prognostic covariates such as sex, T stage, N stage, primary tumor location, and MSI status. Moreover, Immunoscore had the strongest contribution to survival risk for TTR and OS. Importantly, chemotherapy was significantly associated with survival in the high-Immunoscore group for patients with either low-risk or high-risk stage II CC as compared to the low-Immunoscore group (Mlecnik et al., 2020). These studies have paved the path for investigating the prognostic role of Immunoscore in larger prospective studies, and, even more importantly, in randomized clinical trials to evaluate its predictive potential for chemotherapy benefit.

3.2 Palliative setting

Although the backbone of treatment in patients with mCRC has historically been chemotherapy, over the last few decades targeted molecular therapies against EGFR and angiogenic factors have been introduced into daily clinical practice. Furthermore, new treatment options have recently been added to the mCRC armamentarium, including BRAF inhibitors (BRAFi) and anti-programmed cell death 1 (PD-1) antibodies. Historically, and in parallel with drug development, multiple research efforts have been undertaken to discover and implement molecular biomarkers to guide therapeutic strategies. This becomes even more important in today's clinical scenario in which multiple therapeutic options are available, and therefore treatment selection aims not only to improve patient survival, but also to spare patients from unnecessary toxicity and reduce the economic burden of expensive treatments. In this section, we summarize the most relevant milestones achieved in the field of biomarkers for various treatments in mCRC (Fig. 2) and provide insights into some of the important clinical and methodological aspects.

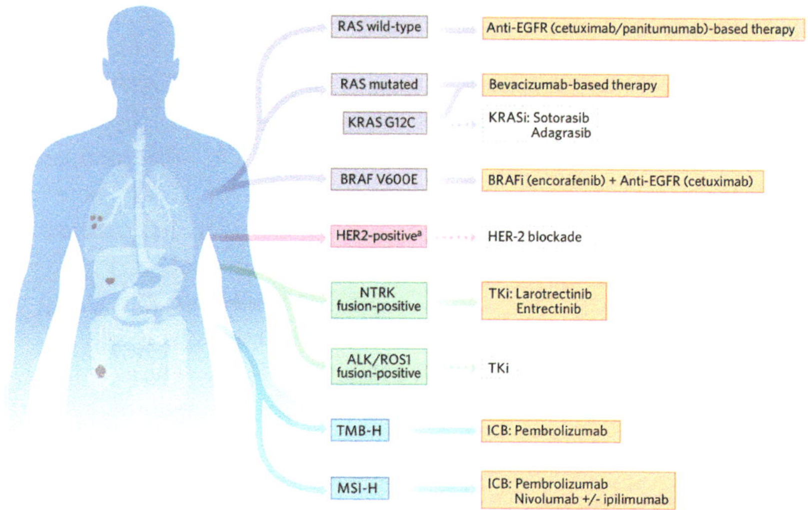

Fig. 2 Predictive biomarkers for metastatic CRC treatment. Summary of currently known molecular tumor alterations that enable improved clinical decision-making regarding use of therapies that are tailored to the metastatic CRC (mCRC) patient. The therapies shown go beyond classical chemotherapeutic agents. Excluding *RAS* mutations for EGFR-targeted therapies, which are considered a negative predictive biomarker, the molecular alterations shown represent positive predictors of benefit with the indicated targeted therapies. Although bevacizumab is considered a useful therapeutic option in combination with chemotherapy in mCRC independent of any particular molecular alteration, currently there is no consistent predictive biomarker to guide bevacizumab use. Although many of these molecular alterations are applicable only to a minority of mCRC patients, collectively these low-prevalence actionable characteristics support a new targeted therapeutic horizon for many patients. Color-coded boxes: yellow, US FDA-approved therapy; gray, not US FDA-approved therapy. [a]*HER2* amplification or overexpression. Abbreviations: BRAFi, BRAF inhibitors; CRC, colorectal cancer; ICB, immune-checkpoint blockade; KRASi, KRAS inhibitors; MSI-H, microsatellite instability high; TKi, tyrosine kinase inhibitors; TMB-H, tumor mutation burden high.

3.2.1 Conventional chemotherapy

The backbone of treatment in patients with mCRC has historically been chemotherapy, and several chemotherapeutic agents are now approved in this setting: fluoropyrimidines (fluorouracil [FU] and capecitabine), oxaliplatin, irinotecan, and, since 2015, trifluridine/tipiracil (TAS-102).

**Fluoropyrimidine-based chemotherapy and TAS-102*—Several studies have investigated the predictive role of fluoropyrimidine metabolic pathway enzymes in response to FU-based therapies. Various retrospective and prospective studies of the role of thymidylate synthase (TS) in fluoropyrimidine-based therapy (primarily FU plus leucovorin) have yielded discordant results. In this regard, multiple studies have shown that low levels of TS expression in metastatic tumor tissues generally correlate with higher overall response rate (ORR) (Aschele et al., 2002; Aschele, Debernardis, Tunesi, Maley, & Sobrero, 2000; Cascinu et al., 2000; Etienne-Grimaldi et al., 2008). Surprisingly, such a correlation was not evident when TS levels were measured in primary tumor tissues (Aschele et al., 2000; Johnston et al., 2003). Similarly, low levels of TS and dihydropyrimidine dehydrogenase (DPD) in metastatic tumor tissues were associated with a favorable response to FU in patients with mCRC (Salonga et al., 2000); however, a subsequent report in 2006 failed to validate these findings (Smorenburg et al., 2006). Likewise, the role of thymidine phosphorylase as a predictive biomarker was investigated, but the results remain inconclusive (Gustavsson et al., 2009; Lindskog, Derwinger, Gustavsson, Falk, & Wettergren, 2014).

In 2009, a meta-analysis of five studies examining a total of 861 patients with mCRC concluded that, compared with MSS patients, MSI-H patients did not achieve a statistically significant better response rate to FU-based chemotherapy (Des Guetz, Uzzan, Nicolas, Schischmanoff, & Morere, 2009). Similarly, a study of the relationship between *MSH2* gene expression and capecitabine efficacy in patients with mCRC revealed that a higher MSH2 expression was associated with a better response (Jensen, Danenberg, Danenberg, & Jakobsen, 2007).

In an attempt to identify noncoding RNA-based predictive biomarkers, a low expression of miR-143 was shown to be associated with improved ORR and progression-free survival (PFS) in patients treated with capecitabine (Simmer et al., 2015). Likewise, low expression of miR-429 correlated with improved 5-year disease-free survival and OS in patients with mCRC treated with FU-based chemotherapy (Dong, Cai, & Li, 2016).

On the basis of the results of the phase III RECOURSE (Study of TAS-102 in Patients With Metastatic Colorectal Cancer Refractory to Standard Chemotherapies) trial, the US FDA-approved TAS-102 for patients with mCRC. In 2017, Suenaga et al. (2017) analyzed genomic DNA extracted from 233 samples from three cohorts: an evaluation cohort of 52 patients who received TAS-102, a validation cohort of 129 patients who received TAS-102, and a control cohort of 52 patients who were treated with regorafenib. Single-nucleotide polymorphisms of genes involved in homologous recombination repair (*ATM*, *BRCA1*, *BRCA2*, *XRCC3*, *FANCD2*, *H2AX*, and *RAD51*), and cell cycle checkpoints (*ATR*, *CHEK1*, *CHEK2*, *CDKN1A*, *TP53*, *CHE1*, *PIN1*, and *PCNA*) were analyzed by PCR-based direct sequencing. Interestingly, patients who harbored any G allele at the *ATM rs609429* locus exhibited improved OS rates when compared with patients with a *C/C* variant (Suenaga et al., 2017).

**Oxaliplatin-based chemotherapy*—Several key genes involved in the nucleotide excision repair pathway have been explored as potential predictive biomarkers for treatment with oxaliplatin. The most notable studies in this area have been on the *ERCC1* gene, but as reported based on results from the MAVERICC (Marker Evaluation for Avastin Research in CRC) trial (NCT01374425), intratumoral *ERCC1* expression failed to predict response to oxaliplatin treatment (Parikh et al., 2019). The *XRCC1* gene, a base excision repair modulator, has also been evaluated as a possible predictive biomarker for oxaliplatin treatment, and a polymorphism in this gene (*XRCC1–839 Arg/Gln* or *Gln/Gln*) correlated with worse ORR to treatment with FU/oxaliplatin (Stoehlmacher et al., 2001).

Several miRNAs have also been assessed for predictive response potential to FOLFOX- or capecitabine plus oxaliplatin-based regimens. In this treatment setting, high miR-625-3p and low miR-148a expression were associated with poor response (Rasmussen et al., 2013; Takahashi et al., 2012), whereas high miR-126 microvessel density was associated with improved PFS (Hansen, Nielsen, Jakobsen, & Sorensen, 2013).

**Irinotecan-based chemotherapy*—The most notable marker studied as a potential predictive biomarker for irinotecan treatment is topoisomerase 1 (TOP1). The first large study in which the predictive power of TOP1 was evaluated used tumor tissue samples from the FOCUS (Fluorouracil, Oxaliplatin, CPT-11: Use and Sequencing; NCT00008060) trial and reported that moderate or high expression of TOP1 was associated with a significant gain in survival after irinotecan-based therapy (Braun et al., 2008). Unfortunately, these findings were not subsequently confirmed in an analysis of samples from 545 patients involved in the CAIRO (Capecitabine, Irinotecan, Oxaliplatin; NCT00312000) study, even though patients received similar treatment regimens and similar analytical approaches were used (Koopman et al., 2009).

A study of the role of genetic polymorphisms in the *TDP1* and *XRCC1* genes in response to irinotecan-based regimens, showed a positive correlation with improved ORR in patients with the *TDP1 IVS12+79G.T* and *XRCC1* GGCC-G/GGCC-G genotypes (Hoskins et al., 2008). Aprataxin, a protein member of the histidine triad superfamily, has also shown potential ability to be used to discriminate responders from nonresponders. In a retrospective cohort of 128 patients treated with irinotecan-based chemotherapy, low expression of aprataxin correlated with improved disease control, PFS, and OS (Dopeso et al., 2010). Although studies of *UGT1A* gene polymorphisms and their predictive value for response to irinotecan treatment have yielded conflicting results (Cecchin et al., 2009; Dias, McKinnon, & Sorich, 2012), higher methylation levels of the *BNIP3* gene in patients treated with irinotecan plus S1 correlated with lower response rates (Hiraki et al., 2010).

3.2.2 Anti-angiogenic drugs

**Bevacizumab*—Since the introduction of bevacizumab for the management of patients with mCRC, substantial efforts have been made to discover predictive biomarkers for this anti-angiogenic drug. To date, the most promising approach evaluated the correlation between VEGF-A levels and clinical benefit from bevacizumab-based chemotherapy. A 2014 study reported higher ORR and improved PFS in patients with a low expression of VEGF-A in primary tumor specimens (Bruhn et al., 2014). Unfortunately, these encouraging results were not confirmed by the MAVERICC trial (Parikh et al., 2019). The possible predictive role of *RAS* was evaluated in a subset of 230 patients with mCRC treated with irinotecan, FU, and leucovorin (IFL) plus placebo, or IFL plus bevacizumab, in a phase III randomized clinical trial. In this study, only the subset of patients who had wild-type *KRAS* showed a significantly higher response rate in the bevacizumab arm (Hurwitz, Yi, Ince, Novotny, & Rosen, 2009).

**Regorafenib*—While most of the studies have explored biomarkers from a radiomic perspective, others have used tissue- and plasma-based approaches. As a potential tissue-based predictive molecular marker for regorafenib, Wong et al. (2015) described a correlation between down-regulation of p53 and phosphorylated-proline-rich AKT substrate and higher PFS and metabolic response, respectively. The authors also evaluated the association between total plasma circulating cfDNA at baseline and major efficacy outcomes and found a significant inverse correlation with PFS . Moreover, patients with detectable *KRAS* mutation in plasma presented a worse PFS .

Regarding *RAS* gene evaluated in circulating cfDNA, a small retrospective study involving 21 CRC patients showed a clinically significant association between reduced rates of *RAS* mutation in plasma, as detected by digital droplet PCR, and improved PFS and OS (Khan et al., 2018). A prognostic score for regorafenib has also been developed based on combinatorial clinico-molecular models and the REBECCA (Regorafenib in

Metastatic Colorectal Cancer: A Cohort Study in the Real-Life Setting) study. This prognostic score included the following parameters that were independently associated with poorer OS: high Eastern Cooperative Oncology Group performance status, a shorter time from initial diagnosis of metastases, an initial regorafenib dose of 160 mg, more than three metastatic sites, presence of liver metastases, and *KRAS* mutations (Adenis et al., 2016).

**Aflibercept and ramucirumab*—A 2015 study of 87 patients with mCRC who were enrolled on the phase II AFFIRM (Study of Aflibercept and Modified FOLFOX6 as First-Line Treatment in Patients With Metastatic Colorectal Cancer) trial and were treated with aflibercept plus mFOLFOX6 reported that high plasma levels of interleukin-8 at baseline, together with increase in these plasma levels during the course of treatment, correlated with shorter PFS (Lambrechts et al., 2015). In 2017, Tabernero et al. (2018) evaluated the predictive potential of VEGF-D in a translational research study based on the RAISE (Ramucirumab versus Placebo in Combination With Second-Line FOLFIRI in Patients With Metastatic Colorectal Carcinoma That Progressed During or After First-Line Therapy With Bevacizumab, Oxaliplatin, and a Fluoropyrimidine) trial. Using an adaptive analysis design, the researchers analyzed plasma proteins from two sets of patients, one exploratory ($n=294$) and one confirmatory ($n=590$), and found that high VEGF-D basal levels significantly correlated with a better PFS and OS in patients treated in the ramucirumab arm. Although VEGF-D is not a genomic or epigenomic biomarker, it is the most compelling study to date in a ramucirumab-treated CRC cohort.

3.2.3 Anti-EGFR drugs: Cetuximab and panitumumab

KRAS mutations in tumor tissues were the first predictive biomarker approved to guide decision-making and determine eligibility for anti-EGFR therapy in patients with mCRC. One of the first studies of KRAS mutations as predictive biomarkers comprised 30 patients with mCRC treated with cetuximab-based regimens. This study showed a significant correlation between the presence of *KRAS* mutations and the lack of response to anti-EGFR therapy (Lievre et al., 2006), an observation that was subsequently validated in a cohort of 427 patients with mCRC treated with panitumumab (Amado et al., 2008). Similarly, the potential of *NRAS* mutation status as a negative response predictor for panitumumab and cetuximab was later confirmed in several clinical trials and meta-analyses (De Roock, Claes, et al., 2010; De Roock, Jonker, et al., 2010;

Douillard et al., 2013; Peeters, Oliner, & Parker, 2010; Sorich et al., 2015). Interestingly, in addition to the lack of efficacy in patients with *RAS*-mutated mCRC, panitumumab or cetuximab also showed reduced efficacy when given in combination with FOLFOX in a retrospective data analysis from two phase III clinical trials (Bokemeyer et al., 2017; Douillard et al., 2013). Such a negative effect for panitumumab or cetuximab when given in conjunction with FU, leucovorin, and irinotecan (FOLFIRI) has not been confirmed to date (Stintzing et al., 2017). In an attempt to translate these tissue-based predictive biomarkers into ctDNA-based liquid biopsy assays, two 2017 studies used an innovative beads, emulsions, amplification, and magnetics (BEAMing) assay and reported promising agreement rates of 89.7% and 93% for the mutational status of RAS between tissue and ctDNA (Grasselli et al., 2017; Vidal et al., 2017).

BRAF mutations have also been evaluated as a predictive biomarker for anti-EGR therapies. The presence of *V600E* mutation within *BRAF* often reflects a poor prognosis in patients with CRC (Yuan et al., 2013). In addition, two meta-analyses reported a lack of benefit in terms of PFS, OS, and ORR when anti-EGFR therapies were combined with standard chemotherapy in a subset of patients harboring BRAF mutations (Pietrantonio et al., 2015; Rowland et al., 2015). Despite these early data suggesting that the presence of the *BRAF V600E* mutation may suggest a lack of response to anti-EGFR-based therapies, there is not yet enough clinical evidence to consider *BRAF* mutational status as a negative predictive biomarker in patients with advanced disease. Moreover, emerging therapeutic strategies that combine anti-EGFR drugs with BRAFi have recently changed the paradigm of how we must understand these molecular alterations from a therapeutic perspective.

In 2014, Manceau et al. (2014) investigated the roles of different miRNAs in response to anti-EGFR therapies. They conducted miRNA profiling in a training set of tissue specimens (33 fresh-frozen [FF] and 35 FFPE samples that were retrospectively collected and 19 prospectively collected FF samples) from 87 patients with mCRC refractory to chemotherapy and treated with anti-EGFR antibodies, and determined that hsa-miR-31-3p expression levels were significantly associated with PFS. They subsequently validated these results in an independent validation cohort consisting of 19 FF and 26 FFPE prospectively collected samples. The percentage of size variation of target lesions by RECIST criteria in the validation series and risk status were significantly associated with expression levels of hsa-miR-31-3p. In addition, the investigators built and validated nomograms to predict PFS based on hsa-miR-31-3p expression. In the same context, Miller-Phillips et al. (2018) evaluated the role of miR-21 as a predictive biomarker for anti-EGFR therapies. Using an RT-qPCR assay, they quantified miR-21 expression in FFPE tumor samples from patients enrolled on the FIRE-3 trial (NCT00433927) patients. The median miR-21 expression within the FIRE-3 population was determined and subsequently used to segment the population into low and high miR-21 expression groups. Patients who had wild-type *RAS* wild-type and low miR-21 expression showed significantly improved ORR and OS when cetuximab, but not bevacizumab, was added to FOLFIRI chemotherapy. In contrast, patients who had wild-type RAS but high miR-21 expression showed no significant difference in ORR and OS between the cetuximab and bevacizumab treatment groups. Similarly, patients with mutated *RAS* and either high or low miR-21 expression showed no significant difference in ORR between cetuximab or bevacizumab added to FOLFIRI. Unfortunately, other correlations between groups and PFS or OS were not significant. In a later study to determine the biological changes underlying these differences in efficacy outcomes, Miller-Phillips et al. (2019) used normalized mRNA microarray data to compare gene expression in the miR-21 low and high groups. This analysis identified 538 genes that were significantly and differentially expressed in patients with wild-type *RAS* after adjustment for multiple testing. Including data from the two groups into single-sample Gene Set Enrichment Analysis showed 23 hallmarks of cancer gene sets significantly differentially enriched, being *KRAS*-signaling higher in the miR-21 high group.

3.2.4 Anti-PD-1 drugs: Pembrolizumab and nivolumab

In May and July of 2017, the US FDA approved the anti-PD-1 therapies pembrolizumab and nivolumab for the treatment of patients with MSI-H mCRC for whom the disease has progressed after treatment with fluoropyrimidine, oxaliplatin, and irinotecan. Almost a year later, in July 2018, a nivolumab plus ipilimumab combination regimen was approved, which opened three novel treatment options for patients with MSI-H or dMMR mCRC (patients with MSI-H or dMMR mCRC represent approximately 5% of all patients with mCRC) (Braun et al., 2008). Although patients with MSI-positive mCRC have worse prognosis, it is thought that they derive clinical benefit from anti-PD-1 therapy because of a large proportion of lymphocytic infiltration and the presence of mutation-associated neoantigens (Giannakis et al., 2016; Le et al., 2017,

2015; Overman et al., 2017, 2018). This exciting discovery has led to universal MSI testing for the management of patients with mCRC.

Not surprisingly, in May 2020 the US FDA-approved pembrolizumab as first-line therapy for patients with MSI-H/dMMR mCRC. This approval was based on the results of the KEYNOTE-177 study (NCT02563002), a multicenter, international, open-label, active-controlled, randomized trial that compared first-line therapy with pembrolizumab vs chemotherapy in 307 patients with MSI-H/dMMR mCRC. This study demonstrated a statistically significant improvement in PFS, with a median PFS of 16.5 months vs 8.2 months for pembrolizumab compared to chemotherapy standard-of-care. Longer-term analysis is needed to assess the effect on OS. Moreover, in June 2020 the US FDA granted accelerated approval to pembrolizumab for the treatment of patients with any unresectable or metastatic solid tumor with high mutational burden (as determined by the FDA-approved test, the FoundationOneCDx assay) whose cancer has progressed after previous treatment and has no satisfactory alternative treatment options (Bersanelli, 2020). Several clinical trials evaluating the combination of anti-PD-1 therapy with chemotherapy are ongoing for previously untreated MSI-H/dMMR mCRC patients with the goal of improving on results from previous studies and further extending survival of these patients. Meanwhile, other different immunotherapeutic approaches are being evaluated for treatment of MSS CRC, which is less responsive to immune-checkpoint inhibition than MSI-H mCRC. Although all of these results represent substantial therapeutic advances in the treatment of mCRC, they also emphasize the growing need for more precise predictive biomarkers to support more rational development of immunotherapies. A more comprehensive understanding of the intersection between genomics, epigenomics, and immunology in mCRC seems essential for meeting this need for new strategies.

Recently, Grasso et al. (2018) reported the results of a large-scale genomic analysis (TCGA, Nurses' Health Study, and Health Professionals Follow-up Study cohorts) involving 1211 primary CRC tumor specimens. Mutations in genes involved in immune modulatory pathways, as well as in the neoantigen-presentation machinery (mainly *B2M* and *HLA*), significantly correlated with MSI-H. Along with *JAK1/2* and *IFN-gamma receptor 1* mutations, similar alterations have been observed in melanoma, non-small cell lung cancer, and CRC and deemed to be genetic drivers of primary or acquired resistance to immune-checkpoint blockade, reflecting their role as a mechanism of adaptive resistance against T-cell tumor infiltration

(Gao et al., 2016; Giannakis et al., 2016; Grasso et al., 2018; Gurjao et al., 2019; Le et al., 2017; Zaretsky et al., 2016). The interaction between somatic alterations and the immune system is complex, as indicated by a recent study in which 11 out of 13 B2M-mutant CRC patients achieved mCRC control with anti-PD-1 or anti-PD-L1 agents, despite the presence of a mutation that, theoretically, conferred primary resistance to immune-checkpoint inhibitors (Middha et al., 2019). On the other hand, for both MSS and MSI-H tumors, active WNT/β-catenin signaling was inversely associated with tumor T-cell infiltration, providing evidence of the existence of an anti-immune response mechanism beyond the MSI profile (Grasso et al., 2018).

3.2.5 BRAF inhibitors: Vemurafenib and encorafenib

BRAF mutations occur in 10–15% of all CRCs and in ~7% of all mCRC (Clarke & Kopetz, 2015; Davies et al., 2002). Although most *BRAF* mutations occur in codon 600 (mainly *BRAF V600E*), which leads to constitutive BRAF kinase activity and sustained MAPK pathway signaling, 2% of mCRCs have atypical *BRAF* mutations that are outside of codon 600, usually in codon 594 (Jones et al., 2017). Surprisingly, although monotherapy with BRAFi has proven effective in the treatment of *BRAF*-mutant melanoma, it was ineffective in *BRAF V600E*-mutant CRCs. Preclinical evidence demonstrated that despite transient inhibition of pERK by BRAFi such as vemurafenib, rapid ERK reactivation occurs through EGFR-mediated activation of RAS and CRAF (Corcoran et al., 2012). Furthermore, the fact that *BRAF V600E*-mutant CRCs express higher levels of pEGFR than do *BRAF*-mutant melanomas, positions them for EGFR-mediated resistance (Corcoran et al., 2012). Collectively, these findings provided rationale to test dual BRAF and EGFR blockade. Results from preclinical studies and early-phase clinical trials, have demonstrated this strategy is feasible and safe, and can potentially improve therapeutic efficacy of BRAFi. Moreover, preclinical studies have suggested that combined inhibition of BRAF and MEK was more effective than dual BRAF and EGFR blockade. This strategy was tested in subsequent phase 1 and phase 2 clinical trials that combined BRAFi with both anti-EGFR monoclonal antibodies and MEK inhibitors (MEKi) (Corcoran et al., 2010, 2012, 2018). Results of these trials led to US FDA approval (in April 2020) of encorafenib, a BRAF tyrosine kinase inhibitor, used in combination with cetuximab for the treatment of adult patients with *BRAF V600E*-mutated mCRC. The efficacy of this combination of drugs was evaluated in the BEACON

CRC study (Kopetz et al., 2019), a phase 3 randomized, active-controlled, open-label, multicenter trial (NCT02928224). In this trial, encorafenib plus cetuximab demonstrated a clinical and statistically significant OS and PFS benefit compared to the control arm of either irinotecan or FOLFIRI plus cetuximab in patients with *BRAF V600E*-mutated mCRC who had progressed on one or two prior regimens. This trial also evaluated the efficacy of triple-therapy with encorafenib, binimetinib (a MEKi), and cetuximab in a second experimental arm, but although this regimen showed an improved OS and PFS compared to the control arm, it was more toxic than the dual BRAF and EGFR blockade and had similar efficacy.

Another BRAFi, vemurafenib, which has more modest clinical activity, was recently included in the NCCN guidelines as a treatment option for patients with *BRAF V600E*-mutated mCRC when used in combination with cetuximab/panitumumab plus irinotecan (Kopetz et al., 2021; National Comprehensive Cancer Network, 2020). Inclusion in the guidelines was based on results of the randomized phase 2 Southwest Oncology Group (SWOG) 1406 trial, in which the triple-therapy (vemurafenib, cetuximab, and irinotecan) demonstrated improved PFS and ORR as compared with cetuximab plus irinotecan (Kopetz et al., 2021). In addition to the previously described regimens, based on the results of a phase 1 study (Corcoran et al., 2018), the NCCN Panel has recommend the combination of dabrafenib (BRAFi) plus trametinib (MEKi) plus either cetuximab or panitumumab as another treatment option beyond the first-line setting for *BRAF V600E*-mutated mCRC (National Comprehensive Cancer Network, 2020).

3.2.6 HER-2 blockade

A large body of evidence, accrued primarily from breast and gastric cancer patients, supports the role of *HER-2* amplification or overexpression as a predictive biomarker for anti-HER-2-based therapies. Therefore, there is renewed interest in evaluating HER-2 as a clinically actionable target in mCRC. Although initial mCRC clinical trials interrogating the anti-HER-2 monoclonal antibody trastuzumab in combination with other chemotherapeutic agents (either FOLFOX or irinotecan) closed early due to lack of patient accrual, mechanistic insights gained from preclinical analyses of *HER-2*-amplified mCRC patient-derived xenografts has led to improved design of new clinical trials (Bertotti et al., 2011; Clark, Niedzwiecki, & Hollis, 2003; Ramanathan et al., 2004). Three phase 2 clinical trials evaluated dual HER-2 blockade in a biomarker-selected subset of heavily pretreated mCRC patients. Study treatments included trastuzumab plus lapatinib (HERACLES trial, NCT03225937), pertuzumab and trastuzumab (MyPathway trial, NCT02091141), or the antibody-drug conjugate trastuzumab deruxtecan (DESTINY-CRC01, NCT03384940). Results of these studies demonstrated an impressive ORR of ~30–45% (Hainsworth et al., 2018; Sartore-Bianchi et al., 2017; Siena et al., 2020). These data have paved the way for development of ongoing phase 2 clinical trials evaluating the efficacy of new anti-HER-2 agents, such as S1613 (NCT03365882), trastuzumab-emtansine (NCT03418558), or tucatinib (NCT03043313) in this clinical scenario comprising ~5% of RAS wild-type mCRC patients (Valtorta et al., 2015). Furthermore, determining the utility of ctDNA analyses in monitoring therapeutic efficacy and in identifying mechanisms of resistance to dual HER-2 blockade is also an attractive area of study (Siravegna et al., 2018).

3.2.7 Tyrosine kinase inhibitors

New drugs that target tyrosine kinase (TK) fusions in genes such as *NRTK1/2/3*, *RET*, *ALK*, and *ROS1* are showing promising preliminary results in phase 1 and 2 clinical trials that include patients with CRC. One agent, LOXO 101 (larotrectinib), is a selective tropomyosin receptor kinase (TRK) inhibitor that demonstrated tumor-agnostic efficacy in patients with *NTRK* fusion-positive malignancies (including four patients with CRC who achieved a partial response) (Drilon et al., 2018). A second agent, entrectinib, an ALK, ROS1, TRKA, TRKB, and TRKC selective inhibitor, demonstrated clinical activity in patients who had fusions in the previously described TK genes (Drilon et al., 2017). Patients who responded to entrectinib included two patients whose mCRC harbored *CAD-ALK* or *LMNA-NTRK1* gene fusions (Amatu et al., 2015; Sartore-Bianchi et al., 2015). Anticipating potential resistance mechanisms to larotrectinib based on evidence from other pan-TK inhibitors, Drilon et al. (2017) developed LOXO-195 (selitrectinib), a potent and selective TRK kinase inhibitor designed to have a molecular structure that would overcome typical TRK resistance mutations. LOXO-195 was initially evaluated in a mCRC patient whose cancer had an *LMNA-NTRK1* rearrangement with a G595R larotrectinib-resistance mutation. This patient successfully achieved a durable partial response (Drilon, Nagasubramanian, et al., 2017). Although the prevalence of rearrangements in TK genes in mCRC patients may be as low as 1.5%, the accelerated development of TK inhibitors offers new hope for some heavily pretreated mCRC patients who have no other therapeutic options

(Pietrantonio et al., 2017). Given these promising results, the US FDA granted accelerated approval to larotrectinib (November 2018) and entrectinib (August 2019) for patients with *NTRK* gene fusion-positive solid tumors without a known acquired resistance mutation. The Committee for Medicinal Products for Human Use of the European Medicines Agency has also recommended the granting of a conditional marketing authorization for larotrectinib (July 2019) and entrectinib (May 2020) for the same indication.

3.2.8 KRAS inhibitors

KRAS is one of the most commonly altered oncogenes in human cancers, and was long considered an undruggable target because of the small size of abnormal KRAS proteins, the presence of few binding sites, and the rapid, tight binding of active KRAS to GTP. However, recent data have suggested that KRAS may be targetable. For example, preliminary data on the activity of AMG510 (sotorasib), a small covalent inhibitor, have shown that it rapidly and irreversibly occupies *KRAS G12C* and extinguishes its activity through a unique interaction with the P2 pocket (Janes et al., 2018). The *KRAS G12C* mutation occurs in ~4% of CRC (Neumann, Zeindl-Eberhart, Kirchner, & Jung, 2009). In a recent phase 1 trial, sotorasib showed encouraging antitumor activity in heavily pretreated patients who had advanced, *KRAS G12C*-mutated solid tumors (Hong et al., 2020). A total of 129 patients were included in this study, 42 of whom had CRC. Within CRC patients, sotorasib treatment yielded an ORR and disease control rate of 7.1% and 73.8%, respectively. The median duration of stable disease was 5.4 months and the median PFS was 4.0 months. Although sotorasib showed promising anticancer activity in patients with heavily pretreated solid tumors bearing the *KRAS G12C* mutation, inconsistency was seen in tumor response between patients with non-small cell lung cancer and those with CRC, which the authors suggested indicated either that *KRAS G12C* is not the dominant oncogenic driver for CRC or that other pathways, such as the WNT or EGFR pathways, mediate oncogenic signaling beyond *KRAS*. These hypotheses are supported by solid preclinical evidence (Amodio et al., 2020; Lee et al., 2018; Xue et al., 2020), and therefore, clinical trials that combine sotorasib with other agents that block additional pathways have already been initiated (i.e., NCT04185883 and NCT04303780). Although many *KRAS G12C* inhibitors in addition to sotorasib are under development, to date only adagrasib, an irreversible covalent inhibitor, has shown promising antitumor activity in *KRAS G12C*-mutated CRC. Furthermore, inhibitors for mutations other than *KRAS G12C* are being developed. For example, initial preclinical data for MRTX1133, a new, first-in-class *KRAS G12D* inhibitor, have demonstrated significant tumor regression in preclinical animal models (Mirati Therapeutics, 2020). Thus, through development of a range of inhibitors, effective means of targeting KRAS are emerging.

3.2.9 Disease monitoring by liquid biopsies

Liquid biopsies have recently emerged as powerful tools for monitoring disease evolution and therapeutic response through the analysis of cfDNA and RNA biomarkers in bodily fluids. One of the basis for this concept of liquid biopsies came from a study that reported that a gradual decrease in CEA levels during chemotherapy was significantly associated with better survival rates (Al-Sarraf, Baker, Talley, Kithier, & Vaitkevicius, 1979). Such a correlation between CEA flare and improved PFS and OS was confirmed a few years later in a subset of 670 patients with mCRC being treated with first-line chemotherapy (Strimpakos et al., 2010). Although CEA is not a CRC-specific biomarker, CEA monitoring in blood, alone or in addition to CA19–9 monitoring, is still commonly performed in routine clinical practice (Stiksma, Grootendorst, & van der Linden, 2014). In addition to this strategy, analyses of circulating tumor cells and endothelial cells in mCRC have been undertaken by several groups (S. J. Cohen et al., 2008; Ronzoni et al., 2010). In 2015, Hansen, Carlsen, Heegaard, Sørensen, and Jakobsen (2015) reported that circulating levels of miRNA-126 in a subset of 68 patients with mCRC were predictive of tumor response to bevacizumab-based chemotherapy. On the other hand, the role of ctDNA in genotyping CRCs and tracking clonal evolution of CRC during and after treatment with anti-EGFR-based regimens has been evaluated (Siravegna et al., 2015). Multiple studies have since reported distinct genetic alterations in ctDNA from patients with primary or acquired resistance to anti-EGFR-based therapies. These alterations were seen in genes such as *KRAS*, *NRAS*, *MET*, *ERBB2*, *FLT3*, *EGFR*, and *MAP2K1* and detected by approaches including droplet digital PCR, BEAMing, and next-generation sequencing methodologies (Siravegna et al., 2015; Van Emburgh et al., 2016). Further demonstrating the potential utility of changes in ctDNA, a massively parallel sequencing-based assay of ctDNA from a prospective cohort of 53 patients with mCRC showed that early changes in ctDNA during first-line standard chemotherapy can predict subsequent radiologic response (J Tie et al., 2015). Similarly, a 2017 study demonstrated a significant correlation between a decrease in RAS mutant clones in blood after 8 weeks of therapy and improved PFS and OS in a cohort of patients treated with regorafenib (Khan et al., 2018).

Intriguingly, although clonal evolution is dynamic, the emergence of drug-resistant clones in circulation increases during treatment and decreases after drug withdrawal. Understanding of this fact has led to novel treatment strategies that are being evaluated in studies such as the RASINTRO (*RAS* Mutations in ctDNA and Anti-EGFR Reintroduction in mCRC) study (NCT03259009) and the CHRONOS (Rechallenge With Panitumumab Driven by *RAS* Dynamic of Resistance) trial (NCT03227926). Both of these studies are evaluating the predictive impact of ctDNA *RAS* mutations on the efficacy of anti-EGFR monotherapy rechallenge in patients with mCRC with wild-type *RAS* whose disease has progressed after anti-EGFR-free chemotherapy. Another recently described approach involved a five-gene methylation panel for monitoring tumor burden in liquid biopsies using a methyl-BEAMing assay (Barault et al., 2018). Using this method to analyze plasma samples from 182 patients with mCRC treated with chemotherapy and/or targeted therapy, revealed a significant correlation between the dynamics of circulating cfDNA methylation markers and ORR and PFS. Using a similar approach, as part of studies presented in previous sections of this chapter, Luo et al. (2020) evaluated the utility of ddPCR analysis of cg10673833 methylation for monitoring disease evolution. Dynamic changes in cg10673833 methylation were consistent with treatment response, and these changes were more pronounced than changes in CEA. Patients who had positive responses to treatment had parallel decreases in cg10673833 methylation as compared with untreated patients, and a further reduction in cg10673833 methylation was observed in patients after surgery. In contrast, patients who had progressive or recurrent disease showed increased methylation of cg10673833.

4. Toxicity-associated biomarkers

4.1 *DPYD* gene

Severe adverse events and side effects as a result of treatment-related toxicity are a major concern for any cancer therapy. One particular concern in the context of CRC is DPD deficiency, an inherited defect in the enzyme that catabolizes fluoropyrimidines and thus increases the risk of toxicity secondary to fluoropyrimidine-based therapies (Amstutz et al., 2018; Diasio, Beavers, & Carpenter, 1988; Mattison, Soong, & Diasio, 2002; Tuchman et al., 1985). DPD deficiency causes a deficit in the metabolism of thymine and uracil, resulting in excessive amounts of uracil and thymine in the blood, urine, and cerebrospinal fluid (Innocenti et al., 2020). It can be inherited as an autosomal recessive trait, such as homozygosity for deleterious _DPYD_ variants. Patients who are homozygous for _DPYD_ variants can develop severe DPD deficiency, which can result in neurologic and other severe defects during infancy. However, patients with homozygous variants may also be asymptomatic. The prevalence of DPD deficiency in the population ranges from 2% to 8% depending on ethnicity (Innocenti et al., 2020). The Clinical Pharmacogenetics Implementation Consortium selected 4 _DPYD_ variants that confer DPD deficiency and for which recommendations exist to aid fluoropyrimidine treatment and dosing decisions: _c.1905*1G>A (*2A), c.2846A>T, c.1679T>G,_ and _c.1129-5923C>G_ (haplotype B3). These patients could be offered alternative regimens or receive dose reductions (Amstutz et al., 2018). Two prospective studies have shown _DPYD_ genotyping followed by tailoring of the fluoropyrimidine dose to be feasible in clinical practice and to improve patient safety and cost effectiveness (Deenen et al., 2016; Henricks et al., 2018, 2019). We note that universal _DPYD_ genotyping before starting fluoropyrimidine treatment remains controversial. For example the _NCCN Panel_ does not currently support this practice, while the _ESMO Clinical Guidelines_, in accordance to the Pharmacovigilance Risk Assessment Committee of the European Medicines Agency statement (March 2020) (EMA, n.d.), strongly recommend _DPYD_ genotyping or DPD phenotyping before initiating fluoropyrimidine-based therapy in the adjuvant setting (Argilés et al., 2020).

4.2 UGT1A1

The UDP-glucuronosyltransferase (UGT) enzyme encoded by _UGT1A1_ is important in the metabolism of irinotecan, a pro-drug that is widely used in the treatment of mCRC. Irinotecan is converted by carboxylesterase enzymes into the active metabolite SN-38. SN-38 produces severe treatment-related toxicities, including early and late forms of diarrhea, dehydration, and severe neutropenia (Innocenti et al., 2004; Package Insert, 2019). SN-38 is inactivated and detoxified by the UGT1A1 enzyme. UGT enzymes are responsible for glucuronidation, which transforms lipophilic metabolites into hydrophilic substances that can be easily excreted from the body. Deficiencies in UGT1A1 can be caused by certain genetic polymorphisms and can result in conditions associated with accumulation of unconjugated hyperbilirubinemias, such as types I and II of the Crigler-Najjar and Gilbert syndromes. Moreover, certain *UGT1A1* polymorphisms can result in a decreased glucuronidation of SN-38, resulting in increased drug levels and thus increased risk for toxicity, although severe irinotecan-related toxicity is not experienced by all patients who have these polymorphisms (Innocenti et al., 2004; Liu, Cheng, Kuang, Liu, & Xu, 2014; O'Dwyer & Catalano, 2006). *UGT1A* is located on chromosome 2q37, which encodes for multiple genes and pseudogenes. Moreover it exists in many alternatively spliced isoforms (van Es et al., 1993). Variants of *UGT1A1* that decrease UGT1A1 enzyme activity can lead to jaundice. Up to 135 genetic variants of *UGT1A1* have been reported; *UGT1A1*1* is the wild-type allele associated with normal enzyme activity (Strassburg, 2008). The most common variant allele is *UGT1A1*28*, which is also associated with drug toxicity. Current literature suggests that one copy of the *UGTA1A*28* allele results in an approximately 35% decrease in transcriptional activity, and two copies (**28/*28*, homozygous) result in approximately a 70% decrease (Barbarino, Haidar, Klein, & Altman, 2014; Lam, 2019). Approximately 10% of the North American population is homozygous for the **28* allele (**28/*28* genotype, also known as *UGT1A1 7/7* genotype) and are at an increased risk of neutropenia following intravenous irinotecan therapy (Hall, Ybazeta, Destro-Bisol, Petzl-Erler, & Di Rienzo, 1999). The rate of severe neutropenia in **28/*28* homozygous patients is as high as 36%, and is strongly associated with a higher hospitalization rate (Etienne-Grimaldi et al., 2015; Shulman et al., 2011). There is less evidence to support a link between *UGT1A1* genotype and irinotecan treatment-related diarrhea, and data is conflicting regarding the influence of *UGT1A* genotype on response to irinotecan therapy (Dias et al., 2012; Evaluation of Genomic Applications in Practice and Prevention EGAPP Working Group, 2009).

Individuals homozygous for another variant allele, *UGT1A1*6*, which is more prevalent in Asian populations, have reduced UGT1A1 enzyme activity and homozygosity appears to be an important predictor of severe toxicity to irinotecan therapy in this population (Dean, 2015).

Moreover, many other genetic markers, including genetic variations in *ABCC1* and *ABCB2*, SLC transporter genes, *TGFB*, and *NR1I2* may influence the risk of irinotecan toxicity (Dean, 2015). On the other hand, emerging data suggests that other *UGT1* variant alleles may have a protective effect. This could be the case for *UGT1A1 c.-1068A>G*, which seems to have a lower risk of irinotecan-induced neutropenia, probably due to an enhanced capacity for glucuronidation (Chen et al., 2015).

5. Future perspectives and conclusion

Over the past several years, many different approaches have been developed and tested for diagnosing and treating CRC. The rising incidence of early-onset CRC, coupled with the typical screening and diagnosis of the disease in individuals of older ages, represents a tremendous challenge for healthcare systems worldwide (Akimoto et al., 2020). Given current evidence and recent advances, the future of CRC early-diagnosis would appear to be in multimodal liquid biopsy approaches that include robust DNA methylation analyses. However, much additional study is needed to identify the ideal diagnostic test as well as a better understanding of the molecular characteristics of the various subsets of CRC and premalignant lesions.

Beyond diagnosis, a detailed understanding of CRC from a multi-omic perspective seems necessary to improve clinical outcomes in the adjuvant and palliative settings. Although risk stratification of CRC based on multi-omic molecular analyses has not yet been implements in the clinic, such a strategy together with in-depth knowledge of the interaction among (*epi*) genetic somatic and germline alterations and the immune system could have an important role in patient selection for adjuvant therapy. Moreover, although current studies have involved only a small number of patients, the growing evidence of the utility of various target genomic alterations such as specific gene mutations and fusions, together with specific genomic phenotypes, are laying the groundwork for developing more precise therapeutic approaches. This is exemplified by, for example, the emerging importance of *NRTK* fusions and *KRAS* mutations as targets for TK and RAS inhibitors, respectively. Moreover, many alterations that were previously identified in other tumor types such as *HER2* amplification in breast cancer, are also showing promise for CRC.

Finally, an understanding of CRC from a phylogenetic perspective and a better characterization of tumor heterogeneity and evolution should be incorporated into clinical guidelines for future CRC screening and therapeutic strategies. Cancer is a complex disease, and using rational tissue sampling, detailed clinical phenotype annotation, and integrative genomics coupled with comprehensive mathematical models of cancer evolution and biomarker shedding should become the basis for data generation and biomarker development (Avanzini et al., 2020; Wheeler et al., 2021). Implementing all of these approaches will require making use of novel high-throughput molecular analytic techniques for comprehensive molecular profiling as well as rethinking and adapting the design of the next generation of clinical trials. Integrating multi-omic approaches with clinical and epidemiologic data using machine learning algorithms will hasten biomarker development, and, most importantly, the decision-making process (Ruiz-Bañobre et al., 2019). To make this feasible in daily clinical practice, it will be necessary to implement user-friendly decision support tools that foster interactive treatment planning based on streamlined, expert-driven genomic-data interpretation and reporting (Tamborero et al., 2020).

Gradually, CRC management is moving from a "one-size-fits-all" approach to a more personalized medicine strategy in which increasing numbers of subsets of patients can obtain long-term survival benefits targeting low-prevalence driver molecular alterations. With this concept in mind, biomarker research is moving forward, bringing together many types of "omics" data and establishing novel liquid biopsy-based strategies to capture temporospatial tumor heterogeneity. The future of CRC diagnosis and treatment is promising as many scientific disciplines come together, generating new knowledge and strategies that will improve patient care and outcomes.

Acknowledgments

We thank Raquel Varela-Fernández and Jared Millar for their assistance in medical illustration.

Funding information

This work was supported by CA72851, CA181572, CA184792 and CA227602 grants from the National Cancer Institute, National Institutes of Health to A.G. In addition, J.R.-B. is supported by Río Hortega fellowship from the Institute of Health Carlos III (CM19/00087) and 2020 TTD Research Grant from the Spanish Cooperative Group for the Treatment of Digestive Tumors (TTD).

References

Adenis, A., de la Fouchardiere, C., Paule, B., Burtin, P., Tougeron, D., Wallet, J., et al. (2016). Survival, safety, and prognostic factors for outcome with Regorafenib in patients with metastatic colorectal cancer refractory to standard therapies: Results from a multicenter study (REBECCA) nested within a compassionate use program. *BMC Cancer, 16*(1), 412. https://doi.org/10.1186/s12885-016-2440-9.

Aigrain, L., Gu, Y., & Quail, M. A. (2016). Quantitation of next generation sequencing library preparation protocol efficiencies using droplet digital PCR assays—A systematic comparison of DNA library preparation kits for Illumina sequencing. *BMC Genomics, 17*(1), 458. https://doi.org/10.1186/s12864-016-2757-4.

Akimoto, N., Ugai, T., Zhong, R., Hamada, T., Fujiyoshi, K., Giannakis, M., et al. (2020). Rising incidence of early-onset colorectal cancer—A call to action. *Nature Reviews Clinical Oncology*. Advance online publication. https://doi.org/10.1038/s41571-020-00445-1.

Alaiyan, B., Ilyayev, N., Stojadinovic, A., Izadjoo, M., Roistacher, M., Pavlov, V., et al. (2013). Differential expression of colon cancer associated transcript1 (CCAT1) along the colonic adenoma-carcinoma sequence. *BMC Cancer*, *13*, 196. https://doi.org/10.1186/1471-2407-13-196.

Al-Sarraf, M., Baker, L., Talley, R. W., Kithier, K., & Vaitkevicius, V. K. (1979). The value of serial carcinoembryonic antigen (CEA) in predicting response rate and survival of patients with gastrointestinal cancer treated with chemotherapy. *Cancer*, *44*(4), 1222–1225.

Amado, R. G., Wolf, M., Peeters, M., Van Cutsem, E., Siena, S., Freeman, D. J., et al. (2008). Wild-type KRAS is required for panitumumab efficacy in patients with metastatic colorectal cancer. *Journal of Clinical Oncology: Official Journal of the American Society of Clinical Oncology*, *26*(10), 1626–1634. https://doi.org/10.1200/JCO.2007.14.7116.

Amatu, A., Somaschini, A., Cerea, G., Bosotti, R., Valtorta, E., Buonandi, P., et al. (2015). Novel CAD-ALK gene rearrangement is drugable by entrectinib in colorectal cancer. *British Journal of Cancer*, *113*, 1730.

Amodio, V., Yaeger, R., Arcella, P., Cancelliere, C., Lamba, S., Lorenzato, A., et al. (2020). EGFR blockade reverts resistance to KRASG12C inhibition in colorectal cancer. *Cancer Discovery*, *10*(8), 1129–1139. https://doi.org/10.1158/2159-8290.CD-20-0187.

Amstutz, U., Henricks, L. M., Offer, S. M., Barbarino, J., Schellens, J. H. M., Swen, J. J., et al. (2018). Clinical pharmacogenetics implementation consortium (CPIC) guideline for dihydropyrimidine dehydrogenase genotype and fluoropyrimidine dosing: 2017 update. *Clinical Pharmacology and Therapeutics*, *103*(2), 210–216. https://doi.org/10.1002/cpt.911.

André, T., Boni, C., Mounedji-Boudiaf, L., Navarro, M., Tabernero, J., Hickish, T., et al. (2004). Oxaliplatin, fluorouracil, and leucovorin as adjuvant treatment for colon cancer. *New England Journal of Medicine*, *350*(23), 2343–2351. https://doi.org/10.1056/NEJMoa032709.

Argilés, G., Tabernero, J., Labianca, R., Hochhauser, D., Salazar, R., Iveson, T., et al. (2020). Localised colon cancer: ESMO clinical practice guidelines for diagnosis, treatment and follow-up[&#x2020]. *Annals of Oncology*, *31*(10), 1291–1305. https://doi.org/10.1016/j.annonc.2020.06.022.

Aschele, C., Debernardis, D., Bandelloni, R., Cascinu, S., Catalano, V., Giordani, P., et al. (2002). Thymidylate synthase protein expression in colorectal cancer metastases predicts for clinical outcome to leucovorin-modulated bolus or infusional 5-fluorouracil but not methotrexate-modulated bolus 5-fluorouracil. *Annals of Oncology: Official Journal of the European Society for Medical Oncology*, *13*(12), 1882–1892.

Aschele, C., Debernardis, D., Tunesi, G., Maley, F., & Sobrero, A. (2000). Thymidylate synthase protein expression in primary colorectal cancer compared with the corresponding distant metastases and relationship with the clinical response to 5-fluorouracil. *Clinical Cancer Research: An Official Journal of the American Association for Cancer Research*, *6*(12), 4797–4802.

Ashktorab, H., Belgrave, K., Hosseinkhah, F., Brim, H., Nouraie, M., Takkikto, M., et al. (2009). Global histone H4 acetylation and HDAC2 expression in colon adenoma and carcinoma. *Digestive Diseases and Sciences*, *54*(10), 2109–2117. https://doi.org/10.1007/s10620-008-0601-7.

Aust, D. E., Sommer, U., & Baretton, G. B. (2012). What can molecular pathology offer for optimal decision making? *Annals of Oncology: Official Journal of the European Society for Medical Oncology*, *23*(Suppl 10), x63–x70. England, September https://doi.org/10.1093/annonc/mds346.

Avanzini, S., Kurtz, D. M., Chabon, J. J., Moding, E. J., Hori, S. S., Gambhir, S. S., et al. (2020). A mathematical model of ctDNA shedding predicts tumor detection size. *Science Advances*, *6*(50), eabc4308. https://doi.org/10.1126/sciadv.abc4308.

Baek, Y. H., Chang, E., Kim, Y. J., Kim, B. K., Sohn, J. H., & Park, D. I. (2009). Stool methylation-specific polymerase chain reaction assay for the detection of colorectal neoplasia in Korean patients. *Diseases of the Colon and Rectum*, *52*(8), 1452–1463. https://doi.org/10.1007/DCR.0b013e3181a79533.

Baker, S. J., Preisinger, A. C., Jessup, J. M., Paraskeva, C., Markowitz, S., Willson, J. K., et al. (1990). p53 gene mutations occur in combination with 17p allelic deletions as late events in colorectal tumorigenesis. *Cancer Research*, *50*(23), 7717–7722.

Ballman, K. V. (2015). Biomarker: Predictive or Prognostic? *Journal of Clinical Oncology: Official Journal of the American Society of Clinical Oncology*, *33*(33), 3968–3971. https://doi.org/10.1200/JCO.2015.63.3651.

Barault, L., Amatu, A., Siravegna, G., Ponzetti, A., Moran, S., Cassingena, A., et al. (2018). Discovery of methylated circulating DNA biomarkers for comprehensive non-invasive monitoring of treatment response in metastatic colorectal cancer. *Gut*, *67*(11), 1995–2005. https://doi.org/10.1136/gutjnl-2016-313372.

Barbarino, J. M., Haidar, C. E., Klein, T. E., & Altman, R. B. (2014). PharmGKB summary: Very important pharmacogene information for UGT1A1. *Pharmacogenetics and Genomics*, *24*(3), 177–183. https://doi.org/10.1097/FPC.0000000000000024.

Barták, B. K., Kalmár, A., Péterfia, B., Patai, Á. V., Galamb, O., Valcz, G., et al. (2017). Colorectal adenoma and cancer detection based on altered methylation pattern of SFRP1, SFRP2, SDC2, and PRIMA1 in plasma samples. *Epigenetics*, *12*(9), 751–763. https://doi.org/10.1080/15592294.2017.1356957.

Benatti, P., Gafa, R., Barana, D., Marino, M., Scarselli, A., Pedroni, M., et al. (2005). Microsatellite instability and colorectal cancer prognosis. *Clinical Cancer Research: An Official Journal of the American Association for Cancer Research*, *11*(23), 8332–8340. https://doi.org/10.1158/1078-0432.CCR-05-1030.

Bergheim, J., Semaan, A., Gevensleben, H., Groening, S., Knoblich, A., Dietrich, J., et al. (2018). Potential of quantitative SEPT9 and SHOX2 methylation in plasmatic circulating cell-free DNA as auxiliary staging parameter in colorectal cancer: A prospective observational cohort study. *British Journal of Cancer*, *118*(9), 1217–1228. https://doi.org/10.1038/s41416-018-0035-8.

Bersanelli, M. (2020). Tumour mutational burden as a driver for treatment choice in resistant tumours (and beyond). *The Lancet Oncology*, *21*(10), 1255–1257. https://doi.org/10.1016/S1470-2045(20)30433-2.

Bertotti, A., Migliardi, G., Galimi, F., Sassi, F., Torti, D., Isella, C., et al. (2011). A molecularly annotated platform of patient-derived xenografts ("xenopatients") identifies HER2 as an effective therapeutic target in cetuximab-resistant colorectal cancer. *Cancer Discovery*, *1*(6), 508–523. https://doi.org/10.1158/2159-8290.CD-11-0109.

Bokemeyer, C., Köhne, C.-H., Ciardiello, F., Lenz, H.-J., Heinemann, V., Klinkhardt, U., et al. (2017). FOLFOX4 plus cetuximab treatment and *RAS* mutations in colorectal cancer. *European Journal of Cancer*, *51*(10), 1243–1252. https://doi.org/10.1016/j.ejca.2015.04.007.

Bovell, L., Shanmugam, C., Katkoori, V. R., Zhang, B., Vogtmann, E., Grizzle, W. E., et al. (2012). miRNAs are stable in colorectal cancer archival tissue blocks. *Frontiers in Bioscience (Elite Edition)*, *4*, 1937–1940. https://doi.org/10.2741/514.

Braun, M. S., Richman, S. D., Quirke, P., Daly, C., Adlard, J. W., Elliott, F., et al. (2008). Predictive biomarkers of chemotherapy efficacy in colorectal cancer: Results from the UK MRC FOCUS trial. *Journal of Clinical Oncology*, *26*(16), 2690–2698. https://doi.org/10.1200/JCO.2007.15.5580.

Bruhn, M. A., Townsend, A. R., Khoon Lee, C., Shivasami, A., Price, T. J., Wrin, J., et al. (2014). Proangiogenic tumor proteins as potential predictive or prognostic biomarkers for bevacizumab therapy in metastatic colorectal cancer. *International Journal of Cancer*, *135*(3), 731–741. https://doi.org/10.1002/ijc.28698.

Carethers, J. M., Smith, E. J., Behling, C. A., Nguyen, L., Tajima, A., Doctolero, R. T., et al. (2004). Use of 5-fluorouracil and survival in patients with microsatellite-unstable colorectal cancer. *Gastroenterology, 126*(2), 394–401. https://doi.org/10.1053/j.gastro.2003.12.023.

Carmona, F. J., Azuara, D., Berenguer-Llergo, A., Fernández, A. F., Biondo, S., de Oca, J., et al. (2013). DNA methylation biomarkers for noninvasive diagnosis of colorectal cancer. *Cancer Prevention Research (Philadelphia, PA.), 6*(7), 656–665. https://doi.org/10.1158/1940-6207.CAPR-12-0501.

Cascinu, S., Catalano, V., Aschele, C., Barni, S., Debernardis, D., Gallo, L., et al. (2000). Immunohistochemical determination of p53 protein does not predict clinical response in advanced colorectal cancer with low thymidylate synthase expression receiving a bolus 5-fluorouracil-leucovorin combination. *Annals of Oncology: Official Journal of the European Society for Medical Oncology, 11*(8), 1053–1056.

Cecchin, E., Innocenti, F., D'Andrea, M., Corona, G., De Mattia, E., Biason, P., et al. (2009). Predictive role of the UGT1A1, UGT1A7, and UGT1A9 genetic variants and their haplotypes on the outcome of metastatic colorectal cancer patients treated with fluorouracil, leucovorin, and irinotecan. *Journal of Clinical Oncology: Official Journal of the American Society of Clinical Oncology, 27*(15), 2457–2465. https://doi.org/10.1200/JCO.2008.19.0314.

Chalabi, M., Fanchi, L. F., Dijkstra, K. K., Van den Berg, J. G., Aalbers, A. G., Sikorska, K., et al. (2020). Neoadjuvant immunotherapy leads to pathological responses in MMR-proficient and MMR-deficient early-stage colon cancers. *Nature Medicine, 26*(4), 566–576. https://doi.org/10.1038/s41591-020-0805-8.

Chang, P.-Y., Chen, C.-C., Chang, Y.-S., Tsai, W.-S., You, J.-F., Lin, G.-P., et al. (2016). MicroRNA-223 and microRNA-92a in stool and plasma samples act as complementary biomarkers to increase colorectal cancer detection. *Oncotarget,* 7(9), 10663–10675. https://doi.org/10.18632/oncotarget.7119.

Chen, X., Gole, J., Gore, A., He, Q., Lu, M., Min, J., et al. (2020). Non-invasive early detection of cancer four years before conventional diagnosis using a blood test. *Nature Communications, 11*(1), 3475. https://doi.org/10.1038/s41467-020-17316-z.

Chen, W.-D., Han, Z. J., Skoletsky, J., Olson, J., Sah, J., Myeroff, L., et al. (2005). Detection in fecal DNA of colon cancer-specific methylation of the nonexpressed vimentin gene. *Journal of the National Cancer Institute, 97*(15), 1124–1132. https://doi.org/10.1093/jnci/dji204.

Chen, S., Laverdiere, I., Tourancheau, A., Jonker, D., Couture, F., Cecchin, E., et al. (2015). A novel UGT1 marker associated with better tolerance against irinotecan-induced severe neutropenia in metastatic colorectal cancer patients. *The Pharmacogenomics Journal, 15*(6), 513–520. https://doi.org/10.1038/tpj.2015.12.

Chen, H., Li, N., Ren, J., Feng, X., Lyu, Z., Wei, L., et al. (2019). Participation and yield of a population-based colorectal cancer screening programme in China. *Gut, 68*(8), 1450–1457. https://doi.org/10.1136/gutjnl-2018-317124.

Church, T. R., Wandell, M., Lofton-Day, C., Mongin, S. J., Burger, M., Payne, S. R., et al. (2014). Prospective evaluation of methylated SEPT9 in plasma for detection of asymptomatic colorectal cancer. *Gut, 63*(2), 317–325. https://doi.org/10.1136/gutjnl-2012-304149.

Clark, J. W., Niedzwiecki, D., & Hollis, D. P. (2003). Phase-II trial of 5-fluororuacil (5-FU), leucovorin (LV), oxaliplatin (Ox), and trastuzumab (T) for patients with metastatic colorectal cancer (CRC) refractory to initial therapy. *Onkologie, 26*, 13–46.

Clarke, C. N., & Kopetz, E. S. (2015). BRAF mutant colorectal cancer as a distinct subset of colorectal cancer: Clinical characteristics, clinical behavior, and response to targeted therapies. *Journal of Gastrointestinal Oncology, 6*(6), 660–667. https://doi.org/10.3978/j.issn.2078-6891.2015.077.

Coakley, M., Garcia-Murillas, I., & Turner, N. C. (2019). Molecular residual disease and adjuvant trial design in solid tumors. *Clinical Cancer Research: An Official Journal of the American Association for Cancer Research, 25*(20), 6026–6034. https://doi.org/10.1158/1078-0432.CCR-19-0152.

Cohen, J. D., Li, L., Wang, Y., Thoburn, C., Afsari, B., Danilova, L., et al. (2018). Detection and localization of surgically resectable cancers with a multi-analyte blood test. *Science, 359*(6378), 926–930. https://doi.org/10.1126/science.aar3247.

Cohen, S. J., Punt, C. J. A., Iannotti, N., Saidman, B. H., Sabbath, K. D., Gabrail, N. Y., et al. (2008). Relationship of circulating tumor cells to tumor response, progression-free survival, and overall survival in patients with metastatic colorectal cancer. *Journal of Clinical Oncology, 26*(19), 3213–3221. https://doi.org/10.1200/JCO.2007.15.8923.

Cohen, R., Taieb, J., Fiskum, J., Yothers, G., Goldberg, R., Yoshino, T., et al. (2020). Microsatellite instability in patients with stage III colon cancer receiving fluoropyrimidine with or without oxaliplatin: An ACCENT pooled analysis of 12 adjuvant trials. *Journal of Clinical Oncology, 20*, 01600. https://doi.org/10.1200/JCO.20.01600.

ColonES Product Sheet. (n.d.). Singlera genomics. https://singleraoncology.com/wp-content/uploads/2018/10/Singlera-ColonES-Product-Sheet-V2018OCT.pdf. Accessed on 6 December 2020.

Corcoran, R. B., Andre, T., Atreya, C. E., Schellens, J. H. M., Yoshino, T., Bendell, J. C., et al. (2018). Combined BRAF, EGFR, and MEK inhibition in patients with BRAFV600E-mutant colorectal cancer. *Cancer Discovery, 8*(4), 428–443. https://doi.org/10.1158/2159-8290.CD-17-1226.

Corcoran, R. B., Dias-Santagata, D., Bergethon, K., Iafrate, A. J., Settleman, J., & Engelman, J. A. (2010). BRAF gene amplification can promote acquired resistance to MEK inhibitors in cancer cells harboring the BRAF V600E mutation. *Science Signaling, 3*(149), ra84. https://doi.org/10.1126/scisignal.2001148.

Corcoran, R. B., Ebi, H., Turke, A. B., Coffee, E. M., Nishino, M., Cogdill, A. P., et al. (2012). EGFR-mediated reactivation of MAPK signaling contributes to insensitivity of BRAF-mutant colorectal cancers to RAF inhibition with vemurafenib. *Cancer Discovery, 2*(3), 227–235. https://doi.org/10.1158/2159-8290.CD-11-0341.

Cortez, M. A., Bueso-Ramos, C., Ferdin, J., Lopez-Berestein, G., Sood, A. K., & Calin, G. A. (2011). MicroRNAs in body fluids—The mix of hormones and biomarkers. *Nature Reviews. Clinical Oncology, 8*(8), 467–477. https://doi.org/10.1038/nrclinonc.2011.76.

Dalerba, P., Sahoo, D., Paik, S., Guo, X., Yothers, G., Song, N., et al. (2016). CDX2 as a prognostic biomarker in stage II and stage III colon cancer. *New England Journal of Medicine, 374*(3), 211–222. https://doi.org/10.1056/NEJMoa1506597.

Davies, H., Bignell, G. R., Cox, C., Stephens, P., Edkins, S., Clegg, S., et al. (2002). Mutations of the BRAF gene in human cancer. *Nature, 417*(6892), 949–954. https://doi.org/10.1038/nature00766.

De Roock, W., Claes, B., Bernasconi, D., De Schutter, J., Biesmans, B., Fountzilas, G., et al. (2010). Effects of KRAS, BRAF, NRAS, and PIK3CA mutations on the efficacy of cetuximab plus chemotherapy in chemotherapy-refractory metastatic colorectal cancer: A retrospective consortium analysis. *The Lancet Oncology, 11*(8), 753–762. https://doi.org/10.1016/S1470-2045(10)70130-3.

De Roock, W., Jonker, D. J., Di Nicolantonio, F., Sartore-Bianchi, A., Tu, D., Siena, S., et al. (2010). Association of KRAS p.G13D mutation with outcome in patients with chemotherapy-refractory metastatic colorectal cancer treated with cetuximab. *JAMA, 304*(16), 1812–1820. https://doi.org/10.1001/jama.2010.1535.

Dean, L. (2015). Irinotecan therapy and UGT1A1 genotype. In V. M. Pratt, H. L. McLeod, W. S. Rubinstein, S. A. Scott, L. C. Dean, B. L. Kattman, & A. J.

Malheiro (Eds.), *Medical Genetics Summaries* National Center for Biotechnology Information (US). Bethesda, MD.

Deenen, M. J., Meulendijks, D., Cats, A., Sechterberger, M. K., Severens, J. L., Boot, H., et al. (2016). Upfront genotyping of DPYD*2A to individualize fluoropyrimidine therapy: A safety and cost analysis. *Journal of Clinical Oncology: Official Journal of the American Society of Clinical Oncology*, *34*(3), 227–234. https://doi.org/10.1200/JCO.2015.63.1325.

Deng, L., Chang, D., Foshaug, R. R., Eisner, R., Tso, V. K., Wishart, D. S., et al. (2017). Development and validation of a high-throughput mass spectrometry based urine metabolomic test for the detection of colonic adenomatous polyps. *Metabolites*, 7(3), 32. https://doi.org/10.3390/metabo7030032.

Des Guetz, G., Uzzan, B., Nicolas, P., Schischmanoff, O., & Morere, J.-F. (2009). Microsatellite instability: A predictive marker in metastatic colorectal cancer? *Targeted Oncology*, *4*(1), 57–62. https://doi.org/10.1007/s11523-008-0103-8.

Dias, M. M., McKinnon, R. A., & Sorich, M. J. (2012). Impact of the UGT1A1*28 allele on response to irinotecan: A systematic review and meta-analysis. *Pharmacogenomics*, *13*(8), 889–899. https://doi.org/10.2217/pgs.12.68.

Diasio, R. B., Beavers, T. L., & Carpenter, J. T. (1988). Familial deficiency of dihydropyrimidine dehydrogenase. Biochemical basis for familial pyrimidinemia and severe 5-fluorouracil-induced toxicity. *The Journal of Clinical Investigation*, *81*(1), 47–51. https://doi.org/10.1172/JCI113308.

Diehl, F., Schmidt, K., Choti, M. A., Romans, K., Goodman, S., Li, M., et al. (2008). Circulating mutant DNA to assess tumor dynamics. *Nature Medicine*, *14*(9), 985–990. https://doi.org/10.1038/nm.1789.

Dienstmann, R., Mason, M. J., Sinicrope, F. A., Phipps, A. I., Tejpar, S., Nesbakken, A., et al. (2017). Prediction of overall survival in stage II and III colon cancer beyond TNM system: A retrospective, pooled biomarker study. *Annals of Oncology: Official Journal of the European Society for Medical Oncology*, *28*(5), 1023–1031. https://doi.org/10.1093/annonc/mdx052.

Dong, S.-J., Cai, X.-J., & Li, S.-J. (2016). The clinical significance of MiR-429 as a predictive biomarker in colorectal cancer patients receiving 5-fluorouracil treatment. *Medical Science Monitor: International Medical Journal of Experimental and Clinical Research*, *22*, 3352–3361.

Dopeso, H., Mateo-Lozano, S., Elez, E., Landolfi, S., Ramos Pascual, F. J., Hernandez-Losa, J., et al. (2010). Aprataxin tumor levels predict response of colorectal cancer patients to irinotecan-based treatment. *Clinical Cancer Research: An Official Journal of the American Association for Cancer Research*, *16*(8), 2375–2382. https://doi.org/10.1158/1078-0432.CCR-09-3275.

Douillard, J.-Y., Oliner, K. S., Siena, S., Tabernero, J., Burkes, R., Barugel, M., et al. (2013). Panitumumab-FOLFOX4 treatment and RAS mutations in colorectal cancer. *The New England Journal of Medicine*, *369*(11), 1023–1034. https://doi.org/10.1056/NEJMoa1305275.

Drilon, A., Laetsch, T. W., Kummar, S., DuBois, S. G., Lassen, U. N., Demetri, G. D., et al. (2018). Efficacy of larotrectinib in TRK fusion–positive cancers in adults and children. *New England Journal of Medicine*, *378*(8), 731–739. https://doi.org/10.1056/NEJMoa1714448.

Drilon, A., Nagasubramanian, R., Blake, J. F., Ku, N., Tuch, B. B., Ebata, K., et al. (2017). A next-generation TRK kinase inhibitor overcomes acquired resistance to prior TRK kinase inhibition in patients with TRK fusion-positive solid tumors. *Cancer Discovery*, 7(9), 63–972.

Drilon, A., Siena, S., Ou, S.-H. I., Patel, M., Ahn, M. J., Lee, J., et al. (2017). Safety and antitumor activity of the multitargeted pan-TRK, ROS1, and ALK inhibitor

entrectinib: Combined results from two phase I trials (ALKA-372-001 and STARTRK-1). *Cancer Discovery*, 7(4), 400–409.
Duran-Sanchon, S., Moreno, L., Augé, J. M., Serra-Burriel, M., Cuatrecasas, M., Moreira, L., et al. (2020). Identification and validation of microRNA profiles in fecal samples for detection of colorectal cancer. *Gastroenterology*, *158*(4), 947–957.e4. https://doi.org/10.1053/j.gastro.2019.10.005.
Ebert, M. P. A., Tänzer, M., Balluff, B., Burgermeister, E., Kretzschmar, A. K., Hughes, D. J., et al. (2012). TFAP2E-DKK4 and chemoresistance in colorectal cancer. *The New England Journal of Medicine*, *366*(1), 44–53. https://doi.org/10.1056/NEJMoa1009473.
Evaluation of Genomic Applications in Practice and Prevention (EGAPP) Working Group. (2009). Recommendations from the EGAPP Working Group: Can UGT1A1 genotyping reduce morbidity and mortality in patients with metastatic colorectal cancer treated with irinotecan? *Genetics in Medicine: Official Journal of the American College of Medical Genetics*, *11*(1), 15–20. https://doi.org/10.1097/GIM.0b013e31818efd9d.
EMA. (n.d.), EMA recommendations on DPD testing prior to treatment with fluorouracil, capecitabine, tegafur and flucytosine. https://www.ema.europa.eu/en/news/ema-recommendations-dpd-testing-prior-treatment-fluorouracil-capecitabine-tegafur-flucytosine. Accessed November.
Erlander, M. G., Ma, X.-J., Kesty, N. C., Bao, L., Salunga, R., & Schnabel, C. A. (2011). Performance and clinical evaluation of the 92-gene real-time PCR assay for tumor classification. *The Journal of Molecular Diagnostics*, *13*(5), 493–503. https://doi.org/10.1016/j.jmoldx.2011.04.004.
Esteller, M. (2011). Non-coding RNAs in human disease. *Nature Reviews Genetics*, *12*(12), 861–874. https://doi.org/10.1038/nrg3074.
Etienne-Grimaldi, M.-C., Boyer, J.-C., Thomas, F., Quaranta, S., Picard, N., Loriot, M.-A., et al. (2015). UGT1A1 genotype and irinotecan therapy: general review and implementation in routine practice. *Fundamental & Clinical Pharmacology*, *29*(3), 219–237. https://doi.org/10.1111/fcp.12117.
Etienne-Grimaldi, M.-C., Formento, J.-L., Francoual, M., Francois, E., Formento, P., Renee, N., et al. (2008). K-Ras mutations and treatment outcome in colorectal cancer patients receiving exclusive fluoropyrimidine therapy. *Clinical Cancer Research: An Official Journal of the American Association for Cancer Research*, *14*(15), 4830–4835. https://doi.org/10.1158/1078-0432.CCR-07-4906.
Fearon, E. R., & Vogelstein, B. (1990). A genetic model for colorectal tumorigenesis. *Cell*, *61*(5), 759–767. https://doi.org/10.1016/0092-8674(90)90186-i.
Feinberg, A. P., & Tycko, B. (2004). The history of cancer epigenetics. *Nature Reviews. Cancer*, *4*(2), 143–153. February https://doi.org/10.1038/nrc1279.
Feinberg, A. P., & Vogelstein, B. (1983). Hypomethylation distinguishes genes of some human cancers from their normal counterparts. *Nature*, *301*(5895), 89–92. https://doi.org/10.1038/301089a0.
Flejou, J.-F., André, T., Chibaudel, B., Scriva, A., Hickish, T., Tabernero, J., et al. (2013). Effect of adding oxaliplatin to adjuvant 5-fluorouracil/leucovorin (5FU/LV) in patients with defective mismatch repair (dMMR) colon cancer stage II and III included in the MOSIAC study. *Journal of Clinical Oncology*, *31*(15_suppl), 3524. https://doi.org/10.1200/jco.2013.31.15_suppl.3524.
Fraga, M. F., Herranz, M., Espada, J., Ballestar, E., Paz, M. F., Ropero, S., et al. (2004). A mouse skin multistage carcinogenesis model reflects the aberrant DNA methylation patterns of human tumors. *Cancer Research*, *64*(16), 5527–5534. https://doi.org/10.1158/0008-5472.CAN-03-4061.
Friedland, S., Pan, J. Y., Watson, D., Chen, Y., Nimgaonkar, A., Gulzar, Z., et al. (2020). A sensitive and quantitative multimodal blood test for the detection of colorectal adenomas

and cancer: Correlation with size and number of polyps. *Journal of Clinical Oncology, 38*-(15_suppl), 1555. https://doi.org/10.1200/JCO.2020.38.15_suppl.1555.

Fu, B., Yan, P., Zhang, S., Lu, Y., Pan, L., Tang, W., et al. (2018). Cell-free circulating methylated SEPT9 for noninvasive diagnosis and monitoring of colorectal cancer. *Disease Markers, 2018*, 6437104. https://doi.org/10.1155/2018/6437104.

Galon, J., Galon, J., Costes, A., Costes, A., Sanchez-Cabo, F., Sanchez-Cabo, F., et al. (2006). Type, density, and location of immune cells within human colorectal tumors predict clinical outcome. *Science, 313*(September), 1960–1964. https://doi.org/10.1126/science.1129139.

Gansauge, M.-T., & Meyer, M. (2013). Single-stranded DNA library preparation for the sequencing of ancient or damaged DNA. *Nature Protocols, 8*(4), 737–748. https://doi.org/10.1038/nprot.2013.038.

Gao, J., Shi, L. Z., Zhao, H., Chen, J., Xiong, L., He, Q., et al. (2016). Loss of IFN-γ pathway genes in tumor cells as a mechanism of resistance to anti-CTLA-4 therapy. *Cell, 167*(2), 397–404.e9. https://doi.org/10.1016/j.cell.2016.08.069.

Gavin, P. G., Colangelo, L. H., Fumagalli, D., Tanaka, N., Remillard, M. Y., Yothers, G., et al. (2012). Mutation profiling and microsatellite instability in stage II and III colon cancer: An assessment of their prognostic and oxaliplatin predictive value. *Clinical Cancer Research, 18*(23), 6531–6541. https://doi.org/10.1158/1078-0432.CCR-12-0605.

Gezer, U., Ustek, D., Yörüker, E. E., Cakiris, A., Abaci, N., Leszinski, G., et al. (2013). Characterization of H3K9me3- and H4K20me3-associated circulating nucleosomal DNA by high-throughput sequencing in colorectal cancer. *Tumour Biology: The Journal of the International Society for Oncodevelopmental Biology and Medicine, 34*(1), 329–336. https://doi.org/10.1007/s13277-012-0554-5.

Gezer, U., Yörüker, E. E., Keskin, M., Kulle, C. B., Dharuman, Y., & Holdenrieder, S. (2015). Histone methylation marks on circulating nucleosomes as novel blood-based biomarker in colorectal cancer. *International Journal of Molecular Sciences, 16*(12), 29654–29662. https://doi.org/10.3390/ijms161226180.

Gharib, E., Nazemalhosseini-Mojarad, E., Baghdar, K., Nayeri, Z., Sadeghi, H., Rezasoltani, S., et al. (2021). Identification of a stool long non-coding RNAs panel as a potential biomarker for early detection of colorectal cancer. *Journal of Clinical Laboratory Analysis, 35*(2), e23601. https://doi.org/10.1002/jcla.23601.

Giannakis, M., Mu, X. J., Shukla, S. A., Qian, Z. R., Cohen, O., Nishihara, R., et al. (2016). Genomic correlates of immune-cell infiltrates in colorectal carcinoma. *Cell Reports, 15*(4), 857–865. https://doi.org/10.1016/j.celrep.2016.03.075.

Glöckner, S. C., Dhir, M., Yi, J. M., McGarvey, K. E., Van Neste, L., Louwagie, J., et al. (2009). Methylation of *TFPI2* in stool DNA: A potential novel biomarker for the detection of colorectal cancer. *Cancer Research, 69*(11), 4691–4699. https://doi.org/10.1158/0008-5472.CAN-08-0142.

Grady, W. M., & Carethers, J. M. (2008). Genomic and epigenetic instability in colorectal cancer pathogenesis. *Gastroenterology, 135*(4), 1079–1099. https://doi.org/10.1053/j.gastro.2008.07.076.

Grasselli, J., Elez, E., Caratù, G., Matito, J., Santos, C., Macarulla, T., et al. (2017). Concordance of blood- and tumor-based detection of RAS mutations to guide anti-EGFR therapy in metastatic colorectal cancer. *Annals of Oncology, 28*(6), 1294–1301.

Grasso, C. S., Giannakis, M., Wells, D. K., Hamada, T., Mu, X. J., Quist, M., et al. (2018). Genetic mechanisms of immune evasion in colorectal cancer. *Cancer Discovery, 8*(6), 730–749. https://doi.org/10.1158/2159-8290.CD-17-1327.

Guastadisegni, C., Colafranceschi, M., Ottini, L., & Dogliotti, E. (2010). Microsatellite instability as a marker of prognosis and response to therapy: A meta-analysis of colorectal cancer survival data. *European Journal of Cancer (Oxford, England: 1990)*, *46*(15), 2788–2798. https://doi.org/10.1016/j.ejca.2010.05.009.

Guinney, J., Dienstmann, R., Wang, X., de Reynies, A., Schlicker, A., Soneson, C., et al. (2015). The consensus molecular subtypes of colorectal cancer. *Nature Medicine*, *21*(11), 1350–1356. https://doi.org/10.1038/nm.3967.

Gurjao, C., Liu, D., Hofree, M., AlDubayan, S. H., Wakiro, I., Su, M.-J., et al. (2019). Intrinsic resistance to immune checkpoint blockade in a mismatch repair-deficient colorectal cancer. *Cancer Immunology Research*, 7(8), 1230–1236. https://doi.org/10.1158/2326-6066.CIR-18-0683.

Gustavsson, B., Kaiser, C., Carlsson, G., Wettergren, Y., Odin, E., Lindskog, E. B., et al. (2009). Molecular determinants of efficacy for 5-FU-based treatments in advanced colorectal cancer: mRNA expression for 18 chemotherapy-related genes. *International Journal of Cancer*, *124*(5), 1220–1226. https://doi.org/10.1002/ijc.23852.

Hainsworth, J. D., Meric-Bernstam, F., Swanton, C., Hurwitz, H., Spigel, D. R., Sweeney, C., et al. (2018). Targeted therapy for advanced solid tumors on the basis of molecular profiles: Results from mypathway, an open-label, phase IIa multiple basket study. *Journal of Clinical Oncology*, *36*(6), 536–542. https://doi.org/10.1200/JCO.2017.75.3780.

Hainsworth, J. D., Rubin, M. S., Spigel, D. R., Boccia, R. V., Raby, S., Quinn, R., et al. (2012). Molecular gene expression profiling to predict the tissue of origin and direct site-specific therapy in patients with carcinoma of unknown primary site: A prospective trial of the sarah cannon research institute. *Journal of Clinical Oncology*, *31*(2), 217–223. https://doi.org/10.1200/JCO.2012.43.3755.

Hall, D., Ybazeta, G., Destro-Bisol, G., Petzl-Erler, M. L., & Di Rienzo, A. (1999). Variability at the uridine diphosphate glucuronosyltransferase 1A1 promoter in human populations and primates. *Pharmacogenetics*, *9*(5), 591–599.

Hansen, T. F., Carlsen, A. L., Heegaard, N. H. H., Sørensen, F. B., & Jakobsen, A. (2015). Changes in circulating microRNA-126 during treatment with chemotherapy and bevacizumab predicts treatment response in patients with metastatic colorectal cancer. *British Journal of Cancer*, *112*(4), 624–629. October 2014 https://doi.org/10.1038/bjc.2014.652.

Hansen, T. F., Nielsen, B. S., Jakobsen, A., & Sorensen, F. B. (2013). Visualising and quantifying angiogenesis in metastatic colorectal cancer: A comparison of methods and their predictive value for chemotherapy response. *Cellular Oncology (Dordrecht)*, *36*(4), 341–350. https://doi.org/10.1007/s13402-013-0139-3.

He, N., Song, L., Kang, Q., Jin, P., Cai, G., Zhou, J., et al. (2018). The pathological features of colorectal cancer determine the detection performance on blood ctDNA. *Technology in Cancer Research & Treatment*, *17*, 1533033818791794. https://doi.org/10.1177/1533033818791794.

He, X., Tan, X., Wang, X., Jin, H., Liu, L., Ma, L., et al. (2014). C-Myc-activated long noncoding RNA CCAT1 promotes colon cancer cell proliferation and invasion. *Tumour Biology: The Journal of the International Society for Oncodevelopmental Biology and Medicine*, *35*(12), 12181–12188. https://doi.org/10.1007/s13277-014-2526-4.

Heinemann, V., Kraemer, N., Buchner, H., Fischer von Weikersthal, L., Decker, T., Kiani, A., et al. (2018). Somatic DNA mutations, tumor mutational burden (TMB), and MSI status: Association with efficacy in patients (pts) with metastatic colorectal cancer (mCRC) of FIRE-3 (AIO KRK-0306). *Journal of Clinical Oncology*, *36*(15_suppl), 3591. https://doi.org/10.1200/JCO.2018.36.15_suppl.3591.

Henricks, L. M., Lunenburg, C. A. T. C., de Man, F. M., Meulendijks, D., Frederix, G. W. J., Kienhuis, E., et al. (2018). DPYD genotype-guided dose individualisation of fluoropyrimidine therapy in patients with cancer: A prospective safety analysis. *The Lancet Oncology*, *19*(11), 1459–1467. https://doi.org/10.1016/S1470-2045(18)30686-7.

Henricks, L. M., Lunenburg, C. A. T. C., de Man, F. M., Meulendijks, D., Frederix, G. W. J., Kienhuis, E., et al. (2019). A cost analysis of upfront DPYD genotype-guided dose individualisation in fluoropyrimidine-based anticancer therapy. *European Journal of Cancer (Oxford, England: 1990)*, *107*, 60–67. https://doi.org/10.1016/j.ejca.2018.11.010.

Herreros-Villanueva, M., Duran-Sanchon, S., Martín, A. C., Pérez-Palacios, R., Vila-Navarro, E., Marcuello, M., et al. (2019). Plasma microRNA signature validation for early detection of colorectal cancer. *Clinical and Translational Gastroenterology*, *10*(1), e00003. https://doi.org/10.14309/ctg.0000000000000003.

Heyn, H., & Esteller, M. (2012). DNA methylation profiling in the clinic: Applications and challenges. *Nature Reviews. Genetics*, *13*(10), 679–692. https://doi.org/10.1038/nrg3270.

Hiraki, M., Kitajima, Y., Nakafusa, Y., Nakamura, J., Hashiguchi, K., Sumi, K., et al. (2010). CpG island methylation of BNIP3 predicts resistance against S-1/CPT-11 combined therapy in colorectal cancer patients. *Oncology Reports*, *23*(1), 191–197.

Hong, D. S., Fakih, M. G., Strickler, J. H., Desai, J., Durm, G. A., Shapiro, G. I., et al. (2020). KRASG12C inhibition with sotorasib in advanced solid tumors. *New England Journal of Medicine*, *383*(13), 1207–1217. https://doi.org/10.1056/NEJMoa1917239.

Hoskins, J. M., Marcuello, E., Altes, A., Marsh, S., Maxwell, T., Van Booven, D. J., et al. (2008). Irinotecan pharmacogenetics: Influence of pharmacodynamic genes. *Clinical Cancer Research*, *14*(6), 1788–1796. https://doi.org/10.1158/1078-0432.CCR-07-1472.

Hurwitz, H. I., Yi, J., Ince, W., Novotny, W. F., & Rosen, O. (2009). The clinical benefit of bevacizumab in metastatic colorectal cancer is independent of K-ras mutation status: analysis of a phase III study of bevacizumab with chemotherapy in previously untreated metastatic colorectal cancer. *The Oncologist*, *14*(1), 22–28. https://doi.org/10.1634/theoncologist.2008-0213.

Hutchins, G., Southward, K., Handley, K., Magill, L., Beaumont, C., Stahlschmidt, J., et al. (2011). Value of mismatch repair, KRAS, and BRAF mutations in predicting recurrence and benefits from chemotherapy in colorectal cancer. *Journal of Clinical Oncology: Official Journal of the American Society of Clinical Oncology*, *29*(10), 1261–1270. https://doi.org/10.1200/JCO.2010.30.1366.

Imperiale, T. F., Ransohoff, D. F., Itzkowitz, S. H., Levin, T. R., Lavin, P., Lidgard, G. P., et al. (2014). Multitarget stool DNA testing for colorectal-cancer screening. *The New England Journal of Medicine*, *370*(14), 1287–1297. https://doi.org/10.1056/NEJMoa1311194.

Innocenti, F., Mills, S. C., Sanoff, H., Ciccolini, J., Lenz, H.-J., & Milano, G. (2020). All you need to know about DPYD genetic testing for patients treated with fluorouracil and capecitabine: A practitioner-friendly guide. *JCO Oncology Practice*, *16*(12), 793–798. https://doi.org/10.1200/OP.20.00553. OP.20.00553.

Innocenti, F., Undevia, S. D., Iyer, L., Chen, P. X., Das, S., Kocherginsky, M., et al. (2004). Genetic variants in the UDP-glucuronosyltransferase 1A1 gene predict the risk of severe neutropenia of irinotecan. *Journal of Clinical Oncology: Official Journal of the American Society of Clinical Oncology*, *22*(8), 1382–1388. https://doi.org/10.1200/JCO.2004.07.173.

Itzkowitz, S. H., Jandorf, L., Brand, R., Rabeneck, L., Schroy, P. C., 3rd, Sontag, S., et al. (2007). Improved fecal DNA test for colorectal cancer screening. *Clinical Gastroenterology and Hepatology: The Official Clinical Practice Journal of the American Gastroenterological Association*, *5*(1), 111–117. https://doi.org/10.1016/j.cgh.2006.10.006.

Janes, M. R., Zhang, J., Li, L.-S., Hansen, R., Peters, U., Guo, X., et al. (2018). Targeting KRAS mutant cancers with a covalent G12C-specific inhibitor. *Cell*, *172*(3), 578–589. e17. https://doi.org/10.1016/j.cell.2018.01.006.

Jensen, L. H., Danenberg, K. D., Danenberg, P. V., & Jakobsen, A. (2007). Predictive value of MSH2 gene expression in colorectal cancer treated with capecitabine. *Clinical Colorectal Cancer*, *6*(6), 433–435. https://doi.org/10.3816/CCC.2007.n.012.

Jin, Z., Sanhueza, C. T., Johnson, B., Nagorney, D. M., Larson, D. W., Mara, K. C., et al. (2018). Outcome of mismatch repair-deficient metastatic colorectal cancer: the mayo clinic experience. *The Oncologist, 23*(9), 1083–1091. https://doi.org/10.1634/theoncologist.2017-0289.

Jo, W.-S., & Carethers, J. M. (2006). Chemotherapeutic implications in microsatellite unstable colorectal cancer. *Cancer Biomarkers: Section A of Disease Markers, 2*(1–2), 51–60. https://doi.org/10.3233/cbm-2006-21-206.

Johnston, P. G., Benson, A. B., 3rd, Catalano, P., Rao, M. S., O'Dwyer, P. J., & Allegra, C. J. (2003). Thymidylate synthase protein expression in primary colorectal cancer: lack of correlation with outcome and response to fluorouracil in metastatic disease sites. *Journal of Clinical Oncology: Official Journal of the American Society of Clinical Oncology, 21*(5), 815–819. https://doi.org/10.1200/JCO.2003.07.039.

Jones, J. C., Renfro, L. A., Al-Shamsi, H. O., Schrock, A. B., Rankin, A., Zhang, B. Y., et al. (2017). (Non-V600) BRAF mutations define a clinically distinct molecular subtype of metastatic colorectal cancer. *Journal of Clinical Oncology: Official Journal of the American Society of Clinical Oncology, 35*(23), 2624–2630. https://doi.org/10.1200/JCO.2016.71.4394.

Jover, R., Zapater, P., Castells, A., Llor, X., Andreu, M., Cubiella, J., et al. (2009). The efficacy of adjuvant chemotherapy with 5-fluorouracil in colorectal cancer depends on the mismatch repair status. *European Journal of Cancer, 45*(3), 365–373. https://doi.org/10.1016/j.ejca.2008.07.016.

Jung, G., Hernández-Illán, E., Moreira, L., Balaguer, F., & Goel, A. (2020). Epigenetics of colorectal cancer: biomarker and therapeutic potential. *Nature Reviews Gastroenterology & Hepatology, 17*(2), 111–130. https://doi.org/10.1038/s41575-019-0230-y.

Kandimalla, R., Xu, J., Link, A., Matsuyama, T., Yamamura, K., Parker, I., et al. (2020). Abstract 1084: EpiPanGI-Dx: A cell-free DNA methylation fingerprint for the early detection of gastrointestinal cancers. *Cancer Research, 80*(16 Suppl), 1084. https://doi.org/10.1158/1538-7445.AM2020-1084.

Kandoth, C., McLellan, M. D., Vandin, F., Ye, K., Niu, B., Lu, C., et al. (2013). Mutational landscape and significance across 12 major cancer types. *Nature, 502*(7471), 333–339. https://doi.org/10.1038/nature12634.

Karczmarski, J., Rubel, T., Paziewska, A., Mikula, M., Bujko, M., Kober, P., et al. (2014). Histone H3 lysine 27 acetylation is altered in colon cancer. *Clinical Proteomics, 11*(1), 24. https://doi.org/10.1186/1559-0275-11-24.

Khan, K., Rata, M., Cunningham, D., Koh, D.-M., Tunariu, N., Hahne, J. C., et al. (2018). Functional imaging and circulating biomarkers of response to regorafenib in treatment-refractory metastatic colorectal cancer patients in a prospective phase II study. *Gut, 67*(8), 1484–1492. https://doi.org/10.1136/gutjnl-2017-314178.

Kim, C. G., Ahn, J. B., Jung, M., Beom, S. H., Kim, C., Kim, J. H., et al. (2016). Effects of microsatellite instability on recurrence patterns and outcomes in colorectal cancers. *British Journal of Cancer, 115*(1), 25–33. https://doi.org/10.1038/bjc.2016.161.

Kisiel, J. B., Yab, T. C., Nazer Hussain, F. T., Taylor, W. R., Garrity-Park, M. M., Sandborn, W. J., et al. (2013). Stool DNA testing for the detection of colorectal neoplasia in patients with inflammatory bowel disease. *Alimentary Pharmacology & Therapeutics, 37*(5), 546–554. https://doi.org/10.1111/apt.12218.

Klingbiel, D., Saridaki, Z., Roth, A. D., Bosman, F. T., Delorenzi, M., & Tejpar, S. (2015). Prognosis of stage II and III colon cancer treated with adjuvant 5-fluorouracil or FOLFIRI in relation to microsatellite status: Results of the PETACC-3 trial. *Annals of Oncology: Official Journal of the European Society for Medical Oncology, 26*(1), 126–132. https://doi.org/10.1093/annonc/mdu499.

Koopman, M., Knijn, N., Richman, S., Seymour, M., Quirke, P., van Tinteren, H., et al. (2009). The correlation between Topoisomerase-I (Topo1) expression and outcome of treatment with capecitabine and irinotecan in advanced colorectal cancer (ACC) patients (pts) treated in the CAIRO study of the Dutch Colorectal Cancer Group (DCCG). *EJC Supplements*, 7(2), 321–322. https://doi.org/10.1016/S1359-6349(09)71098-5.

Kopetz, S., Grothey, A., Yaeger, R., Van Cutsem, E., Desai, J., Yoshino, T., et al. (2019). Encorafenib, binimetinib, and cetuximab in BRAF V600E-mutated colorectal cancer. *New England Journal of Medicine*, *381*(17), 1632–1643. https://doi.org/10.1056/NEJMoa1908075.

Kopetz, S., Guthrie, K. A., Morris, V. K., Lenz, H.-J., Magliocco, A. M., Maru, D., et al. (2021). Randomized trial of irinotecan and cetuximab with or without vemurafenib in BRAF-mutant metastatic colorectal cancer (SWOG S1406). *Journal of Clinical Oncology*, *39*(4), 285–294. JCO.20.01994 https://doi.org/10.1200/JCO.20.01994.

Laird, P. W. (2003). The power and the promise of DNA methylation markers. *Nature Reviews. Cancer*, *3*(4), 253–266. https://doi.org/10.1038/nrc1045.

Lam, Y. W. F. (2019). In Y. W. F. Lam, S. A. B. T.-P, & S. E. Scott (Eds.), *Chapter 1—Principles of pharmacogenomics: Pharmacokinetic, pharmacodynamic, and clinical implications* (pp. 1–53). Academic Press. https://doi.org/10.1016/B978-0-12-812626-4.00001-2.

Lambrechts, D., Thienpont, B., Thuillier, V., Sagaert, X., Moisse, M., Peuteman, G., et al. (2015). Evaluation of efficacy and safety markers in a phase II study of metastatic colorectal cancer treated with aflibercept in the first-line setting. *British Journal of Cancer*, *113*(7), 1027–1034. https://doi.org/10.1038/bjc.2015.329.

Lanza, G., Gafà, R., Santini, A., Maestri, I., Guerzoni, L., & Cavazzini, L. (2006). Immunohistochemical test for MLH1 and MSH2 expression predicts clinical outcome in stage II and III colorectal cancer patients. *Journal of Clinical Oncology: Official Journal of the American Society of Clinical Oncology*, *24*(15), 2359–2367. https://doi.org/10.1200/JCO.2005.03.2433.

Le, D. T., Durham, J. N., Smith, K. N., Wang, H., Bartlett, B. R., Aulakh, L. K., et al. (2017). Mismatch repair deficiency predicts response of solid tumors to PD-1 blockade. *Science (New York, N.Y.)*, *357*(6349), 409–413. https://doi.org/10.1126/science.aan6733.

Le, D. T., Uram, J. N., Wang, H., Bartlett, B. R., Kemberling, H., Eyring, A. D., et al. (2015). PD-1 blockade in tumors with mismatch-repair deficiency. *New England Journal of Medicine*, *372*(26), 2509–2520. https://doi.org/10.1056/NEJMoa1500596.

Lee, S.-K., Jeong, W.-J., Cho, Y.-H., Cha, P.-H., Yoon, J.-S., Ro, E. J., et al. (2018). β-Catenin-RAS interaction serves as a molecular switch for RAS degradation via GSK3β. *EMBO Reports*, *19*(12), e46060. https://doi.org/10.15252/embr.201846060.

Li, M., Chen, W.-D., Papadopoulos, N., Goodman, S. N., Bjerregaard, N. C., Laurberg, S., et al. (2009). Sensitive digital quantification of DNA methylation in clinical samples. *Nature Biotechnology*, *27*(9), 858–863. https://doi.org/10.1038/nbt.1559.

Liao, X., Morikawa, T., Lochhead, P., Imamura, Y., Kuchiba, A., Yamauchi, M., et al. (2012). Prognostic role of PIK3CA mutation in colorectal cancer: Cohort study and literature review. *Clinical Cancer Research: An Official Journal of the American Association for Cancer Research*, *18*(8), 2257–2268. https://doi.org/10.1158/1078-0432.CCR-11-2410.

Lievre, A., Bachet, J.-B., Le Corre, D., Boige, V., Landi, B., Emile, J.-F., et al. (2006). KRAS mutation status is predictive of response to cetuximab therapy in colorectal cancer. *Cancer Research*, *66*(8), 3992–3995. https://doi.org/10.1158/0008-5472.CAN-06-0191.

Lindskog, E. B., Derwinger, K., Gustavsson, B., Falk, P., & Wettergren, Y. (2014). Thymidine phosphorylase expression is associated with time to progression in patients with metastatic colorectal cancer. *BMC Clinical Pathology*, *14*, 25. https://doi.org/10.1186/1472-6890-14-25.

Liu, X., Cheng, D., Kuang, Q., Liu, G., & Xu, W. (2014). Association of UGT1A1*28 polymorphisms with irinotecan-induced toxicities in colorectal cancer: A meta-analysis in

Caucasians. *The Pharmacogenomics Journal, 14*(2), 120–129. https://doi.org/10.1038/tpj.2013.10.

Liu, M. C., Klein, E., Hubbell, E., Maddala, T., Aravanis, A. M., Beausang, J. F., et al. (2018). Plasma cell-free DNA (cfDNA) assays for early multi-cancer detection: The circulating cell-free genome atlas (CCGA) study. *Annals of Oncology, 29*, viii14. https://doi.org/10.1093/annonc/mdy269.048.

Liu, H.-N., Liu, T.-T., Wu, H., Chen, Y.-J., Tseng, Y.-J., Yao, C., et al. (2018). Serum microRNA signatures and metabolomics have high diagnostic value in colorectal cancer using two novel methods. *Cancer Science, 109*(4), 1185–1194. https://doi.org/10.1111/cas.13514.

Liu, M. C., Oxnard, G. R., Klein, E. A., Swanton, C., Seiden, M. V., Liu, M. C., et al. (2020). Sensitive and specific multi-cancer detection and localization using methylation signatures in cell-free DNA. *Annals of Oncology, 31*(6), 745–759. https://doi.org/10.1016/j.annonc.2020.02.011.

Liu, G.-H., Zhou, Z.-G., Chen, R., Wang, M.-J., Zhou, B., Li, Y., et al. (2013). Serum miR-21 and miR-92a as biomarkers in the diagnosis and prognosis of colorectal cancer. *Tumour Biology: The Journal of the International Society for Oncodevelopmental Biology and Medicine, 34*(4), 2175–2181. https://doi.org/10.1007/s13277-013-0753-8.

Lothe, R. A., Peltomäki, P., Meling, G. I., Aaltonen, L. A., Nyström-Lahti, M., Pylkkänen, L., et al. (1993). Genomic instability in colorectal cancer: Relationship to clinicopathological variables and family history. *Cancer Research, 53*(24), 5849–5852.

Lu, J., Getz, G., Miska, E. A., Alvarez-Saavedra, E., Lamb, J., Peck, D., et al. (2005). MicroRNA expression profiles classify human cancers. *Nature, 435*(7043), 834–838. https://doi.org/10.1038/nature03702.

Lu, H., Huang, S., Zhang, X., Wang, D., Zhang, X., Yuan, X., et al. (2014). DNA methylation analysis of SFRP2, GATA4/5, NDRG4 and VIM for the detection of colorectal cancer in fecal DNA. *Oncology Letters, 8*(4), 1751–1756. https://doi.org/10.3892/ol.2014.2413.

Luo, Y., Yu, M., & Grady, W. M. (2014). Field cancerization in the colon: A role for aberrant DNA methylation? *Gastroenterology Report, 2*(1), 16–20. https://doi.org/10.1093/gastro/got039.

Luo, H., Zhao, Q., Wei, W., Zheng, L., Yi, S., Li, G., et al. (2020). Circulating tumor DNA methylation profiles enable early diagnosis, prognosis prediction, and screening for colorectal cancer. *Science Translational Medicine, 12*(524), eaax7533. https://doi.org/10.1126/scitranslmed.aax7533.

Ma, X.-J., Patel, R., Wang, X., Salunga, R., Murage, J., Desai, R., et al. (2006). Molecular classification of human cancers using a 92-gene real-time quantitative polymerase chain reaction assay. *Archives of Pathology & Laboratory Medicine, 130*(4), 465–473. https://doi.org/10.1043/1543-2165(2006)130[465,MCOHCU]2.0.CO;2.

Malathi, N., Mythili, S., & Vasanthi, H. R. (2014). Salivary diagnostics: A brief review. *ISRN Dentistry, 2014*, 158786. https://doi.org/10.1155/2014/158786.

Manceau, G., Imbeaud, S., Thiébaut, R., Liébaert, F., Fontaine, K., Rousseau, F., et al. (2014). Hsa-miR-31-3p expression is linked to progression-free survival in patients with KRAS wild-type metastatic colorectal cancer treated with anti-EGFR therapy. *Clinical Cancer Research: An Official Journal of the American Association for Cancer Research, 20*(12), 3338–3347. https://doi.org/10.1158/1078-0432.CCR-13-2750.

Mann, B., Gelos, M., Siedow, A., Hanski, M. L., Gratchev, A., Ilyas, M., et al. (1999). Target genes of beta-catenin-T cell-factor/lymphoid-enhancer-factor signaling in human colorectal carcinomas. *Proceedings of the National Academy of Sciences of the United States of America, 96*(4), 1603–1608. https://doi.org/10.1073/pnas.96.4.1603.

Margonis, G. A., Buettner, S., Wagner, D., McVey, J., Andreatos, N., Beer, A., et al. (2018). Microsatellite instability in resectable colorectal liver metastasis: An international multi-institutional analysis. *Journal of Clinical Oncology*, *36*(4_suppl), 220. https://doi.org/10.1200/JCO.2018.36.4_suppl.220.

Mattison, L. K., Soong, R., & Diasio, R. B. (2002). Implications of dihydropyrimidine dehydrogenase on 5-fluorouracil pharmacogenetics and pharmacogenomics. *Pharmacogenomics*, *3*(4), 485–492. https://doi.org/10.1517/14622416.3.4.485.

McCarthy, A. J., Serra, S., & Chetty, R. (2019). Traditional serrated adenoma: An overview of pathology and emphasis on molecular pathogenesis. *BMJ Open Gastroenterology*, *6*(1), e000317. https://doi.org/10.1136/bmjgast-2019-000317.

Meiri, E., Mueller, W. C., Rosenwald, S., Zepeniuk, M., Klinke, E., Edmonston, T. B., et al. (2012). A second-generation microRNA-based assay for diagnosing tumor tissue origin. *The Oncologist*, *17*(6), 801–812. https://doi.org/10.1634/theoncologist.2011-0466.

Merok, M. A., Ahlquist, T., Røyrvik, E. C., Tufteland, K. F., Hektoen, M., Sjo, O. H., et al. (2013). Microsatellite instability has a positive prognostic impact on stage II colorectal cancer after complete resection: results from a large, consecutive Norwegian series. *Annals of Oncology: Official Journal of the European Society for Medical Oncology*, *24*(5), 1274–1282. https://doi.org/10.1093/annonc/mds614.

Middha, S., Yaeger, R., Shia, J., Stadler, Z. K., King, S., Guercio, S., et al. (2019). Majority of B2M-mutant and -deficient colorectal carcinomas achieve clinical benefit from immune checkpoint inhibitor therapy and are microsatellite instability-high. *JCO Precision Oncology*, *3*, 1–14. https://doi.org/10.1200/PO.18.00321.

Miller-Phillips, L., Fischer von Weikersthal, L., Kaiser, F., Al-Batran, S.-E., Heintges, T., Neureiter, D., et al. (2018). Association of microRNA-21 (miR-21) with efficacy of cetuximab (cet) and bevacizumab (bev) in patients with metastatic colorectal cancer (mCRC) within the FIRE-3 study (AIO KRK-0306). *Annals of Oncology*, *29*, viii39. https://doi.org/10.1093/annonc/mdy269.122.

Miller-Phillips, L., Heinemann, V., Stahler, A., Fischer von Weikersthal, L., Kaiser, F., Al-Batran, S.-E., et al. (2019). Association of microRNA-21 with efficacy of cetuximab in RAS wild-type patients in the FIRE-3 study (AIO KRK-0306) and microRNA-21's influence on gene expression in the EGFR signaling pathway. *Journal of Clinical Oncology*, *37*(15_suppl), 3593. https://doi.org/10.1200/JCO.2019.37.15_suppl.3593.

Mirati Therapeutics, 2020, Accessed on 6 December. https://ir.mirati.com/news-releases/news-details/2020/Mirati-Therapeutics-Reports-Investigational-Adagrasib-MRTX849-Preliminary-Data-Demonstrating-Tolerability-and-Durable-Anti-Tumor-Activity-as-well-as-Initial-MRTX1133-Preclinical-Data/default.aspx.

Mlecnik, B., Bifulco, C., Bindea, G., Marliot, F., Lugli, A., Lee, J. J., et al. (2020). Multicenter international society for immunotherapy of cancer study of the consensus immunoscore for the prediction of survival and response to chemotherapy in stage III colon cancer. *Journal of Clinical Oncology*, *38*(31), 3638–3651. https://doi.org/10.1200/JCO.19.03205.

Moran, S., Martinez-Cardus, A., Sayols, S., Musulen, E., Balana, C., Estival-Gonzalez, A., et al. (2016). Epigenetic profiling to classify cancer of unknown primary: A multicentre, retrospective analysis. *The Lancet. Oncology*, *17*(10), 1386–1395. https://doi.org/10.1016/S1470-2045(16)30297-2.

Murcia, O., Jover, R., Egoavil, C., Perez-Carbonell, L., Juárez, M., Hernández-Illán, E., et al. (2018). TFAP2E methylation and expression status does not predict response to 5-FU-based chemotherapy in colorectal cancer. *Clinical Cancer Research: An Official Journal of the American Association for Cancer Research*, *24*(12), 2820–2827. https://doi.org/10.1158/1078-0432.CCR-17-2940.

Nakazawa, T., Kondo, T., Ma, D., Niu, D., Mochizuki, K., Kawasaki, T., et al. (2012). Global histone modification of histone H3 in colorectal cancer and its precursor lesions. *Human Pathology*, *43*(6), 834–842. https://doi.org/10.1016/j.humpath.2011.07.009.

National Comprehensive Cancer Network. (2020). Colon cancer (Version 4.2020). https://www.nccn.org/professionals/physician_gls/pdf/colon.pdf. Accessed December 6, 2020.

Ned, R. M., Melillo, S., & Marrone, M. (2011). Fecal DNA testing for colorectal cancer screening: The ColoSure™ test. *PLoS Currents*, *3*, RRN1220. https://doi.org/10.1371/currents.RRN1220.

Neumann, J., Zeindl-Eberhart, E., Kirchner, T., & Jung, A. (2009). Frequency and type of KRAS mutations in routine diagnostic analysis of metastatic colorectal cancer. *Pathology, Research and Practice*, *205*(12), 858–862. https://doi.org/10.1016/j.prp.2009.07.010.

Ng, E. K. O., Chong, W. W. S., Jin, H., Lam, E. K. Y., Shin, V. Y., Yu, J., et al. (2009). Differential expression of microRNAs in plasma of patients with colorectal cancer: A potential marker for colorectal cancer screening. *Gut*, *58*(10), 1375–1381. https://doi.org/10.1136/gut.2008.167817.

Nguyen, L. H., Goel, A., & Chung, D. C. (2020). Pathways of colorectal carcinogenesis. *Gastroenterology*, *158*(2), 291–302. https://doi.org/10.1053/j.gastro.2019.08.059.

Nissan, A., Stojadinovic, A., Mitrani-Rosenbaum, S., Halle, D., Grinbaum, R., Roistacher, M., et al. (2012). Colon cancer associated transcript-1: A novel RNA expressed in malignant and pre-malignant human tissues. *International Journal of Cancer*, *130*(7), 1598–1606. https://doi.org/10.1002/ijc.26170.

O'Dwyer, P. J., & Catalano, R. B. (2006). Uridine diphosphate glucuronosyltransferase (UGT) 1A1 and irinotecan: Practical pharmacogenomics arrives in cancer therapy. *Journal of Clinical Oncology: Official Journal of the American Society of Clinical Oncology*, *24*(28), 4534–4538. https://doi.org/10.1200/JCO.2006.07.3031.

Ojala, K. A., Kilpinen, S. K., & Kallioniemi, O. P. (2011). Classification of unknown primary tumors with a data-driven method based on a large microarray reference database. *Genome Medicine*, *3*(9), 63. https://doi.org/10.1186/gm279.

Overman, M. J., Lonardi, S., Leone, F., McDermott, R. S., Morse, M. A., Wong, K. Y. M., et al. (2017). Nivolumab in patients with DNA mismatch repair deficient/microsatellite instability high metastatic colorectal cancer: Update from CheckMate 142. *Journal of Clinical Oncology*, *35*(4_suppl), 519. https://doi.org/10.1200/JCO.2017.35.4_suppl.519.

Overman, M. J., Lonardi, S., Wong, K. Y. M., Lenz, H.-J., Gelsomino, F., Aglietta, M., et al. (2018). Durable clinical benefit with nivolumab plus ipilimumab in DNA mismatch repair–deficient/microsatellite instability–high metastatic colorectal cancer. *Journal of Clinical Oncology*, *36*(8), 773–779. https://doi.org/10.1200/JCO.2017.76.9901.

Oxnard, G. R., Klein, E. A., Seiden, M. V., Hubbell, E., Venn, O., Jamshidi, A., et al. (2019). LBA77—Simultaneous multi-cancer detection and tissue of origin (TOO) localization using targeted bisulfite sequencing of plasma cell-free DNA (cfDNA). *Annals of Oncology*, *30*, v912. https://doi.org/10.1093/annonc/mdz394.074.

Ozawa, T., Matsuyama, T., Toiyama, Y., Takahashi, N., Ishikawa, T., Uetake, H., et al. (2017). CCAT1 and CCAT2 long noncoding RNAs, located within the 8q.24.21 "gene desert", serve as important prognostic biomarkers in colorectal cancer. *Annals of Oncology: Official Journal of the European Society for Medical Oncology*, *28*(8), 1882–1888. https://doi.org/10.1093/annonc/mdx248.

Package Insert. (2019). Camptosar® (Irinotecan) Injection, intravenous infusion. Available at: https://www.accessdata.fda.gov/drugsatfda_docs/label/2019/020571s050l bl.pdf. Accessed November 16, 2020.

Pagès, F., Mlecnik, B., Marliot, F., Bindea, G., Ou, F.-S., Bifulco, C., et al. (2018). International validation of the consensus Immunoscore for the classification of colon cancer: A prognostic and accuracy study. *The Lancet*, *391*(10135), 2128–2139. https://doi.org/10.1016/S0140-6736(18)30789-X.

Parikh, A. R., Lee, F.-C., Yau, L., Koh, H., Knost, J., Mitchell, E. P., et al. (2019). MAVERICC, a randomized, biomarker-stratified, phase II study of mFOLFOX6-bevacizumab versus FOLFIRI-bevacizumab as first-line chemotherapy in metastatic colorectal cancer. *Clinical Cancer Research: An Official Journal of the American Association for Cancer Research, 25*(10), 2988–2995. https://doi.org/10.1158/1078-0432.CCR-18-1221.

Park, S.-K., Baek, H. L., Yu, J., Kim, J. Y., Yang, H.-J., Jung, Y. S., et al. (2017). Is methylation analysis of SFRP2, TFPI2, NDRG4, and BMP3 promoters suitable for colorectal cancer screening in the Korean population? *Intestinal Research, 15*(4), 495–501. https://doi.org/10.5217/ir.2017.15.4.495.

Peeters, M., Oliner, K. S., & Parker, A. (2010). Use of massively parallel, next-generation sequencing to identify gene mutations beyond KRAS that predict response to panitumumab in a randomized, phase 3, monotherapy study of metastatic colorectal cancer (mCRC). In *American Association for Cancer Research 101st Annual Meeting. Washington, DC, USA. April 17–2.*

Peng, Q., Zhang, X., Min, M., Zou, L., Shen, P., & Zhu, Y. (2017). The clinical role of microRNA-21 as a promising biomarker in the diagnosis and prognosis of colorectal cancer: A systematic review and meta-analysis. *Oncotarget, 8*(27), 44893–44909. https://doi.org/10.18632/oncotarget.16488.

Pietrantonio, F., Di Nicolantonio, F., Schrock, A. B., Lee, J., Tejpar, S., Sartore-Bianchi, A., et al. (2017). ALK, ROS1, and NTRK rearrangements in metastatic colorectal cancer. *Journal of the National Cancer Institute, 109*(12), djx089. https://doi.org/10.1093/jnci/djx089.

Pietrantonio, F., Petrelli, F., Coinu, A., Di Bartolomeo, M., Borgonovo, K., Maggi, C., et al. (2015). Predictive role of BRAF mutations in patients with advanced colorectal cancer receiving cetuximab and panitumumab: A meta-analysis. *European Journal of Cancer (Oxford, England: 1990), 51*(5), 587–594. https://doi.org/10.1016/j.ejca.2015.01.054.

Popat, S. (2004). Systematic review of microsatellite instability and colorectal cancer prognosis. *Journal of Clinical Oncology, 23*(3), 609–618. https://doi.org/10.1200/JCO.2005.01.086.

Powell, S. M., Zilz, N., Beazer-Barclay, Y., Bryan, T. M., Hamilton, S. R., Thibodeau, S. N., et al. (1992). APC mutations occur early during colorectal tumorigenesis. *Nature, 359*(6392), 235–237. https://doi.org/10.1038/359235a0.

Quintero, E., Castells, A., Bujanda, L., Cubiella, J., Salas, D., Lanas, Á., et al. (2012). Colonoscopy versus fecal immunochemical testing in colorectal-cancer screening. *New England Journal of Medicine, 366*(8), 697–706. https://doi.org/10.1056/NEJMoa1108895.

Rajagopalan, H., Bardelli, A., Lengauer, C., Kinzler, K. W., Vogelstein, B., & Velculescu, V. E. (2002). Tumorigenesis: RAF/RAS oncogenes and mismatch-repair status. *Nature, 418*(6901), 934. https://doi.org/10.1038/418934a.

Ramanathan, R. K., Hwang, J. J., Zamboni, W. C., Sinicrope, F. A., Safran, H., Wong, M. K., et al. (2004). Low overexpression of HER-2/neu in advanced colorectal cancer limits the usefulness of trastuzumab (Herceptin) and irinotecan as therapy. A phase II trial. *Cancer Investigation, 22*(6), 858–865.

Rapado-González, Ó., Majem, B., Álvarez-Castro, A., Díaz-Peña, R., Abalo, A., Suárez-Cabrera, L., et al. (2019). A novel saliva-based miRNA signature for colorectal cancer diagnosis. *Journal of Clinical Medicine, 8*(12), 2029. https://doi.org/10.3390/jcm8122029.

Rasmussen, M. H., Jensen, N. F., Tarpgaard, L. S., Qvortrup, C., Romer, M. U., Stenvang, J., et al. (2013). High expression of microRNA-625-3p is associated with poor response to first-line oxaliplatin based treatment of metastatic colorectal cancer. *Molecular Oncology*, 7(3), 637–646. https://doi.org/10.1016/j.molonc.2013.02.016.

Reinert, T., Henriksen, T. V., Christensen, E., Sharma, S., Salari, R., Sethi, H., et al. (2019). Analysis of plasma cell-free DNA by ultradeep sequencing in patients with stages I to III colorectal cancer. *JAMA Oncology*, *5*(8), 1124–1131. https://doi.org/10.1001/jamaoncol.2019.0528.

Ribic, C. M., Sargent, D. J., Moore, M. J., Thibodeau, S. N., French, A. J., Goldberg, R. M., et al. (2003). Tumor microsatellite-instability status as a predictor of benefit from fluorouracil-based adjuvant chemotherapy for colon cancer. *The New England Journal of Medicine*, *349*(3), 247–257. https://doi.org/10.1056/NEJMoa022289.

Ronzoni, M., Manzoni, M., Mariucci, S., Loupakis, F., Brugnatelli, S., Bencardino, K., et al. (2010). Circulating endothelial cells and endothelial progenitors as predictive markers of clinical response to bevacizumab-based first-line treatment in advanced colorectal cancer patients. *Annals of Oncology: Official Journal of the European Society for Medical Oncology*, *21*(12), 2382–2389. https://doi.org/10.1093/annonc/mdq261.

Rosenfeld, N., Aharonov, R., Meiri, E., Rosenwald, S., Spector, Y., Zepeniuk, M., et al. (2008). MicroRNAs accurately identify cancer tissue origin. *Nature Biotechnology*, *26*(4), 462–469. https://doi.org/10.1038/nbt1392.

Rowland, A., Dias, M. M., Wiese, M. D., Kichenadasse, G., McKinnon, R. A., Karapetis, C. S., et al. (2015). Meta-analysis of BRAF mutation as a predictive biomarker of benefit from anti-EGFR monoclonal antibody therapy for RAS wild-type metastatic colorectal cancer. *British Journal of Cancer*, *112*(12), 1888–1894. https://doi.org/10.1038/bjc.2015.173.

Ruiz-Bañobre, J., & Goel, A. (2019). DNA mismatch repair deficiency and immune checkpoint inhibitors in gastrointestinal cancers. *Gastroenterology*, *156*(4), 890–903. https://doi.org/10.1053/j.gastro.2018.11.071.

Ruiz-Bañobre, J., Kandimalla, R., & Goel, A. (2019). Predictive biomarkers in metastatic colorectal cancer: A systematic review. *JCO Precision Oncology*, *3*, 1–17. https://doi.org/10.1200/PO.18.00260.

Ruiz-Bañobre, J., Roy, R., Alustiza Fernández, M., Murcia, Ó., Jover, R., Pera, M., et al. (2020). Clinical significance of a microRNA signature for the identification and predicting prognosis in colorectal cancers with mucinous differentiation. *Carcinogenesis*, *41*(11), 1498–1506. https://doi.org/10.1093/carcin/bgaa097.

Salonga, D., Danenberg, K. D., Johnson, M., Metzger, R., Groshen, S., Tsao-Wei, D. D., et al. (2000). Colorectal tumors responding to 5-fluorouracil have low gene expression levels of dihydropyrimidine dehydrogenase, thymidylate synthase, and thymidine phosphorylase. *Clinical Cancer Research*, *6*(4), 1322–1327.

Samowitz, W. S., Curtin, K., Ma, K. N., Schaffer, D., Coleman, L. W., Leppert, M., et al. (2001). Microsatellite instability in sporadic colon cancer is associated with an improved prognosis at the population level. *Cancer Epidemiology, Biomarkers & Prevention: A Publication of the American Association for Cancer Research, Cosponsored by the American Society of Preventive Oncology*, *10*(9), 917–923.

Santini, D., Loupakis, F., Vincenzi, B., Floriani, I., Stasi, I., Canestrari, E., et al. (2008). High concordance of KRAS status between primary colorectal tumors and related metastatic sites: Implications for clinical practice. *The Oncologist*, *13*(12), 1270–1275. https://doi.org/10.1634/theoncologist.2008-0181.

Sargent, D. J., Marsoni, S., Monges, G., Thibodeau, S. N., Labianca, R., Hamilton, S. R., et al. (2010). Defective mismatch repair as a predictive marker for lack of efficacy of fluorouracil-based adjuvant therapy in colon cancer. *Journal of Clinical Oncology: Official Journal of the American Society of Clinical Oncology*, *28*(20), 3219–3226. https://doi.org/10.1200/JCO.2009.27.1825.

Sartore-Bianchi, A., Ardini, E., Bosotti, R., Amatu, A., Valtorta, E., Somaschini, A., et al. (2015). Sensitivity to entrectinib associated with a novel LMNA-NTRK1 gene fusion in metastatic colorectal cancer. *Journal of the National Cancer Institute*, *108*(1), djv306. https://doi.org/10.1093/jnci/djv306.

Sartore-Bianchi, A., Trusolino, L., Martino, C., Bencardino, K., Lonardi, S., Bergamo, F., et al. (2017). Dual-targeted therapy with trastuzumab and lapatinib in treatment-refractory, *KRAS* codon 12/13 wild-type, HER2-positive metastatic colorectal cancer (HERACLES): A proof-of-concept, multicentre, open-label, phase 2 trial. *The Lancet Oncology*, *17*(6), 738–746. https://doi.org/10.1016/S1470-2045(16)00150-9.

Sazanov, A. A., Kiselyova, E. V., Zakharenko, A. A., Romanov, M. N., & Zaraysky, M. I. (2017). Plasma and saliva miR-21 expression in colorectal cancer patients. *Journal of Applied Genetics*, *58*(2), 231–237. https://doi.org/10.1007/s13353-016-0379-9.

Shulman, K., Cohen, I., Barnett-Griness, O., Kuten, A., Gruber, S. B., Lejbkowicz, F., et al. (2011). Clinical implications of UGT1A1*28 genotype testing in colorectal cancer patients. *Cancer*, *117*(14), 3156–3162. https://doi.org/10.1002/cncr.25735.

Siegel, R. L., Miller, K. D., Goding Sauer, A., Fedewa, S. A., Butterly, L. F., Anderson, J. C., et al. (2020). Colorectal cancer statistics, 2020. *CA: A Cancer Journal for Clinicians*, *70*(3), 145–164. https://doi.org/10.3322/caac.21601.

Siena, S., Di Bartolomeo, M., Raghav, K. P. S., Masuishi, T., Loupakis, F., Kawakami, H., et al. (2020). A phase II, multicenter, open-label study of trastuzumab deruxtecan (T-DXd; DS-8201) in patients (pts) with HER2-expressing metastatic colorectal cancer (mCRC): DESTINY-CRC01. *Journal of Clinical Oncology*, *38*(15_suppl), 4000. https://doi.org/10.1200/JCO.2020.38.15_suppl.4000.

Simmer, F., Venderbosch, S., Dijkstra, J. R., Vink-Borger, E. M., Faber, C., Mekenkamp, L. J., et al. (2015). MicroRNA-143 is a putative predictive factor for the response to fluoropyrimidine-based chemotherapy in patients with metastatic colorectal cancer. *Oncotarget*, *6*(26), 22996–23007. https://doi.org/10.18632/oncotarget.4035.

Sinicrope, F. A., Foster, N. R., Thibodeau, S. N., Marsoni, S., Monges, G., Labianca, R., et al. (2011). DNA mismatch repair status and colon cancer recurrence and survival in clinical trials of 5-fluorouracil-based adjuvant therapy. *Journal of the National Cancer Institute*, *103*(11), 863–875. https://doi.org/10.1093/jnci/djr153.

Sinicrope, F. A., Mahoney, M. R., Smyrk, T. C., Thibodeau, S. N., Warren, R. S., Bertagnolli, M. M., et al. (2013). Prognostic impact of deficient DNA mismatch repair in patients with stage III colon cancer from a randomized trial of FOLFOX-based adjuvant chemotherapy. *Journal of Clinical Oncology: Official Journal of the American Society of Clinical Oncology*, *31*(29), 3664–3672. https://doi.org/10.1200/JCO.2013.48.9591.

Sinicrope, F. A., Yoon, H. H., Mahoney, M. R., Nelson, G. D., Thibodeau, S. N., Goldberg, R. M., et al. (2014). Overall survival result and outcomes by KRAS, BRAF, and DNA mismatch repair in relation to primary tumor site in colon cancers from a randomized trial of adjuvant chemotherapy: NCCTG (Alliance) N0147. *Journal of Clinical Oncology*, *32*(15_suppl), 3525. https://doi.org/10.1200/jco.2014.32.15_suppl.3525.

Siravegna, G., Lazzari, L., Crisafulli, G., Sartore-Bianchi, A., Mussolin, B., Cassingena, A., et al. (2018). Radiologic and genomic evolution of individual metastases during HER2 blockade in colorectal cancer. *Cancer Cell*, *34*(1), 148–162.e7. https://doi.org/10.1016/j.ccell.2018.06.004.

Siravegna, G., Mussolin, B., Buscarino, M., Corti, G., Cassingena, A., Crisafulli, G., et al. (2015). Clonal evolution and resistance to EGFR blockade in the blood of colorectal cancer patients. *Nature Medicine*, *21*, 795.

Smorenburg, C. H., Peters, G. J., van Groeningen, C. J., Noordhuis, P., Smid, K., van Riel, A. M. G. H., et al. (2006). Phase II study of tailored chemotherapy for advanced colorectal cancer with either 5-fluouracil and leucovorin or oxaliplatin and irinotecan based on the expression of thymidylate synthase and dihydropyrimidine dehydrogenase. *Annals of Oncology: Official Journal of the European Society for Medical Oncology*, *17*(1), 35–42. https://doi.org/10.1093/annonc/mdj046.

Song, L., Jia, J., Peng, X., Xiao, W., & Li, Y. (2017). The performance of the SEPT9 gene methylation assay and a comparison with other CRC screening tests: A meta-analysis. *Scientific Reports*, 7(1), 3032. https://doi.org/10.1038/s41598-017-03321-8.

Song, L., Peng, X., Li, Y., Xiao, W., Jia, J., Dong, C., et al. (2017). The SEPT9 gene methylation assay is capable of detecting colorectal adenoma in opportunistic screening. *Epigenomics*, *9*(5), 599–610. https://doi.org/10.2217/epi-2016-0146.

Sorich, M. J., Wiese, M. D., Rowland, A., Kichenadasse, G., McKinnon, R. A., & Karapetis, C. S. (2015). Extended RAS mutations and anti-EGFR monoclonal antibody survival benefit in metastatic colorectal cancer: A meta-analysis of randomized, controlled trials. *Annals of Oncology: Official Journal of the European Society for Medical Oncology*, *26*(1), 13–21. https://doi.org/10.1126/scitranslmed.3002442.

Stiksma, J., Grootendorst, D. C., & van der Linden, P. W. G. (2014). CA 19-9 as a marker in addition to CEA to monitor colorectal cancer. *Clinical Colorectal Cancer*, *13*(4), 239–244. https://doi.org/10.1016/j.clcc.2014.09.004.

Stintzing, S., Miller-Phillips, L., Modest, D. P., Fischer von Weikersthal, L., Decker, T., Kiani, A., et al. (2017). Impact of BRAF and RAS mutations on first-line efficacy of FOLFIRI plus cetuximab versus FOLFIRI plus bevacizumab: Analysis of the FIRE-3 (AIO KRK-0306) study. *European Journal of Cancer*, *79*, 50–60. https://doi.org/10.1016/j.ejca.2017.03.023.

Stoehlmacher, J., Ghaderi, V., Iobal, S., Groshen, S., Tsao-Wei, D., Park, D., et al. (2001). A polymorphism of the XRCC1 gene predicts for response to platinum based treatment in advanced colorectal cancer. *Anticancer Research*, *21*(4B), 3075–3079.

Strassburg, C. P. (2008). Pharmacogenetics of Gilbert's syndrome. *Pharmacogenomics*, *9*(6), 703–715. https://doi.org/10.2217/14622416.9.6.703.

Strimpakos, A. S., Cunningham, D., Mikropoulos, C., Petkar, I., Barbachano, Y., & Chau, I. (2010). The impact of carcinoembryonic antigen flare in patients with advanced colorectal cancer receiving first-line chemotherapy. *Annals of Oncology: Official Journal of the European Society for Medical Oncology*, *21*(5), 1013–1019. https://doi.org/10.1093/annonc/mdp449.

Suenaga, M., Schirripa, M., Cao, S., Zhang, W., Yang, D., Murgioni, S., et al. (2017). Genetic variants of DNA repair-related genes predict efficacy of TAS-102 in patients with refractory metastatic colorectal cancer. *Annals of Oncology: Official Journal of the European Society for Medical Oncology*, *28*(5), 1015–1022. https://doi.org/10.1093/annonc/mdx035.

Sveen, A., Kopetz, S., & Lothe, R. A. (2020). Biomarker-guided therapy for colorectal cancer: Strength in complexity. *Nature Reviews. Clinical Oncology*, *17*(1), 11–32. https://doi.org/10.1038/s41571-019-0241-1.

Svoboda, M., Slyskova, J., Schneiderova, M., Makovicky, P., Bielik, L., Levy, M., et al. (2014). HOTAIR long non-coding RNA is a negative prognostic factor not only in primary tumors, but also in the blood of colorectal cancer patients. *Carcinogenesis*, *35*(7), 1510–1515. https://doi.org/10.1093/carcin/bgu055.

Swanton, C., Venn, O., Aravanis, A., Hubbell, E., Maddala, T., Beausang, J. F., et al. (2018). Prevalence of clonal hematopoiesis of indeterminate potential (CHIP) measured by an ultra-sensitive sequencing assay: Exploratory analysis of the Circulating Cancer Genome Atlas (CCGA) study. *Journal of Clinical Oncology*, *36*(15_suppl), 12003. https://doi.org/10.1200/JCO.2018.36.15_suppl.12003.

Tabernero, J., Hozak, R. R., Yoshino, T., Cohn, A. L., Obermannova, R., Bodoky, G., et al. (2018). Analysis of angiogenesis biomarkers for ramucirumab efficacy in patients with metastatic colorectal cancer from RAISE, a global, randomized, double-blind, phase III study. *Annals of Oncology*, *29*(3), 602–609.

Takahashi, M., Cuatrecasas, M., Balaguer, F., Hur, K., Toiyama, Y., Castells, A., et al. (2012). The clinical significance of MiR-148a as a predictive biomarker in patients with advanced colorectal cancer. *PLoS One*, 7(10), e46684. https://doi.org/10.1371/journal.pone.0046684.

Tamborero, D., Dienstmann, R., Rachid, M. H., Boekel, J., Baird, R., Braña, I., et al. (2020). Support systems to guide clinical decision-making in precision oncology: The Cancer Core Europe Molecular Tumor Board Portal. *Nature Medicine, 26*(7), 992–994. https://doi.org/10.1038/s41591-020-0969-2.

Tang, D., Liu, J., Wang, D., Yu, H., Li, Y., & Zhang, J. (2011). Diagnostic and prognostic value of the methylation status of secreted frizzled-related protein 2 in colorectal cancer. *Clinical and Investigative Medicine. Medecine Clinique et Experimentale, 34*(2), E88–E95. https://doi.org/10.25011/cim.v34i1.15105.

Tänzer, M., Balluff, B., Distler, J., Hale, K., Leodolter, A., Röcken, C., et al. (2010). Performance of epigenetic markers SEPT9 and ALX4 in plasma for detection of colorectal precancerous lesions. *PLoS One, 5*(2), e9061. https://doi.org/10.1371/journal.pone.0009061.

Thibodeau, S. N., Bren, G., & Schaid, D. (1993). Microsatellite instability in cancer of the proximal colon. *Science (New York, N.Y.), 260*(5109), 816–819. https://doi.org/10.1126/science.8484122.

Tie, J., Cohen, J. D., Lo, S. N., Wang, Y., Li, L., Christie, M., et al. (2021). Prognostic significance of postsurgery circulating tumor DNA in nonmetastatic colorectal cancer: Individual patient pooled analysis of three cohort studies. *International Journal of Cancer, 148*(4), 1014–1026. https://doi.org/10.1002/ijc.33312.

Tie, J., Cohen, J. D., Wang, Y., Christie, M., Simons, K., Lee, M., et al. (2019). Circulating tumor DNA analyses as markers of recurrence risk and benefit of adjuvant therapy for stage III colon cancer. *JAMA Oncology, 5*(12), 1710–1717. https://doi.org/10.1001/jamaoncol.2019.3616.

Tie, J., Cohen, J. D., Wang, Y., Li, L., Christie, M., Simons, K., et al. (2019). Serial circulating tumour DNA analysis during multimodality treatment of locally advanced rectal cancer: A prospective biomarker study. *Gut, 68*(4), 663–671. https://doi.org/10.1136/gutjnl-2017-315852.

Tie, J., Kinde, I., Wang, Y., Wong, H. L., Roebert, J., Christie, M., et al. (2015). Circulating tumor DNA as an early marker of therapeutic response in patients with metastatic colorectal cancer. *Annals of Oncology: Official Journal of the European Society for Medical Oncology, 26*(8), 1715–1722. https://doi.org/10.1093/annonc/mdv177.

Tie, J., Wang, Y., Tomasetti, C., Li, L., Springer, S., Kinde, I., et al. (2016). Circulating tumor DNA analysis detects minimal residual disease and predicts recurrence in patients with stage II colon cancer. *Science Translational Medicine, 8*(346), 346ra92. https://doi.org/10.1126/scitranslmed.aaf6219.

Toiyama, Y., Takahashi, M., Hur, K., Nagasaka, T., Tanaka, K., Inoue, Y., et al. (2013). Serum miR-21 as a diagnostic and prognostic biomarker in colorectal cancer. *Journal of the National Cancer Institute, 105*(12), 849–859. https://doi.org/10.1093/jnci/djt101.

Tougeron, D., Mouillet, G., Trouilloud, I., Lecomte, T., Coriat, R., Aparicio, T., et al. (2016). Efficacy of adjuvant chemotherapy in colon cancer with microsatellite instability: A large multicenter AGEO study. *Journal of the National Cancer Institute, 108*(7). https://doi.org/10.1093/jnci/djv438.

Toyota, M., Ahuja, N., Ohe-Toyota, M., Herman, J. G., Baylin, S. B., & Issa, J. P. (1999). CpG island methylator phenotype in colorectal cancer. *Proceedings of the National Academy of Sciences of the United States of America, 96*(15), 8681–8686. https://doi.org/10.1073/pnas.96.15.8681.

Tran, B., Kopetz, S., Tie, J., Gibbs, P., Jiang, Z.-Q., Lieu, C. H., et al. (2011). Impact of BRAF mutation and microsatellite instability on the pattern of metastatic spread and prognosis in metastatic colorectal cancer. *Cancer, 117*(20), 4623–4632. https://doi.org/10.1002/cncr.26086.

Tuchman, M., Stoeckeler, J. S., Kiang, D. T., O'Dea, R. F., Ramnaraine, M. L., & Mirkin, B. L. (1985). Familial pyrimidinemia and pyrimidinuria associated with severe fluorouracil toxicity. *New England Journal of Medicine, 313*(4), 245–249. https://doi.org/10.1056/NEJM198507253130407.

Uratani, R., Toiyama, Y., Kitajima, T., Kawamura, M., Hiro, J., Kobayashi, M., et al. (2016). Diagnostic potential of cell-free and exosomal microRNAs in the identification of patients with high-risk colorectal adenomas. *PLoS One, 11*(10), e0160722. https://doi.org/10.1371/journal.pone.0160722.

USPSTF. US Preventive Services Task Force et al. (2020). Screening for colorectal cancer: US preventive services task force recommendation statement. https://uspreventiveservicestaskforce.org/uspstf/draft-recommendation/colorectal-cancer-screening3#ful. https://doi.org/10.3978/j.issn.2223-4683.2012.07.04.

Valtorta, E., Martino, C., Sartore-Bianchi, A., Penaullt-Llorca, F., Viale, G., Risio, M., et al. (2015). Assessment of a HER2 scoring system for colorectal cancer: Results from a validation study. *Modern Pathology: An Official Journal of the United States and Canadian Academy of Pathology, Inc, 28*(11), 1481–1491. https://doi.org/10.1038/modpathol.2015.98.

Van Emburgh, B. O., Arena, S., Siravegna, G., Lazzari, L., Crisafulli, G., Corti, G., et al. (2016). Acquired RAS or EGFR mutations and duration of response to EGFR blockade in colorectal cancer. *Nature Communications*, 7, 13665. https://doi.org/10.1038/ncomms13665.

van Es, H. H., Bout, A., Liu, J., Anderson, L., Duncan, A. M., Bosma, P., et al. (1993). Assignment of the human UDP glucuronosyltransferase gene (UGT1A1) to chromosome region 2q37. *Cytogenetics and Cell Genetics, 63*(2), 114–116. https://doi.org/10.1159/000133513.

Varadhachary, G. R., Spector, Y., Abbruzzese, J. L., Rosenwald, S., Wang, H., Aharonov, R., et al. (2011). Prospective gene signature study using microRNA to identify the tissue of origin in patients with carcinoma of unknown primary. *Clinical Cancer Research: An Official Journal of the American Association for Cancer Research, 17*(12), 4063–4070. https://doi.org/10.1158/1078-0432.CCR-10-2599.

Venderbosch, S., Nagtegaal, I. D., Maughan, T. S., Smith, C. G., Cheadle, J. P., Fisher, D., et al. (2014). Mismatch repair status and BRAF mutation status in metastatic colorectal cancer patients: A pooled analysis of the CAIRO, CAIRO2, COIN, and FOCUS studies. *Clinical Cancer Research: An Official Journal of the American Association for Cancer Research, 20*(20), 5322–5330. https://doi.org/10.1158/1078-0432.CCR-14-0332.

Vidal, J., Muinelo, L., Dalmases, A., Jones, F., Edelstein, D., Iglesias, M., et al. (2017). Plasma ctDNA RAS mutation analysis for the diagnosis and treatment monitoring of metastatic colorectal cancer patients. *Annals of Oncology, 28*(6), 1325–1332. https://doi.org/10.1093/annonc/mdx125.

Vilar, E., & Gruber, S. B. (2010). Microsatellite instability in colorectal cancer-the stable evidence. *Nature Reviews. Clinical Oncology*, 7(3), 153–162. https://doi.org/10.1038/nrclinonc.2009.237.

Volinia, S., Calin, G. A., Liu, C.-G., Ambs, S., Cimmino, A., Petrocca, F., et al. (2006). A microRNA expression signature of human solid tumors defines cancer gene targets. *Proceedings of the National Academy of Sciences of the United States of America, 103*(7), 2257–2261. https://doi.org/10.1073/pnas.0510565103.

Wheeler, D. A., Takebe, N., Hinoue, T., Hoadley, K. A., Cardenas, M. F., Hamilton, A. M., et al. (2021). Molecular features of cancers exhibiting exceptional responses to treatment. *Cancer Cell, 39*(1), 38–53.e7. https://doi.org/10.1016/j.ccell.2020.10.015.

Wong, A. L. A., Lim, J. S. J., Sinha, A., Gopinathan, A., Lim, R., Tan, C.-S., et al. (2015). Tumour pharmacodynamics and circulating cell free DNA in patients with refractory colorectal carcinoma treated with regorafenib. *Journal of Translational Medicine, 13*, 57. https://doi.org/10.1186/s12967-015-0405-4.

Wu, D., Zhou, G., Jin, P., Zhu, J., Li, S., Wu, Q., et al. (2016). Detection of colorectal cancer using a simplified SEPT9 gene methylation assay is a reliable method for opportunistic screening. *The Journal of Molecular Diagnostics*, *18*(4), 535–545. https://doi.org/10.1016/j.jmoldx.2016.02.005.

Xi, Y., Formentini, A., Nakajima, G., Kornmann, M., & Ju, J. (2008). Validation of biomarkers associated with 5-fluorouracil and thymidylate synthase in colorectal cancer. *Oncology Reports*, *19*(1), 257–262. https://doi.org/10.3892/or.19.1.257.

Xi, Y., Nakajima, G., Schmitz, J. C., Chu, E., & Ju, J. (2006). Multi-level gene expression profiles affected by thymidylate synthase and 5-fluorouracil in colon cancer. *BMC Genomics*, 7, 68. https://doi.org/10.1186/1471-2164-7-68.

Xue, J. Y., Zhao, Y., Aronowitz, J., Mai, T. T., Vides, A., Qeriqi, B., et al. (2020). Rapid non-uniform adaptation to conformation-specific KRAS(G12C) inhibition. *Nature*, *577*(7790), 421–425. https://doi.org/10.1038/s41586-019-1884-x.

Yan, H. H. N., Lai, J. C. W., Ho, S. L., Leung, W. K., Law, W. L., Lee, J. F. Y., et al. (2017). RNF43 germline and somatic mutation in serrated neoplasia pathway and its association with BRAF mutation. *Gut*, *66*(9), 1645–1656. https://doi.org/10.1136/gutjnl-2016-311849.

Yang, X., Han, H., De Carvalho, D. D., Lay, F. D., Jones, P. A., & Liang, G. (2014). Gene body methylation can alter gene expression and is a therapeutic target in cancer. *Cancer Cell*, *26*(4), 577–590. https://doi.org/10.1016/j.ccr.2014.07.028.

Yau, T. O., Tang, C.-M., Harriss, E. K., Dickins, B., & Polytarchou, C. (2019). Faecal microRNAs as a non-invasive tool in the diagnosis of colonic adenomas and colorectal cancer: A meta-analysis. *Scientific Reports*, *9*(1), 9491. https://doi.org/10.1038/s41598-019-45570-9.

Yuan, Z.-X., Wang, X.-Y., Qin, Q.-Y., Chen, D.-F., Zhong, Q.-H., Wang, L., et al. (2013). The prognostic role of BRAF mutation in metastatic colorectal cancer receiving anti-EGFR monoclonal antibodies: A meta-analysis. *PLoS One*, *8*(6), e65995. https://doi.org/10.1371/journal.pone.0065995.

Zaanan, A., Fléjou, J.-F., Emile, J.-F., Des, G. G., Cuilliere-Dartigues, P., Malka, D., et al. (2011). Defective mismatch repair status as a prognostic biomarker of disease-free survival in stage III colon cancer patients treated with adjuvant FOLFOX chemotherapy. *Clinical Cancer Research: An Official Journal of the American Association for Cancer Research*, *17*(23), 7470–7478. https://doi.org/10.1158/1078-0432.CCR-11-1048.

Zaretsky, J. M., Garcia-Diaz, A., Shin, D. S., Escuin-Ordinas, H., Hugo, W., Hu-Lieskovan, S., et al. (2016). Mutations associated with acquired resistance to PD-1 blockade in melanoma. *New England Journal of Medicine*, *375*(9), 819–829. https://doi.org/10.1056/NEJMoa1604958.

Zhang, H., Zhu, Y.-Q., Wu, Y.-Q., Zhang, P., & Qi, J. (2014). Detection of promoter hypermethylation of Wnt antagonist genes in fecal samples for diagnosis of early colorectal cancer. *World Journal of Gastroenterology*, *20*(20), 6329–6335. https://doi.org/10.3748/wjg.v20.i20.6329.

Zhao, W., Song, M., Zhang, J., Kuerban, M., & Wang, H. (2015). Combined identification of long non-coding RNA CCAT1 and HOTAIR in serum as an effective screening for colorectal carcinoma. *International Journal of Clinical and Experimental Pathology*, *8*(11), 14131–14140.

Zheng, Z., Liebers, M., Zhelyazkova, B., Cao, Y., Panditi, D., Lynch, K. D., et al. (2014). Anchored multiplex PCR for targeted next-generation sequencing. *Nature Medicine*, *20*(12), 1479–1484. https://doi.org/10.1038/nm.3729.

The Cancer Genome Atlas (TCGA) Research Network. (2012). Comprehensive molecular characterization of human colon and rectal cancer. *Nature*, *487*(7407), 330–337. https://doi.org/10.1038/nature11252.

CHAPTER EIGHT

Field carcinogenesis for risk stratification of colorectal cancer

Dionne Rebello[a], Elliott Rebello[b], Matthew Custodio[c], Xixi Xu[c], Sanil Gandhi[d], and Hemant K. Roy[a,e,*]

[a]Department of Gastroenterology, Boston University Medical Center, Boston, MA, United States
[b]Warren Alpert Medical School of Brown University, Providence, RI, United States
[c]Department of Internal Medicine, Boston University Medical Center, Boston, MA, United States
[d]Boston University, Boston, MA, United States
[e]Department of Internal Medicine, Baylor College of Medicine, Houston, TX, United States
*Corresponding author: e-mail address: hemant.roy@bcm.edu

Contents

Advances in Cancer Research, Volume 151
ISSN 0065-230X
https://doi.org/10.1016/bs.acr.2021.03.004

Abstract

Colorectal cancer (CRC) is characterized by genetic-environmental interplay leading to diffuse changes in the entire colonic mucosa (field carcinogenesis or field of injury) and to a pro-neoplastic genetic/epigenetic/physiological milieu. The clinical consequences are increased risk of synchronous and metachronous neoplasia. Factors such as genetics, race, ethnicity, age, and socioeconomic status are thought to influence neoplasia development. Here, we explore the potential improvement to CRC screening through exploiting field carcinogenesis, with particular focus on racial disparities and chemoprevention strategies. Also, we discuss future directions for field carcinogenesis/risk stratification using molecular and novel biophotonic techniques for personalized CRC screening.

1. Field carcinogenesis and colorectal cancer risk

The concept of field effect (a.k.a. field defect, field carcinogenesis, field of injury) has a myriad of definitions and has evolved over time. It was first coined by Slaughter in 1953 when he noted the high rate of metachronous cancers in patients whose oral cancers were successfully resected (Slaughter, Southwick, & Smejkal, 1953). The molecular basis for this has been supported by observations from Sidransky et al. that p53 mutations occur diffusely in the microscopically normal mucosa of the oral cavity of patients with head and neck cancer, predicting recurrent disease (Brennan et al., 1995). This exemplifies the concept of field of injury, where exogenous factors such as tobacco smoke and alcohol lead to diffuse genomic injury which can then result in second primaries (Braakhuis, Tabor, Kummer, Leemans, & Brakenhoff, 2003).

The field of injury can occur throughout areas exposed by toxins (Steiling, Ryan, Brody, & Spira, 2008). However, there is some evidence that this may extend to neighboring organs that share similar tissue types and exposures. This is the concept of extended field carcinogenesis espoused by Kopelovich (Kopelovich et al., 1999). Again, tobacco smoke represents the most well-studied example and may be an explanation for the high risk of lung and esophageal cancers in patients with head and neck cancer. Teleologically one could argue that our genetic makeup allows us to be more or less susceptible to the neoplastic effects of the exposures (leveraging the fact that only ~10%–20% of significant smokers go on to develop lung cancer). Interactions between exposures and the genetic makeup of the epithelium cell in the right host may promote carcinogenesis (Platz et al., 2000). Indeed, in the bronchial epithelium, there are several epigenetic events that continue even after smoking cessation, which is in keeping with the continued elevated risk of lung cancer even in former smokers (Steiling et al., 2008).

In the colon, etiological exposures play a major role which along with lifestyle factors are believed to account for 30%–50% of CRC risk (Erdrich, Zhang, Giovannucci, & Willett, 2015). Risk factors for CRC include things that are likely to target the fecal stream and hence the colonic mucosa, such as high fat diet, smoking and other aspects of the "exposome." However, there are also endogenous factors such as obesity, lack of physical activity, diabetes etc. that may not directly impact the fecal stream. Lochhead and colleagues extended this to an etiological field carcinogenesis assessing the impact of not only exogenous factors (diet, smoking) but also endogenous factors (Lochhead et al., 2015). The etiological field carcinogenesis concept highlights the impact of genetic-exogenous interactions (G × E), which incorporates genetic and epigenetic changes in the field along with other events including microbiome changes and inflammation (Dekker, Tanis, Vleugels, Kasi, & Wallace, 2019). This goes hand-in-hand with molecular events, such as overexpression of cyclooxygenase (COX)-2, which is a hallmark of field carcinogenesis. This leads to pro-inflammatory prostaglandins such as prostaglandin E2 (PGE_2). Furthermore, the field effect concept shifts the focus of attention from somatic, genetic, and epigenetic alterations to the influence of etiologic factors that might predispose to acquisition of a myriad of pathological molecular alterations.

The inclusion of microbiome changes is critical given emerging appreciation of its role in CRC risk. Indeed, there is unequivocal data from both clinical and mouse studies on the importance of the microbiome, either specific species (increased *Fusobacterium nucleatum* and enterotoxigenic *Bacteroides fragilis* and decreasing antineoplastic short chain fatty acid generating species) or more high-level changes in microbiome complexity. The mechanisms of action of microbiome alterations in CRC risk include generation of secondary bile acids (e.g., deoxycholate) known to be pro-carcinogenic in the colon. The microbiome can potentially lead to inflammation, changing Treg function, etc. Alterations in the microbiome can modify the metabolomic milieu. Indeed, obesity has been shown to exert at least part of its pro-neoplastic effect through modulation of these factors. Other CRC risk factors such as diet and exercise have been shown to impact the microbiome in pro- and anti-neoplastic directions, respectively (Song, Chan, & Sun, 2020).

A consequence of the microbiome is that both the mucosa and submucosa may be impacted in field carcinogenesis. From the mucosal perspective, while most colonocytes are short lived, if intestinal stem cells become genetically or epigenetically altered in the base, fixation of changes may result. Altered intestinal stem cells can lead to new crypts with mutations built in (niche succession) (Patel, Tripathi, Gopalakrishnan, Williams, & Arasaradnam, 2015). These abnormal crypts can create small clonal patches in the mucosa via crypt fission and may provide a potential explanation for the non-homogeneity noted in field carcinogenesis (e.g., as evident by immunohistochemical detection of crypt-restricted cytochrome C oxidase subunit I) (Payne et al., 2005). Another way of fixing the changes in the colon is through the stromal micro-environment. Dotto reviewed the importance of the stroma in field carcinogenesis as providing the "bad soil" for epithelial cells (Dotto, 2014). The crosstalk between stromal fibroblasts and epithelial cells can occur via signals from the micro-environment inducing gene expression, mitogenic kinases and reactive oxygen species. These can lead to changes that become permanent in cells such as stromal fibroblasts. This, in turn, can provide long-term pro-neoplastic signals to the epithelium establishing long-term risk. The stroma provides the "bad soil" component and works hand-in-hand with epithelial cells in field carcinogenesis (Dotto, 2014).

While there is also evidence that the tumor itself can secrete cytokines and have a local effect, we will focus on the more important "field of injury concept" for this chapter (Steiling et al., 2008). In this paradigm, genetic and exogenous stimuli interact at the mucosal level enabling the accumulation of genetic/epigenetic events that increase the propensity towards neoplastic transformation (Baba et al., 2016). The occurrence of tumors is stochastic, requiring key genetic events such as mutations in/loss of critical genes, such as truncation of the adenomatous polyposis coli (APC) gene or silencing (via methylation) of the mismatch repair gene hMLH-1. The concept that detecting this "condemned mucosa" through a variety of techniques may offer opportunities for risk stratification of the entire colon.

2. Biomarkers of field carcinogenesis

In order to exploit field carcinogenesis for clinical use, marker selection is critical. The nature of the markers varies, and includes biochemical, immunohistochemical, cellular, genomic, epigenetic, and microvascular/

microarchitectural alterations. Here we will focus on the evidence for various schemata that capture metabolic and optically detectable markers of field carcinogenesis (Backman & Roy, 2011).

2.1 Cellular

The colonic crypt is dynamic with a lifecycle of approximately 5–7 days. Cells originate from the bottom of the crypt (where stem cells reside) to the top (where cells are sloughed off into the fecal stream generally by apoptosis). Thus, the proliferative compartment is generally in the bottom third of the crypt whereas the apoptotic compartment is at the top third. As would be expected, the balance between cell proliferation and apoptosis are biomarkers of cancer risk (Khosraviani, Williamson, & Hamilton, 1996).

2.2 Proliferation

Proliferation in cells can be assessed immunohistochemically through PCNA (proliferating cell nuclear antigen) or Ki-67 assays. Alternatively, in live cells, incorporation of the nucleotide analogue BrdU or tritiated thymidine can be used. Numerous studies indicate that patients harboring adenomas have diffuse hyperproliferation in the microscopically normal colonic mucosa. For instance, Ponz De Leon and colleagues noted that in the normal mucosa from patients harboring neoplasia, expansion of the proliferative compartment occurred, especially at the top of the crypt (Ponz de Leon et al., 1988). Moreover, Anti and colleagues addressed proliferation as a predictor of adenoma recurrence in a cohort of post-polypectomy patients followed serially with colonoscopy for 2 years. There was increased mean proliferative indices in the upper crypt in the 22 patients with recurring polyps as compared to the 33 who were polyp-free (Anti et al., 1993). In addition, induction of rectal proliferation has been noted in polyp recurrence patients who have a strong family history (Rooney, Robinson, Clarke, Hardcastle, & Armitage, 1993) and hereditary syndromes such as hereditary nonpolyposis colorectal cancer (HNPCC) (Green et al., 1998). However, while proliferation is clearly increased in field carcinogenesis, its diagnostic ability has been limited.

2.3 Apoptosis

Apoptosis or programmed cell death is important in removing cells from the top of the crypt. It can be measured by a variety of techniques, including light microscopy (membrane blebbing, ballooning nuclei etc.), immunohistochemical methods such as DNA fragmentation as detected by the TUNEL

(Terminal deoxynucleotidyl transferase dUTP nick end labeling), protease activity assays (e.g., cleaved caspase 3 or neoepitopes including M30), and cell membrane events (annexin V assays). Suppression of apoptosis has been thought to be one of the earliest mechanisms in colon carcinogenesis allowing the otherwise short-lived colonocytes to accumulate the genetic/epigenetic events necessary for neoplastic transformation. Studies have shown that patients with colonic neoplasia have a marked imbalance with both increased proliferation as discussed above and decreased apoptosis (Khosraviani et al., 1996). Bernstein and colleagues noted that bile salt-induced apoptosis in the left colon (20 cm from anal verge) was suppressed in patients with a history of CRC (Bernstein et al., 1999). Intriguingly, this assay noted the hallmark patchiness in field carcinogenesis (Payne et al., 2005). The role of apoptosis as a predictor of future neoplasia was assessed by Keku and colleagues using a cohort of 257 patients who had rectal biopsies. Using TUNEL assays, these investigators noted that lower apoptosis rates were associated with a higher risk of concurrent adenomas ($P=0.0001$) (Keku et al., 2008). Additionally, those within the highest tertile of apoptosis on initial rectal biopsy had a much lower rate of adenoma on follow-up colonoscopy, as compared to those within the lowest tertile. Furthermore, subjects with high apoptosis were less likely to have an adenoma at follow-up (OR, 0.25; 95% CI, 0.09–0.65) (Keku et al., 2008). Rectal apoptosis has been a standard marker in chemoprevention studies in conjunction with proliferation rates (Keller et al., 1999).

2.4 Genomic

Given the cellular changes in the normal mucosa, it is logical to predict that genomics must serve as the underpinnings. Chen and colleagues evaluated 15 gene expression signatures in the microscopically normal mucosa in the MIN mouse (murine model of familial adenomatous polyposis) (Chen et al., 2004). They observed that expression of two genes, encoding osteopontin and COX2, were >20-fold increased in polyps, and 5–20-fold increased in the histologically normal mucosa. In human cancer patients, both COX2 and osteopontin were overexpressed in cancerous tissues by >20-fold as compared to mucosa in samples obtained from resections for benign indications) (Chen et al., 2004). Importantly, in the uninvolved mucosa from these patients, expression was markedly higher than in benign resections (5–20-fold) and there was no evidence of decay as distance from tumor increased. This is consistent with the field of injury concept. This gene panel was used in first degree family members with CRC to demonstrate distinct patterns vs controls (Hao et al., 2005), and was extended to patients harboring adenomas with similar findings (Hao et al., 2005). Importantly, this was from the distal colon (rectosigmoid) suggesting that rectal swabs may be useful as a minimally intrusive approach to risk assessment.

2.5 Methylation

From an epigenetic perspective, methylation of CpG islands remains a common modality of silencing of gene expression. For many years, it was appreciated that during colon carcinogenesis there was a global hypomethylation of the genome. Later it was discovered that there was gene specific hypermethylation (Okugawa, Grady, & Goel, 2015). Currently it is believed that global hypomethylation might result in chromosomal instability, whereas hypermethylation is an important modality for silencing tumor suppressor genes and is mediated through alterations in DNA methyltransferases (DNMT-1). Altered methylation leads to a distinct molecular phenotype such as the CpG island methylator phenotype (CIMP). Most microsatellite unstable CRCs originate via hMLH1 silencing. Furthermore, methylation is the hallmark of the serrated polyp pathway, which is an alternative pathway for CRC development. Methylation is widely used as a biomarker for CRC and has achieved acceptance in the clinical armamentarium. For instance, methylation of septin 9 (Church et al., 2014) in the blood is FDA approved as a screening modality, as is fecal DNA (Cologuard) (Imperiale et al., 2014) which utilizes several methylation biomarkers. While methylation is common in many putative tumor suppressor genes (p16, MGMT-1, estrogen receptor etc.) (Baba et al., 2016), concern has been raised that hypermethylation in general, including some tumor suppressor genes, may also occur in an age-dependent fashion (Wang et al., 2020).

Despite these concerns, several groups have shown that methylation in the distal colon can serve as a biomarker for risk of colonic neoplasia (Cesaroni, Powell, & Sapienza, 2014; Leclerc et al., 2013). For example, Cesaroni and colleagues applied an 18 gene methylation signature in 24 patients with and without cancer. In the training set this had 96% sensitivity and 100% specificity. Using only 7 genes enabled classifying 44 of the 48 patients in the validation set correctly (Cesaroni et al., 2014). Paun evaluated specifically rectal methylation biomarkers (both hyper and hypomethylation) using 3 genes (NELL1, APC, and p14), and obtained an area under receiver operator character curve of 0.6661. The methylation was influenced by age and smoking (Paun et al., 2010).

2.6 Proteomic

As would be anticipated, changes have also been exploited at the protein level. Polley and colleagues performed two-dimensional gel electrophoresis and mass spectroscopy on microscopically normal mucosa from 13 healthy subjects, 9 patients with adenomatous polyps, and 9 with cancer (Polley et al., 2006). Systematic comparisons yielded 839 spots that significantly differed among the groups. Protein analysis yielded 61 proteins that differed, which included a variety of cytokeratins.

Daniel and colleagues performed individual studies on several protein biomarkers including TGF-alpha, MSH2 and hMLH-1 (Daniel et al., 2009). They evaluated microscopically normal mucosal biopsies from adenoma-harboring patients and controls (negative colonoscopies) for TGF- alpha, hMLH-1 and hMSH2 evaluated by automated immunohistochemistry (sample sizes for cases/controls was 29/31, 51/154 and 51/154, respectively). Rectal TGF-alpha was 51% higher in cases versus controls ($P=0.05$). Intriguingly, TGF-alpha mirrored known risk factors (decreased with putative chemopreventive agents such as NSAIDS or estrogens and increased in patients with CRC family history) (Daniel et al., 2009). For hMLH1, cases were 49% lower in the ascending colon ($P=0.03$) but not in other areas consonant with the predilection of mismatch repair protein deficient tumors within the right colon (Sidelnikov, Bostick, Flanders, Long, & Seabrook, 2009). hMSH2, another DNA mismatch repair protein, was similarly lower in the right colon of cases than controls and this was mirrored in the rectum (Sidelnikov et al., 2009).

2.7 Karyometry

Alberts and colleagues have pioneered the application of karyometry to field carcinogenesis in the colon. Karyometry is a technique to measure nuclear chromatin patterns using a computerized measure of the spatial and statistical relationships between nuclear pixel gray values. The investigators observed great efficacy for the technique in skin and breast cancer, and have applied it to colonic field carcinogenesis. In a study of patients harboring neoplasia (either adenomas or carcinomas), they evaluated 95 karyometric features in normal mucosa and noted striking chromatin alterations (Alberts et al., 2007). Importantly, the magnitude of the differences in the microscopically normal mucosa was comparable regardless of distance from lesions, arguing for the field of injury component and against the alternative explanation of tumor impact upon surrounding mucosa (Alberts et al., 2007). Further supporting the role of field carcinogenesis is the demonstration that rectal mucosal karyometry can predict recurrence of adenomas (Ranger-Moore et al., 2005).

2.8 MicroRNAs

Pioneering work by Goel and colleagues among others has demonstrated the power of microRNAs as biomarkers for colon carcinogenesis, with a panel showing good diagnostic performance in blood and stool (Balaguer et al., 2010; Link et al., 2010). MicroRNA levels can be regulated by a variety of factors including methylation, which has been a common mechanism for field carcinogenesis. In this regard, Grady et al. noted methylation-induced silencing of hsa-miR-342, a microRNA encoded within an intron of the gene EVL, in both CRC tissues and the uninvolved mucosa as a clear demonstration of field carcinogenesis (Grady et al., 2008). Kunte et al. noted the same in a genetic and carcinogenic animal model (Kunte et al., 2012). Importantly, in the carcinogen model, microRNAs predicted the onset of future neoplasia suggesting value for risk stratification (Kunte et al., 2012).

2.9 Biochemical

There have been a number of biochemical markers of field carcinogenesis. These include protein kinase C (PKC), a family of proteins that govern a myriad of processes. McGarrity and Peiffer performed PKC activity assays on microscopically normal mucosa from 12 patients with negative colonoscopies vs 12 with CRC (McGarrity & Peiffer, 1994). Similarly, given the importance of polyamines in colon carcinogenesis, ornithine decarboxylase (ODC) activity assays were conducted in microscopically normal mucosa of 23 patients with adenomas vs 22 who had normal colonoscopies (Hixson et al., 1993). The adenoma-harboring patients' activity levels were 2.5-fold higher. The relevance to carcinogenesis has been supported by several studies of the chemopreventive efficacy of the ODC inhibitor DFMO (Meyskens et al., 2008). Another approach utilized galactose oxidase-Schiff testing in rectal mucus for screening. Shamsuddin assessed 137 samples of mucus from cases and controls and had 100% specificity and 96% sensitivity (Vucenik, Gotovac, Druzijanic, & Shamsuddin, 2001). As would be predicted by field carcinogenesis, the majority of patients remained positive even after cancer resection.

2.10 Aberrant crypt foci (ACF)

ACF are morphological lesions that were first noted using methylene blue tissue staining in carcinogen-treated rat models. These abnormal crypts have altered luminal openings, thickened epithelia, and are larger than normal adjacent crypts. In humans, they appear to have a predilection for the rectal mucosa. Takayama and colleagues developed a modality where they sprayed the rectum with methylene blue followed by magnification colonoscopy on 171 patients without neoplasia, 131 harboring adenomas, and 48 with CRC (Takayama et al., 1998). While ACF tended to increase with age, overall there was a striking increase associated with the presence of neoplasia (rectal ACF numbers were 1, 5, and 26.5, respectively). Furthermore, there were significant ($P < 0.001$) correlations between the number of ACF, the presence of dysplastic foci, the size of the foci, and the number of adenomas. While there have been a number of discordant studies, by and large, most are supportive of ACF as a marker of field carcinogenesis even though the diagnostic power is suboptimal for clinical applications.

3. Metabolic markers

Metabolic changes are considered one of the hallmarks of malignant transformation (Hanahan & Weinberg, 2011) and can provide the requisite anabolic underpinnings for the hyperproliferative mucosa noted in field carcinogenesis. With regard to potential clinical correlates, metabolic diseases are well-established to predispose to colonic neoplasia. For instance, a number of studies have linked obesity to ~30%–70% increased risk of CRC, albeit greater in men than women (Bardou, Barkun, & Martel, 2013). There does appear to be a site difference as well (e.g., a weaker risk of rectal cancer in women). The etiologies are incompletely understood, but suggestions include inflammation, insulin resistance, adipokines, leptin, the microbiome etc. Similarly, diabetes might underlie increased risk with differences in gender and site-specificity. Overbeek and colleagues did a study with ~55,000 diabetics matched with ~215,000 controls (Overbeek et al., 2019). While the increased CRC risk in diabetics was similar in men and women (~1.3-old), in men the predominant effect was in the distal colon (OR 1.42, 95% CI 1.08–1.88), whereas in women it was in the proximal (OR 1.58, 95% CI 1.13–2.19). For rectal cancer, no statistically significant risk was observed in either men or women (Overbeek et al., 2019). The biological mechanisms are also not understood, but there has been copious data on insulin growth factor (IGF) as a biomarker for colon cancer (Wei et al., 2005). Even metabolic syndrome has been shown to increase risk. In a cross-sectional study of 5707 participants, those with metabolic syndrome had ORs for colonic adenomas and carcinomas of 1.76 and 1.92, respectively (Milano et al., 2019). Keku and colleagues noted that insulin resistance as measured by biochemical methods correlated with risk of concurrent adenomas and inversely correlated with rectal apoptosis. Finally, treatment with the anti-diabetic drug metformin had striking chemopreventive effects in a meta-analysis of randomized control trials (Higurashi et al., 2016). This supports the importance of metabolic changes in early colon carcinogenesis.

From a diagnostic perspective, fatty acid synthase (FASN), the driver of the lipidomic switch, was analyzed (Cruz et al., 2014). FASN expression progressively increased ~2–4-fold in both a genetic and carcinogen-induced rat model. Furthermore, FASN was significantly overexpressed (~1.5–2-fold) in rectal biopsies from patients harboring adenomas compared with those with no adenomas. Overall, the performance of rectal FASN was excellent (area under ROC curve of 0.81).

With regards to mechanism, as is well-established, induction of COX-2 occurs in the premalignant mucosa and is thought to have pleiotropic, pro-neoplastic effects. COX-2 drives production of a variety of factors including PGE2, a central bioactive lipid implicated in early carcinogenesis. Also, others have shown that hydroxyprostaglandin dehydrogenase is a marker of field carcinogenesis (Fink et al., 2014). Comprehensive mechanistic investigations were performed by Manna and colleagues into the molecular basis of metabolic early colon carcinogenesis, utilizing mass spectroscopy and gene expression in both animal models and clinical specimens. They noted that changes in a variety of metabolites were consistent with metabolic reprogramming (Manna et al., 2014).

One recent thrust is the role of the Warburg effect in field carcinogenesis. Championed almost a century ago, the Warburg effect is the concept that tumors preferentially use glycolysis instead of oxidative phosphorylation even in normo-oxemic states (Liberti & Locasale, 2016). While this is clearly inefficient for energy generation, the hypothesis is that increased glucose is used by cancer cells as a carbon source for anabolism necessary for cell proliferation. Specifically, the extra carbon is critical for de novo synthesis of lipids, proteins and nucleotides (Manna et al., 2014). Furthermore, the Warburg effect leads to modulation of various signaling pathways that foster neoplastic transformation.

Detection of these early metabolic changes in premalignant colonic tissue has been largely under-studied to date. However, Cruz et al. provided novel evidence of early Warburg effect-like changes in metabolism of premalignant colorectal mucosa, yielding potential for the development of metabolic biomarkers (Cruz et al., 2017). They evaluated the expression of 4 markers of glycolysis and 3 markers of mitochondrial changes in premalignant colonic tissue in two validated rat models (AOM-treated and PIRC), as well as in human rectal tissue. Among markers of glycolysis, they examined HIF1α, GLUT1, PKM2 and LDHA. HIF1α is a transcription factor that induces the expression of genes involved in increasing aerobic glycolysis and angiogenesis, and has been found to be overexpressed in CRC tissues (Chen, Pore, Behrooz, Ismail-Beigi, & Maity, 2001; Furlan et al., 2007; Rasheed, McDonald, Northover, & Guenther, 2008; Zhdanov, Okkelman, Collins, Melgar, & Papkovsky, 2015). GLUT1, or SCL2A1, is a glucose transporter that is known to increase both carcinogenesis and hypoxia (Chen et al., 2001). GLUT1 expression has been found to be increased in CRC tissue and its level of expression correlates with poor prognosis in CRC (Shen, Arbman, Olsson, & Sun, 2011). PKM2 is an isoform of pyruvate kinase, which catalyzes ATP and pyruvate production in glycolysis; this particular isoform is seen mostly in cancer cells and is induced by HIF1a (Zhdanov et al., 2015). PKM2 is overexpressed in CRC and its levels of expression correlate with TNM stage and poorer prognosis (Cui & Shi, 2015; Dekker et al., 2019; Nimri, Saadi, Peri, Yehuda-Shnaidman, & Schwartz, 2015; Song et al., 2020). Finally, lactate dehydrogenase (LDHA), an enzyme that catalyzes pyruvate fermentation into lactate and generates NAD^+ for glycolysis, is expressed at increased level in CRC tissue (Patel et al., 2015; Wang et al., 2015). Expression of all of these markers is elevated in the rectal mucosa of patients harboring precancerous lesions, as compared to patients without such lesions based upon screening colonoscopy.

In examining the expression of markers of alterations in mitochondrial function, Cruz et al. evaluated mitochondrial dynamics (DRP1 and OPA1), biogenesis (PGC1-α), uncoupling (UCP2), DNA (mtDNA), and copy number (mtND1). DRP1 and OPA1 promote mitochondrial fission and fusion respectively, allowing for increased cellular proliferation and oxidative capacity (Archer, 2013; Boland, Chourasia, & Macleod, 2013; Dotto, 2014; Payne et al., 2005). PGC1-α promotes increased numbers of mitochondria, which has been observed in CRC (Baba et al., 2016; Backman & Roy, 2011; Kaldma et al., 2014; Khosraviani et al., 1996;

Payne et al., 2005; Ussakli et al., 2013). UCP2 is a mitochondrial uncoupling protein whose expression levels are increased in adenomas and CRC in a stage-dependent manner (Anti et al., 1993; Baffy, 2010; Kuai, Ji, & Zhang, 2010; Ponz de Leon et al., 1988). Finally, mtND1 is a mitochondrial gene, which along with mtDNA copy number is known to be elevated in CRC and correlates with CRC risk (Thyagarajan, Wang, Barcelo, Koh, & Yuan, 2012; Ussakli et al., 2013). As with the glycolytic markers, all of these markers exhibit statistically significant elevated levels of expression in the rectal mucosa of patients harboring precancerous lesions as compared to patients lacking such lesions.

4. Current clinical applications

By definition, field carcinogenesis is a determinant of progression to CRC, and increases risk of both synchronous and metachronous neoplasia. Clinically, this is already well-established and acts as the basis of practice for CRC screening and diagnosis. Examination of the distal colon alone with flexible sigmoidoscopy has been shown to decrease the risk of both distal and proximal cancers (Chung, Hakim, Siddiqui, & Cash, 2020; Holme et al., 2017). Furthermore, current guidelines recommend shorter intervals for surveillance colonoscopy in patients who have had previous lesions removed (Lieberman et al., 2012).

The utilization of field carcinogenesis for CRC risk stratification provides significant advantages relative to current standards. First, field carcinogenesis is the milieu in which genetic and epigenetic events occur, leading to a mucosa predisposed to neoplastic development. Given this, assessing field carcinogenesis may be a surrogate for CRC risk, possibly at a point prior to development of an adenoma. Second, since field carcinogenesis involves colorectal mucosa diffusely, its detection can be performed by evaluation of readily available, uninvolved tissue within the field. This can be achieved conveniently, without an endoscope, and possibly without bowel purge, by utilizing the rectal mucosa, addressing major barriers to CRC screening such as bowel preparation, fear, cost, and logistics (Jones, Devers, Kuzel, & Woolf, 2010; Keku et al., 2008; Keller et al., 1999; Muthukrishnan, Arnold, & James, 2019). Ultimately, it lends itself to a promising future for CRC screening, surveillance, diagnosis, and even treatment.

Much of clinical implementation of the concept of field carcinogenesis is leveraged by both flexible sigmoidoscopy and colonoscopy. The overarching theme is that adenomas can serve as a marker for field carcinogenesis and hence provide insight into clinical management decisions.

5. Flexible sigmoidoscopy (FS) and field carcinogenesis

FS epitomizes field carcinogenesis in that detection of polyps in the distal colon is well established to be a harbinger of lesions throughout the organ. FS examines the lower half of the large intestine, and requires no sedation and a more limited bowel preparation than other structural examinations. FS also has the capacity to remove polyps as for biopsy necessary.

In a systemic review, using the sigmoid-descending colon junction as the extent of examination, Lewis et al. noted an OR for proximal neoplasia of 2.36 (95% CI, 1.30–4.29). However, there was clearly a significant amount of isolated proximal neoplasia (Lewis et al., 2003) indicating that the sensitivity of distal lesions was suboptimal. There are several studies that have demonstrated efficacy of FS in cancer prevention and mortality.

Data from United States was available through the PLCO (prostate, lung, colon and ovarian cancer) cohort. This cohort of 154,900 trial subjects between the ages of 55–74 randomized to usual care (no screening) or FS (at baseline with repeat in 3–5 years), with a median of ~12 years follow up. The uptake of FS was 84% at baseline and 54% undergoing second procedure (Schoen et al., 2012). Significant reductions were observed in the relative risk of both distal (OR 0.71; 95% CI 0.64–0.80; $P < 0.001$) and proximal CRC (OR 0.86; 95% CI, 0.76–0.97; $P = 0.01$). However, CRC mortality reduction was noted only for disease in the distal colon (OR 0.50; 95% CI, 0.38–0.64), but not for that in proximal colon (OR 0.97; 95% CI, 0.77–1.22; $P = 0.81$).

Schoenfeld and colleagues made a powerful observation regarding the impact of gender on FS outcomes from a VA Cooperative Study 380 (Schoenfeld et al., 2005). It is well-established that women tend to be diagnosed with CRC in their right colon, whereas men are diagnosed predominantly in the left colon. The diagnostic yield of FS was calculated by estimating the proportion of patients with advanced neoplasia whose lesions would have been identified if they had undergone FS alone (either located in distal colon or sentinel adenoma that would have mandated follow up colonoscopy). Performance of FS was estimated to be much lower in women than in men (35.2% vs 66%; $P < 0.001$) (Schoenfeld et al., 2005).

The power of field carcinogenesis as a risk stratification tool was underscored by a study led by Atkin and colleagues in the United Kingdom, who examined once-in-a-lifetime screening with FS in subjects age 55–64. In this design, ~50,000 patients assigned to FS and ~100,000 controls were followed for over a decade. In the intention-to-treat analyses (71% of those assigned to FS actually underwent the procedure), the CRC incidence was reduced by 23% (HR 0.77; 95% CI, 0.70–0.84) and mortality by 31% (HR 0.69; 95% CI, 0.59–0.82). The incidence of distal CRC (rectal and sigmoid colon) was reduced by 50% (HR 0.50; 95% CI, 0.42–0.59). Calculation of the number needed to screen during the study period to prevent one CRC case and one death was 191 (95% CI, 145–277) and 489 (343–852), respectively (Atkin et al., 2010).

Holme and colleagues summarized data from a pooled analysis of randomized trials with FS focusing on gender and age. They identified 115,139 patients randomized to screening and 172,789 to usual care; follow-up was 10.5–12.1 years. The relative risk of CRC incidence was decreased in both men (HR 0.76; 95% CI, 0.70–0.83) and women (HR, 0.83; 95% CI 0.75–0.92). No difference in the effect of screening was seen among men by age (dichotomized at age 60). In women, however, FS decreased incidence of CRC in those <60 (RR 0.71; 95% CI, 0.59–0.84), but not in those >60 versus (RR 0.92; 95% CI 0.80–1.02) (Holme, Loberg, Kalager, Bretthauer, & Hoff, 2018).

While the issues with FS are many, at least in the United States, the point is moot since over the last several decades FS has been replaced by colonoscopy. About 2.5% of adults at the age of CRC screening undergo FS in the recommended interval, compared with 60% for colonoscopy (National Center for Health Statistics (US), 2017). Importantly, however, the fact that FS has efficacy in both distal and proximal neoplasia underscores the potential power of field carcinogenesis for risk stratification (National Center for Health Statistics (US), 2017).

6. Colonoscopy: Post-polypectomy surveillance

Since colonoscopy entails visualization of the entire colon, at first pass one might not think field carcinogenesis is relevant. However, it is integral to management of patients in two scenarios: adenomas found at colonoscopy, and a completely negative colonoscopy.

Adenomatous polyps are marker of field carcinogenesis. The concept of condemned mucosa portends metachronous lesions. There is strong data supporting the notion that the severity of neoplasia mirrors the strength of field carcinogenesis, and hence risk of metachronous neoplasia.

Lieberman and colleagues reviewed records from a Veterans Administration database of patients age 50–75 (1171 with neoplasia and 501 without) who had a repeat colonoscopy done a median of 5.5 years after the initial procedure. The outcome was advanced neoplasia (defined by size >1 cm, high-grade dysplasia, or >25% villous features). In those whose baseline colonoscopy showed no neoplasia, 2.4% had advanced neoplasia on follow-up. If the baseline colonoscopy showed 1–2 non-advanced adenomas, the relative risk of advanced neoplasia was 1.92 (95% CI, 0.83–4.42). However, if the baseline colonoscopy identified either advanced neoplasia or >2 non-advanced lesions, there was a 5.4–6.9-fold increase in risk of advanced adenoma on follow-up. These data was largely responsible for the multigroup recommendation in 2012 that those with 1–2 adenomas should undergo repeat colonoscopy in 5–10 years, while those with ≥3 non-advanced adenomas or any advanced adenomas should repeat in 3 years (Lieberman et al., 2012).

The durability of field carcinogenesis has been an issue. Specifically, the question arises: If a follow-up colonoscopy shows no neoplasia after a first one with neoplasia, does that mean that field carcinogenesis has been ameliorated? This was addressed by Robertson and colleagues who examined a cohort of 564 patients from clinical trials that had colonoscopies at baseline, at 3 years, and at 6 years. For comparison, if a low-risk adenoma was identified at baseline, and no adenomas were identified at first follow-up, the risk of advanced adenoma on third colonoscopy was 4.9%. However, if the second colonoscopy had advanced neoplasia (regardless of findings at baseline), the rate of advanced adenomas was 18.2%–20% at the third. Importantly, if the first exam showed high-risk features, but the second showed low-risk or no adenomas, the probability of an advanced adenoma at the third colonoscopy was still quite elevated, i.e., 12.3%–13.6%. Thus, while adenoma development is a stochastic event within a permissive field, if a person has one, they remain at higher risk for future lesions (Robertson et al., 2009). This is bolstered by a report from Sullivan et al. who identified participants with 3 or more colonoscopies and found advanced adenomas in 114 of 891 participants (Sullivan et al., 2020). When compared to subjects without adenomas on both prior colonoscopies, those with advanced adenoma on first exam remained at higher risk for future neoplasia, with 3, 5 and 7 year risk increases of 3.41-, 3.14-, and 2.89-fold.

While numerous lines of evidence those have indicated that those with previous adenomas are at risk of future neoplasia (metachronous lesions), the crucial issue is to accurately quantify risk so as to prevent future CRCs while avoiding more procedures than is necessary (i.e., overutilization). Fundamentally, the prevalence of adenomas based on colonoscopy studies can range as high as 50%–60%, whereas the lifetime risk of CRC in the U.S. is ~4%. Thus, the vast majority of adenomas may not progress to cancer. While previous surveillance recommendations have been made based on advanced adenoma recurrence, the latest guidelines also take into account CRC risk. For most criteria, aside from advanced adenomas on initial endoscopy, there is data indicating that surveillance endoscopy decreases CRC incidence but not necessarily CRC mortality, though the quality is low (Gupta et al., 2020). However, for other non-advanced adenomas, such as non-advanced sessile serrated polyps, the benefit of surveillance colonoscopy on CRC is uncertain. Specifically, Gupta and colleagues reviewed a meta-analysis of 10,139 patients across 8 surveillance studies. The proportion of patients with metachronous advanced adenomas on surveillance was slightly higher in patients whose baseline colonoscopy showed non-advanced adenomas as compared to those who were adenoma-free (4.9% vs 3.3%) (Gupta et al., 2020). On the other hand, if the baseline finding was an advanced adenoma, the risk was markedly higher (17.3%). Thus, field carcinogenesis severity as marked by the presence of advanced adenomas, portends a higher risk of future neoplasia. Given these findings and a focus on CRC, the most recent US Multi-Society Task Force on Colorectal Cancer recommended extending intervals for 1–2 non-advanced adenomas from 5 to 10 years to 7–10 years; intervals for 3–4 non-advanced adenomas were extended from 3 years to 3–5 years). The outcome for the first surveillance colonoscopy determines the recommendations for the second surveillance colonoscopy, except if the incident (baseline) colonoscopy showed advanced neoplasia. In the latter case, repeat surveillance colonoscopy should be every 5 years (Gupta et al., 2020).

7. Negative colonoscopy follow-up

The converse of using adenoma detection as a marker for the presence of a field effect, and thus more frequent follow-up, is the situation where no adenoma is detected. Does that indicate the lack of field carcinogenesis and hence a long-term projection of low risk? For instance, Singh and colleagues did an administrative review of a cohort of 35,975 patients from Manitoba with negative colonoscopy. The standardized incidence ratios (SIRs) of CRC in this cohort as compared to the general provincial population was 0.28 (95% CI, 0.09–0.65) at 10 years (Singh, Turner, Xue, Targownik, &

Bernstein, 2006). Those CRCs that occurred had a predilection for right versus left colon (47% vs 28%; $P < 0.001$) consonant with the demonstration that colonoscopy is less protective for proximal CRC.

There are numerous other high-quality, large scale studies indicating that a negative colonoscopy connotes a long-term risk reduction of CRC. However, colonoscopy is operator- dependent in that its quality, as gauged by parameters such as the adenoma detection rate, varying widely among trained endoscopists. Pelonis and colleagues evaluated 165,887 individuals who had a negative colonoscopy for an average 17.4 years and noted that CRC incidence and mortality were 72% and 81% lower, respectively, than in the general population. They dichotomized quality by endoscopists having adenoma detection rate of at least 20% (Pilonis et al., 2020). Those with a high-quality exam (defined by cecal intubation rate, good bowel preparation, and endoscopist adenoma detection rate > 20%) had half the risk of CRC incidence (SIR 0.16 vs 0.32) and mortality (SMR 0.10 vs 0.22) as compared to those undergoing a low-quality exam. Only high-quality colonoscopy achieved a long-term dramatic decrease in CRC incidence and mortality, underscoring the potential of using colonoscopy for long-term (potentially even once in a lifetime) risk stratification.

8. Field carcinogenesis in inflammatory bowel disease (IBD)

CRC is one of the most feared complications of chronic inflammatory bowel disease (IBD). The prevalence has been decreasing with better medical therapies, but it is still believed to occur in ~18% for those with colitis for 30 years. The biology of CRC is quite distinct from sporadic disease from a molecular genetic perspective. Chronic inflammation is believed to induce oxidative stress leading to DNA damage, and resulting in activation of proto-oncogenes and inhibition of tumor suppressor genes. The consequences are chronically multifocal precancerous and cancerous changes or patchy limited-field changes (Baker, Salk, Brentnall, & Risques, 2018). For instance, p53 mutations are one of the earliest events in IBD-CRC but one of the latest in sporadic-CRC progression. In a landmark article, Brentall and colleagues noted that p53 mutation occurred in the flat mucosa (invisible) (Brentnall et al., 1994). For many years, given concerns about the flat invisible dysplasia and the practice of random biopsies of the colon (~30–40), any significant dysplasia led to a recommendation of colectomy, given a high risk of the presence of undiagnosed high-grade dysplasia or carcinoma. More recently, this approach has changed with the advent of high-definition colonoscopy with or without chromo-endoscopy, where most lesions can now be detected; still, many recommend concurrent random biopsies assessing for invisible dysplasia as a marker for field carcinogenesis.

With regards to markers of field carcinogenesis in IBD, while the biology is quite different, there are striking parallels:

1. *Proliferation*—There has been limited work on the role of proliferation. A study by Biasco and colleagues looked separately at proliferation in rectal biopsies from patients with ulcerative colitis (UC) or adenomas using tritiated thymidine to determine labeling frequency in the top 40% of crypts. In controls, it was 0.04 ± 0.01, which was increased in both UC and adenoma patients to 0.16 ± 0.02 and 0.10 ± 0.01, respectively (Biasco et al., 1990). While this does not specifically compare UC patients with and without neoplasia, it suggests that proliferation may be a useful marker, although inflammation and repair may be confounders.
2. *MicroRNAs*—Toiyama and colleagues evaluated microRNA methylation by bisulfite pyrosequencing of DNA in rectal biopsies from 238 patients with UC (including 17 with dysplasia, and 69 with CRC) (Toiyama et al., 2017). The analysis showed that methylation of MIR1, MIR9, MIR124 and MIR137 discriminated UC patients with and without dysplasia or CRC with an area under receiver operator characteristic curve (AUROC) of 0.81. In an independent validation dataset, the AUROC was comparable (0.79).
3. *Proteomics*—May and colleagues examined rectal biopsies from UC patients with or without dysplasia using accurate mass and time methods with ion intensity-based label-free proteomics (May et al., 2011). Among the differentially expressed proteins, CPS1 showed a statistically significant difference in immunohistochemical staining between those with and without dysplasia. Furthermore, rectal CPS1 staining predicted UC dysplasia and/or cancer in the colon with 87% sensitivity and 45%, specificity.
4. *Methylation*—Toiyama et al. determined if aberrant DNA methylations can discriminate UC patients with and without concurrent dysplasia. In the training set, 187 differentially methylated regions were identified which was pared down to 11 using the elastic net classification algorithm (Toiyama et al., 2018). This 11 gene classifier was from non-neoplastic, differentially methylated regions (DMR) in rectal tissues from UC-CRC. The 11 chosen DMRs could discriminate UC patients with or without

CRC in the training set (area under the curve, 0.96) and the validation set (area under the curve, 0.81).

5. *ACF:* ACF—Seventy-six patients with long-standing UC were prospectively screened by colonoscopy with chromoendoscopy-guided endomicroscopy in the distal rectum (Freire et al., 2014). The ability of rectal ACF to predict concomitant dysplasia was excellent, with 100% sensitivity and 65.2% specificity.
6. *Metabolic*—Ussakli and colleagues examined the role of mitochondria in UC tumorigenesis (Ussakli et al., 2013). Nondysplastic colon biopsies from UC patients with high-grade dysplasia or cancer (termed progressors) and from UC patients who are dysplasia-free (nonprogressors) were assayed for the mitochondrial protein cytochrome C oxidase. For six additional progressors, nondysplastic and dysplastic biopsies were analyzed for the oxidase as well as additional mitochondrial proteins (PGC1α) and mitochondrial DNA (mtDNA) copy number. The nondysplastic biopsies of UC progressors showed statistically significant reduction in oxidase expression and mitochondrial loss as compared to that in nonprogressors, consistent with a Warburg effect in the uninvolved mucosa of UC patients with dysplasia or cancer.
7. *Biophotonic*—Uttam and colleagues developed a novel light-scattering technology, denoted 3-dimensional (3D) nanoscale nuclear architecture mapping (nanoNAM) (Uttam et al., 2019), to analyze the rectal mucosa of 103 patients with IBD colitis. The nanoNAM-based structural analysis discriminated between patients with and without neoplasia with an AUC of 0.87 (sensitivity = 81%, specificity = 82%). Importantly, this approach was able to estimate likelihood of future onset of neoplasia.

9. CRC racial disparities and field carcinogenesis

As discussed in depth in other chapters, there are important racial differences in CRC incidence and mortality. While there are a myriad of potential etiologies underlying such disparities, including access to care and concurrent risk factors (e.g., diabetes and obesity), there is evidence for at least some biological underpinnings, including mutational predilection (Guda et al., 2015). Field carcinogenesis may differ by race, based on interval cancer rates (i.e., cancers that occur after a previously negative colonoscopy). Fedewa and colleagues assessed rates of interval cancers for Blacks versus Whites among 2735 CRC cases. The probability of interval CRC by the end of follow-up was 7.1% in Black persons and 5.8% in White persons

(HR 1.31, 95% CI 1.13–1.51). While interval cancer rates may have both an endoscopist component (e.g., missed lesions) and a biological component (e.g., newly formed neoplasia), adjusting for endoscopist polyp detection rates did not ameliorate these findings. This suggests a more pronounced field carcinogenesis in Blacks (Fedirko et al., 2009). A number of groups have evaluated DNA methylation changes in tumors. For instance, Ashktorab and colleagues identified some distinctively hypomethylated genes in Blacks with CRC (Ashktorab et al., 2020). Busch et al. compared interactions of lifestyle factors and tumor DNA methylation, and noted distinct racial patterns in variables such as fruit and folate consumption (Busch, Galanko, Sandler, Goel, & Keku, 2018). With regard to uninvolved mucosa, Alonso and colleagues evaluated methylation of the O^6-methylguanine-DNA methyltransferase (MGMT) and matrix metalloproteinase (ADAMTS14) genes in normal mucosa of CRC patients, and found that methylation was higher in the elderly and in African Americans ($P=0.00001$), suggesting a racial predisposition (Alonso et al., 2015). Intriguingly, in a similar vein, Deval et al. evaluated epigenetic aging in 88 African Americans vs 40 European Americans, and observed that the former manifested age acceleration in the right colon when compared to the left colon (1.51 years $P=0.001$), whereas the latter showed age deceleration (1.93 years) in the right colon when compared to the left colon (Devall et al., 2020). In addition, African-Americans had a distinct pattern of hypermethylation in the right colon, as noted by epigenomic analysis. This may explain, at least partially, the predilection for earlier age of onset and right sided disease in African Americans (Zisman, Nickolov, Brand, Gorchow, & Roy, 2006).

There is emerging recognition of the importance of chromatin remodeling in carcinogenesis. Cohesins are known to impact gene expression through modulation of chromatin looping. Stromal antigen-1 (SA-1) plays an important role. Wali and colleagues first demonstrated that SA-1 mRNA was lost in CRC, which was more pronounced in Blacks than Whites. With regard to field carcinogenesis, they observed that in the endoscopically normal rectal mucosa, SA-1 expression was suppressed in patients with concurrent adenomas (Wali et al., 2016). It is noteworthy that while the percent decrease in SA-1 was equivalent in Blacks and Whites, the baseline expression in Blacks was half that in Whites. In order to elucidate the mechanisms for this decrease, Datta and colleagues analyzed single nucleotide polymorphism (SNP) arrays, and noted one site in particular *(rs 34,149,860)* that was associated with both a decrease in SA-1 mRNA expression and increased risk of colon neoplasia (Datta et al., 2018). Importantly, this SNP was found almost exclusively in Black patients. From a mechanistic perspective, the SNP created a new binding site for miR-29b, leading to degradation of SA-1. These findings may provide a mechanism for the more dramatic loss of the SA-1 tumor suppressor in Blacks when compared to Whites.

10. Optically detectable markers of field carcinogenesis

Biophotonics applies optical techniques to probing tissue structure. Various techniques have been applied in endoscopy, albeit the focus of most work has been on optical biopsy (i.e., ascertaining whether lesions are neoplastic or benign). There are a myriad of techniques that numerous groups have developed to examine the micro- and nano-structures within the cell and to quantify disturbances. The Backman lab has pioneered development of various technologies to probe different structural facets of field carcinogenesis (Backman & Roy, 2013). One advantage of this approach is that while colon carcinogenesis is heterogeneous (i.e., there are four distinct consensus molecular subtypes, distinct epigenetic pathways, etc.), the structural aspects (chromatin nano-architecture, microvascular blood content, stroma, etc.) appear to be common among diverse neoplasia (Guinney et al., 2015). Though the physics involved in biophotonics is beyond the scope of this chapter, general methods include (Backman & Roy, 2013):

1. *Polarization-gated spectroscopy (PGS)*—This targets mucosal microcirculation. There is a pathognomonic light absorption signature for hemoglobin, making it an attractive target. This allows obtaining quantitative information about the microvasculature in tissue samples by analyzing the characteristic absorption/reflection spectra of red blood cells (RBCs). Given that single scattering events retain polarization whereas multiple events do not, the co-polarization versus cross-polarization signals can be exploited for depth selectivity. PGS can glean information about hemoglobin content, oxygenation status, and vessel size (Wali et al., 2005).
2. *Low-coherence enhanced backscattering spectroscopy (LEBS)*—Enhanced backscattering is an optical phenomenon that can provide quantitative information about the nanoscale composition of the epithelium. Enhanced backscattering, otherwise known as coherent backscattering, is a manifestation of the self-interference of light propagating in biological tissues, which leads to an enhanced scattering peak in the backward direction (i.e., direction opposite to that of the light beam incident on the tissue surface). In LEBS, both epithelial cells and the mucosal/

submucosal tissue microenvironment are investigated via scattering of light shone onto colonic mucosa. Macromolecules (including DNA and collagen) and small organelles of the colonic lamina propria are sensed by LEBS to provide a perceptible scattering signature, thus enabling subcellular scale detection beyond what is possible with conventional light microscopy. Analyzing these signatures via spectroscopy elucidates many key features of the mucosal ultrastructure, including macromolecular density and spatial packing. LEBS directly measures spectroscopic variables (e.g., correlation function shape, average amplitude, and length scale of mass density variations) whose magnitudes correlate with physiological ultrastructural alterations. Practically, LEBS-detected field carcinogenesis alterations identify disorganization of macromolecules (e.g., chromatin and extracellular collagen). LEBS peak characteristics (height, width, slope, auto-correlation, etc.) provide quantitative insights within length scales of 100–800 nm. It is important to note that LEBS provides data on tissue samples of area $\sim$1 mm^2 so it averages information from a large number of cells and the stroma.

3. *Partial-wave spectroscopic microscopy (PWS)*—This technology harnesses information from one-dimensional propagating waves in different parts of a cell. PWS measures the disorder strength (Ld) of the intracellular architecture that is proportional to the variance and the correlation length (e.g., the effective size of intracellular nanostructures) of the spatial fluctuations of local mass density. PWS provides data on individual cells, dividing each cell into $\sim$60,000 pixels and measuring the pixel-to-pixel variability in refractive index. This can be superimposed on cell visualizations so that the status of specific cell compartments can be interrogated.

11. Biological basis of biophotonic signatures

11.1 Micro-circulation

There is an increase in pericryptal micro-circulation during colon carcinogenesis, termed early increase in blood supply (EIBS) (Wali et al., 2005). This reflects the increased energy demand during field carcinogenesis coupled with metabolic changes, such as those of the Warburg effect. PGS measurements have been corroborated with other techniques, including confocal endomicroscopy, Western blot analysis of hemoglobin in colonic mucosal scrapings, and immunohistochemical analysis of microvessel density. Tiwari et al. showed that these changes are mirrored by induction of genes involved in control of micro-circulation (Tiwari et al., 2011). Intriguingly, Ruderman and colleagues noted that targeting the angiotensin II receptor with losartan not only blocked EIBS but also suppressed neoplasia, implicating EIBS in adenoma formation (Ruderman et al., 2018).

11.2 Chromatin architecture

Given changes in gene expression, it is expected that chromatin architecture will be altered. Globally, heterochromatin is more compacted and denser than euchromatin, which is more open and thus amenable to transcription. Even locally, there can be differences in silenced (compacted) genes versus more accessible chromatin. Modulators of gene expression, and hence chromatin architecture, include histone acetylation. Stypula and colleagues showed that histone deacetylase inhibitors (HDACs) modulate the PWS marker (Stypula-Cyrus et al., 2013). Almassahla and colleagues showed changes in transcription, which was evaluated through alterations in PWS (Almassalha et al., 2017). Importantly, PWS examines nuclear changes in carcinogenesis from the perspective of higher-order nuclear topology, as opposed to the typical molecular viewpoint. The individual biomolecular pathways which underlie and directly incite the cancer state are multitudinous and varied. The focus of PWS microscopy on chromatin shape and compaction lends insight into whether and how fundamental molecular mechanisms converge into specific higher-order chromatin structures (a genomic folding code) as a truly causative agent in carcinogenesis.

11.3 Cytoskeleton

Cytoskeletal structural alterations are a hallmark of carcinogenesis. Using the microtubule inhibitor colchicine, Mutyal and colleagues were able to reverse the neoplastic changes in premalignant mucosa in azoxymethane-treated rats, as measured by LEBS, suggesting that these cytoskeletal changes contributed to the LEBS signatures (Mutyal et al., 2013). They also conducted focused array studies to show that a number of cytoskeletal proteins are involved.

11.4 Stroma

Both LEBS and PGS sample deeper structures. As discussed in the field carcinogenesis section, there is evidence that stroma may play a role in determining epithelial neoplasia. Spicer and colleagues utilized spectral imaging technology to investigate matrix crosslinking in a novel Matrigel system. They noted that increased collagen crosslinking imbued cells with aggressive features and a number of modulators such as matrix metalloproteases and lysyl oxidase-like 4 (Spicer et al., 2016).

12. PGS

EIBS was demonstrated to be increased in both the azoxymethane-treated rat and the MIN mouse (multiple intestinal neoplasia, the murine model of familial adenomatous polyposis) at timepoints that precede neoplasia. Indeed, at the earliest timepoint, the increased blood supply had a sensitivity of 94%, a specificity of 96%, and a positive predictive value of 96.8% for future neoplasia. This was followed up with a study of endoscopically normal mucosa from 37 patients, comparing those with no neoplasia to those harboring advanced adenomas elsewhere in their colons. Patients with advanced adenomas had a ~3-fold increased number of RBCs in their non-neoplastic mucosa ($P < 0.02$) (Wali et al., 2005).

The next step was to convert to an endoscopically compatible probe. This was successfully accomplished with a ~2 mm diameter probe that was used to map the colons of 222 patients, including the endoscopically normal cecum, the mid-transverse colon, and the rectum (Roy et al., 2008). If a polyp was present, readings were taken from the polyp along with 10- and 30-cm areas surrounding the lesion. Consistent with ex vivo studies, EIBS was most pronounced in the mucosa as compared to the submucosa, and was detectable within 30 cm of adenomas. EIBS correlated with adenoma size. Thus, there seemed to be two EIBS components in the colons of patients harboring adenomas—a component in the mucosa on top of a diffuse component in the submucosa.

Gomes and colleagues focused entirely on the rectal mucosa to evaluate the ability of EIBS to serve as a risk stratification tool. EIBS was evaluated using PGS in 216 patients undergoing screening colonoscopy (Gomes et al., 2009). Comparing patients with advanced adenomas to those without neoplasia showed that there was a ~50% induction in EIBS in the former, regardless of adenoma location. Using oxyhemoglobin, the area under the ROC curve was 0.88 (83% sensitivity, 82% specificity) for advanced adenomas. One intriguing phenomenon was the observation of a rectal EIBS-gender interaction, in that a dramatically better prediction of proximal neoplasia was observed in women. Rectal EIBS (oxyhemoglobin) had excellent performance characteristics (sensitivity, 100%; specificity, 76.8%; positive predictive value, 33%; and negative predictive value, 100%).

13. LEBS

LEBS initially was developed with a table-top instrument on ex vivo tissue. This was validated in both the azoxymethane-treated rat and the MIN mouse. Pilot human studies ($N = 63$) centered on one parameter, spectral slope, which was shown to be altered in the microscopically normal mucosa in proximity to the adenomas (Roy et al., 2006). To investigate the potential clinical translatability of this approach, biopsies were obtained from the normal-appearing rectal mucosa of 219 patients undergoing colonoscopy (Roy et al., 2009). LEBS signals were recorded, and four parameters characterizing the LEBS signal were linearly combined into a single marker. The LEBS signal parameters generally mirrored neoplasia progression from patients with no neoplasia, to those with 5–9 mm sized adenomas to advanced adenomas. The composite LEBS marker calculated from the LEBS signal paralleled adenoma risk status (ANOVA $P < 0.001$), and was independent of other CRC risk factors, such as benign colonic findings or clinically unimportant lesions (diminutive adenomas, hyperplastic polyps). For advanced adenomas, the LEBS marker had a sensitivity of 100%, specificity of 80%, and area under the receiver operator characteristic curve of 0.895. Leave-one-out cross-validation on an independent data set from 51 patients showed that the AUC was comparable (0.845), supporting the robustness of the findings.

This was transitioned into a fiberoptic probe by Radosevich et al. (2015). The probe diameter (3.4 mm) precluded use through a standard colonoscope, so that it was inserted either directly or via the ancillary channel of a colonoscope. Ten readings from the rectum were taken (each reading required 250 milliseconds) and thus added approximately 1 min to the procedure. There was a total of 619 patients recruited, and two LEBS parameters were assessed: the optical reduced scattering coefficient and mass density distribution factor analysis. These markers were combined utilizing logistic regression and the combined rectal LEBS marker had a sensitivity of 88% and specificity of 72% for advanced adenomas. The rectal LEBS was not confounded by location of the adenomas and worked comparably with patients harboring ≥ 3 non-advanced adenomas. Blinded and cross-validated test performance characteristics showed 88% sensitivity and 72% specificity. The independent validation set had a comparable AUROC at 0.82. Pilot data from unprepped patients was equivalent, suggesting that the rectal LEBS approach may be viable for population screening (e.g., in the primary care clinic).

14. PWS

The clinical approach with PWS involves brushings of the rectum, done endoscopically or with an anoscope/cytological brush. These are applied to microscope slides and fixed. The PWS instrument, in conjunction with scanning brightfield components, allows targeting of the nucleus, supranuclear areas, cytoplasm etc.

The PWS parameter Ld was shown to be able to accurately distinguish the intestinal mucosa of MIN mice prior to tumorigenesis from age-matched wildtype mice (Subramanian et al., 2008). The initial clinical data was obtained from hemicolectomy samples. The uninvolved mucosa of neoplastic resections (4 cm from tumor) exhibited a dramatically higher PWS (~10-fold) than that from non-neoplastic resections. Intriguingly, evaluating the tumor per se was only slightly greater (~12-fold greater than control brushings). Damania and colleagues obtained and evaluated rectal brushings from 146 patients, including those with negative colonoscopy, with diminutive adenomas, with advanced adenomas, and with germline HNPCC mutations (with or without concurrent adenocarcinoma) (Damania et al., 2012). The investigators noted a progressive increase in Ld which mirrored risk of CRC. Specifically, patients with advanced adenoma, germline HNPCC mutations, and invasive CRC had a ~2, ~3 and ~4 fold increase, respectively, in Ld with high diagnostic performance (AUROC 0.85, 0.89 and 0.92). The Ld cut-point from the training set was optimized for advanced adenomas, and prospectively applied to an independent validation cohort enriched for these cases (14 controls, and 25 with advanced adenomas). The diagnostic characteristics for the validation set (sensitivity = 74% and specificity = 83%) mirrored performance for the training set, supporting the robustness of the data.

Gladstein and colleagues furthered the work by evaluating the ability of rectal Ld to be a determinant of long-term risk in a cohort of 191 patients, including a significant number of subjects with previous history of colonic neoplasia. PWS was taken from rectal mucosa, and cataloged by surveillance colonoscopy (no lesion, low-risk adenoma and high-risk adenoma) and history of neoplasia (low-risk or high-risk). Both a history of previous advanced adenomas and current adenomas appeared to be reflected in rectal Ld; This is consonant with data indicating that even after a negative surveillance colonoscopy, there is long-term elevation in risk, as discussed above. Importantly, when rectal Ld was plotted against a literature-derived 5-year risk of advanced neoplasia, the correlation was surprisingly strong (r squared = 0.95) (Rasheed et al., 2008). This indicates that rectal Ld is a robust marker for both concurrent and future neoplasia.

15. Problems in implementing field carcinogenesis in clinical practice

Biologically, the concept of field carcinogenesis (field of injury) is robust and utilized clinically in some scenarios (e.g., post-polypectomy surveillance); however, there are potential caveats that need to be considered before there can be wider implementation for patient care. The major limitation is that it provides a probability of concurrent/future neoplasia, and does not definitely identify the lesion. The occurrence of neoplasia is a stochastic event requiring the inactivation of APC, hMLH-1, or other tumor suppressor genes. These are more likely to occur in a pro-neoplastic milieu that is reflected by field carcinogenesis. However, it is conceivable that some lower risk fields might develop neoplastic changes, whereas some higher risk milieus might not. Moreover, the impact of biomarkers on the biological heterogeneity of polyps (tubular adenomas versus serrated) or subtypes of CRC is unclear. It is possible that different biomarkers may perform better with each other. In addition, there is evidence that field carcinogenesis may be somewhat patchy, requiring adequate sampling of the rectal mucosa. These and other issues need to be kept in mind when developing biomarkers for field carcinogenesis and translation to patient care.

16. Chemoprevention via field carcinogenesis biomarkers assessed by biophotonics

Chemoprevention of colorectal neoplasia epitomizes field changes in that effective agents can suppress neoplasia throughout the colon. While there has been a myriad of agents that have been proven to prevent colon cancer, clinical implementation has been stymied by lack of intermediate biomarkers to both perform clinical validation and to personalize therapy (Umezawa et al., 2019). For clinical validation, the standard approach has been adenoma recurrence (Katona & Weiss, 2020). However, this is fraught with issues, including the need to follow a large number of patients for a minimum of 3–5 years and perform surveillance colonoscopy (with latest guidelines not meeting the necessity for clinical care). This is extraordinarily laborious and expensive. These factors are sometimes confounded by colonoscopic quality metrics (e.g., varying adenoma detection rates). Furthermore, while they are biologically the most relevant endpoints, the vast majority of adenomas never progress to CRC. Thus, other intermediate biomarkers would allow more rapid translation. As an example, from an individual perspective, chemoprevention with aspirin is well-established (and approved by USPSTF) (Bibbins-Domingo & Force, 2016); however, it works in only ~30%–50% of the population, though the entire population is at risk. Moreover, the dosage is not clear. Baron et al. showed that 81 mg of aspirin is superior to 325 mg (Baron et al., 2003), whereas Chan et al. demonstrated aspirin to have dose-dependent effects that are likely determined by various factors, including numerous pharmacogenomic influences (Chan et al., 2005). Thus, having a minimally intrusive modality to assess efficacy would be of critical importance.

In this regard, rectal markers have been widely used for chemoprevention. These include proliferation and apoptosis rates. As previously discussed, rectal biomarkers correlate with synchronous and metachronous neoplasia. With aspirin and NSAIDs, numerous studies in both sporadic and familial adenomatous polyposis patients have shown the importance of proliferation suppression and/or apoptosis induction, although there are discordant studies. Moreover, correlations with adenoma regression are unclear. On the other hand, rectal mucosal proliferation was not changed with metformin (Zell et al., 2020), whereas a randomized controlled trial showed significant benefit (Higurashi et al., 2016). With regard to rectal ACF, rodent models have shown that these are one of the most sensitive biomarkers of CRC. Takayama and colleagues evaluated 4 normal, 6 adenoma and 1 CRC patient treated for 10–12 months with sulindac (100 mg tid) (Takayama et al., 1998). The ACFs, as detected by magnification colonoscopy, decreased by ~80%–90%. On the other hand, in a subset from a randomized placebo-controlled trial of celecoxib (200 or 400 mg) for adenoma prevention (3 year), rectal ACF was at baseline at 8–12 months in 45 patients. In this nested sub-study, there was no significant modulation of ACF by celecoxib versus placebo ($P = 0.77$), which was juxtaposed by robust reduction in advanced adenomas (relative risks was 0.34–0.43).

Given the promise of spectral markers of field carcinogenesis for risk stratification, LEBS was evaluated as an intermediate biomarker for chemopreventive response. In a randomized placebo-controlled double-blinded trial, patients with a history of adenomas were randomized to aspirin (325 mg) or placebo (Roy et al., 2017). Rectal biopsies were taken at baseline colonoscopy for LEBS evaluation (via a tabletop instrument). At 3 months, FS was performed, another rectal biopsy was taken for repeat LEBS analysis, and the differences between the initial and final readings for two LEBS parameter, i.e., spectral slope (SPEC) and fractal dimension (FRAC), were determined. Previous studies had shown that SPEC decreases whereas FRAC increases in colonic field carcinogenesis. Thirty-six patients completed the study in each arm (placebo and aspirin-treated) (Roy et al., 2017). With regards to SPEC, in the aspirin group there was an increase (less neoplastic direction) that bordered on statistical significance when compared to placebo ($P=0.055$). For FRAC, there was a decrease in aspirin-treated patients (less neoplastic), although again, this did not achieve statistical significance relative to placebo ($P=0.17$). Measurement of established biomarkers such as prostaglandin E2 (PGE2) showed that the aspirin-induced increase in SPEC and decrease in FRAC was strongly correlated with reduced PGE2 (r squared of -0.55 and 0.57 respectively, $P=0.01$); there was no correlation in the placebo group. Finally, spectral markers were compared with pharmacogenomics markers. Uridine diphosphate glucuronosyltransferase (1A6UGT1A6) catalyzes glucuronidation of aspirin and thereby inactivates it. A variant allele of the UGP1A6 gene leads to slower degradation of aspirin. Earlier, Chan and colleagues, using the Nurses Health Study, found that patients with the variant, but not the wildtype, genotype had decreased colon polyps upon treatment with aspirin (Chan, Tranah, Giovannucci, Hunter, & Fuchs, 2005). In the LEBS trial (Roy et al., 2017), SPEC and FRAC both changed in an anti-neoplastic direction only in patients with the variant genotype, further supporting the potential chemopreventive relevance of the spectral markers.

17. Conclusions/future potential

The role of field carcinogenesis in the development of colorectal neoplasia is biologically robust and clinically utilized. The evidence for field carcinogenesis in CRC is unequivocal, and there is a myriad of putative biomarkers that provide biological insights into the earliest events in colon carcinogenesis. While bench-to-bedside transition has been disappointing thus far, the adenoma remains a stalwart of clinical risk stratification (e.g., post-polypectomy surveillance or paradigms with FS), underscoring the need for new and novel biomarkers.

The approach of using a minimally intrusive modality (i.e., rectal interrogation) is quite attractive for population screening. This would offer the possibility to personalize screening since approaches to date have been largely "one size fits all," leading to inefficient screening (i.e., overuse of colonoscopy). If advanced adenoma removal is considered the goal (colon cancer prevention rather than screening), then <10% of colonoscopies are fruitful. This is particularly critical given the resource limitations and the fact that a sizeable proportion of the population do not undergo screening. This has been exacerbated by the need for moving screening initiation to address the rising incidence of early onset CRC. Furthermore, this could potentially be utilized for other higher risk groups, including African Americans, those with family history of CRC, those who are obese, etc. Finally, given the need for precision medicine during chemoprevention, field carcinogenesis represents an attractive possible solution.

Issues in the future include identifying and designing biomarkers that are robust, cost-effective and easy to implement in a primary care setting (e.g., during annual physical exam). Biophotonics may be particularly compelling given that it meets all of these criteria. Evaluating structural approaches may be potentially superior to molecular approaches, due to the well-established genetic/epigenetic heterogeneity during carcinogenesis.

References

Alberts, D. S., Einspahr, J. G., Krouse, R. S., Prasad, A., Ranger-Moore, J., Hamilton, P., et al. (2007). Karyometry of the colonic mucosa. *Cancer Epidemiology, Biomarkers & Prevention, 16*(12), 2704–2716. https://doi.org/10.1158/1055-9965.EPI-07-0595.

Almassalha, L. M., Tiwari, A., Ruhoff, P. T., Stypula-Cyrus, Y., Cherkezyan, L., Matsuda, H., et al. (2017). The global relationship between chromatin physical topology, fractal structure, and gene expression. *Scientific Reports*, 7, 41061. https://doi.org/10.1038/srep41061.

Alonso, S., Dai, Y., Yamashita, K., Horiuchi, S., Dai, T., Matsunaga, A., et al. (2015). Methylation of MGMT and ADAMTS14 in normal colon mucosa: biomarkers of a field defect for cancerization preferentially targeting elder African-Americans. *Oncotarget, 6*(5), 3420–3431. https://doi.org/10.18632/oncotarget.2852.

Anti, M., Marra, G., Armelao, F., Percesepe, A., Ficarelli, R., Ricciuto, G. M., et al. (1993). Rectal epithelial cell proliferation patterns as predictors of adenomatous colorectal polyp recurrence. *Gut, 34*(4), 525–530. https://doi.org/10.1136/gut.34.4.525.

Archer, S. L. (2013). Mitochondrial dynamics—Mitochondrial fission and fusion in human diseases. *The New England Journal of Medicine, 369*(23), 2236–2251. https://doi.org/10.1056/NEJMra1215233.

Ashktorab, H., Washington, K., Zarnogi, S., Shakoori, A., Varma, S., Lee, E., et al. (2020). Determination of distinctive hypomethylated genes in African American colorectal neoplastic lesions. *Therapeutic Advances in Gastroenterology, 13*, 1756284820905482. https://doi.org/10.1177/1756284820905482. .

Atkin, W. S., Edwards, R., Kralj-Hans, I., Wooldrage, K., Hart, A. R., Northover, J. M., et al. (2010). Once-only flexible sigmoidoscopy screening in prevention of colorectal cancer: A multicentre randomised controlled trial. *Lancet*, *375*(9726), 1624–1633. https://doi.org/10.1016/S0140-6736(10)60551-X.

Baba, Y., Ishimoto, T., Kurashige, J., Iwatsuki, M., Sakamoto, Y., Yoshida, N., et al. (2016). Epigenetic field cancerization in gastrointestinal cancers. *Cancer Letters*, *375*(2), 360–366. https://doi.org/10.1016/j.canlet.2016.03.009.

Backman, V., & Roy, H. K. (2011). Light-scattering technologies for field carcinogenesis detection: A modality for endoscopic prescreening. *Gastroenterology*, *140*(1), 35–41. https://doi.org/10.1053/j.gastro.2010.11.023.

Backman, V., & Roy, H. K. (2013). Advances in biophotonics detection of field carcinogenesis for colon cancer risk stratification. *Journal of Cancer*, *4*(3), 251–261. https://doi.org/10.7150/jca.5838.

Baffy, G. (2010). Uncoupling protein-2 and cancer. *Mitochondrion*, *10*(3), 243–252. https://doi.org/10.1016/j.mito.2009.12.143.

Baker, K. T., Salk, J. J., Brentnall, T. A., & Risques, R. A. (2018). Precancer in ulcerative colitis: The role of the field effect and its clinical implications. *Carcinogenesis*, *39*(1), 11–20. https://doi.org/10.1093/carcin/bgx117.

Balaguer, F., Link, A., Lozano, J. J., Cuatrecasas, M., Nagasaka, T., Boland, C. R., et al. (2010). Epigenetic silencing of miR-137 is an early event in colorectal carcinogenesis. *Cancer Research*, *70*(16), 6609–6618. https://doi.org/10.1158/0008-5472.CAN-10-0622.

Bardou, M., Barkun, A. N., & Martel, M. (2013). Obesity and colorectal cancer. *Gut*, *62*(6), 933–947. https://doi.org/10.1136/gutjnl-2013-304701.

Baron, J. A., Cole, B. F., Sandler, R. S., Haile, R. W., Ahnen, D., Bresalier, R., et al. (2003). A randomized trial of aspirin to prevent colorectal adenomas. *The New England Journal of Medicine*, *348*(10), 891–899. https://doi.org/10.1056/NEJMoa021735.

Bernstein, C., Bernstein, H., Garewal, H., Dinning, P., Jabi, R., Sampliner, R. E., et al. (1999). A bile acid-induced apoptosis assay for colon cancer risk and associated quality control studies. *Cancer Research*, *59*(10), 2353–2357.

Biasco, G., Paganelli, G. M., Miglioli, M., Brillanti, S., Di Febo, G., Gizzi, G., et al. (1990). Rectal cell proliferation and colon cancer risk in ulcerative colitis. *Cancer Research*, *50*(4), 1156–1159.

Bibbins-Domingo, K., & Force, USPST. (2016). Aspirin use for the primary prevention of cardiovascular disease and colorectal cancer: U.S. preventive services task force recommendation statement. *Annals of Internal Medicine*, *164*(12), 836–845. https://doi.org/10.7326/M16-0577.

Boland, M. L., Chourasia, A. H., & Macleod, K. F. (2013). Mitochondrial dysfunction in cancer. *Frontiers in Oncology*, *3*, 292. https://doi.org/10.3389/fonc.2013.00292.

Braakhuis, B. J., Tabor, M. P., Kummer, J. A., Leemans, C. R., & Brakenhoff, R. H. (2003). A genetic explanation of Slaughter's concept of field cancerization: Evidence and clinical implications. *Cancer Research*, *63*(8), 1727–1730.

Brennan, J. A., Mao, L., Hruban, R. H., Boyle, J. O., Eby, Y. J., Koch, W. M., et al. (1995). Molecular assessment of histopathological staging in squamous-cell carcinoma of the head and neck. *The New England Journal of Medicine*, *332*(7), 429–435. https://doi.org/10.1056/NEJM199502163320704.

Brentnall, T. A., Crispin, D. A., Rabinovitch, P. S., Haggitt, R. C., Rubin, C. E., Stevens, A. C., et al. (1994). Mutations in the p53 gene: An early marker of neoplastic progression in ulcerative colitis. *Gastroenterology*, *107*(2), 369–378. https://doi.org/10.1016/0016-5085(94)90161-9.

Busch, E. L., Galanko, J. A., Sandler, R. S., Goel, A., & Keku, T. O. (2018). Lifestyle factors, colorectal tumor methylation, and survival among African Americans and European Americans. *Scientific Reports*, *8*(1), 9470. https://doi.org/10.1038/s41598-018-27738-x.

Cesaroni, M., Powell, J., & Sapienza, C. (2014). Validation of methylation biomarkers that distinguish normal colon mucosa of cancer patients from normal colon mucosa of patients without cancer. *Cancer Prevention Research (Philadelphia, Pa.)*, 7(7), 717–726. https://doi.org/10.1158/1940-6207.CAPR-13-0407.

Chan, A. T., Giovannucci, E. L., Meyerhardt, J. A., Schernhammer, E. S., Curhan, G. C., & Fuchs, C. S. (2005). Long-term use of aspirin and nonsteroidal anti-inflammatory drugs and risk of colorectal cancer. *Journal of the American Medical Association*, *294*(8), 914–923. https://doi.org/10.1001/jama.294.8.914.

Chan, A. T., Tranah, G. J., Giovannucci, E. L., Hunter, D. J., & Fuchs, C. S. (2005). Genetic variants in the UGT1A6 enzyme, aspirin use, and the risk of colorectal adenoma. *Journal of the National Cancer Institute*, *97*(6), 457–460. https://doi.org/10.1093/jnci/dji066.

Chen, L. C., Hao, C. Y., Chiu, Y. S., Wong, P., Melnick, J. S., Brotman, M., et al. (2004). Alteration of gene expression in normal-appearing colon mucosa of APC(min) mice and human cancer patients. *Cancer Research*, *64*(10), 3694–3700. https://doi.org/10.1158/0008-5472.CAN-03-3264.

Chen, C., Pore, N., Behrooz, A., Ismail-Beigi, F., & Maity, A. (2001). Regulation of glut1 mRNA by hypoxia-inducible factor-1. Interaction between H-ras and hypoxia. *The Journal of Biological Chemistry*, *276*(12), 9519–9525. https://doi.org/10.1074/jbc.M010144200.

Chung, S. W., Hakim, S., Siddiqui, S., & Cash, B. D. (2020). Update on flexible sigmoidoscopy, computed tomographic colonography, and capsule colonoscopy. *Gastrointestinal Endoscopy Clinics of North America*, *30*(3), 569–583. https://doi.org/10.1016/j.giec.2020.02.009.

Church, T. R., Wandell, M., Lofton-Day, C., Mongin, S. J., Burger, M., Payne, S. R., et al. (2014). Prospective evaluation of methylated SEPT9 in plasma for detection of asymptomatic colorectal cancer. *Gut*, *63*(2), 317–325. https://doi.org/10.1136/gutjnl-2012-304149.

Cruz, M. D., Ledbetter, S., Chowdhury, S., Tiwari, A. K., Momi, N., Wali, R. K., et al. (2017). Metabolic reprogramming of the premalignant colonic mucosa is an early event in carcinogenesis. *Oncotarget*, *8*(13), 20543–20557. https://doi.org/10.18632/oncotarget.16129.

Cruz, M. D., Wali, R. K., Bianchi, L. K., Radosevich, A. J., Crawford, S. E., Jepeal, L., et al. (2014). Colonic mucosal fatty acid synthase as an early biomarker for colorectal neoplasia: Modulation by obesity and gender. *Cancer Epidemiology, Biomarkers & Prevention*, *23*(11), 2413–2421. https://doi.org/10.1158/1055-9965.EPI-14-0026.

Cui, R., & Shi, X. Y. (2015). Expression of pyruvate kinase M2 in human colorectal cancer and its prognostic value. *International Journal of Clinical and Experimental Pathology*, *8*(9), 11393–11399.

Damania, D., Roy, H. K., Subramanian, H., Weinberg, D. S., Rex, D. K., Goldberg, M. J., et al. (2012). Nanocytology of rectal colonocytes to assess risk of colon cancer based on field cancerization. *Cancer Research*, *72*(11), 2720–2727. https://doi.org/10.1158/0008-5472.CAN-11-3807.

Daniel, C. R., Bostick, R. M., Flanders, W. D., Long, Q., Fedirko, V., Sidelnikov, E., et al. (2009). TGF-alpha expression as a potential biomarker of risk within the normal-appearing colorectal mucosa of patients with and without incident sporadic adenoma. *Cancer Epidemiology, Biomarkers & Prevention*, *18*(1), 65–73. https://doi.org/10.1158/1055-9965.EPI-08-0732.

Datta, S., Sherva, R. M., De La Cruz, M., Long, M. T., Roy, P., Backman, V., et al. (2018). Single nucleotide polymorphism facilitated down-regulation of the cohesin stromal antigen-1: Implications for colorectal cancer racial disparities. *Neoplasia*, *20*(3), 289–294. https://doi.org/10.1016/j.neo.2018.01.003.

Dekker, E., Tanis, P. J., Vleugels, J. L. A., Kasi, P. M., & Wallace, M. B. (2019). Colorectal cancer. *Lancet*, *394*(10207), 1467–1480. https://doi.org/10.1016/S0140-6736(19)32319-0.

Devall, M., Sun, X., Yuan, F., Cooper, G. S., Willis, J., Weisenberger, D. J., et al. (2020). Racial disparities in epigenetic aging of the right vs left colon. *Journal of the National Cancer Institute*. https://doi.org/10.1093/jnci/djaa206.

Dotto, G. P. (2014). Multifocal epithelial tumors and field cancerization: Stroma as a primary determinant. *The Journal of Clinical Investigation*, *124*(4), 1446–1453. https://doi.org/10.1172/JCI72589.

Erdrich, J., Zhang, X., Giovannucci, E., & Willett, W. (2015). Proportion of colon cancer attributable to lifestyle in a cohort of US women. *Cancer Causes & Control*, *26*(9), 1271–1279. https://doi.org/10.1007/s10552-015-0619-z.

Fedirko, V., Bostick, R. M., Flanders, W. D., Long, Q., Sidelnikov, E., Shaukat, A., et al. (2009). Effects of vitamin d and calcium on proliferation and differentiation in normal colon mucosa: A randomized clinical trial. *Cancer Epidemiology, Biomarkers & Prevention*, *18*(11), 2933–2941. https://doi.org/10.1158/1055-9965.EPI-09-0239.

Fink, S. P., Yamauchi, M., Nishihara, R., Jung, S., Kuchiba, A., Wu, K., et al. (2014). Aspirin and the risk of colorectal cancer in relation to the expression of 15-hydroxyprostaglandin dehydrogenase (HPGD). *Science Translational Medicine*, *6*(233), 233re2. https://doi.org/10.1126/scitranslmed.3008481.

Freire, P., Figueiredo, P., Cardoso, R., Manuel Donato, M., Ferreira, M., Mendes, S., et al. (2014). Predictive value of rectal aberrant crypt foci for intraepithelial neoplasia in ulcerative colitis—A cross-sectional study. *Scandinavian Journal of Gastroenterology*, *49*(10), 1219–1229. https://doi.org/10.3109/00365521.2014.951390.

Furlan, D., Sahnane, N., Carnevali, I., Cerutti, R., Uccella, S., Bertolini, V., et al. (2007). Up-regulation and stabilization of HIF-1alpha in colorectal carcinomas. *Surgical Oncology*, *16*(Suppl. 1), S25–S27. https://doi.org/10.1016/j.suronc.2007.10.014.

Gomes, A. J., Roy, H. K., Turzhitsky, V., Kim, Y., Rogers, J. D., Ruderman, S., et al. (2009). Rectal mucosal microvascular blood supply increase is associated with colonic neoplasia. *Clinical Cancer Research*, *15*(9), 3110–3117. https://doi.org/10.1158/1078-0432.CCR-08-2880.

Grady, W. M., Parkin, R. K., Mitchell, P. S., Lee, J. H., Kim, Y. H., Tsuchiya, K. D., et al. (2008). Epigenetic silencing of the intronic microRNA hsa-miR-342 and its host gene EVL in colorectal cancer. *Oncogene*, *27*(27), 3880–3888. https://doi.org/10.1038/onc.2008.10.

Green, S. E., Chapman, P., Burn, J., Burt, A. D., Bennett, M., Appleton, D. R., et al. (1998). Colonic epithelial cell proliferation in hereditary non-polyposis colorectal cancer. *Gut*, *43*(1), 85–92. https://doi.org/10.1136/gut.43.1.85.

Guda, K., Veigl, M. L., Varadan, V., Nosrati, A., Ravi, L., Lutterbaugh, J., et al. (2015). Novel recurrently mutated genes in African American colon cancers. *Proceedings of the National Academy of Sciences of the United States of America*, *112*(4), 1149–1154. https://doi.org/10.1073/pnas.1417064112.

Guinney, J., Dienstmann, R., Wang, X., de Reynies, A., Schlicker, A., Soneson, C., et al. (2015). The consensus molecular subtypes of colorectal cancer. *Nature Medicine*, *21*(11), 1350–1356. https://doi.org/10.1038/nm.3967.

Gupta, S., Lieberman, D., Anderson, J. C., Burke, C. A., Dominitz, J. A., Kaltenbach, T., et al. (2020). Recommendations for follow-up after colonoscopy and polypectomy: A consensus update by the US multi-society task force on colorectal cancer. *Gastroenterology*, *158*(4), 1131–1153.e5. https://doi.org/10.1053/j.gastro.2019.10.026.

Hanahan, D., & Weinberg, R. A. (2011). Hallmarks of cancer: The next generation. *Cell*, *144*(5), 646–674. https://doi.org/10.1016/j.cell.2011.02.013.

Hao, C. Y., Moore, D. H., Chiu, Y. S., Wong, P., Bennington, J. L., Smith, A. P., et al. (2005). Altered gene expression in normal colonic mucosa of individuals with polyps of the colon. *Diseases of the Colon and Rectum*, *48*(12), 2329–2335. https://doi.org/10.1007/s10350-005-0153-2.

Hao, C. Y., Moore, D. H., Wong, P., Bennington, J. L., Lee, N. M., & Chen, L. C. (2005). Alteration of gene expression in macroscopically normal colonic mucosa from individuals with a family history of sporadic colon cancer. *Clinical Cancer Research*, *11*(4), 1400–1407. https://doi.org/10.1158/1078-0432.CCR-04-1942.

Higurashi, T., Hosono, K., Takahashi, H., Komiya, Y., Umezawa, S., Sakai, E., et al. (2016). Metformin for chemoprevention of metachronous colorectal adenoma or polyps in post-polypectomy patients without diabetes: A multicentre double-blind, placebo-controlled, randomised phase 3 trial. *The Lancet Oncology*, *17*(4), 475–483. https://doi.org/10.1016/S1470-2045(15)00565-3.

Hixson, L. J., Garewal, H. S., McGee, D. L., Sloan, D., Fennerty, M. B., Sampliner, R. E., et al. (1993). Ornithine decarboxylase and polyamines in colorectal neoplasia and mucosa. *Cancer Epidemiology, Biomarkers & Prevention*, *2*(4), 369–374.

Holme, O., Loberg, M., Kalager, M., Bretthauer, M., & Hoff, G. (2018). Long-term effectiveness of sigmoidoscopy screening in women and men. *Annals of Internal Medicine*, *169*(9), 663–664. https://doi.org/10.7326/L18-0512.

Holme, O., Schoen, R. E., Senore, C., Segnan, N., Hoff, G., Loberg, M., et al. (2017). Effectiveness of flexible sigmoidoscopy screening in men and women and different age groups: Pooled analysis of randomised trials. *BMJ*, *356*, i6673. https://doi.org/10.1136/bmj.i6673.

Imperiale, T. F., Ransohoff, D. F., Itzkowitz, S. H., Levin, T. R., Lavin, P., Lidgard, G. P., et al. (2014). Multitarget stool DNA testing for colorectal-cancer screening. *The New England Journal of Medicine*, *370*(14), 1287–1297. https://doi.org/10.1056/NEJMoa1311194.

Jones, R. M., Devers, K. J., Kuzel, A. J., & Woolf, S. H. (2010). Patient-reported barriers to colorectal cancer screening: A mixed-methods analysis. *American Journal of Preventive Medicine*, *38*(5), 508–516. https://doi.org/10.1016/j.amepre.2010.01.021.

Kaldma, A., Klepinin, A., Chekulayev, V., Mado, K., Shevchuk, I., Timohhina, N., et al. (2014). An in situ study of bioenergetic properties of human colorectal cancer: The regulation of mitochondrial respiration and distribution of flux control among the components of ATP synthasome. *The International Journal of Biochemistry & Cell Biology*, *55*, 171–186. https://doi.org/10.1016/j.biocel.2014.09.004.

Katona, B. W., & Weiss, J. M. (2020). Chemoprevention of colorectal cancer. *Gastroenterology*, *158*(2), 368–388. https://doi.org/10.1053/j.gastro.2019.06.047.

Keku, T. O., Amin, A., Galanko, J., Martin, C., Schliebe, B., & Sandler, R. S. (2008). Apoptosis in normal rectal mucosa, baseline adenoma characteristics, and risk of future adenomas. *Cancer Epidemiology, Biomarkers & Prevention*, *17*(2), 306–310. https://doi.org/10.1158/1055-9965.EPI-07-0066.

Keller, J. J., Offerhaus, G. J., Polak, M., Goodman, S. N., Zahurak, M. L., Hylind, L. M., et al. (1999). Rectal epithelial apoptosis in familial adenomatous polyposis patients treated with sulindac. *Gut*, *45*(6), 822–828. https://doi.org/10.1136/gut.45.6.822.

Khosraviani, K., Williamson, K., & Hamilton, P. (1996). Apoptosis and proliferation: The balance. *Gastroenterology*, *110*(4), 1323–1324. https://doi.org/10.1053/gast.1996.v110.agast961323b.

Kopelovich, L., Henson, D. E., Gazdar, A. F., Dunn, B., Srivastava, S., Kelloff, G. J., et al. (1999). Surrogate anatomic/functional sites for evaluating cancer risk: An extension of the field effect. *Clinical Cancer Research*, *5*(12), 3899–3905.

Kuai, X. Y., Ji, Z. Y., & Zhang, H. J. (2010). Mitochondrial uncoupling protein 2 expression in colon cancer and its clinical significance. *World Journal of Gastroenterology*, *16*(45), 5773–5778. https://doi.org/10.3748/wjg.v16.i45.5773.

Kunte, D. P., DelaCruz, M., Wali, R. K., Menon, A., Du, H., Stypula, Y., et al. (2012). Dysregulation of microRNAs in colonic field carcinogenesis: Implications for screening. *PLoS One*, 7(9), e45591. https://doi.org/10.1371/journal.pone.0045591.

Leclerc, D., Levesque, N., Cao, Y., Deng, L., Wu, Q., Powell, J., et al. (2013). Genes with aberrant expression in murine preneoplastic intestine show epigenetic and expression changes in normal mucosa of colon cancer patients. *Cancer Prevention Research (Philadelphia, Pa.)*, *6*(11), 1171–1181. https://doi.org/10.1158/1940-6207.CAPR-13-0198.

Lewis, J. D., Ng, K., Hung, K. E., Bilker, W. B., Berlin, J. A., Brensinger, C., et al. (2003). Detection of proximal adenomatous polyps with screening sigmoidoscopy: A systematic review and meta-analysis of screening colonoscopy. *Archives of Internal Medicine*, *163*(4), 413–420. https://doi.org/10.1001/archinte.163.4.413.

Liberti, M. V., & Locasale, J. W. (2016). The Warburg effect: How does it benefit cancer cells? *Trends in Biochemical Sciences*, *41*(3), 211–218. https://doi.org/10.1016/j.tibs.2015.12.001.

Lieberman, D. A., Rex, D. K., Winawer, S. J., Giardiello, F. M., Johnson, D. A., & Levin, T. R. (2012). Guidelines for colonoscopy surveillance after screening and polypectomy: A consensus update by the US multi-society task force on colorectal cancer. *Gastroenterology*, *143*(3), 844–857. https://doi.org/10.1053/j.gastro.2012.06.001.

Link, A., Balaguer, F., Shen, Y., Nagasaka, T., Lozano, J. J., Boland, C. R., et al. (2010). Fecal MicroRNAs as novel biomarkers for colon cancer screening. *Cancer Epidemiology, Biomarkers & Prevention*, *19*(7), 1766–1774. https://doi.org/10.1158/1055-9965.EPI-10-0027.

Lochhead, P., Chan, A. T., Nishihara, R., Fuchs, C. S., Beck, A. H., Giovannucci, E., et al. (2015). Etiologic field effect: Reappraisal of the field effect concept in cancer predisposition and progression. *Modern Pathology*, *28*(1), 14–29. https://doi.org/10.1038/modpathol.2014.81.

Manna, S. K., Tanaka, N., Krausz, K. W., Haznadar, M., Xue, X., Matsubara, T., et al. (2014). Biomarkers of coordinate metabolic reprogramming in colorectal tumors in mice and humans. *Gastroenterology*, *146*(5), 1313–1324. https://doi.org/10.1053/j.gastro.2014.01.017.

May, D., Pan, S., Crispin, D. A., Lai, K., Bronner, M. P., Hogan, J., et al. (2011). Investigating neoplastic progression of ulcerative colitis with label-free comparative proteomics. *Journal of Proteome Research*, *10*(1), 200–209. https://doi.org/10.1021/pr100574p.

McGarrity, T. J., & Peiffer, L. P. (1994). Protein kinase C activity as a potential marker for colorectal neoplasia. *Digestive Diseases and Sciences*, *39*(3), 458–463. https://doi.org/10.1007/BF02088328.

Meyskens, F. L., Jr., McLaren, C. E., Pelot, D., Fujikawa-Brooks, S., Carpenter, P. M., Hawk, E., et al. (2008). Difluoromethylornithine plus sulindac for the prevention of sporadic colorectal adenomas: A randomized placebo-controlled, double-blind trial. *Cancer Prevention Research (Philadelphia, Pa.)*, *1*(1), 32–38. https://doi.org/10.1158/1940-6207.CAPR-08-0042.

Milano, A., Bianco, M. A., Buri, L., Cipolletta, L., Grossi, E., Rotondano, G., et al. (2019). Metabolic syndrome is a risk factor for colorectal adenoma and cancer: A study in a white population using the harmonized criteria. *Therapeutic Advances in Gastroenterology*, *12*, 1756284819867839. https://doi.org/10.1177/1756284819867839.

Muthukrishnan, M., Arnold, L. D., & James, A. S. (2019). Patients' self-reported barriers to colon cancer screening in federally qualified health center settings. *Preventive Medical Reports*, *15*, 100896. https://doi.org/10.1016/j.pmedr.2019.100896.

Mutyal, N. N., Radosevich, A., Tiwari, A. K., Stypula, Y., Wali, R., Kunte, D., et al. (2013). Biological mechanisms underlying structural changes induced by colorectal field carcinogenesis measured with low-coherence enhanced backscattering (LEBS) spectroscopy. *PLoS One*, *8*(2), e57206. https://doi.org/10.1371/journal.pone.0057206.

National Center for Health Statistics (US). (2017). *Health, United States, 2016: With chartbook on long-term trends in health*. Hyattsville (MD): Health, United States.

Nimri, L., Saadi, J., Peri, I., Yehuda-Shnaidman, E., & Schwartz, B. (2015). Mechanisms linking obesity to altered metabolism in mice colon carcinogenesis. *Oncotarget*, *6*(35), 38195–38209. https://doi.org/10.18632/oncotarget.5561.

Okugawa, Y., Grady, W. M., & Goel, A. (2015). Epigenetic alterations in colorectal cancer: Emerging biomarkers. *Gastroenterology*, *149*(5), 1204–1225.e12. https://doi.org/10.1053/j.gastro.2015.07.011.

Overbeek, J. A., Kuiper, J. G., van der Heijden, A., Labots, M., Haug, U., Herings, R. M. C., et al. (2019). Sex- and site-specific differences in colorectal cancer risk among people with type 2 diabetes. *International Journal of Colorectal Disease*, *34*(2), 269–276. https://doi.org/10.1007/s00384-018-3191-7.

Patel, A., Tripathi, G., Gopalakrishnan, K., Williams, N., & Arasaradnam, R. P. (2015). Field cancerisation in colorectal cancer: A new frontier or pastures past? *World Journal of Gastroenterology*, *21*(13), 3763–3772. https://doi.org/10.3748/wjg.v21.i13.3763.

Paun, B. C., Kukuruga, D., Jin, Z., Mori, Y., Cheng, Y., Duncan, M., et al. (2010). Relation between normal rectal methylation, smoking status, and the presence or absence of colorectal adenomas. *Cancer*, *116*(19), 4495–4501. https://doi.org/10.1002/cncr.25348.

Payne, C. M., Holubec, H., Bernstein, C., Bernstein, H., Dvorak, K., Green, S. B., et al. (2005). Crypt-restricted loss and decreased protein expression of cytochrome C oxidase subunit I as potential hypothesis-driven biomarkers of colon cancer risk. *Cancer Epidemiology, Biomarkers & Prevention*, *14*(9), 2066–2075. https://doi.org/10.1158/1055-9965.EPI-05-0180.

Pilonis, N. D., Bugajski, M., Wieszczy, P., Franczyk, R., Didkowska, J., Wojciechowska, U., et al. (2020). Long-term colorectal cancer incidence and mortality after a single negative screening colonoscopy. *Annals of Internal Medicine*, *173*(2), 81–91. https://doi.org/10.7326/M19-2477.

Platz, E. A., Willett, W. C., Colditz, G. A., Rimm, E. B., Spiegelman, D., & Giovannucci, E. (2000). Proportion of colon cancer risk that might be preventable in a cohort of middle-aged US men. *Cancer Causes & Control*, *11*(7), 579–588. https://doi.org/10.1023/a:1008999232442.

Polley, A. C., Mulholland, F., Pin, C., Williams, E. A., Bradburn, D. M., Mills, S. J., et al. (2006). Proteomic analysis reveals field-wide changes in protein expression in the morphologically normal mucosa of patients with colorectal neoplasia. *Cancer Research*, *66*(13), 6553–6562. https://doi.org/10.1158/0008-5472.CAN-06-0534.

Ponz de Leon, M., Roncucci, L., Di Donato, P., Tassi, L., Smerieri, O., Amorico, M. G., et al. (1988). Pattern of epithelial cell proliferation in colorectal mucosa of normal subjects and of patients with adenomatous polyps or cancer of the large bowel. *Cancer Research*, *48*(14), 4121–4126.

Radosevich, A. J., Mutyal, N. N., Eshein, A., Nguyen, T. Q., Gould, B., Rogers, J. D., et al. (2015). Rectal optical markers for in vivo risk stratification of premalignant colorectal lesions. *Clinical Cancer Research*, *21*(19), 4347–4355. https://doi.org/10.1158/1078-0432.CCR-15-0136.

Ranger-Moore, J., Frank, D., Lance, P., Alberts, D., Yozwiak, M., Bartels, H. G., et al. (2005). Karyometry in rectal mucosa of patients with previous colorectal adenomas. *Analytical and Quantitative Cytology and Histology*, *27*(3), 134–142.

Rasheed, S., McDonald, P. J., Northover, J. M., & Guenther, T. (2008). Angiogenesis and hypoxic factors in colorectal cancer. *Pathology, Research and Practice*, *204*(7), 501–510. https://doi.org/10.1016/j.prp.2008.03.005.

Robertson, D. J., Burke, C. A., Welch, H. G., Haile, R. W., Sandler, R. S., Greenberg, E. R., et al. (2009). Using the results of a baseline and a surveillance colonoscopy to predict recurrent adenomas with high-risk characteristics. *Annals of Internal Medicine*, *151*(2), 103–109. https://doi.org/10.7326/0003-4819-151-2-200907210-00007.

Rooney, P. S., Robinson, M. H., Clarke, P. A., Hardcastle, J. D., & Armitage, N. C. (1993). Individuals with a strong family history of colorectal cancer demonstrate abnormal rectal mucosal proliferation. *The British Journal of Surgery*, *80*(2), 249–251. https://doi.org/10.1002/bjs.1800800246.

Roy, H. K., Gomes, A., Turzhitsky, V., Goldberg, M. J., Rogers, J., Ruderman, S., et al. (2008). Spectroscopic microvascular blood detection from the endoscopically normal colonic mucosa: Biomarker for neoplasia risk. *Gastroenterology*, *135*(4), 1069–1078. https://doi.org/10.1053/j.gastro.2008.06.046.

Roy, H. K., Kim, Y. L., Liu, Y., Wali, R. K., Goldberg, M. J., Turzhitsky, V., et al. (2006). Risk stratification of colon carcinogenesis through enhanced backscattering spectroscopy analysis of the uninvolved colonic mucosa. *Clinical Cancer Research*, *12*(3 Pt. 1), 961–968. https://doi.org/10.1158/1078-0432.CCR-05-1605.

Roy, H. K., Turzhitsky, V., Kim, Y., Goldberg, M. J., Watson, P., Rogers, J. D., et al. (2009). Association between rectal optical signatures and colonic neoplasia: Potential applications for screening. *Cancer Research*, *69*(10), 4476–4483. https://doi.org/10.1158/0008-5472.CAN-08-4780.

Roy, H. K., Turzhitsky, V., Wali, R., Radosevich, A. J., Jovanovic, B., Della'Zanna, G., et al. (2017). Spectral biomarkers for chemoprevention of colonic neoplasia: A placebo-controlled double-blinded trial with aspirin. *Gut*, *66*(2), 285–292. https://doi.org/10.1136/gutjnl-2015-309996.

Ruderman, S., Eshein, A., Valuckaite, V., Dougherty, U., Almoghrabi, A., Gomes, A., et al. (2018). Early increase in blood supply (EIBS) is associated with tumor risk in the azoxymethane model of colon cancer. *BMC Cancer*, *18*(1), 814. https://doi.org/10.1186/s12885-018-4709-7.

Schoen, R. E., Pinsky, P. F., Weissfeld, J. L., Yokochi, L. A., Church, T., Laiyemo, A. O., et al. (2012). Colorectal-cancer incidence and mortality with screening flexible sigmoidoscopy. *The New England Journal of Medicine*, *366*(25), 2345–2357. https://doi.org/10.1056/NEJMoa1114635.

Schoenfeld, P., Cash, B., Flood, A., Dobhan, R., Eastone, J., Coyle, W., et al. (2005). Colonoscopic screening of average-risk women for colorectal neoplasia. *The New England Journal of Medicine*, *352*(20), 2061–2068. https://doi.org/10.1056/NEJMoa042990.

Shen, Y. M., Arbman, G., Olsson, B., & Sun, X. F. (2011). Overexpression of GLUT1 in colorectal cancer is independently associated with poor prognosis. *The International Journal of Biological Markers*, *26*(3), 166–172. https://doi.org/10.5301/JBM.2011.8550.

Sidelnikov, E., Bostick, R. M., Flanders, W. D., Long, Q., & Seabrook, M. E. (2009). Colorectal mucosal expression of MSH2 as a potential biomarker of risk for colorectal neoplasms. *Cancer Epidemiology, Biomarkers & Prevention*, *18*(11), 2965–2973. https://doi.org/10.1158/1055-9965.EPI-09-0519.

Singh, H., Turner, D., Xue, L., Targownik, L. E., & Bernstein, C. N. (2006). Risk of developing colorectal cancer following a negative colonoscopy examination: Evidence for a 10-year interval between colonoscopies. *Journal of the American Medical Association*, *295*(20), 2366–2373. https://doi.org/10.1001/jama.295.20.2366.

Slaughter, D. P., Southwick, H. W., & Smejkal, W. (1953). Field cancerization in oral stratified squamous epithelium; clinical implications of multicentric origin. *Cancer*, *6*(5), 963–968. https://doi.org/10.1002/1097-0142(195309)6:5<963::aid-cncr2820060515->3.0.co;2-q.

Song, M., Chan, A. T., & Sun, J. (2020). Influence of the gut microbiome, diet, and environment on risk of colorectal cancer. *Gastroenterology*, *158*(2), 322–340. https://doi.org/10.1053/j.gastro.2019.06.048.

Spicer, G. L., Azarin, S. M., Yi, J., Young, S. T., Ellis, R., Bauer, G. M., et al. (2016). Detection of extracellular matrix modification in cancer models with inverse spectroscopic optical coherence tomography. *Physics in Medicine and Biology*, *61*(19), 6892–6904. https://doi.org/10.1088/0031-9155/61/19/6892.

Steiling, K., Ryan, J., Brody, J. S., & Spira, A. (2008). The field of tissue injury in the lung and airway. *Cancer Prevention Research (Philadelphia, Pa.)*, *1*(6), 396–403. https://doi.org/10.1158/1940-6207.CAPR-08-0174.

Stypula-Cyrus, Y., Damania, D., Kunte, D. P., Cruz, M. D., Subramanian, H., Roy, H. K., et al. (2013). HDAC up-regulation in early colon field carcinogenesis is involved in cell tumorigenicity through regulation of chromatin structure. *PLoS One*, *8*(5), e64600. https://doi.org/10.1371/journal.pone.0064600.

Subramanian, H., Pradhan, P., Liu, Y., Capoglu, I. R., Li, X., Rogers, J. D., et al. (2008). Optical methodology for detecting histologically unapparent nanoscale consequences of genetic alterations in biological cells. *Proceedings of the National Academy of Sciences of the United States of America*, *105*(51), 20118–20123. https://doi.org/10.1073/pnas.0804723105.

Sullivan, B. A., Redding, T. S., Hauser, E. R., Gellad, Z. F., Qin, X., Gupta, S., et al. (2020). High-risk adenomas at screening colonoscopy remain predictive of future high-risk adenomas despite an intervening negative colonoscopy. *The American Journal of Gastroenterology*, *115*(8), 1275–1282. https://doi.org/10.14309/ajg.0000000000000677.

Takayama, T., Katsuki, S., Takahashi, Y., Ohi, M., Nojiri, S., Sakamaki, S., et al. (1998). Aberrant crypt foci of the colon as precursors of adenoma and cancer. *The New England Journal of Medicine*, *339*(18), 1277–1284. https://doi.org/10.1056/NEJM199810293391803.

Thyagarajan, B., Wang, R., Barcelo, H., Koh, W. P., & Yuan, J. M. (2012). Mitochondrial copy number is associated with colorectal cancer risk. *Cancer Epidemiology, Biomarkers & Prevention*, *21*(9), 1574–1581. https://doi.org/10.1158/1055-9965.EPI-12-0138-T.

Tiwari, A. K., Crawford, S. E., Radosevich, A., Wali, R. K., Stypula, Y., Kunte, D. P., et al. (2011). Neo-angiogenesis and the premalignant micro-circulatory augmentation of early colon carcinogenesis. *Cancer Letters*, *306*(2), 205–213. https://doi.org/10.1016/j.canlet.2011.03.008.

Toiyama, Y., Okugawa, Y., Kondo, S., Okita, Y., Araki, T., Kusunoki, K., et al. (2018). Comprehensive analysis identifying aberrant DNA methylation in rectal mucosa from ulcerative colitis patients with neoplasia. *Oncotarget*, *9*(69), 33149–33159. https://doi.org/10.18632/oncotarget.26032.

Toiyama, Y., Okugawa, Y., Tanaka, K., Araki, T., Uchida, K., Hishida, A., et al. (2017). A panel of methylated microRNA biomarkers for identifying high-risk patients with ulcerative colitis-associated colorectal cancer. *Gastroenterology*, *153*(6), 1634–1646.e8. https://doi.org/10.1053/j.gastro.2017.08.037.

Umezawa, S., Higurashi, T., Komiya, Y., Arimoto, J., Horita, N., Kaneko, T., et al. (2019). Chemoprevention of colorectal cancer: Past, present, and future. *Cancer Science*, *110*(10), 3018–3026. https://doi.org/10.1111/cas.14149.

Ussakli, C. H., Ebaee, A., Binkley, J., Brentnall, T. A., Emond, M. J., Rabinovitch, P. S., et al. (2013). Mitochondria and tumor progression in ulcerative colitis. *Journal of the National Cancer Institute*, *105*(16), 1239–1248. https://doi.org/10.1093/jnci/djt167.

Uttam, S., Hashash, J. G., LaFace, J., Binion, D., Regueiro, M., Hartman, D. J., et al. (2019). Three-dimensional nanoscale nuclear architecture mapping of rectal biopsies detects colorectal neoplasia in patients with inflammatory bowel disease. *Cancer Prevention Research (Philadelphia, Pa.)*, *12*(8), 527–538. https://doi.org/10.1158/1940-6207.CAPR-19-0024.

Vucenik, I., Gotovac, J., Druzijanic, N., & Shamsuddin, A. M. (2001). Usefulness of galactose oxidase-Schiff test in rectal mucus for screening of colorectal malignancy. *Anticancer Research, 21*(2B), 1247–1255.

Wali, R. K., Momi, N., Dela Cruz, M., Calderwood, A. H., Stypula-Cyrus, Y., Almassalha, L., et al. (2016). Higher order chromatin modulator cohesin SA1 is an early biomarker for colon carcinogenesis: Race-specific implications. *Cancer Prevention Research (Philadelphia, Pa.), 9*(11), 844–854. https://doi.org/10.1158/1940-6207.CAPR-16-0054.

Wali, R. K., Roy, H. K., Kim, Y. L., Liu, Y., Koetsier, J. L., Kunte, D. P., et al. (2005). Increased microvascular blood content is an early event in colon carcinogenesis. *Gut, 54*(5), 654–660. https://doi.org/10.1136/gut.2004.056010.

Wang, T., Maden, S. K., Luebeck, G. E., Li, C. I., Newcomb, P. A., Ulrich, C. M., et al. (2020). Dysfunctional epigenetic aging of the normal colon and colorectal cancer risk. *Clinical Epigenetics, 12*(1), 5. https://doi.org/10.1186/s13148-019-0801-3.

Wang, J., Wang, H., Liu, A., Fang, C., Hao, J., & Wang, Z. (2015). Lactate dehydrogenase A negatively regulated by miRNAs promotes aerobic glycolysis and is increased in colorectal cancer. *Oncotarget, 6*(23), 19456–19468. https://doi.org/10.18632/oncotarget.3318.

Wei, E. K., Ma, J., Pollak, M. N., Rifai, N., Fuchs, C. S., Hankinson, S. E., et al. (2005). A prospective study of C-peptide, insulin-like growth factor-I, insulin-like growth factor binding protein-1, and the risk of colorectal cancer in women. *Cancer Epidemiology, Biomarkers & Prevention, 14*(4), 850–855. https://doi.org/10.1158/1055-9965.EPI-04-0661.

Zell, J. A., McLaren, C. E., Morgan, T. R., Lawson, M. J., Rezk, S., Albers, C. G., et al. (2020). A phase IIa trial of metformin for colorectal cancer risk reduction among individuals with history of colorectal adenomas and elevated body mass index. *Cancer Prevention Research (Philadelphia, Pa.), 13*(2), 203–212. https://doi.org/10.1158/1940-6207.CAPR-18-0262.

Zhdanov, A. V., Okkelman, I. A., Collins, F. W., Melgar, S., & Papkovsky, D. B. (2015). A novel effect of DMOG on cell metabolism: Direct inhibition of mitochondrial function precedes HIF target gene expression. *Biochimica et Biophysica Acta, 1847*(10), 1254–1266. https://doi.org/10.1016/j.bbabio.2015.06.016.

Zisman, A. L., Nickolov, A., Brand, R. E., Gorchow, A., & Roy, H. K. (2006). Associations between the age at diagnosis and location of colorectal cancer and the use of alcohol and tobacco: Implications for screening. *Archives of Internal Medicine, 166*(6), 629–634. https://doi.org/10.1001/archinte.166.6.629.

CHAPTER NINE

Modeling colorectal cancers using multidimensional organoids

Ibrahim M. Sayed[a,†,§], Amer Ali Abd El-Hafeez[b,‡,§], Priti P. Maity[b,§], Soumita Das[a,c,f,*], and Pradipta Ghosh[b,c,d,e,f,*]

[a]Department of Pathology, University of California, San Diego, CA, United States
[b]Department of Cellular and Molecular Medicine, University of California, San Diego, CA, United States
[c]Rebecca and John Moore Comprehensive Cancer Center, University of California, San Diego, CA, United States
[d]Department of Medicine, University of California, San Diego, CA, United States
[e]Veterans Affairs Medical Center, San Diego, CA, United States
[f]HUMANOID Center of Research Excellence (CoRE), University of California, San Diego, CA, United States
*Corresponding authors: e-mail address: sodas@health.ucsd.edu; prghosh@health.ucsd.edu

Contents

† Department of Medical Microbiology and Immunology, Faculty of Medicine, Assiut University, Asyut, Egypt.
‡ Pharmacology and Experimental Oncology Unit, Cancer Biology Department, National Cancer Institute, Cairo University, Cairo, Egypt.
§ Equal contribution.

Advances in Cancer Research, Volume 151
ISSN 0065-230X
https://doi.org/10.1016/bs.acr.2021.02.005

Abstract

Organoids have revolutionized cancer research as highly adaptable models that enable an array of experimental techniques to interrogate tissue morphology and function. Because they preserve the genetic, phenotypic, and behavioral traits of their source tissue, organoids have gained traction as the most relevant models for drug discovery, tracking therapeutic response and for personalized medicine. As organoids are indisputably becoming a mainstay of cancer research, this review specifically addresses how colon-derived organoids can be perfected as multidimensional, scalable, reproducible models of healthy, pre-neoplastic and neoplastic conditions of the colon and for use in high-throughput "Phase-0" human clinical trials-in-a-dish.

1. Introduction

Colorectal cancer (CRC) is the most frequently diagnosed heterogeneous disease in the digestive tract; it is the second most common cause of cancer-related death in the United States and the third leading cause worldwide (Siegel et al., 2020; Yan et al., 2020). Most deaths from CRCs can be attributed to cancer recurrence after initial treatment, which presents as distant metastases in secondary sites such as the liver or lung. Although the total global incidence of CRCs has dropped in the last decade, the incidence of early-onset colorectal cancer (EOCRC), defined as CRCs in patients aged 50 or younger, has been on the rise (Mauri et al., 2019). If left untreated, the 5-year survival rate of patients with metastatic CRC can be as low as 5% (Engel et al., 2020).

Like most cancers, CRCs are comprised of a group of molecularly heterogeneous subtypes, each characterized by a range of genomic and epigenomic alterations and distinct tumor drivers. Such heterogeneity makes the standard one-size-fits-all approach to CRC-treatment ineffective. Another major challenge is that the pathogenic mechanisms of some CRCs, such as EOCRCs, still remain poorly understood and insights into their molecular characterization remain scant. Without such insights, the development of relevant preclinical models also remains unfulfilled. Thus, the need for comprehensive studies that delineate the underlying molecular basis for disease pathogenesis using physiologically relevant human preclinical model systems is both urgent and unmet. Such studies could not only reveal disease mechanisms but also ensure that the translational potential of the discoveries remains high.

Since the ability to culture human intestinal organoids has been realized, several teams have established a library of patient-derived CRC organoid lines, encompassing different histological subtypes and clinical stages, as preclinical models for downstream genomic profiling (Sato et al., 2011; Schütte et al., 2017; van de Wetering et al., 2015; Weeber et al., 2015; Yan et al., 2020). Given the fact that sporadic CRCs arising through the classical adenoma-carcinoma sequence, with mutational activation of the *WNT* signaling pathway in ~80% of CRCs, the rationalized addition of *WNT*-3A and R-spondin1 in the culture medium can support normal organoid growth, and their withdrawal can prevent overgrowth. These insights have increased the success rate of tumor organoid establishment (van de Wetering et al., 2015). As organoids are indisputably becoming a mainstay of cancer research, we dedicate this review to not only highlight the challenges in modeling CRC initiation and progression and summarize what has been accomplished so far but also outline strategies that can help us realize the full potential of organoid-based CRC research by building multidimensional, scalable, reproducible models of healthy, pre-neoplastic and neoplastic conditions of the colon.

2. Challenges in modeling CRC initiation and progression

2.1 Complex etiopathology

While the etiopathology of CRC is not well known, it is assumed that CRCs are initiated by a cascade of DNA mutations within epithelial cells (Allen & Sears, 2019). It is also considered that some external stimuli like bacteria, and fungal toxins, incorporation of exoDNA into epithelial DNA and micro-environmental factors may alter the genetic and epigenetic signature of the cellular genome and may propel early to intermediate to late steps neoplastic progression of this disease (Allen & Sears, 2019; Bardhan & Liu, 2013; Chan & Giovannucci, 2010). Metastasis of CRCs, on the other hand, is fueled mostly by epigenetic mechanisms, believed to be of multi-factorial triggers, the exact nature of which is less well understood (Bardhan & Liu, 2013). Epigenetic alterations that have been observed include DNA methylation, histone modifications, chromatin remodeling and non-coding RNAs, affect every aspect of tumor development from initiation to metastasis (Ju et al., 2011; Vaiopoulos, Athanasoula, & Papavassiliou, 2014). Without a complete understanding of the triggers or sequence of these events, their cause-effect relationships, or if they occur concomitantly and synergize, it is difficult to model such progressive events *ex vivo*. Informed models of tumor initiation as well as metastatic progression have not been realized, and whatever exists remains rudimentary at best.

2.2 Molecular subtypes

CRC is a heterogeneous disease; its subtypes carry distinct prognostic and predictive information that is associated with different patient outcomes (Inamura, 2018). CRCs are classified on the basis of their comprehensive molecular characterizations into The Cancer Genome Atlas (TCGA) and consensus molecular subtype (CMS). TCGA is further categorized into hypermutated and non-hypermutated CRC subtypes according to their DNA copy number, exome sequences, promoter methylation, messenger RNA as well as microRNA expression patterns. For CMS classification, gene expression profiles and TCGA data are used together to generate a consensus molecular subtypes of CRCs (Budinska et al., 2013; Marisa et al., 2013; Sadanandam et al., 2013). CRC subtypes include CMS1 (MSI immune, 14%), CMS2 (canonical, 37%), CMS3 (metabolic, 13%) and CMS4 (mesenchymal, 23%); they are categorized based on whether they are microsatellite proficient (i.e., stable; MSS) or deficient (i.e., unstable; MSI), and whether they have a CpG island methylator phenotype (CIMP) or somatic copy number alteration (SCNA) phenotype. CRCs are also categorized into distinct subtypes by considering their key molecular features and clinical outcomes. Based on their distinct clinicopathological features with regard to *MSI, CIMP, BRAF and KRAS* mutation status five distinct subtypes of CRCs have been recognized (Inamura, 2018; Phipps et al., 2015). Type 1 and 2 subtypes demonstrate a serrated morphology, type 3 reflects an alternate serrated pathway with origins in *KRAS*-mutated adenomas, type 4 subtype represents the origin of CRC from traditional adenoma-carcinoma sequence and type 5 subtype indicates possible Lynch syndrome with a high prevalence of CRC family history (Guinney et al., 2015). Patients with CRC subtypes 1 and 5 (MSI high) have the most favorable outcomes, whereas those with subtype 2 (MSS/MSI-low, *CIMP*-positive, *BRAF*-mutated, *KRAS*-mutation–negative) have the highest mortality (Jass, 2007; Phipps et al., 2015). The broad spectrum in genetic changes and mutations in various subtypes have created a gap between cancer genetics and patient treatment. These circumstances necessitate the creation of large genotyped phenotyped biobanks to generate reproducible findings. In this aspect, patient-derived organoids can help create a "living biobank," which can minimize the gap between the disease and the experimental model, enhance the precision and personalization in the process of biomarker/drug discovery.

2.3 Lacking models for intermediate (pre-neoplastic) lesions

CRCs are the end results of a multistage process that is slow, and spans over years, as aberrant crypts eventually turn into malignant polyps. Because about 85% of CRCs begin as pre-cancerous polyps, the early detection and removal of polyps can notably reduce the risk of cancer (Gopalappa, Aydogan-Cremaschi, Das, & Orcun, 2011). But polyps are not all one and the same; several subtypes of polyps exist, e.g., hyperplastic polyps (HPs), sessile serrated adenoma/polyps (SSAPs) and traditional serrated adenoma (TSA) (Snover, Jass, Fenoglio-Preiser, & Batts, 2005). It is known that HPs polyps represent very low malignant potential where SSA/P and TSA both correspond as a precursor of cancer via the serrated neoplastic pathway (O'Brien et al., 2006; Spring et al., 2006; Uraoka et al., 2015). The SSA/P polyps are also recognized as precursors of sporadic colorectal cancer with findings of microsatellite instability and aberrant DNA hypermethylation (Bardhan & Liu, 2013; Goldstein, 2006; Noffsinger, 2009; Simon, 2016). However, the process of initiation and progression of polyps and the risk factors that propel such progression have not been comprehensively modeled to understand what constitutes such risk and why some polyps progress, and others do not (Harrington, Wei, Suriawinata, Mackenzie, & Hassanpour, 2020). Despite advances in the modeling and understanding of CRCs, precise preclinical ex-vivo models for investigating pre-neoplastic states (polyps) are lacking.

3. Conventional CRC models

3.1 2D monolayers of CRC cell lines

Human 2D CRC cell line grown as monolayer is the most popular conventional CRC model that allows easy assessment of a variety of tumor cell phenotypes, e.g., cell viability, cell cycle, colony formation, migration, invasion, metastasis, and survival and apoptosis (Kannen et al., 2012; Nautiyal et al., 2011; Young, Ordonez, & Clarke, 2013). Moreover, the 2D monolayers have been used to study the efficacy of anti-cancer therapies, identification and validation of drug targets, and targeted drug delivery. For example, The role of *JNK* signaling in cell migration and invasion was studied in the COLO-205 cell line (Zhang et al., 2016), while the SW480 cell line was used to investigate β-catenin as a novel target for cancer therapy (Li, Zhou, Ji, & Luo, 2016). The popularity of the 2D monolayer model stems from multiple factors such as their low cost, high reproducibility, and their ease of creation and propagation (Joseph, Malindisa, & Ntwasa, 2018). However, this model has serious drawbacks, i.e., they do not recapitulate the tumor cell microenvironment and lack dimensionality (Porter, Murray, & McLean, 2020; Ronen, Hayat, & Akalin, 2019). Other disadvantages include (a) the molecular profiles of CRC cell lines often differ significantly from human tumors; (b) cancer cell lines are subjected to the risk of contamination and mislabeling, and genetic drifts during prolonged cultures; (c) lack of dimensionality leading to unnatural adhesion forces; and (d) lack the tumor-stroma and tumor-immune cell interactions (Domcke, Sinha, Levine, Sander, & Schultz, 2013; Huang, Liu, Zheng, & Shen, 2017).

3.2 3D tumor spheroids of CRC cell lines

3D tumor spheroids, developed by either growing tumor cells within 3D scaffolds or as multicellular spheroids, recapitulate the *in vivo* tumor microenvironment and they have been developed to study the interactions between tumor and stromal cells as well as tumor cells and the extracellular matrix. This model mitigates some of the disadvantages of the 2D monolayer model, namely, as it is well documented to regain intrinsic properties and to better mimic the *in vivo* situation due to the added third dimension (Dolznig et al., 2011; Fischbach et al., 2009). Over the past few years, the 3D tumor spheroids have been preferred over 2D monolayers to study tumor cell metabolism, gene expression studies, and for tracking response to anti-tumor treatments (Desoize & Jardillier, 2000; Takagi et al., 2007). This model is advantaged by its simplicity, high degrees of reproducibility, and the ability to coculture with other cell types. Major disadvantages of this model are the poor structural organization and poor supply of oxygen and nutrients to the spheroid core (Porter et al., 2020). Moreover, the 2D and the 3D models have been shown to differentially activate signaling pathways, especially in the *mTOR, AKT, S6K* signaling in response to the therapeutic agents (Riedl et al., 2017). These apparent inconsistencies and major disadvantages require researchers to use pre-clinical animal models for human CRC research to confirm findings, forcing us one step backward (in terms of species).

3.3 Animal models

The main goal of modeling human CRC in animals is to understand the molecular etiology, pathology, and progression of CRC (Oliveira, Abrantes, Tralhão, & Botelho, 2020). This model has been used to translate and confirm the hypotheses derived from cell models (Johnson & Fleet, 2013). Moreover, it is used to study the precise molecular mechanisms as well as to test the possible potential preventative and therapeutic strategies for CRC (Oliveira et al., 2020). There five major strategies for the development of potential animal models for CRCs, including: (i) spontaneous intestinal cancers in various animal species including dogs, sheep and mice; (ii) chemically or environmentally induced cancers in rodents; (iii) mutagen-induced germline mutation models, including APC^{min} mouse and F344-Pirc Rat; (iv) genetically modified mice; and (v) implantation of tumor cells into mice, either syngeneic murine cells (Morikawa et al., 1988; Naito, von Eschenbach, Giavazzi, & Fidler, 1986) or xenotransplantation of human CRC tissues/cells (Johnson & Fleet, 2013; Oliveira et al., 2020). Although these animal models mitigate a lot of the disadvantages of the 2D and 3D models, there are still several weaknesses, for example, (i) the gut environment (diet, toxins, microbes) is non-human, and hence, the innate immune response in the gut lining will lack humanness; (ii) the ability to study both the initial and late stages (metastasis) of CRCs is limited because the models either allow studying initiation (e.g., spontaneous tumor models) or progression (syngeneic or xenograft models), but not both; (iii) The need for extensive resources, high costs related to animal care and breeding and ethical concerns (Johnson & Fleet, 2013; Porter et al., 2020). Furthermore, the human xenograft model, which requires the use of immune deficient mice has one major flaw that is unique to this model, in that, it lacks a key component in cancer progression, i.e., the host immune response to a growing tumor. Therefore, an optimal model that can be uniformly used to study both the early and late stages of CRCs is still lacking.

4. The development of patient-derived colorectal organoids as 3D models for studying CRCs

Recent technological innovation in stem cell research has had a significant impact specifically on CRC research, enabling the development of 3D colorectal organoid models and their optimized growth conditions which bypass some of the disadvantages of the conventional models (Du et al., 2020; Ghosh et al., 2020; Pleguezuelos-Manzano et al., 2020; Sayed, Chakraborty, et al., 2020; Sayed, Sahan, et al., 2020; Sayed, Suarez, et al., 2020; Yuki, Cheng, Nakano, & Kuo, 2020). Although most of the published papers used the murine *APC*^{min} colon-derived tissue to grow organoids, others (Sato et al., 2011) and we Ghosh et al. (2020), Sayed, Chakraborty, et al. (2020), Sayed, Sahan, et al. (2020), Sayed, Suarez, et al. (2020) have shown that the long-term culture of colorectal organoids from healthy subjects and patients with inflammatory bowel disease (IBD) (d'Aldebert et al., 2020; Sayed, Chakraborty, et al., 2020; Sayed, Sahan, et al., 2020; Sayed, Suarez, et al., 2020), adenoma and colorectal cancers (Bruun et al., 2020; Sato et al., 2011). These organoids spurred an unprecedented growth of associated technologies, all of which have highlighted the limitless potential of this human pre-clinical model system (see Fig. 1).

4.1 Standardized methodologies for culturing 3D organoids, enhancing reproducibility

The optimization of the 3D organoids' culture media has had an enormous impact on CRC research. Sato et al. had successfully established mouse intestinal crypt-villus organoids from single sorted *LGR5*$^{+}$ stem cells using *WNT* 3a, *R*-spondin and *N*oggin (WRN) containing culture media and 3D serum-free extracellular matrix such as Matrigel (Sato et al., 2009). The composition of WRN media and other constituents required for organoid culture are listed in Table 1. This media supports the growth of *LGR5*$^{+}$ stem cells into crypt-villus-forming epithelial domains containing all the differentiated cell types that are present in the colon epithelium *in vivo*. It is notable that these organoids can be maintained for long-term in culture despite the absence of a mesenchymal niche (more than 8 months) (Sato et al., 2009).

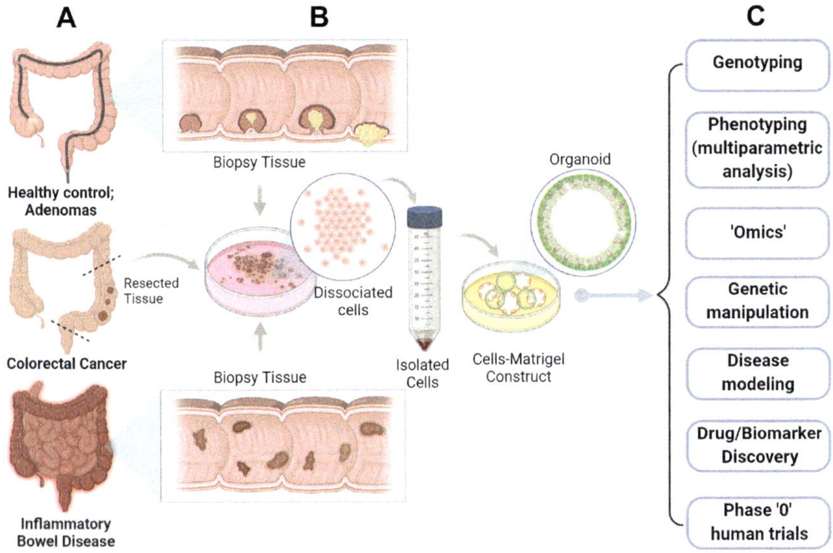

Fig. 1 The limitless potential of colorectal organoids. Organoids can be generated from virtually all colon tissues representing diverse pathophysiologic states of the colon (A), e.g., normal healthy mucosa and pre-neoplastic states such as diverse types of colon polyps (e.g., tubular, villous and serrated adenomas) (top), colorectal cancers (middle), and finally, inflammatory conditions, e.g., inflammatory bowel diseases (IBD), check-point colitis, etc., as long as the tissue can be preserved to serve as a source of adult stem cells. (B) Schematic outlines key steps in the process of organoid generation. Briefly, fresh tissue or viable frozen tissue is dissociated via enzymatic and mechanical methods to release the stem cells. Dissociated cells are subsequently embedded within the extracellular matrix (Matrigel). Under the right growth conditions, which includes a delicate ratio of growth factors (*WNT*, *R-spondin* and *Noggin*), 3D organoids are formed, which contain all cell types normally found in the colonic epithelium. These adult stem-cell-derived organoids can be passaged, expanded and biobanked. (C) The organoid model system can be exploited for several applications such as multiparametric analysis, genetic manipulation, drug and biomarker discovery and for carrying out phase "0" clinical trials in humans to assess not just drug toxicity/safety but also efficacy.

The addition of nicotinamide, A83-01 (an *ACTIVIN/NODAL/TGF-β* pathway inhibitor), and SB202190 (a p38 *MAPK* inhibitor) to the WRN media allows the long-term culture of organoids derived from the human small intestine and colon tissues and mouse *APC*-deficient adenomas, but also from human colorectal cancer cells and metaplastic epithelia from regions of Barrett's esophagus (Jung et al., 2011; Sato et al., 2011) Table 1. The use of such conditioned media (instead of custom media with cocktails) has been shown to improve reproducibility (VanDussen, Sonnek, & Stappenbeck, 2019). With the optimization of *R-spondin*-based culture conditions,

Table 1 Composition of the culture media required from the growth of healthy and CRC colon organoid.

Component required for organoid culture	Function	References
Wnt	• Important for crypt proliferation	Kim et al. (2005) and Pinto, Gregorieff, Begthel, and Clevers (2003)
R-spondins	• Upregulate the *WNT* signaling pathway through a specific effect on the Lgr receptors driving the crypt hyperplasia *in vivo*	de Lau et al. (2011)
Noggin	• Inhibitor for bone morphogenetic protein (BMP) signaling pathway • Helps in the expansion of crypt numbers	Haramis et al. (2004)
Epidermal growth factor (EGF)	• Important for stem cell proliferation and inhibit apoptosis	Dignass and Sturm (2001)
Rho-kinase inhibitor Y-27632	• Inhibits anoikis in isolated embryonic stem cells • Decreases cell death	Watanabe et al. (2007)
Notch-agonistic peptide	• Notch/gamma-secretase inhibition turns proliferative cells in intestinal crypts and adenomas into goblet cells	Li et al. (1998) and van Es et al. (2005)
Extracellular Matrix (Matrigel)	• Rich with laminin (α1 and α2) • Support the intestinal epithelial growth through laminin enrichment at the crypt base	Sasaki, Giltay, Talts, Timpl, and Talts (2002)
Nicotinamide	• Inhibitors to ROCK kinase • Promotes Cell Survival and Differentiation • Required for long-term growth of human derived organoids and organoids from CRC patients, mouse *APC*-deficient adenomas, and Barrett's esophagus	Jung et al. (2011), Meng et al. (2018), and Sato et al. (2011)

Table 1 Composition of the culture media required from the growth of healthy and CRC colon organoid.—cont'd

Component required for organoid culture	Function	References
A83–01	• Potent inhibitor of activin receptor-like kinase (*ALK*) including *ALK5* (type I transforming growth factor-β receptor), *ALK4* (type IB activin receptor), and *ALK7* (type I *NODAL receptor*) • Inhibits *TGF-β*-induced epithelial-to-mesenchymal transition • Allow long-term culture of organoid from CRC patients and *APC* deficient adenoma	Jung et al. (2011), Sato et al. (2011), and Wu et al. (2017)
SB202190	• Inhibits p38 of mitogen-activated protein kinase (*MAPK*) • Decreases the degradation of *EGF* receptor • Increase plating efficiency and allows long-term culture of organoid from CRC patients and *APC* deficient adenoma	Frey, Dise, Edelblum, and Polk, (2006), Jung et al. (2011), and Sato et al. (2011)
Gastrin	• Not always required • Increase the culture efficiency	Jung et al. (2011) and Sato et al. (2011)

organoid technology has been expanded to enable several other human carcinomas, including gastric (Nanki et al., 2018), breast (Sachs et al., 2018), ovarian (Kopper et al., 2019), lung (Sachs et al., 2019), and liver cancers (Broutier et al., 2017); spurring the notion that if there are $LGR5^{+}$ adult tissue stem cells, establishment of organoids is feasible. Importantly, the organoids derived from human CRC samples, especially tissues that represent the adenoma-carcinoma transition stage, can grow without the niche factors *WNT*, *R-spondin*, *EGF*, and *Noggin*; this is because, the mutational background of these samples ensure that the associated downstream signaling pathways are constitutively activated (Drost et al., 2015; Fujii et al., 2016; Matano et al., 2015). Recently, it has been shown that

Fibroblast growth factor 2 (*FGF2*) is essential and necessary for the self-renewal of colon cancer organoids (Otte et al., 2019) especially to grow organoids from tumor metastases.

Thus, the past decade has seen a concerted push toward the standardization and simplification of protocols for organoid isolation and expansion, and most groups converge on the fundamentals (i.e., what to add). However, like most external manipulations in cell biology, we emphasize that "*less is more*" and that such a restrained approach which avoids the addition of too many/too much external growth factors might be best in order to preserve the epigenetic and phenotypic properties of the organoid closest to its tissue of origin.

4.2 Generation of a living organoid biobank from CRC patients

The importance of building patient tumor-derived organoid biobank has been documented before with other cancers, e.g., primary pancreatic ductal adenocarcinoma (Huang, Holtzinger, et al., 2015; Seino et al., 2018; Tiriac et al., 2018), breast cancer (Sachs et al., 2018), prostate cancer (Gao et al., 2014), and primary liver cancer (Broutier et al., 2017). These efforts not only spurred the discovery of novel pathways and actionable targets, but also personalize anti-cancer therapeutics to match the molecular phenotype. In the case of CRCs, efforts to build similar biobank(s) have been reported previously (van de Wetering et al., 2015; Weeber et al., 2015); van de Wetering and colleagues established patient-derived organoids from 20 patients using tumor tissues that represented major CRC molecular subtypes and matching adjacent healthy regions (van de Wetering et al., 2015). The tumor organoids recapitulated the mutation spectra found in the parent CRC tissue (van de Wetering et al., 2015), making them ideal for use in high-throughput drug screening assays to not just facilitate personalized therapy but also exploit for genomic and functional studies on the individual patient (van de Wetering et al., 2015). Weeber and colleagues showed that organoids reflect the mutation load of original metastases from which they were derived, capturing ~90% of the somatic mutations seen in the patient, as determined by sequencing of 14 patient-derived organoids (PDOs) for 1977 cancer-relevant genes (Weeber et al., 2015). Similarly, Fujii et al. demonstrated that organoids ($n=55$ PDOs) preserve the histopathological grade and differentiation status of their parental tumors in vitro and develop tumors after implantation to the immunodeficient mice (Fujii et al., 2016). Moreover, the OncoTrack consortium generated a large biobank of 106 tumors, 35 organoids, and 59 xenografts from CRC patients (stages I–IV) ($n = 106$ PDOs), and developed a preclinical platform generating a compendium of drug sensitivity data more than 4000 assays for evaluation of 16 anti-cancer drugs (Schütte et al., 2017). By combining the molecular profiles and drug sensitivity patterns, they could identify biomarkers to predict sensitivity to the EGFR inhibitor cetuximab (Schütte et al., 2017).

These key proof-of-concept studies demonstrate that a well-annotated organoid biobank accurately reflecting the patient's tumor and its unique biology are useful for drug and biomarker discovery, for strategizing personalized treatment plans, and could even serve as platforms for carrying out phase "0" pre-clinical human clinical trials for novel targets/compounds (Fig. 1). The key, however, is to have proper metadata collection, as complete as possible, which not just includes information such as age, gender, source (biopsy or surgical resection), clinical staging, prior treatment history (chemotherapy or targeted), location and histology (moderately differentiated or poorly differentiated), but also mutation load estimates, MSI/MSS status, and molecular makeup. In other words, biobanks would benefit from diligent collection of PDO-associated metadata, which is any and all information that can aid in the selection of therapeutics and formulation of treatment plans.

5. The coculture models with CRC organoids and its benefits in CRC research

5.1 Coculture of CRC organoids with the immune cells

Somatic mutations in the cancer cells have the potential to encode immunogenic antigens, named neoantigens (Le et al., 2015). The host immune response produced against the neoantigens may be insufficient to neutralize them, leading to immune escape and survival of cancer cells. To overcome this challenge, patient's immune cells could be expanded *in vitro* to generate large quantities of activated immune cells for *in vivo* use, and this is the principle of cancer immunotherapy (Chen & Mellman, 2017; Sharma & Allison, 2015). Immune cells can be expanded *in vitro* in coculture with organoids (see Fig. 2). Nozaki and colleagues successfully established a coculture model between mouse intestinal epithelial organoids and intraepithelial lymphocytes (IELs) that allows efficient expansion and motility analysis of IELs in the presence of IL-2, IL-7, and IL-15 (Nozaki et al., 2016). IELs showed a fourfold increase in both αβT and γδT IELs for 2 weeks and were

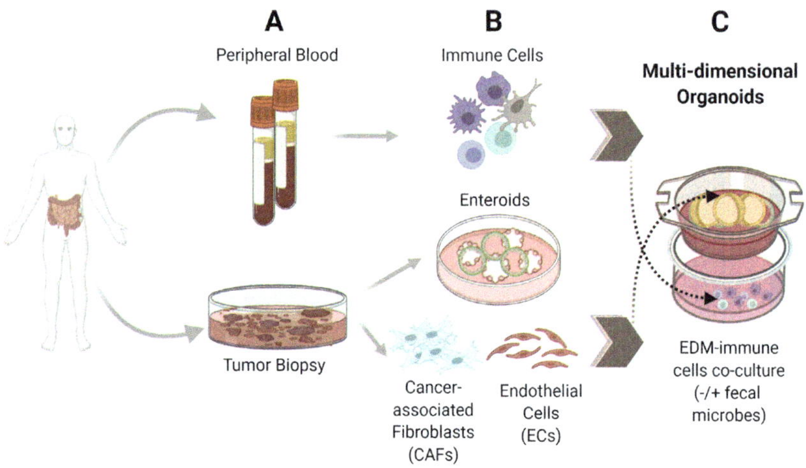

Fig. 2 Building multi-dimensional organoids that emulate the complexity of the gut. Schematic displays the isogenic cell components derived from the same patient that can be used to build multidimensional organoid-based CRC models. Blood and tumor tissue collected from the same patient (A) can be used as a source of immune cells (B, top) or tumor organoids and tumor-associated stromal cells (B, bottom). Coculture models can then be assembled to preserve the physiologic orientation of tissue (i.e., basolateral side of the tumor epithelium facing stroma and immune cells; see C). Complexity can be further increased with the addition of fecal microbes from the same patient. Although not shown here, components of the enteric nervous system can also be incorporated into the coculture where desired. Such an understanding-by-modeling approach is critical to creating wholesome models for testing drug efficacy or for testing fundamental research hypotheses in the setting of complex cell types.

efficiently maintained within and outside of organoids (Nozaki et al., 2016). Finnberg and colleagues generated 3D-tumoroids, organoids derived from CRC patients, cocultured them with the patient's peripherally and tumor-derived immune cells *in vitro* to mimic the tumor microenvironment and assess the tumor response to chemotherapy such as 5-fluorouracil (Finnberg et al., 2017). Similarly, Neal and colleagues used the air-liquid interface method to develop a coculture of organoids, derived from more than 100 human biopsies and mouse tumors, and native embedded immune cells (T, B, and NK cells and macrophages) to recapitulate the tumor microenvironment (Neal et al., 2018). They also found that human and murine derived organoids successfully modeled immune checkpoint blockade with anti-PD-1- and/or anti-PD-L1 expanding and activating tumor antigen-specific $CD8^+$ T lymphocytes and eliciting tumor cytotoxicity (Neal et al., 2018). Furthermore, Dijkstra and colleagues showed that cocultures of autologous tumor organoids, derived from mismatch repair-deficient CRC patients and non-small-cell lung cancer, and peripheral blood lymphocytes were efficient in enrichment and induction the tumor activity of the circulating T cells that specifically kill autologous tumor organoids (Dijkstra et al., 2018). Moreover, analysis of immune cells from human mammary ductal epithelial organoids revealed the presence of T lymphocytes subset, Vδ2+ T cells, which produced antitumor cytokine IFNγ and efficiently killed breast carcinoma cells in the presence of FDA—approved drugs bisphosphonate drugs (Zumwalde et al., 2016). In addition, the development of reverse-engineered thymus-derived organoids can be applied in regenerative medicine and solid organ transplantation as they provide a physiological thymic microenvironment that can be a source for tumor-specific lymphocytes (Tajima, Pradhan, Trucco, & Fan, 2016).

The tumor immune microenvironment (TME) can be generated using CRC organoids by three approaches: (a) reconstituted models, in which immune cells can be cocultured with organoids; (b) native TME models, in which the intrinsic immune microenvironment of tumors are preserved when establishing tumor organoids; (c) air–liquid interface (ALI) culture models, in which tissue fragments containing both tumor cells and immune components are embedded in collagen gels (Yuki et al., 2020). All these models are promising platforms for use in interrogating immunotherapeutic strategies.

5.2 Organoid models to assess the contribution of microbes in cancer initiation and progression

Microbes have been implicated in both initiation and progression of cancers, as exemplified by the oncopathogen that increases the risk of gastric cancers, *Helicobacter pylori*. Cocultures of microbes and organoids are useful to assess the mechanisms and risk factors for cancer development. Bartfeld and colleagues first developed a 3D organoid culture of human gastric stem cells and challenged with *H. pylori* by microinjecting the bacteria into the lumen; they noted a strong inflammatory response that was characterized by an induction of IL-8 and activation of NF-κB (Bartfeld et al., 2015). Since then, several groups used similar coculture models of human gastric organoids with *H. pylori* to understand the host epithelial cell responses to the microbe and the microbial response to host metabolites and host signaling pathways (Bartfeld et al., 2015; Holokai et al., 2019; Huang, Sweeney, et al., 2015; Sebrell et al., 2019). We recently used gastric organoid-derived monolayers and challenged them with *H. pylori* and observed altered DNA repair pathways, specifically, the base excision repair process, followed by induction of inflammatory response and oxidative DNA damage

(Sayed, Sahan, et al., 2020); the latter are key cellular events that have been implicated in gastric cancer initiation and progression.

Besides cocultures of *H. pylori* and gastric organoids, murine gallbladder organoids have been cocultured with the enteric pathogen, *Salmonella* (Scanu et al., 2015); this study showed that the microbe can induce host signaling pathways that culminate in the amplification of *c-MYC* through the activation of *MAPK* and *AKT* signaling pathways, and that these events were permissive to cellular transformation associated with gall bladder carcinoma.

In the case of CRCs, it is now widely accepted that gut microbes contribute to CRC progression through various mechanisms such as the induction of chronic inflammatory state and the production of toxic metabolites, which in turn affects stem cell precursors (Tsilimigras, Fodor, & Jobin, 2017). The well-known CRC associated microbes are *Fusobacterium nucleatum* (*Fn*), enterotoxigenic *Bacteroides fragilis* (ETBF), and colibactin producing *E. coli* (Alhinai, Walton, & Commane, 2019). The anaerobe *F. nucleatum* resides in the oral cavity and plays a role in CRC progression (Kostic et al., 2013, 2012), possibly by activating *β-catenin* and *Wnt* signaling and upregulates inflammatory responses (Rubinstein et al., 2013). In addition, high levels of *F. nucleatum* in CRC patients' tissues was associated with poor patient outcome and recurrence of post-chemotherapy (Yu et al., 2017). Much like *H. pylori, F. nucleatum* also induces DNA damage through the effect on the DNA repair pathway (Geng, Zhang, Lu, Zhang, & Pan, 2020). Recently, using 3D colon-derived organoids and polarized 2D epithelial cells derived from healthy humans, we have shown that *F. nucleatum* induces DNA damage that can influence CRC initiation and/or progression (Sayed, Chakraborty, et al., 2020).

The enterotoxigenic *Bacteroides fragilis* (ETBF) is another example of a CRC-associated microbe that secretes a metalloprotease toxin, which causes human inflammatory diarrhea (Sears et al., 2008). ETBF has been shown to enhance CRC progression in *APC* min mice through immunologic mechanisms via selective activation to the colonic signal transducer and activator of transcription-3 (Stat3) and T-helper 17 (Th17), causing inflammation-induced cancer (Wu et al., 2009). In addition, purified *Bacteriodes fragilis* toxin (BFT) upregulates the enzyme spermine oxidase (SMO) in the colon epithelial cells resulting in the accumulation of reactive oxygen species (ROS) and the induction of γ-H_2AX, a marker of DNA damage (Goodwin et al., 2011).

E. coli with the polyketide synthase genomic island (pks^{+} *E. coli*) are also associated with CRC and responsible for the synthesis of colibactin (genotoxin). Colibactin producing *E. coli* NC101 strain induces colon carcinogenesis through induction of DNA double-strand breaks triggering genomic instability (Arthur et al., 2012; Cuevas-Ramos et al., 2010; Nougayrède et al., 2006). A recent study done by Dejea and colleagues reported that patients with familial adenomatous polyposis (FAP) harbor colonic biofilms containing tumorigenic bacteria composed of colibactin producing *E. coli* and ETBF (Dejea, Fathi, & Craig, 2018). Adherent invasive *E. coli*, associated with Crohn's disease, increases the risk of CRC, especially in IBD patients through induction of chronic inflammation and accumulation of DNA damage (Martin et al., 2004).

Taken together, these studies have converged on a common overall mechanism, one that involves infection, chronic inflammation, leading to DNA damage coupled with altered core DNA repair mechanisms in the epithelium, which in turn triggers aberrant cell behavior and immune responses (Alhinai et al., 2019). Despite these insights, whether CRC-associated microbes can "*cause*" CRCs has been met with a healthy dose of criticism. One plausible hypothesis is that they may simply preferentially colonize cancerous lesions because of growth advantage (Coleman & Nunes, 2016; Wassenaar, 2018).

6. The advantage of the organoid model in the CRC research

6.1 Models to interrogate the precise contributions of host genetics and microbes in CRC initiation and propagation

It is conceivable that the 3D colon organoid and polarized epithelial cell model that we already developed for use in coculture assays (Sayed, Chakraborty, et al., 2020; Sayed, Sahan, et al., 2020; Sayed, Suarez, et al., 2020) could be systemically infected with each strain to assess the impact of microbes on CRC initiation. Colon organoids isolated from patients with hereditary polyposis syndromes (derived from uninvolved mucosal tissue that is at risk, as well as from adenomatous tissue) or those from patients with sporadic adenomas and/or CRCs (Fig. 3A) could be cocultured with the aforementioned CRC-associated bacterial strains (in the preceding paragraphs) or even fecal microbiota isolated from the same patient (Fig. 3C).

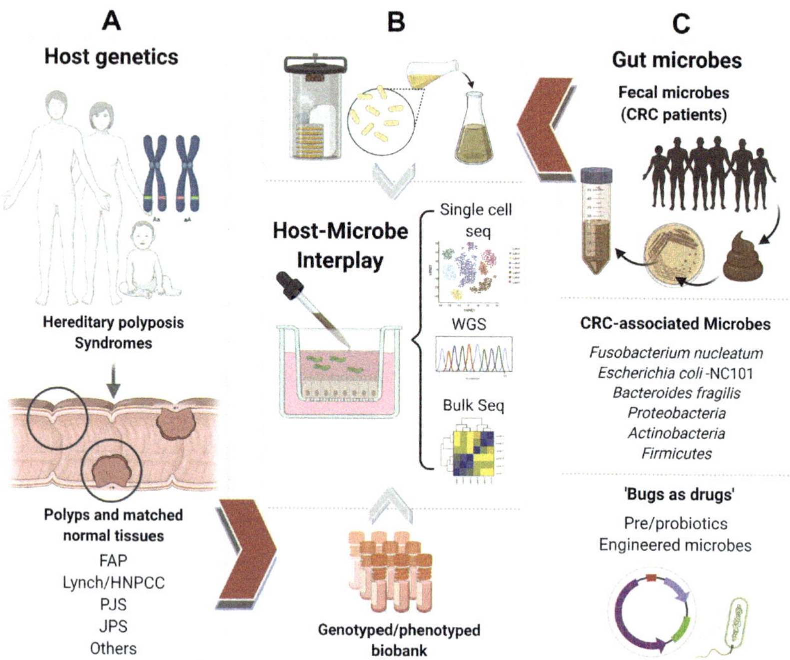

Fig. 3 Patient-derived organoids can enable an objective assessment of the relative contributions of host genetics and microbes toward CRC initiation, reveal the potential use of microbes as therapeutics. Schematic displays how patient-derived organoids (PDOs) from uninvolved normal mucosal biopsies and involved adenomatous regions of colons representing diverse CRC syndromes (A), e.g., Familial adenomatous polyposis (FAP), Lynch or Hereditary nonpolyposis colorectal cancer (HNPCC), Peutz-Jeghers syndrome (PJS), Juvenile polyposis syndrome (JPS) of known genotypes can be used to represent host contributions in organoid-microbe coculture models in panel (B). PDOs can also help interrogate and assess the impact of both harmful and beneficial gut microbes (C). Fecal microbes collected from high-risk patients with CRCs/excessive polyposis (C, top) can be cultured using aerobic or anaerobic conditions (B, top) prior to use in coculture studies with normal PDOs. Bacterial strains associated with CRCs (C, middle) can similarly be used in cocultures with PDOs with known genotypes. Pre- and probiotics and engineered microbes (C, bottom) can be similarly tested for their beneficial effects on PDOs. In each coculture model, the PDOs can be objectively assessed in an unbiased way using "omics"- and computational approaches (B, middle) to measure the impact of host genetics, microbes, individually and when combined.

Such a study model, that is systematically designed to interrogate the 2-way interactions between the host genetics and the gut microbe (either fecal slurry or specific strains) could help objectively assess the exact nature of and the relative contributions of the complex interplay between the microbes and the host in CRC initiation and progression (Fig. 3B).

Such a powerful well-controlled "reductionistic" model could be paired with unbiased "omics"-based analyses (e.g., whole genome or bulk or single-cell sequencing of the transcriptome; Fig. 3B) to ask numerous interesting and exciting questions. For example, what sequence of events may be triggered by microbes in the healthy vs the vulnerable mucosa, and which microbes may be able to "shift" the transcriptome toward CRCs. On the flip side, this very same coculture model could be exploited to detect, interrogate and harness for therapeutic purposes the beneficial effect of probiotics and prebiotics that may shift the cellular processes and transcriptomics away from CRCs.

6.2 Genome editing and genetic cancer modeling in organoids

Colon-derived organoids have proven to be the perfect models for controlled and sequential genetic manipulation to mimic what is currently known to occur during CRC initiation and progression. CRISPR/Cas9 genome-editing system has enabled researchers to introduce mutations in the tumor suppressor genes *APC*, *SMAD4*, and *TP53*, and/or the oncogenes *KRAS/BRAF* and/or *PIK3CA/PTEN*, either alone, or in combination, either sequentially and/or randomly into the normal human intestinal epithelium. Using such an approach, Matano et al. reported that the organoids harbored all five mutations, i.e., *APC, SMAD4, TP53, KRAS* and *PIK3CA*, could grow independently of niche factors (Matano et al., 2015). These isogenic genetically manipulated organoids formed tumors after implantation under the kidney in mice, and they formed micrometastases containing tumor initiating cells after injection in the mice spleen, but neither progressed to form distant metastases in the liver. These findings suggested that the mutations may be sufficient for stem cell maintenance in the hostile tumor microenvironment, but not for metastatic progression (Matano et al., 2015). In addition, Drost et al. reported that loss of *APC* and *TP53* are key drivers of chromosomal instability and aneuploidy that lead to CRC (Drost et al., 2015). Using the CRISPR/Cas9 technique for the generation of quadruple mutant organoids (mutant for *APC, P53, KRAS,* and *SMAD4*), they showed that the mutant organoids could grow independently of all stem-cell-niche factors and tolerate the presence of the P53 stabilizer nutlin-3 (Drost et al., 2015). Once again, these mutant organoids grew independently of niche growth factors from the culture medium, and they could grow as tumors with features of invasive carcinoma upon xenotransplantation into mice (Drost et al., 2015). Furthermore, Fumagalli et al. studied the genetic dissection of CRC progression by orthotopic transplantation of human colon organoids engineered to harbor different combinations of CRC-associated mutations (Fumagalli et al., 2017). They showed that the sequential accumulation of oncogenic mutations in critical signaling pathways, e.g., the *WNT*, *EGFR*, *P53*, and *TGF-β/SMAD* mediates efficient tumor growth, migration, and metastatic colonization (Fumagalli et al., 2017). Besides, Verissimo and colleagues developed normal organoids and tumor organoids with oncogenic *KRAS* mutation introduced via CRISPR and used them to evaluate the therapeutic potential of *RAS*-pathway inhibitors and for use in drug screening (Verissimo et al., 2016). These studies revealed that the presence of mutant *RAS* correlated with resistance to the targeted therapies and that such resistance stemmed from a drug-induced transient cell-cycle arrest rather than cell death (Verissimo et al., 2016). In another study aimed at exploring the origin of cancer-associated mutational signatures (Drost & van Boxtel, 2017), the authors deleted essential DNA repair genes in human colon organoids via CRISPR/Cas9 and found that mutation accumulation in organoids deficient in the mismatch repair gene *MLH1* resembles the mutation profiles recorded in mismatch repair-deficient CRCs. By contrast, mutation accumulation in organoids deficient in the base excision repair protein *NTHL1* resembles mutation previously reported in a breast cancer cohort (Drost & van Boxtel, 2017). Moreover, genome editing enabled the characterization of cancer-associated genes and identified the entire spectrum of tumor progression and metastasis (Roper et al., 2017). Using gene-editing of the *APC* and *TP53* tumor suppressor genes in colon epithelial cells, followed by orthotopic transplantation of *APC*-edited colon organoids, Roper and colleagues reported that $APC\Delta/\Delta$;Kras$^{G12D/+}$;$TP53\Delta/\Delta$ (AKP) mouse colon organoids and human CRC organoids engrafted in the distal colon and metastasized to the liver suggesting that a high-grade dysplasia developed with two driver mutation combinations (Roper et al., 2017).

6.3 CRC organoids as platforms for drug screening

Tumor cell lines grown as monolayers in 2D have been used as models for tumor research and drug screening for a long time (Barretina et al., 2012; Sharma, Haber, & Settleman, 2010). However, gathering evidence suggests that the response of 2D models to drugs often do not mirror the drug response in clinical patients (Liu et al., 2020; Xia, Li, He, Aji, & Gao, 2019). Patient-derived xenograft (PDX) is another model used in cancer research for drug screening and it has advantages over 2D cell lines of being preserving the molecular and morphological characteristics and revealing the heterogenicity of the primary tumor (DeRose et al., 2011; Liu et al., 2017; Wang et al., 2019). However, the PDX model lacks the immune system, and tumor evolution occurred in PDX (Liu et al., 2020). Cancer organoid provided an alternative strategy for drug screening which has advantages over 2D models and PDX models in terms of stability, fidelity, and the possibility of coculture with the immune cells, and hence the cancer organoid is a promising tool for preclinical drug screening and personalized therapy (Kondo & Inoue, 2019; Yang et al., 2019). In addition, the development of organoid biobanks from CRC tissues and adjacent healthy tissue enables high-throughput drug screening and detection of gene-drug associations (van de Wetering et al., 2015). Using CRC organoids, van de Wetering and colleagues screened 83 compound library including drugs in clinical use ($n=25$), chemotherapeutics ($n=10$), drugs in clinical trials ($n=29$), and experimental anti-cancer compounds ($n=29$). They showed that a single organoid culture showed high sensitivity to *WNT* secretion inhibitors (porcupine) was mutated for ring finger protein 43 (RNF43), the negative *WNT* feedback regulator, rather than in *APC* (van de Wetering et al., 2015). Vlachogiannis and colleagues compared the relationship of drug responses between patients and their metastatic gastrointestinal tumor organoids. They assessed 55 drugs that are in phase I–III studies or currently approved for clinical use and concluded the beneficial application of tumor organoids in precision cancer medicine (Vlachogiannis et al., 2018). Moreover, Schütte et al. developed a preclinical platform with more than 4000 assays testing the drug sensitivity of 16 clinical drugs using CRC patient organoids (Schütte et al., 2017). Analysis of drug sensitivity patterns identified biomarkers such as a signature outperforming *RAS/RAF* mutations that could predict the sensitivity to the *EGFR* inhibitor cetuximab (Schütte et al., 2017).

6.4 Organoids aid in personalizing anti-cancer therapies

Because patient-derived tumor organoids recapitulate the original biopsies that they were derived from (Weeber et al., 2015), the development of living organoid biobanks that represent all CRC subtypes could be ideal models for testing markers of treatment resistance (Schütte et al., 2017) or for optimizing personalized anti-cancer therapies. For example, using organoids as mainstay models, it was established that CRCs harboring *KRAS/BRAF* mutations are resistant to targeted therapies (Karapetis et al., 2008; Loupakis et al., 2009).

Reversal of the mutation overcomes the resistance and reinstates sensitivity to anti-cancer therapies (Zhai et al., 2017). Mechanistically, resistance was attributed to the ability of *KRAS* mutations to allow tumor cells to arrest cell cycle when exposed to combination therapies (*EGFR/MEK* inhibitors and/or *EGRF/ERK* inhibitors), and persist (Verissimo et al., 2016); these therapy-resisting persistors rapidly restarted tumor growth when the treatment was stopped (Verissimo et al., 2016). Interestingly, the addition of a low dose of navitoclax, a clinically tested *BCL2/BCLXL* inhibitor, to the *EGFR/MEK/ERK* pathway through the targeted combinational therapies primed the *RAS* mutant CRC organoids into apoptosis, indicating the strong potential of patient-derived CRC organoid libraries in evaluating pathway inhibitors and drug combinations in a preclinical setting (Verissimo et al., 2016). Another prospective clinical study demonstrated that PDOs could predict response to chemotherapy in metastatic CRC patients (Ooft & Weeber, 2019). PDOs were used for evaluation of sensitivity to a standard of care chemotherapy in CRC, and they predicted response of the biopsied lesion in more than 80% of patients treated with irinotecan-based therapies without misclassifying patients who could have treatment benefits (Ooft & Weeber, 2019). However, the PDOs failed to predict the outcome for treatment with 5-fluorouracil plus oxaliplatin, suggesting that the prediction was specific to irinotecan-based chemotherapy (Ooft & Weeber, 2019). Moreover, a rectal cancer organoid platform was used to study individual responses to chemoradiation (Ganesh, Wu, O'Rourke, & Szeglin, 2019); the latter were generated from primary, metastatic, and recurrent tumors ($n = 65$) and subsequently used to assess the responses of individuals to chemoradiation. The response of tumoroids to clinically relevant chemotherapy and radiation treatment were heterogeneous but correlated with the clinical responses noted in individual patients' tumors (Ganesh et al., 2019). Taken together, all these studies generally agree that PDOs recapitulate the patient tumor phenotype and genotype, and therefore, they are ideal models for enhancing both precision and personalization in CRC treatment.

6.5 Impact of CRC organoids on clinical trials

Vlachogiannis and colleagues carried out phase I/II clinical trials to evaluate the clinical value of PDOs collected from metastatic, heavily pretreated CRC and gastroesophageal cancer patients ($n = 71$) in the evaluation of the treatment response (Vlachogiannis et al., 2018). Using 3D screening assays and 55 agents under assessment in phase I–III clinical trials or already clinically acceptable standard of care, the drug responses in cancer organoids mirrored the clinical response in the corresponding patients with 100% sensitivity and 93% specificity. These results provide proof-of-concept that PDOs can recapitulate patient responses in the clinic and may become crucial pre-clinical models for early phase clinical trials and for personalizing therapeutics in the clinic (Vlachogiannis et al., 2018). Another study done by Yao and colleagues used PDOs to predict chemoradiation responses in locally advanced rectal cancer (LARC) patients treated with neoadjuvant chemoradiation who were enrolled in a concomitant phase III clinical trial (Yao et al., 2020). The chemoradiation responses in patients showed 84.43% accuracy, 78.01% sensitivity, and 91.97% specificity, compared to the responses in PDOs, indicating that the PDOs predict LARC patient responses in the clinic and could be a helpful prognostic tool in the personalization of management for rectal cancer (Yao et al., 2020).

7. Current challenges in the organoid model

Although the organoid model is a relevant system that physiologically mimics the *in vivo* environment, some challenges are still associated with this system. Standalone, the organoid model lacks the multicellular nature of the tumor microenvironment, such as the tumor-stromal fibroblasts, vascular endothelial cells, immune cells, and microbial factors. To overcome this challenge, coculture techniques have been developed, featuring immune cells (Finnberg et al., 2017; Neal et al., 2018), and/or fibroblasts (Öhlund, Handly-Santana, Biffi, Elyada, & Almeida, 2017). However, the coculture of the organoids with multiple elements at the same time to mimic the tumor microenvironment remains challenging. Also, the success rate for organoid development from metastatic CRC or from patients undergoing therapy is low, probably due to the low amount of live stem cells found in these biopsies. Besides, fresh biopsies are sometimes not available. Even when tissues are available, organoid biobanking may not be as cost effective in most laboratories unless they have the necessary infrastructure, dedicated funding and expertise that is found only in a handful research laboratories around the world (Drost & Clevers, 2018; Ji & Wu, 2020). In addition, some components in the organoid media such as matrigel and fetal calf serum may affect the experimental readouts, for example, in the setting of drug screening (Drost & Clevers, 2018; Ji & Wu, 2020). Researchers have tried to solve this problem by replacing matrigel with synthetic matrices that support the growth of mouse and human intestinal organoids without interference in the experiment results (Gjorevski et al., 2016). Also, researchers have tried to use serum-free *WNT* media by replacing fetal calf serum with water-soluble ligands for Frizzled/low-density lipoprotein receptor-related protein 5 (LRP5) or LRP6 heterodimerizes that act as surrogate *WNT* agonists and can support the growth of many humanoid organoids (Janda et al., 2017). Alternatively, stabilization of *WNT* protein, using phospholipids and cholesterol, enables the serum-free culture of human organ stem cells (Tüysüz et al., 2017). Also, adequate characterization of each PDO, including whole-exome sequencing, copy number assessment, RNA sequencing, and drug sensitivity tests, is highly recommended for achieving assay specificity, sensitivity, and reproducibility (Ji & Wu, 2020).

8. The next-generation CRC organoids

The traditional 3D culture method for growing 3D organoids depends on the addition of growth factors and extracellular matrix. However, the existing culture systems have reduced the capacity to recapitulate the complex and dynamic microenvironment of a developing organ and organogenesis. Therefore, stem cell researchers, bioengineers, and physicists have collaboratively developed advanced *in vitro* technologies for organoid research such as organ-on-a-chip and 3D bioprinting of organoid culture system that combine dimensionality and multicellularity with physical forces (Clevers, 2017; Fan, Demirci, & Chen, 2019; Park, Georgescu, & Huh, 2019).

8.1 Organ-on-a-chip

Organs-on-a-chip is microfabricated cell culture device designed to model the functional units of human organs *in vitro* (Bhatia & Ingber, 2014; Esch, Bahinski, & Huh, 2015). This chip can be used for the development of organoid models because it overcomes the challenges associated with the traditional 3D cultures, such as microenvironmental control (both, physical and chemical), modeling tissue–tissue and multiorgan interactions, and reducing the variability (Park et al., 2019). These advantages are expanded below:

1 First, organs-on-a-chip mimics tissues perfused by blood vessels, simulated through the use of a microfluidic platform, which mimics the vascularization of organoids and recapitulates the in vivo tumor microenvironment

with regard to nutrient supply, hypoxia, etc. (Hasan, Paul, Memic, & Khademhosseini, 2015; Kim, Lee, Chung, & Jeon, 2013; Wang et al., 2016).

2 Second, these chips are associated with mechanically actuatable microengineered platforms that can generate and apply *in vivo*–like mechanical forces to organoids such as rhythmic contraction of the stomach (Lee et al., 2018).

3 Third, the application of continuous flow in the chip leads to the expansion of endothelial progenitors within the organoids; such expansion is critical for the structural and functional development of several organoid models such as kidney, pancreas, and the intestines (Homan, Gupta, & Kroll, 2019; Tao et al., 2019; Workman et al., 2018).

4 Fourth, organs-on-a-chip provides an automated digital microfluidic platform with an array of electrodes that enables the automated culture of 3D organoid in media droplets, which provides a producible and precise method and reduces the variability compared to the traditional 3D culture system (Au, Chamberlain, Mahesh, Sefton, & Wheeler, 2014; Yin et al., 2016). Moreover, some culture platforms, especially for multiorgan culture, include biosensing elements to permit continuous screening of organoids behavior that will minimize the variability. For example, a multiorgan-on-a-chip device integrated with multi-sensors for long-term monitoring of cardiac, liver organoids, and primary hepatic spheroids during drug treatment (Zhang, Aleman, Shin, & Kilic, 2017).

5 Fifth, the organs-on-a-chip platform also enables the coculture of different cell/tissue types and organs, e.g., a vascularized liver organoid-on-a-chip is a multicompartment microdevice in which hepatic cells are cocultured with human endothelial cells in an ECM hydrogel (Jin et al., 2018). Jin and colleagues used a microfluidic array to coculture stem cell-derived liver, intestinal, and stomach organoids that were maintained in separate compartments but were communicated via rocker-induced media flow between culture chambers (Jin et al., 2018). This multi-organ model was used to study the complex biology and enterohepatic circulation of bile acids (Jin et al., 2018). Similarly, Skardal and colleagues developed microengineered heart-lung-liver model to reveal how cardiotoxicity of bleomycin is mediated due to the cytokine-associated cross-talk between the lung and the heart tissues (Skardal et al., 2017). In another study, a multiorgan microfluidic chip comprised of four organs, lung, brain, liver, and bone was developed to study the metastasis of primary lung cancer (Xu et al., 2016).

Due to the previously mentioned advantages, brain and/or liver-on-a-chip are promising tools as a preclinical model for drug screening and discovery (Wang et al., 2018; Zhu et al., 2017). Also, organoids-on-a-chip can be used for the generation of patient- and population-specific disease models, which mimics the disease phenotypes *in vivo* and hence an excellent tool for personalized medicine (Clevers, 2017). Besides, organoids-on-a-chip can be used in regenerative medicine through the transplanting *in vitro* expanded organoids into animals to repair damaged organs (Nakamura & Sato, 2018).

8.2 3D-bioprinting of organoid culture

3D bioprinting is a biofabrication process that generates biological components, including ECM scaffolds, cells, and growth factors in 3D (Horváth et al., 2015; Wu et al., 2016). This technology closely mimics the human *in vivo* microenvironment and has the advantages of being high-precision, high-accuracy, and high-throughput (Knowlton, Yenilmez, & Tasoglu, 2016; Peng, Unutmaz, & Ozbolat, 2016). The standard 3D organoid culture method depends on the manual mixing of cells with ECM substrates in certain ratios before gelling or manually laying the cells on the preformed ECM gel, which could lead to a different size, morphology, distribution, and cell types within the formed organoids (Reid et al., 2016; Yonemura, 2014). Such variation could affect the results of experiments and/or make experiments uninterpretable, or findings irreproducible (Reid, Mollica, Bruno, & Sachs, 2018; Reid et al., 2016; Yonemura, 2014). 3D bioprinting technology overcomes this limitation in the traditional 3D culture method through the use of a computer-aided process that precisely allows the deposition of the cells layer-by-layer in the organoids to confirm the experimental consistency (Zhou et al., 2013). Reid and colleagues successfully adapted a low-cost 3D-printer for precise cell placement and used this system for printing human induced pluripotent stem cells (hiPSCs) within 3D hydrogels (Reid et al., 2016). This system was subsequently used to generate large 3D mammary organoids in hydrogels (Reid et al., 2018). This printer was later used as a platform for mechanistic analysis of tumoroids and chimeric mammary organoids to study the microenvironmental control of breast cancer (Reid & Palmer, 2019). In the context of CRCs, Rios de la Rosa and colleagues developed and validated a biofabrication method to model the early-stage CRC tumor monitoring two biomarkers (*CD44* and *HIF-1α*) (Rios de la Rosa, Wubetu, Tirelli, & Tirella, 2018). Using alginate hydrogels and HCT-116 cells, they successfully biofabricated the 3D model and monitored two biomarkers (*CD44 and HIF-1α*) during tumor development. This model mimics the tumor environment and closely represents the heterogeneous tumor mass and has distinct advantages over the traditional 3D culture system such as spheroid and monolayer. Therefore, this high throughput 3D system is a promising tool for preclinical drug testing in CRC (Rios de la Rosa et al., 2018).

9. Conclusion

Patient-derived organoids have immense applications in CRC research as they retain the molecular, cellular, and histological complexity and genetic heterogeneity of the individual patients after long-term culture and can be used for the personalizing optimal treatment strategies. Technological advancement in organoid-based studies, especially those that allow their use in the high throughput format, with higher reproducibility, and quality control is expected to spur more accurate modeling of disease, organ replacement and drug testing. Approximately 15% of oncology drugs proceed from Phase I to FDA approval (DiMasi, Reichert, Feldman, & Malins, 2013; Hay, Thomas, Craighead, Economides, & Rosenthal, 2014). The success rate may improve if we can include pre-clinical human models, e.g., patient-specific organoids with cutting edge organ-on-a-chip and 3D bioprinting models before the candidate drugs enter Phase I. The addition of all the necessary cell types and the microbiome population of the same patient will open new avenues in the detection, measurement, tracking of and tackling (through therapeutics) the risk of CRC initiation and progression. Breakthroughs in cancer research in the coming years are expected to include the development of novel treatment strategies that program either the tumor cell, or its environment, or the host immune system and/or some combination of them all; organoids are bound to be big players in that journey. Their use is expected to help maintain our focus on what is relevant and real in humans, and thereby improve precision in patient care.

Acknowledgment

This work was supported, partly by National Institute of Health Grants: R01 AI141630, CA238042, and CA100768 (to P.G.); DK107585, AG069689, C3 Padre Pedal MCC pilot grant (to S.D.); and UG3TR002968 (to P.G. and S.D.). A.A.A. is supported by an NIH-funded Cancer Therapeutics Training Program (CT2, T32 CA121938).

References

Alhinai, E. A., Walton, G. E., & Commane, D. M. (2019). The role of the gut microbiota in colorectal cancer causation. *International Journal of Molecular Sciences*, *20*(21), 5295. https://doi.org/10.3390/ijms20215295.

Allen, J., & Sears, C. L. (2019). Impact of the gut microbiome on the genome and epigenome of colon epithelial cells: Contributions to colorectal cancer development. *Genome Medicine*, *11*(1), 11. https://doi.org/10.1186/s13073-019-0621-2.

Arthur, J. C., Perez-Chanona, E., Mühlbauer, M., Tomkovich, S., Uronis, J. M., Fan, T. J., et al. (2012). Intestinal inflammation targets cancer-inducing activity of the microbiota. *Science*, *338*(6103), 120–123. https://doi.org/10.1126/science.1224820.

Au, S. H., Chamberlain, M. D., Mahesh, S., Sefton, M. V., & Wheeler, A. R. (2014). Hepatic organoids for microfluidic drug screening. *Lab on a Chip*, *14*(17), 3290–3299. https://doi.org/10.1039/c4lc00531g.

Bardhan, K., & Liu, K. (2013). Epigenetics and colorectal cancer pathogenesis. *Cancers (Basel)*, *5*(2), 676–713. https://doi.org/10.3390/cancers5020676.

Barretina, J., Caponigro, G., Stransky, N., Venkatesan, K., Margolin, A. A., Kim, S., et al. (2012). The cancer cell line encyclopedia enables predictive modelling of anticancer drug sensitivity. *Nature*, *483*(7391), 603–607. https://doi.org/10.1038/nature11003.

Bartfeld, S., Bayram, T., van de Wetering, M., Huch, M., Begthel, H., Kujala, P., et al. (2015). In vitro expansion of human gastric epithelial stem cells and their responses to bacterial infection. *Gastroenterology*, *148*(1), 126–136. e126. https://doi.org/10.1053/j.gastro.2014.09.042.

Bhatia, S. N., & Ingber, D. E. (2014). Microfluidic organs-on-chips. *Nature Biotechnology*, *32*(8), 760–772. https://doi.org/10.1038/nbt.2989.

Broutier, L., Mastrogiovanni, G., Verstegen, M. M., Francies, H. E., Gavarró, L. M., Bradshaw, C. R., et al. (2017). Human primary liver cancer-derived organoid cultures for disease modeling and drug screening. *Nature Medicine*, *23*(12), 1424–1435. https://doi.org/10.1038/nm.4438.

Bruun, J., Kryeziu, K., Eide, P. W., Moosavi, S. H., Eilertsen, I. A., Langerud, J., et al. (2020). Patient-derived organoids from multiple colorectal cancer liver metastases reveal moderate intra-patient pharmacotranscriptomic heterogeneity. *Clinical Cancer Research*, *26*(15), 4107–4119. https://doi.org/10.1158/1078-0432.CCR-19-3637.

Budinska, E., Popovici, V., Tejpar, S., D'Ario, G., Lapique, N., Sikora, K. O., et al. (2013). Gene expression patterns unveil a new level of molecular heterogeneity in colorectal cancer. *The Journal of Pathology*, *231*(1), 63–76. https://doi.org/10.1002/path.4212.

Chan, A. T., & Giovannucci, E. L. (2010). Primary prevention of colorectal cancer. *Gastroenterology*, *138*(6), 2029–2043. e2010. https://doi.org/10.1053/j.gastro.2010.01.057.

Chen, D. S., & Mellman, I. (2017). Elements of cancer immunity and the cancer-immune set point. *Nature*, *541*(7637), 321–330. https://doi.org/10.1038/nature21349.

Clevers, H. (2017). Advances in organoid technology: Hans Clevers, Madeline Lancaster, and Takanori Takebe. *Cell Stem Cell*, *20*, 759–762.

Coleman, O. I., & Nunes, T. (2016). Role of the microbiota in colorectal cancer: Updates on microbial associations and therapeutic implications. *Biores Open Access*, *5*(1), 279–288. https://doi.org/10.1089/biores.2016.0028.

Cuevas-Ramos, G., Petit, C. R., Marcq, I., Boury, M., Oswald, E., & Nougayrède, J. P. (2010). Escherichia coli induces DNA damage in vivo and triggers genomic instability in mammalian cells. *Proceedings of the National Academy of Sciences of the United States of America*, *107*(25), 11537–11542. https://doi.org/10.1073/pnas.1001261107.

d'Aldebert, E., Quaranta, M., Sebert, M., Bonnet, D., Kirzin, S., Portier, G., et al. (2020). Characterization of human colon organoids from inflammatory bowel disease patients. *Frontiers in Cell and Development Biology*, *8*, 363. https://doi.org/10.3389/fcell.2020.00363.

de Lau, W., Barker, N., Low, T. Y., Koo, B. K., Li, V. S., Teunissen, H., et al. (2011). Lgr5 homologues associate with Wnt receptors and mediate R-spondin signalling. *Nature, 476*(7360), 293–297. https://doi.org/10.1038/nature10337.

Dejea, C. M., Fathi, P., & Craig, J. M. (2018). Patients with familial adenomatous polyposis harbor colonic biofilms containing tumorigenic bacteria. *Science, 359*(6375), 592–597. https://doi.org/10.1126/science.aah3648.

DeRose, Y. S., Wang, G., Lin, Y. C., Bernard, P. S., Buys, S. S., Ebbert, M. T., et al. (2011). Tumor grafts derived from women with breast cancer authentically reflect tumor pathology, growth, metastasis and disease outcomes. *Nature Medicine, 17*(11), 1514–1520. https://doi.org/10.1038/nm.2454.

Desoize, B., & Jardillier, J.-C. (2000). Multicellular resistance: a paradigm for clinical resistance? *Critical Reviews in Oncology/Hematology, 36*(2–3), 193–207. https://doi.org/10.1016/s1040-8428(00)00086-x.

Dignass, A. U., & Sturm, A. (2001). Peptide growth factors in the intestine. *European Journal of Gastroenterology & Hepatology, 13*(7), 763–770. https://doi.org/10.1097/00042737-200107000-00002.

Dijkstra, K. K., Cattaneo, C. M., Weeber, F., Chalabi, M., van de Haar, J., Fanchi, L. F., et al. (2018). Generation of tumor-reactive T cells by co-culture of peripheral blood lymphocytes and tumor organoids. *Cell, 174*(6), 1586–1598. e1512. https://doi.org/10.1016/j.cell.2018.07.009.

DiMasi, J. A., Reichert, J. M., Feldman, L., & Malins, A. (2013). Clinical approval success rates for investigational cancer drugs. *Clinical Pharmacology and Therapeutics, 94*(3), 329–335. https://doi.org/10.1038/clpt.2013.117.

Dolznig, H., Rupp, C., Puri, C., Haslinger, C., Schweifer, N., Wieser, E., et al. (2011). Modeling colon adenocarcinomas in vitro: A 3D co-culture system induces cancer-relevant pathways upon tumor cell and stromal fibroblast interaction. *The American Journal of Pathology, 179*(1), 487–501. https://doi.org/10.1016/j.ajpath.2011.03.015.

Domcke, S., Sinha, R., Levine, D. A., Sander, C., & Schultz, N. (2013). Evaluating cell lines as tumour models by comparison of genomic profiles. *Nature Communications, 4*(2126). https://doi.org/10.1038/ncomms3126.

Drost, J., & Clevers, H. (2018). Organoids in cancer research. *Nature Reviews Cancer, 18*(7), 407–418. https://doi.org/10.1038/s41568-018-0007-6.

Drost, J., & van Boxtel, R. (2017). Use of CRISPR-modified human stem cell organoids to study the origin of mutational signatures in cancer. *Science, 358*(6360), 234–238. https://doi.org/10.1126/science.aao3130.

Drost, J., van Jaarsveld, R. H., Ponsioen, B., Zimberlin, C., van Boxtel, R., Buijs, A., et al. (2015). Sequential cancer mutations in cultured human intestinal stem cells. *Nature, 521*(7550), 43–47. https://doi.org/10.1038/nature14415.

Du, Y., Li, X., Niu, Q., Mo, X., Qui, M., Ma, T., et al. (2020). Development of a miniaturized 3D organoid culture platform for ultra-high-throughput screening. *Journal of Molecular Cell Biology, 12*(8), 630–643. https://doi.org/10.1093/jmcb/mjaa036.

Engel, R. M., Chan, W. H., Nickless, D., Hlavca, S., Richards, E., Kerr, G., et al. (2020). Patient-derived colorectal cancer organoids upregulate revival stem cell marker genes following chemotherapeutic treatment. *Journal of Clinical Medicine, 9*(1), 128. https://doi.org/10.3390/jcm9010128.

Esch, E. W., Bahinski, A., & Huh, D. (2015). Organs-on-chips at the frontiers of drug discovery. *Nature Reviews. Drug Discovery, 14*(4), 248–260. https://doi.org/10.1038/nrd4539.

Fan, H., Demirci, U., & Chen, P. (2019). Emerging organoid models: Leaping forward in cancer research. *Journal of Hematology & Oncology, 12*(1), 142. https://doi.org/10.1186/s13045-019-0832-4.

Finnberg, N. K., Gokare, P., Lev, A., Grivennikov, S. I., MacFarlane, A. W., Campbell, K. S., et al. (2017). Application of 3D tumoroid systems to define immune and cytotoxic therapeutic responses based on tumoroid and tissue slice culture molecular signatures. *Oncotarget*, *8*(40), 66747–66757. https://doi.org/10.18632/oncotarget.19965.

Fischbach, C., Kong, H. J., Hsiong, S. X., Evangelista, M. B., Yuen, W., & Mooney, D. J. (2009). Cancer cell angiogenic capability is regulated by 3D culture and integrin engagement. *Proceedings of the National Academy of Sciences*, *106*(2), 399–404.

Frey, M. R., Dise, R. S., Edelblum, K. L., & Polk, D. B. (2006). p38 kinase regulates epidermal growth factor receptor downregulation and cellular migration. *The EMBO Journal*, *25*(24), 5683–5692. https://doi.org/10.1038/sj.emboj.7601457.

Fujii, M., Shimokawa, M., Date, S., Takano, A., Matano, M., Nanki, K., et al. (2016). A colorectal tumor organoid library demonstrates progressive loss of niche factor requirements during tumorigenesis. *Cell Stem Cell*, *18*(6), 827–838. https://doi.org/10.1016/j.stem.2016.04.003.

Fumagalli, A., Drost, J., Suijkerbuijk, S. J., van Boxtel, R., de Ligt, J., Offerhaus, G. J., et al. (2017). Genetic dissection of colorectal cancer progression by orthotopic transplantation of engineered cancer organoids. *Proceedings of the National Academy of Sciences*, *114*(12), E2357–e2364. https://doi.org/10.1073/pnas.1701219114.

Ganesh, K., Wu, C., O'Rourke, K. P., & Szeglin, B. C. (2019). A rectal cancer organoid platform to study individual responses to chemoradiation. *Nature Medicine*, *25*(10), 1607–1614. https://doi.org/10.1038/s41591-019-0584-2.

Gao, D., Vela, I., Sboner, A., Iaquinta, P. J., Karthaus, W. R., Gopalan, A., et al. (2014). Organoid cultures derived from patients with advanced prostate cancer. *Cell*, *159*(1), 176–187. https://doi.org/10.1016/j.cell.2014.08.016.

Geng, F., Zhang, Y., Lu, Z., Zhang, S., & Pan, Y. (2020). Fusobacterium nucleatum caused DNA damage and promoted cell proliferation by the Ku70/p53 pathway in oral cancer cells. *DNA and Cell Biology*, *39*(1), 144–151. https://doi.org/10.1089/dna.2019.5064.

Ghosh, P., Swanson, L., Sayed, I. M., Mittal, Y., Lim, B. B., Ibeawuchi, S. R., et al. (2020). The stress polarity signaling (SPS) pathway serves as a marker and a target in the leaky gut barrier: Implications in aging and cancer. *Life Science Alliance*, *3*(3), e201900481. https://doi.org/10.26508/lsa.201900481.

Gjorevski, N., Sachs, N., Manfrin, A., Giger, S., Bragina, M. E., Ordóñez-Morán, P., et al. (2016). Designer matrices for intestinal stem cell and organoid culture. *Nature*, *539*(7630), 560–564. https://doi.org/10.1038/nature20168.

Goldstein, N. S. (2006). Serrated pathway and APC (conventional)-type colorectal polyps: Molecular-morphologic correlations, genetic pathways, and implications for classification. *American Journal of Clinical Pathology*, *125*(1), 146–153.

Goodwin, A. C., Destefano Shields, C. E., Wu, S., Huso, D. L., Wu, X., Murray-Stewart, T. R., et al. (2011). Polyamine catabolism contributes to enterotoxigenic Bacteroides fragilis-induced colon tumorigenesis. *Proceedings of the National Academy of Sciences of the United States of America*, *108*(37), 15354–15359. https://doi.org/10.1073/pnas.1010203108.

Gopalappa, C., Aydogan-Cremaschi, S., Das, T. K., & Orcun, S. (2011). Probability model for estimating colorectal polyp progression rates. *Health Care Management Science*, *14*(1), 1–21. https://doi.org/10.1007/s10729-010-9138-3.

Guinney, J., Dienstmann, R., Wang, X., de Reynies, A., Schlicker, A., Soneson, C., et al. (2015). The consensus molecular subtypes of colorectal cancer. *Nature Medicine*, *21*(11), 1350–1356. https://doi.org/10.1038/nm.3967.

Haramis, A. P., Begthel, H., van den Born, M., van Es, J., Jonkheer, S., Offerhaus, G. J., et al. (2004). De novo crypt formation and juvenile polyposis on BMP inhibition in mouse intestine. *Science*, *303*(5664), 1684–1686. https://doi.org/10.1126/science.1093587.

Harrington, L. X., Wei, J. W., Suriawinata, A. A., Mackenzie, T. A., & Hassanpour, S. (2020). Predicting colorectal polyp recurrence using time-to-event analysis of medical records. *AMIA Joint Summits on Translational Science Proceedings*, *2020*, 211–220.

Hasan, A., Paul, A., Memic, A., & Khademhosseini, A. (2015). A multilayered microfluidic blood vessel-like structure. *Biomedical Microdevices*, *17*(5), 88. https://doi.org/10.1007/s10544-015-9993-2.

Hay, M., Thomas, D. W., Craighead, J. L., Economides, C., & Rosenthal, J. (2014). Clinical development success rates for investigational drugs. *Nature Biotechnology*, *32*(1), 40–51. https://doi.org/10.1038/nbt.2786.

Holokai, L., Chakrabarti, J., Broda, T., Chang, J., Hawkins, J. A., Sundaram, N., et al. (2019). Increased programmed death-ligand 1 is an early epithelial cell response to helicobacter pylori infection. *PLoS Pathogens*, *15*(1), e1007468. https://doi.org/10.1371/journal.ppat.1007468.

Homan, K. A., Gupta, N., & Kroll, K. T. (2019). Flow-enhanced vascularization and maturation of kidney organoids in vitro. *Nature Methods*, *16*(3), 255–262. https://doi.org/10.1038/s41592-019-0325-y.

Horváth, L., Umehara, Y., Jud, C., Blank, F., Petri-Fink, A., & Rothen-Rutishauser, B. (2015). Engineering an in vitro air-blood barrier by 3D bioprinting. *Scientific Reports*, *5*, 7974. https://doi.org/10.1038/srep07974.

Huang, L., Holtzinger, A., Jagan, I., BeGora, M., Lohse, I., Ngai, N., et al. (2015). Ductal pancreatic cancer modeling and drug screening using human pluripotent stem cell- and patient-derived tumor organoids. *Nature Medicine*, *21*(11), 1364–1371. https://doi.org/10.1038/nm.3973.

Huang, Y., Liu, Y., Zheng, C., & Shen, C. (2017). Investigation of cross-contamination and misidentification of 278 widely used tumor cell lines. *PLoS One*, *12*(1), e0170384. https://doi.org/10.1371/journal.pone.0170384.

Huang, J. Y., Sweeney, E. G., Sigal, M., Zhang, H. C., Remington, S. J., Cantrell, M. A., et al. (2015). Chemodetection and destruction of host urea allows helicobacter pylori to locate the epithelium. *Cell Host & Microbe*, *18*(2), 147–156. https://doi.org/10.1016/j.chom.2015.07.002.

Inamura, K. (2018). Colorectal cancers: An update on their molecular pathology. *Cancers (Basel)*, *10*(1), 26. https://doi.org/10.3390/cancers10010026.

Janda, C. Y., Dang, L. T., You, C., Chang, J., de Lau, W., Zhong, Z. A., et al. (2017). Surrogate Wnt agonists that phenocopy canonical Wnt and β-catenin signalling. *Nature*, *545*(7653), 234–237. https://doi.org/10.1038/nature22306.

Jass, J. R. (2007). Classification of colorectal cancer based on correlation of clinical, morphological and molecular features. *Histopathology*, *50*(1), 113–130. https://doi.org/10.1111/j.1365-2559.2006.02549.x.

Ji, D.-B., & Wu, A.-W. (2020). Organoid in colorectal cancer: Progress and challenges. *Chinese Medical Journal*, *133*(16), 1971–1977. https://doi.org/10.1097/cm9.0000000000000882.

Jin, Y., Kim, J., Lee, J. S., Min, S., Kim, S., Ahn, D.-H., et al. (2018). Vascularized liver organoids generated using induced hepatic tissue and dynamic liver-specific microenvironment as a drug testing platform. *Advanced Functional Materials*, *28*(37), 1801954. https://doi.org/10.1002/adfm.201801954.

Johnson, R. L., & Fleet, J. C. (2013). Animal models of colorectal cancer. *Cancer and Metastasis Reviews*, *32*(1–2), 39–61.

Joseph, J. S., Malindisa, S. T., & Ntwasa, M. (2018). *Two-dimensional (2D) and three-dimensional (3D) cell culturing in drug discovery*. IntechOpen.

Ju, H. X., An, B., Okamoto, Y., Shinjo, K., Kanemitsu, Y., Komori, K., et al. (2011). Distinct profiles of epigenetic evolution between colorectal cancers with and without metastasis. *The American Journal of Pathology*, *178*(4), 1835–1846. https://doi.org/10.1016/j.ajpath.2010.12.045.

Jung, P., Sato, T., Merlos-Suárez, A., Barriga, F. M., Iglesias, M., Rossell, D., et al. (2011). Isolation and in vitro expansion of human colonic stem cells. *Nature Medicine*, *17*(10), 1225–1227. https://doi.org/10.1038/nm.2470.

Kannen, V., Hintzsche, H., Zanette, D. L., Silva, W. A., Jr., Garcia, S. B., Waaga-Gasser, A. M., et al. (2012). Antiproliferative effects of fluoxetine on colon cancer cells and in a colonic carcinogen mouse model. *PLoS One*, 7(11), e50043. https://doi.org/10.1371/journal.pone.0050043.

Karapetis, C. S., Khambata-Ford, S., Jonker, D. J., O'Callaghan, C. J., Tu, D., Tebbutt, N. C., et al. (2008). K-ras mutations and benefit from cetuximab in advanced colorectal cancer. *The New England Journal of Medicine*, *359*(17), 1757–1765. https://doi.org/10.1056/NEJMoa0804385.

Kim, K. A., Kakitani, M., Zhao, J., Oshima, T., Tang, T., Binnerts, M., et al. (2005). Mitogenic influence of human R-spondin1 on the intestinal epithelium. *Science*, *309*(5738), 1256–1259. https://doi.org/10.1126/science.1112521.

Kim, S., Lee, H., Chung, M., & Jeon, N. L. (2013). Engineering of functional, perfusable 3D microvascular networks on a chip. *Lab on a Chip*, *13*(8), 1489–1500. https://doi.org/10.1039/c3lc41320a.

Knowlton, S., Yenilmez, B., & Tasoglu, S. (2016). Towards single-step biofabrication of organs on a chip via 3D printing. *Trends in Biotechnology*, *34*(9), 685–688. https://doi.org/10.1016/j.tibtech.2016.06.005.

Kondo, J., & Inoue, M. (2019). Application of cancer organoid model for drug screening and personalized therapy. *Cell*, *8*(5). https://doi.org/10.3390/cells8050470.

Kopper, O., de Witte, C. J., Lõhmussaar, K., Valle-Inclan, J. E., Hami, N., Kester, L., et al. (2019). An organoid platform for ovarian cancer captures intra- and interpatient heterogeneity. *Nature Medicine*, *25*(5), 838–849. https://doi.org/10.1038/s41591-019-0422-6.

Kostic, A. D., Chun, E., Robertson, L., Glickman, J. N., Gallini, C. A., Michaud, M., et al. (2013). Fusobacterium nucleatum potentiates intestinal tumorigenesis and modulates the tumor-immune microenvironment. *Cell Host & Microbe*, *14*(2), 207–215. https://doi.org/10.1016/j.chom.2013.07.007.

Kostic, A. D., Gevers, D., Pedamallu, C. S., Michaud, M., Duke, F., Earl, A. M., et al. (2012). Genomic analysis identifies association of fusobacterium with colorectal carcinoma. *Genome Research*, *22*(2), 292–298. https://doi.org/10.1101/gr.126573.111.

Le, D. T., Uram, J. N., Wang, H., Bartlett, B. R., Kemberling, H., Eyring, A. D., et al. (2015). PD-1 blockade in tumors with mismatch-repair deficiency. *The New England Journal of Medicine*, *372*(26), 2509–2520. https://doi.org/10.1056/NEJMoa1500596.

Lee, K. K., McCauley, H. A., Broda, T. R., Kofron, M. J., Wells, J. M., & Hong, C. I. (2018). Human stomach-on-a-chip with luminal flow and peristaltic-like motility. *Lab on a Chip*, *18*(20), 3079–3085. https://doi.org/10.1039/c8lc00910d.

Li, L., Milner, L. A., Deng, Y., Iwata, M., Banta, A., Graf, L., et al. (1998). The human homolog of rat Jagged1 expressed by marrow stroma inhibits differentiation of 32D cells through interaction with Notch1. *Immunity*, *8*(1), 43–55. https://doi.org/10.1016/s1074-7613(00)80457-4.

Li, K., Zhou, Z. Y., Ji, P. P., & Luo, H. S. (2016). Knockdown of β-catenin by siRNA influences proliferation, apoptosis and invasion of the colon cancer cell line SW480. *Oncology Letters*, *11*(6), 3896–3900. https://doi.org/10.3892/ol.2016.4481.

Liu, J. F., Palakurthi, S., Zeng, Q., Zhou, S., Ivanova, E., Huang, W., et al. (2017). Establishment of patient-derived tumor xenograft models of epithelial ovarian cancer for preclinical evaluation of novel therapeutics. *Clinical Cancer Research*, *23*(5), 1263–1273. https://doi.org/10.1158/1078-0432.ccr-16-1237.

Liu, C., Qin, T., Huang, Y., Li, Y., Chen, G., & Sun, C. (2020). Drug screening model meets cancer organoid technology. *Translational Oncology*, *13*(11), 100840. https://doi.org/10.1016/j.tranon.2020.100840.

Loupakis, F., Ruzzo, A., Cremolini, C., Vincenzi, B., Salvatore, L., Santini, D., et al. (2009). KRAS codon 61, 146 and BRAF mutations predict resistance to cetuximab plus irinotecan in KRAS codon 12 and 13 wild-type metastatic colorectal cancer. *British Journal of Cancer, 101*(4), 715–721. https://doi.org/10.1038/sj.bjc.6605177.

Marisa, L., de Reyniès, A., Duval, A., Selves, J., Gaub, M. P., Vescovo, L., et al. (2013). Gene expression classification of colon cancer into molecular subtypes: Characterization, validation, and prognostic value. *PLoS Medicine, 10*(5), e1001453. https://doi.org/10.1371/journal.pmed.1001453.

Martin, H. M., Campbell, B. J., Hart, C. A., Mpofu, C., Nayar, M., Singh, R., et al. (2004). Enhanced Escherichia coli adherence and invasion in Crohn's disease and colon cancer. *Gastroenterology, 127*(1), 80–93. https://doi.org/10.1053/j.gastro.2004.03.054.

Matano, M., Date, S., Shimokawa, M., Takano, A., Fujii, M., Ohta, Y., et al. (2015). Modeling colorectal cancer using CRISPR-Cas9-mediated engineering of human intestinal organoids. *Nature Medicine, 21*(3), 256–262. https://doi.org/10.1038/nm.3802.

Mauri, G., Sartore-Bianchi, A., Russo, A. G., Marsoni, S., Bardelli, A., & Siena, S. (2019). Early-onset colorectal cancer in young individuals. *Molecular Oncology, 13*(2), 109–131. https://doi.org/10.1002/1878-0261.12417.

Meng, Y., Ren, Z., Xu, F., Zhou, X., Song, C., Wang, V. Y., et al. (2018). Nicotinamide promotes cell survival and differentiation as kinase inhibitor in human pluripotent stem cells. *Stem Cell Reports, 11*(6), 1347–1356. https://doi.org/10.1016/j.stemcr.2018.10.023.

Morikawa, K., Walker, S. M., Nakajima, M., Pathak, S., Jessup, J. M., & Fidler, I. J. (1988). Influence of organ environment on the growth, selection, and metastasis of human colon carcinoma cells in nude mice. *Cancer Research, 48*(23), 6863–6871. Retrieved from https://www.ncbi.nlm.nih.gov/pubmed/2846163.

Naito, S., von Eschenbach, A. C., Giavazzi, R., & Fidler, I. J. (1986). Growth and metastasis of tumor cells isolated from a human renal cell carcinoma implanted into different organs of nude mice. *Cancer Research, 46*(8), 4109–4115. Retrieved from https://www.ncbi.nlm.nih.gov/pubmed/3731078.

Nakamura, T., & Sato, T. (2018). Advancing intestinal organoid technology toward regenerative medicine. *Cellular and Molecular Gastroenterology and Hepatology, 5*(1), 51–60. https://doi.org/10.1016/j.jcmgh.2017.10.006.

Nanki, K., Toshimitsu, K., Takano, A., Fujii, M., Shimokawa, M., Ohta, Y., et al. (2018). Divergent routes toward Wnt and R-spondin niche independency during human gastric carcinogenesis. *Cell, 174*(4), 856–869. e817. https://doi.org/10.1016/j.cell.2018.07.027.

Nautiyal, J., Banerjee, S., Kanwar, S. S., Yu, Y., Patel, B. B., Sarkar, F. H., et al. (2011). Curcumin enhances dasatinib-induced inhibition of growth and transformation of colon cancer cells. *International Journal of Cancer, 128*(4), 951–961. https://doi.org/10.1002/ijc.25410.

Neal, J. T., Li, X., Zhu, J., Giangarra, V., Grzeskowiak, C. L., Ju, J., et al. (2018). Organoid modeling of the tumor immune microenvironment. *Cell, 175*(7), 1972–1988. e1916. https://doi.org/10.1016/j.cell.2018.11.021.

Noffsinger, A. E. (2009). Serrated polyps and colorectal cancer: New pathway to malignancy. *Annual Review of Pathology, 4*, 343–364. https://doi.org/10.1146/annurev.pathol.4.110807.092317.

Nougayrède, J. P., Homburg, S., Taieb, F., Boury, M., Brzuszkiewicz, E., Gottschalk, G., et al. (2006). Escherichia coli induces DNA double-strand breaks in eukaryotic cells. *Science, 313*(5788), 848–851. https://doi.org/10.1126/science.1127059.

Nozaki, K., Mochizuki, W., Matsumoto, Y., Matsumoto, T., Fukuda, M., Mizutani, T., et al. (2016). Co-culture with intestinal epithelial organoids allows efficient expansion and motility analysis of intraepithelial lymphocytes. *Journal of Gastroenterology, 51*(3), 206–213. https://doi.org/10.1007/s00535-016-1170-8.

O'Brien, M. J., Yang, S., Mack, C., Xu, H., Huang, C. S., Mulcahy, E., et al. (2006). Comparison of microsatellite instability, CpG island methylation phenotype, BRAF and KRAS status in serrated polyps and traditional adenomas indicates separate pathways to distinct colorectal carcinoma end points. *The American Journal of Surgical Pathology, 30*(12), 1491–1501. https://doi.org/10.1097/01.pas.0000213313.36306.85.

Öhlund, D., Handly-Santana, A., Biffi, G., Elyada, E., & Almeida, A. S. (2017). Distinct populations of inflammatory fibroblasts and myofibroblasts in pancreatic cancer. *The Journal of Experimental Medicine, 214*(3), 579–596. https://doi.org/10.1084/jem.20162024.

Oliveira, R. C., Abrantes, A. M., Tralhão, J. G., & Botelho, M. F. (2020). The role of mouse models in colorectal cancer research—The need and the importance of the orthotopic models. *Animal Models and Experimental Medicine, 3*(1), 1–8. https://doi.org/10.1002/ame2.12102.

Ooft, S. N., & Weeber, F. (2019). Patient-derived organoids can predict response to chemotherapy in metastatic colorectal cancer patients. *Science Translational Medicine, 11*(513), eaay2574. https://doi.org/10.1126/scitranslmed.aay2574.

Otte, J., Dizdar, L., Behrens, B., Goering, W., Knoefel, W. T., Wruck, W., et al. (2019). FGF signalling in the self-renewal of colon cancer organoids. *Scientific Reports, 9*(1), 17365. https://doi.org/10.1038/s41598-019-53907-7.

Park, S. E., Georgescu, A., & Huh, D. (2019). Organoids-on-a-chip. *Science, 364*(6444), 960–965. https://doi.org/10.1126/science.aaw7894.

Peng, W., Unutmaz, D., & Ozbolat, I. T. (2016). Bioprinting towards physiologically relevant tissue models for pharmaceutics. *Trends in Biotechnology, 34*(9), 722–732. https://doi.org/10.1016/j.tibtech.2016.05.013.

Phipps, A. I., Limburg, P. J., Baron, J. A., Burnett-Hartman, A. N., Weisenberger, D. J., Laird, P. W., et al. (2015). Association between molecular subtypes of colorectal cancer and patient survival. *Gastroenterology, 148*(1), 77–87. e72. https://doi.org/10.1053/j.gastro.2014.09.038.

Pinto, D., Gregorieff, A., Begthel, H., & Clevers, H. (2003). Canonical Wnt signals are essential for homeostasis of the intestinal epithelium. *Genes & Development, 17*(14), 1709–1713. https://doi.org/10.1101/gad.267103.

Pleguezuelos-Manzano, C., Puschhof, J., van den Brink, S., Geurts, V., Beumer, J., & Clevers, H. (2020). Establishment and culture of human intestinal organoids derived from adult stem cells. *Current Protocols in Immunology, 130*(1), e106. https://doi.org/10.1002/cpim.106.

Porter, R. J., Murray, G. I., & McLean, M. H. (2020). Current concepts in tumour-derived organoids. *British Journal of Cancer, 123*(8), 1209–1218.

Reid, J. A., Mollica, P. A., Bruno, R. D., & Sachs, P. C. (2018). Consistent and reproducible cultures of large-scale 3D mammary epithelial structures using an accessible bioprinting platform. *Breast Cancer Research, 20*(1), 122. https://doi.org/10.1186/s13058-018-1045-4.

Reid, J. A., Mollica, P. A., Johnson, G. D., Ogle, R. C., Bruno, R. D., & Sachs, P. C. (2016). Accessible bioprinting: Adaptation of a low-cost 3D-printer for precise cell placement and stem cell differentiation. *Biofabrication, 8*(2), 025017. https://doi.org/10.1088/1758-5090/8/2/025017.

Reid, J. A., & Palmer, X. L. (2019). A 3D bioprinter platform for mechanistic analysis of tumoroids and chimeric mammary organoids. *Scientific Reports, 9*(1), 7466. https://doi.org/10.1038/s41598-019-43922-z.

Riedl, A., Schlederer, M., Pudelko, K., Stadler, M., Walter, S., Unterleuthner, D., et al. (2017). Comparison of cancer cells in 2D vs 3D culture reveals differences in AKT–mTOR–S6K signaling and drug responses. *Journal of Cell Science, 130*(1), 203–218. https://doi.org/10.1242/jcs.188102.

Rios de la Rosa, J. M., Wubetu, J., Tirelli, N., & Tirella, A. (2018). Colorectal tumor 3D in vitro models: Advantages of biofabrication for the recapitulation of early stages of tumour development. *Biomedical Physics & Engineering Express*, *4*(4), 045010. https://doi.org/10.1088/2057-1976/aac1c9.

Ronen, J., Hayat, S., & Akalin, A. (2019). Evaluation of colorectal cancer subtypes and cell lines using deep learning. *Life Science Alliance*, *2*(6), e201900517. https://doi.org/10.26508/lsa.201900517.

Roper, J., Tammela, T., Cetinbas, N. M., Akkad, A., Roghanian, A., Rickelt, S., et al. (2017). In vivo genome editing and organoid transplantation models of colorectal cancer and metastasis. *Nature Biotechnology*, *35*(6), 569–576. https://doi.org/10.1038/nbt.3836.

Rubinstein, M. R., Wang, X., Liu, W., Hao, Y., Cai, G., & Han, Y. W. (2013). Fusobacterium nucleatum promotes colorectal carcinogenesis by modulating E-cadherin/β-catenin signaling via its FadA Adhesin. *Cell Host & Microbe*, *14*(2), 195–206. https://doi.org/10.1016/j.chom.2013.07.012.

Sachs, N., de Ligt, J., Kopper, O., Gogola, E., Bounova, G., Weeber, F., et al. (2018). A living biobank of breast cancer organoids captures disease heterogeneity. *Cell*, *172*(1-2), 373–386. e310. https://doi.org/10.1016/j.cell.2017.11.010.

Sachs, N., Papaspyropoulos, A., Zomer-van Ommen, D. D., Heo, I., Böttinger, L., Klay, D., et al. (2019). Long-term expanding human airway organoids for disease modeling. *EMBO Journal*, *38*(4). https://doi.org/10.15252/embj.2018100300, e100300.

Sadanandam, A., Lyssiotis, C. A., Homicsko, K., Collisson, E. A., Gibb, W. J., Wullschleger, S., et al. (2013). A colorectal cancer classification system that associates cellular phenotype and responses to therapy. *Nature Medicine*, *19*(5), 619–625. https://doi.org/10.1038/nm.3175.

Sasaki, T., Giltay, R., Talts, U., Timpl, R., & Talts, J. F. (2002). Expression and distribution of laminin alpha1 and alpha2 chains in embryonic and adult mouse tissues: An immunochemical approach. *Experimental Cell Research*, *275*(2), 185–199. https://doi.org/10.1006/excr.2002.5499.

Sato, T., Stange, D. E., Ferrante, M., Vries, R. G., Van Es, J. H., Van den Brink, S., et al. (2011). Long-term expansion of epithelial organoids from human colon, adenoma, adenocarcinoma, and Barrett's epithelium. *Gastroenterology*, *141*(5), 1762–1772. https://doi.org/10.1053/j.gastro.2011.07.050.

Sato, T., Vries, R. G., Snippert, H. J., van de Wetering, M., Barker, N., Stange, D. E., et al. (2009). Single Lgr5 stem cells build crypt-villus structures in vitro without a mesenchymal niche. *Nature*, *459*(7244), 262–265. https://doi.org/10.1038/nature07935.

Sayed, I. M., Chakraborty, A., Abd El-Hafeez, A. A., Sharma, A., Sahan, A. Z., Huang, W. J. M., et al. (2020). The DNA glycosylase NEIL2 suppresses fusobacterium-infection--induced inflammation and DNA damage in colonic epithelial cells. *Cells*, *9*(9), 1980. https://doi.org/10.3390/cells9091980.

Sayed, I. M., Sahan, A. Z., Venkova, T., Chakraborty, A., Mukhopadhyay, D., Bimczok, D., et al. (2020). Helicobacter pylori infection down-regulates the DNA glycosylase NEIL2, resulting in increased genome damage and inflammation in gastric epithelial cells. *The Journal of Biological Chemistry*, *295*(32), 11082–11098. https://doi.org/10.1074/jbc.RA119.009981.

Sayed, I. M., Suarez, K., Lim, E., Singh, S., Pereira, M., Ibeawuchi, S. R., et al. (2020). Host engulfment pathway controls inflammation in inflammatory bowel disease. *The FEBS Journal*, *287*(18), 3967–3988. https://doi.org/10.1111/febs.15236.

Scanu, T., Spaapen, R. M., Bakker, J. M., Pratap, C. B., Wu, L. E., Hofland, I., et al. (2015). Salmonella manipulation of host signaling pathways provokes cellular transformation associated with gallbladder carcinoma. *Cell Host & Microbe*, *17*(6), 763–774. https://doi.org/10.1016/j.chom.2015.05.002.

Schütte, M., Risch, T., Abdavi-Azar, N., Boehnke, K., Schumacher, D., Keil, M., et al. (2017). Molecular dissection of colorectal cancer in pre-clinical models identifies biomarkers predicting sensitivity to EGFR inhibitors. *Nature Communications*, *8*, 14262. https://doi.org/10.1038/ncomms14262.

Sears, C. L., Islam, S., Saha, A., Arjumand, M., Alam, N. H., Faruque, A. S., et al. (2008). Association of enterotoxigenic Bacteroides fragilis infection with inflammatory diarrhea. *Clinical Infectious Diseases*, *47*(6), 797–803. https://doi.org/10.1086/591130.

Sebrell, T. A., Hashimi, M., Sidar, B., Wilkinson, R. A., Kirpotina, L., Quinn, M. T., et al. (2019). A novel gastric spheroid co-culture model reveals chemokine-dependent recruitment of human dendritic cells to the gastric epithelium. *Cellular and Molecular Gastroenterology and Hepatology*, *8*(1), 157–171. e153. https://doi.org/10.1016/j.jcmgh.2019.02.010.

Seino, T., Kawasaki, S., Shimokawa, M., Tamagawa, H., Toshimitsu, K., Fujii, M., et al. (2018). Human pancreatic tumor organoids reveal loss of stem cell niche factor dependence during disease progression. *Cell Stem Cell*, *22*(3), 454–467. e456. https://doi.org/10.1016/j.stem.2017.12.009.

Sharma, P., & Allison, J. P. (2015). The future of immune checkpoint therapy. *Science*, *348*(6230), 56–61. https://doi.org/10.1126/science.aaa8172.

Sharma, S. V., Haber, D. A., & Settleman, J. (2010). Cell line-based platforms to evaluate the therapeutic efficacy of candidate anticancer agents. *Nature Reviews. Cancer*, *10*(4), 241–253. https://doi.org/10.1038/nrc2820.

Siegel, R. L., Miller, K. D., Goding Sauer, A., Fedewa, S. A., Butterly, L. F., Anderson, J. C., et al. (2020). Colorectal cancer statistics, 2020. *CA: A Cancer Journal for Clinicians*, *70*(3), 145–164. https://doi.org/10.3322/caac.21601.

Simon, K. (2016). Colorectal cancer development and advances in screening. *Clinical Interventions in Aging*, *11*, 967–976. https://doi.org/10.2147/cia.s109285.

Skardal, A., Murphy, S. V., Devarasetty, M., Mead, I., Kang, H. W., Seol, Y. J., et al. (2017). Multi-tissue interactions in an integrated three-tissue organ-on-a-chip platform. *Scientific Reports*, 7(1), 8837. https://doi.org/10.1038/s41598-017-08879-x.

Snover, D. C., Jass, J. R., Fenoglio-Preiser, C., & Batts, K. P. (2005). Serrated polyps of the large intestine: A morphologic and molecular review of an evolving concept. *American Journal of Clinical Pathology*, *124*(3), 380–391. https://doi.org/10.1309/v2ep-tplj-rb3f-ghjl.

Spring, K. J., Zhao, Z. Z., Karamatic, R., Walsh, M. D., Whitehall, V. L., Pike, T., et al. (2006). High prevalence of sessile serrated adenomas with BRAF mutations: A prospective study of patients undergoing colonoscopy. *Gastroenterology*, *131*(5), 1400–1407. https://doi.org/10.1053/j.gastro.2006.08.038.

Tajima, A., Pradhan, I., Trucco, M., & Fan, Y. (2016). Restoration of thymus function with bioengineered thymus organoids. *Current Stem Cell Reports*, *2*(2), 128–139. https://doi.org/10.1007/s40778-016-0040-x.

Takagi, A., Watanabe, M., Ishii, Y., Morita, J., Hirokawa, Y., Matsuzaki, T., et al. (2007). Three-dimensional cellular spheroid formation provides human prostate tumor cells with tissue-like features. *Anticancer Research*, *27*(1A), 45–53.

Tao, T., Wang, Y., Chen, W., Li, Z., Su, W., Guo, Y., et al. (2019). Engineering human islet organoids from iPSCs using an organ-on-chip platform. *Lab on a Chip*, *19*(6), 948–958. https://doi.org/10.1039/c8lc01298a.

Tiriac, H., Belleau, P., Engle, D. D., Plenker, D., Deschenes, A., Somerville, T. D. D., et al. (2018). Organoid profiling identifies common responders to chemotherapy in pancreatic cancer. *Cancer Discovery*, *8*(9), 1112–1129. https://doi.org/10.1158/2159-8290.CD-18-0349.

Tsilimigras, M. C., Fodor, A., & Jobin, C. (2017). Carcinogenesis and therapeutics: The microbiota perspective. *Nature Microbiology*, *2*, 17008. https://doi.org/10.1038/nmicrobiol.2017.8.

Tüysüz, N., van Bloois, L., van den Brink, S., Begthel, H., Verstegen, M. M., Cruz, L. J., et al. (2017). Lipid-mediated Wnt protein stabilization enables serum-free culture of human organ stem cells. *Nature Communications*, *8*, 14578. https://doi.org/10.1038/ncomms14578.

Uraoka, T., Higashi, R., Horii, J., Harada, K., Hori, K., Okada, H., et al. (2015). Prospective evaluation of endoscopic criteria characteristic of sessile serrated adenomas/polyps. *Journal of Gastroenterology*, *50*(5), 555–563. https://doi.org/10.1007/s00535-014-0999-y.

Vaiopoulos, A. G., Athanasoula, K., & Papavassiliou, A. G. (2014). Epigenetic modifications in colorectal cancer: Molecular insights and therapeutic challenges. *Biochimica et Biophysica Acta*, *1842*(7), 971–980. https://doi.org/10.1016/j.bbadis.2014.02.006.

van de Wetering, M., Francies, H. E., Francis, J. M., Bounova, G., Iorio, F., Pronk, A., et al. (2015). Prospective derivation of a living organoid biobank of colorectal cancer patients. *Cell*, *161*(4), 933–945. https://doi.org/10.1016/j.cell.2015.03.053.

van Es, J. H., van Gijn, M. E., Riccio, O., van den Born, M., Vooijs, M., Begthel, H., et al. (2005). Notch/gamma-secretase inhibition turns proliferative cells in intestinal crypts and adenomas into goblet cells. *Nature*, *435*(7044), 959–963. https://doi.org/10.1038/nature03659.

VanDussen, K. L., Sonnek, N. M., & Stappenbeck, T. S. (2019). L-WRN conditioned medium for gastrointestinal epithelial stem cell culture shows replicable batch-to-batch activity levels across multiple research teams. *Stem Cell Research*, *37*, 101430. https://doi.org/10.1016/j.scr.2019.101430.

Verissimo, C. S., Overmeer, R. M., Ponsioen, B., Drost, J., Mertens, S., Verlaan-Klink, I., et al. (2016). Targeting mutant RAS in patient-derived colorectal cancer organoids by combinatorial drug screening. *eLife*, *5*, e18489. https://doi.org/10.7554/eLife.18489.

Vlachogiannis, G., Hedayat, S., Vatsiou, A., Jamin, Y., Fernández-Mateos, J., Khan, K., et al. (2018). Patient-derived organoids model treatment response of metastatic gastrointestinal cancers. *Science*, *359*(6378), 920–926. https://doi.org/10.1126/science.aao2774.

Wang, Z., Fu, S., Zhao, J., Zhao, W., Shen, Z., Wang, D., et al. (2019). Transbronchoscopic patient biopsy-derived xenografts as a preclinical model to explore chemorefractory-associated pathways and biomarkers for small-cell lung cancer. *Cancer Letters*, *440-441*, 180–188. https://doi.org/10.1016/j.canlet.2018.10.014.

Wang, X., Phan, D. T., Sobrino, A., George, S. C., Hughes, C. C., & Lee, A. P. (2016). Engineering anastomosis between living capillary networks and endothelial cell-lined microfluidic channels. *Lab on a Chip*, *16*(2), 282–290. https://doi.org/10.1039/c5lc01050k.

Wang, Y., Wang, H., Deng, P., Chen, W., Guo, Y., Tao, T., et al. (2018). In situ differentiation and generation of functional liver organoids from human iPSCs in a 3D perfusable chip system. *Lab on a Chip*, *18*(23), 3606–3616. https://doi.org/10.1039/c8lc00869h.

Wassenaar, T. M. (2018). E. coli and colorectal cancer: A complex relationship that deserves a critical mindset. *Critical Reviews in Microbiology*, *44*(5), 619–632. https://doi.org/10.1080/1040841X.2018.1481013.

Watanabe, K., Ueno, M., Kamiya, D., Nishiyama, A., Matsumura, M., Wataya, T., et al. (2007). A ROCK inhibitor permits survival of dissociated human embryonic stem cells. *Nature Biotechnology*, *25*(6), 681–686. https://doi.org/10.1038/nbt1310.

Weeber, F., van de Wetering, M., Hoogstraat, M., Dijkstra, K. K., Krijgsman, O., Kuilman, T., et al. (2015). Preserved genetic diversity in organoids cultured from biopsies of human colorectal cancer metastases. *Proceedings of the National Academy of Sciences of the United States of America*, *112*(43), 13308–13311. https://doi.org/10.1073/pnas.1516689112.

Workman, M. J., Gleeson, J. P., Troisi, E. J., Estrada, H. Q., Kerns, S. J., Hinojosa, C. D., et al. (2018). Enhanced utilization of induced pluripotent stem cell-derived human intestinal organoids using microengineered chips. *Cellular and Molecular Gastroenterology and Hepatology*, *5*(4), 669–677. e662. https://doi.org/10.1016/j.jcmgh.2017.12.008.

Wu, S., Rhee, K. J., Albesiano, E., Rabizadeh, S., Wu, X., Yen, H. R., et al. (2009). A human colonic commensal promotes colon tumorigenesis via activation of T helper type 17 T cell responses. *Nature Medicine*, *15*(9), 1016–1022. https://doi.org/10.1038/nm.2015.

Wu, Z., Su, X., Xu, Y., Kong, B., Sun, W., & Mi, S. (2016). Bioprinting three-dimensional cell-laden tissue constructs with controllable degradation. *Scientific Reports*, *6*, 24474. https://doi.org/10.1038/srep24474.

Wu, Y., Tran, T., Dwabe, S., Sarkissyan, M., Kim, J., Nava, M., et al. (2017). A83-01 inhibits TGF-β-induced upregulation of Wnt3 and epithelial to mesenchymal transition in HER2-overexpressing breast cancer cells. *Breast Cancer Research and Treatment*, *163*(3), 449–460. https://doi.org/10.1007/s10549-017-4211-y.

Xia, X., Li, F., He, J., Aji, R., & Gao, D. (2019). Organoid technology in cancer precision medicine. *Cancer Letters*, *457*, 20–27. https://doi.org/10.1016/j.canlet.2019.04.039.

Xu, Z., Li, E., Guo, Z., Yu, R., Hao, H., Xu, Y., et al. (2016). Design and construction of a multi-organ microfluidic chip mimicking the in vivo microenvironment of lung cancer metastasis. *ACS Applied Materials & Interfaces*, *8*(39), 25840–25847. https://doi.org/10.1021/acsami.6b08746.

Yan, H. H. N., Siu, H. C., Ho, S. L., Yue, S. S. K., Gao, Y., Tsui, W. Y., et al. (2020). Organoid cultures of early-onset colorectal cancers reveal distinct and rare genetic profiles. *Gut*, *69*(12), 2165–2179. https://doi.org/10.1136/gutjnl-2019-320019.

Yang, L., Yang, S., Li, X., Li, B., Li, Y., Zhang, X., et al. (2019). Tumor organoids: From inception to future in cancer research. *Cancer Letters*, *454*, 120–133. https://doi.org/10.1016/j.canlet.2019.04.005.

Yao, Y., Xu, X., Yang, L., Zhu, J., Wan, J., Shen, L., et al. (2020). Patient-derived organoids predict chemoradiation responses of locally advanced rectal cancer. *Cell Stem Cell*, *26*(1), 17–26. e16. https://doi.org/10.1016/j.stem.2019.10.010.

Yin, X., Mead, B. E., Safaee, H., Langer, R., Karp, J. M., & Levy, O. (2016). Engineering stem cell organoids. *Cell Stem Cell*, *18*(1), 25–38. https://doi.org/10.1016/j.stem.2015.12.005.

Yonemura, S. (2014). Differential sensitivity of epithelial cells to extracellular matrix in polarity establishment. *PLoS One*, *9*(11), e112922. https://doi.org/10.1371/journal.pone.0112922.

Young, M., Ordonez, L., & Clarke, A. R. (2013). What are the best routes to effectively model human colorectal cancer? *Molecular Oncology*, 7(2), 178–189. https://doi.org/10.1016/j.molonc.2013.02.006.

Yu, T., Guo, F., Yu, Y., Sun, T., Ma, D., Han, J., et al. (2017). Fusobacterium nucleatum promotes chemoresistance to colorectal cancer by modulating autophagy. *Cell*, *170*(3), 548–563. e516. https://doi.org/10.1016/j.cell.2017.07.008.

Yuki, K., Cheng, N., Nakano, M., & Kuo, C. J. (2020). Organoid models of tumor immunology. *Trends in Immunology*, *41*(8), 652–664. https://doi.org/10.1016/j.it.2020.06.010.

Zhai, Z., Yu, X., Yang, B., Zhang, Y., Zhang, L., Li, X., et al. (2017). Colorectal cancer heterogeneity and targeted therapy: Clinical implications, challenges and solutions for treatment resistance. *Seminars in Cell & Developmental Biology*, *64*, 107–115. https://doi.org/10.1016/j.semcdb.2016.08.033.

Zhang, Y. S., Aleman, J., Shin, S. R., & Kilic, T. (2017). Multisensor-integrated organs-on--chips platform for automated and continual in situ monitoring of organoid behaviors. *Proceedings of the National Academy of Sciences of the United States of America*, *114*(12), E2293–e2302. https://doi.org/10.1073/pnas.1612906114.

Zhang, Y., Lin, L., Jin, Y., Lin, Y., Cao, Y., & Zheng, C. (2016). Overexpression of WNT5B promotes COLO 205 cell migration and invasion through the JNK signaling pathway. *Oncology Reports*, *36*(1), 23–30. https://doi.org/10.3892/or.2016.4772.

Zhou, Y., Arai, T., Horiguchi, Y., Ino, K., Matsue, T., & Shiku, H. (2013). Multiparameter analyses of three-dimensionally cultured tumor spheroids based on respiratory activity and comprehensive gene expression profiles. *Analytical Biochemistry*, *439*(2), 187–193. https://doi.org/10.1016/j.ab.2013.04.020.

Zhu, Y., Wang, L., Yu, H., Yin, F., Wang, Y., Liu, H., et al. (2017). In situ generation of human brain organoids on a micropillar array. *Lab on a Chip*, *17*(17), 2941–2950. https://doi.org/10.1039/c7lc00682a.

Zumwalde, N. A., Haag, J. D., Sharma, D., Mirrielees, J. A., Wilke, L. G., Gould, M. N., et al. (2016). Analysis of immune cells from human mammary ductal epithelial organoids reveals Vδ2+ T cells that efficiently target breast carcinoma cells in the presence of bisphosphonate. *Cancer Prevention Research (Philadelphia, Pa.)*, *9*(4), 305–316. https://doi.org/10.1158/1940-6207.capr-15-0370-t.

CHAPTER TEN

Unraveling the genomic landscape of colorectal cancer through mutational signatures

Marcos Díaz-Gay[a,b,c] and Ludmil B. Alexandrov[a,b,c,*]

[a]Department of Cellular and Molecular Medicine, UC San Diego, La Jolla, CA, United States
[b]Department of Bioengineering, UC San Diego, La Jolla, CA, United States
[c]Moores Cancer Center, UC San Diego, La Jolla, CA, United States
*Corresponding author: e-mail address: l2alexandrov@health.ucsd.edu

Contents

Abstract

Colorectal cancer, along with most other cancer types, is driven by somatic mutations. Characteristic patterns of somatic mutations, known as mutational signatures, arise as a result of the activities of different mutational processes. Mutational signatures have diverse origins, including exogenous and endogenous sources. In the case of colorectal cancer, the analysis of mutational signatures has elucidated specific signatures for classically associated DNA repair deficiencies, namely mismatch repair (leading to microsatellite instability), base excision repair (due to *MUTYH* or *NTHL1* mutations), and polymerase proofreading (due to *POLE* and *POLD1* exonuclease domain mutations). Additional signatures also play a role in colorectal cancer, including those related to normal aging and those associated with gut microbiota, as well as a number of signatures with unknown etiologies. This chapter provides an overview of the current knowledge of mutational signatures, with a focus on colorectal cancer and on the recently reported signatures in physiologically normal and inflammatory bowel disease-affected somatic colon tissues.

Advances in Cancer Research, Volume 151
ISSN 0065-230X
https://doi.org/10.1016/bs.acr.2021.03.003

1. Introduction

Colorectal cancer (CRC), as all types of human cancer, is the result of a Darwinian evolution in which a single cell has evaded growth-control mechanisms and started to divide uncontrollably (Yates & Campbell, 2012). In the majority of cases, this evolutionary trajectory follows a stepwise process in which a normal somatic colorectal cell first converts into a pre-cancerous lesion (*e.g.*, colorectal adenoma) and, subsequently, evolves into an invasive carcinoma which, in turn, may metastasize to distant organs (Vogelstein et al., 1988). A different carcinogenic pathway is found for nearly 15% of CRC cases, arising from a different type of precursor lesions, known as serrated lesions, with profound histopathological and molecular differences than traditional adenomas (Nguyen, Goel, & Chung, 2020). The metamorphism from a normally functioning colorectal cell to an invasive carcinoma is fueled by the acquisition of somatic mutations (Stratton, Campbell, & Futreal, 2009). The majority of somatic mutations found in a cancer genome are considered *passenger mutations* as they do not play a direct role in tumorigenesis; in contrast, a small number of somatic mutations are believed to be *driver mutations* as they are positively selected across cancer genomes and have been shown to confer a clonal growth advantage (Stratton, 2011). The advent of next generation sequencing technologies (Lappalainen, Scott, Brandt, & Hall, 2019) provided a unique opportunity to perform a detailed examination of the repertoire of passenger mutations. Although not directly responsible for neoplastic development, passenger mutations offered an opportunity to elucidate and understand the mutational processes operative in cancer genomes. Specific patterns of somatic mutations, termed *mutational signatures*, are imprinted by the different mutational processes operative throughout the lineage of a cancer cell (Stratton, 2011). Thus, the final catalog of somatic mutations of a cancer can be examined as a historical record bearing the activities of mutational processes operative throughout the lifetime of the individual in which the cancer arose (Stratton et al., 2009). Mutational processes are characterized by an interplay among DNA damage, DNA repair, and DNA replication (Helleday, Eshtad, & Nik-Zainal, 2014). These processes can be triggered by environmental exposures (*e.g.*, ultraviolet light exposure) and lifestyle choices (*e.g.*, tobacco smoking) as well as by infidelity or failure of internal cellular mechanisms (Volkova et al., 2020) or the use of anticancer therapeutics (Pich et al., 2019).

Mutational signatures can be extracted from next generation sequencing data using unsupervised machine learning approaches (Alexandrov, Nik-Zainal, Wedge, Campbell, & Stratton, 2013). Although mutational signatures can be evaluated using different mutation types, signatures were originally explored using single base substitutions (SBS). An intuitive trinucleotide classification is available for SBS, considering both the affected base-pair and its immediate 5′ and 3′ sequence context (*e.g.*, TpCpG mutating to TpApG; mutated base is underlined). Since there are 6 different base substitutions (considering the pyrimidine base of the Watson-Crick base-pair) and 4 possible bases before and 4 possible bases after a substitution, this results in a classification of 96 mutation subtypes (Alexandrov, Nik-Zainal, Wedge, Aparicio, et al., 2013). Although this intuitive classification is not available for other types of somatic mutations, such as short insertions and deletions (indels), copy number variants or large genomic rearrangements, recent efforts have been made to standardize the categorization of other types of somatic mutations (Alexandrov et al., 2020; Nik-Zainal et al., 2016). The transition to other mutation classes could benefit certain cancer types where large-scale somatic mutations play an important role as in the case of CRC, which has a high level of chromosomal instability (Carethers & Jung, 2015; Dienstmann et al., 2017). Moreover, extended contexts beyond trinucleotide classification for SBS signatures (*e.g.*, by considering pentanucleotides) have recently shown potential for specific signature analyses (Alexandrov et al., 2020). This improved resolution could be also applicable for other variant classes and potentially standardized in future pan-cancer analyses as a quality control check for novel signatures (Rogozin et al., 2018).

Since the conception of mutational signatures, an extensive catalog of computational tools has been developed for analyzing signatures from next generation sequencing data (Baez-Ortega & Gori, 2019; Grolleman, Díaz-Gay, Franch-Exposito, Castellvi-Bel, & de Voer, 2019; Omichessan, Severi, & Perduca, 2019). The first algorithm was originally developed in MATLAB (Alexandrov, Nik-Zainal, Wedge, Campbell, & Stratton, 2013) and allowed the extraction of *de novo* mutational signatures by applying nonnegative matrix factorization. *De novo* extraction of mutational signatures is commonly intended for large datasets, allowing the discovery of novel mutational processes operative in the analyzed samples. In contrast, applying mutational signatures in a clinical setting requires the ability to assess signatures and their activities in a single patient. This approach is known as signature *refitting* analysis. Although the lack of statistical power of this alternative methodology prevents the deciphering of novel signatures, the detection of a known set of reference mutational signatures could result in an improved clinical decision-making (Baez-Ortega & Gori, 2019; Rosenthal, McGranahan, Herrero, Taylor, & Swanton, 2016). Importantly, the reduced computational complexity of signature *refitting* analysis also allowed streamlining the implementation and developing easy to use applications (Díaz-Gay et al., 2018).

Benefitting from the large amount of cancer sequencing data produced in the last years, the reference set of mutational signatures was developed and provided to users within the Catalogue of Somatic Mutations in Cancer (COSMIC) database (Tate et al., 2019). Novel reference signatures have been added as the number of examined cancer genomes has increased over the last 8 years. In the case of SBS mutational signatures, four signatures were originally reported in a cohort of 21 breast cancers (Alexandrov, Nik-Zainal, Wedge, Campbell, & Stratton, 2013; Nik-Zainal et al., 2012), then 21 signatures were elucidated after the first pan-cancer analysis across 7042 cases of 30 different cancer types (COSMIC v1 released in 2013) (Alexandrov, Nik-Zainal, Wedge, Aparicio, et al., 2013) and, subsequently, 30 signatures were detected in 12,000 cancer cases (COSMIC v2 released in 2015) (Alexandrov et al., 2015; Nik-Zainal et al., 2016). The most recent update of COSMIC encompasses 54 SBS signatures from the analysis of 23,829 cancer cases (COSMIC v3.1 released in June 2020) (Alexandrov et al., 2020; Boot et al., 2020; Lee-Six et al., 2019; Li et al., 2020; Pleguezuelos-Manzano et al., 2020). Importantly, since COSMIC v3, reference sets of doublet base substitution (DBS) and indel (ID) signatures have been reported from different cancer types, resulting in a total of 83 mutational signatures (54 SBS, 11 DBS, and 18 ID) in COSMIC v3.1, available at https://cancer.sanger.ac.uk/cosmic/signatures.

The repertoire of mutational signatures found in CRC encompasses 25 signatures of SBSs, 9 signatures of DBSs, and 5 signatures of IDs (Fig. 1). These signatures vary from ones associated with failure of endogenous molecular processes to ones attributed to clock-like mutagenesis due to aging. Mutational signatures found in CRC are also consistent with prior observations of hypermutated and ultra-hypermutated CRC samples. Specifically, hypermutated CRC tumors with microsatellite instability (MSI) account for 17% of CRC cancers, while ultra-hypermutated tumors with *POLE* deficiency are found in approximately 3% of CRC patients (Campbell et al., 2017; Cancer Genome Atlas Network et al., 2012; Shinbrot et al., 2014). Specific mutational signatures have been attributed to failed DNA mismatch repair

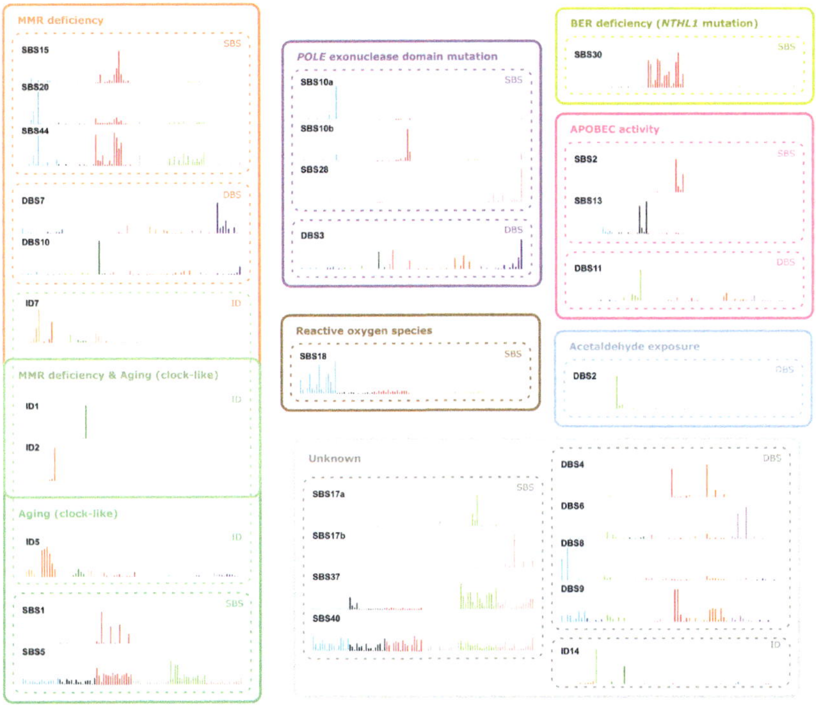

Fig. 1 Mutational profiles of mutational signatures observed in CRC samples from the PCAWG consortium. Signatures are organized by their etiologies. SBS signatures contributing to more than 5 CRC samples with more than 10 mutations are shown. Different profiles are displayed according to the mutation class, including 96 subtypes in the case of SBS, 83 subtypes in the case of ID, and 78 subtypes in the case of DBS (further information about the mutational profiles can be found at https://cancer.sanger.ac.uk/cosmic/signatures).

(MMR) as well as to deficiency of polymerase epsilon (*POLE*) due to mutations in its exonuclease domain (Alexandrov et al., 2020; Alexandrov, Nik-Zainal, Wedge, Aparicio, et al., 2013) (Fig. 1). These potent signatures generally contribute tens to hundreds of thousands of somatic mutations in a sample, thus, giving rise to the hypermutated and ultra-hypermutated phenotypes. Conversely, the most prevalent mutational signatures in CRC have been the ones associated with aging. These signatures reflect clock-like mutational processes generating somatic mutations throughout the cancer cell's lineage, which starts with the first division of the fertilized egg (Alexandrov et al., 2015) (Fig. 1). In most cases, the numbers of somatic mutations generated by these aging signatures in any cancer sample ranges approximately between 1 and 10000 mutations (Alexandrov et al., 2020) and, while they do not give rise to a hypermutated phenotype, these clock-like signatures are ubiquitously found in all cancer samples across the complete spectrum of human neoplasia (Alexandrov et al., 2020). In recent years, additional internal cellular mechanisms, such as other DNA repair pathway deficiencies, reactive oxygen species, activity of APOBEC deaminases, and even an interplay with the gut microbiota, have been linked to specific mutational signatures. A comprehensive analysis of the most up-to-date distribution of mutational signatures in CRC has been summarized for this chapter based on the recently published data from the Pan-Cancer Analysis of Whole Genomes (PCAWG) Consortium (ICGC/TCGA Pan-Cancer Analysis of Whole Genomes Consortium, 2020). The PCAWG dataset includes 2780 whole-genome sequenced patients from different cancer types; these data were further combined with 1865 whole-genome and 19,184 whole-exome sequenced cancers from previous studies to assemble the most recent and largest pan-cancer analysis of mutational signature (Alexandrov et al., 2020). In regard to CRC, a total of 1185 samples were considered, including sequencing data for 73 whole-genome (6%) and 1112 whole-exome (94%) sequenced colorectal cancers. In all cases, these analyses reported mutational signatures of every variant class by considering only high-confidence samples: at least 15 somatic mutations and a signature reconstruction accuracy above 0.90. Further, CRC patients have been generally stratified based on their tumor mutational burden (TMB) for single base substitutions by fitting a Gaussian mixture model, resulting in significant differences in operative mutational signatures among the three TMB groups (Fig. 2).

Recently, the role of mutational signatures as a tool for identifying CRC predisposition syndromes has been proposed, given the mutational signature specificity in certain DNA repair pathways deficiencies (Ma, Setton, Lee, Riaz, & Powell, 2018; Van Hoeck, Tjoonk, van Boxtel, & Cuppen, 2019). Germline mutations affecting genes belonging to these pathways are responsible for well-known hereditary syndromes, as in the case of MMR genes (*MLH1*, *MSH2*, *MSH6*, *PMS2*) for Lynch syndrome or base excision repair genes *MUTYH* and *NTHL1* for *MUTYH*- (MAP) and *NTHL1*-associated polyposis (NAP), respectively (Lynch, Snyder, Shaw, Heinen, & Hitchins, 2015; Valle et al., 2019). In fact, both Lynch syndrome and MAP were originally identified according to their distinctive genetic alterations (Al-Tassan et al., 2002; Lynch & Krush, 1971; Lynch et al., 2015), now captured by specific mutational signatures. In these hereditary cases, mutational signatures could be used as a diagnostic tool, as it was the case for *NTHL1*-mutated patients and their link with a high level of

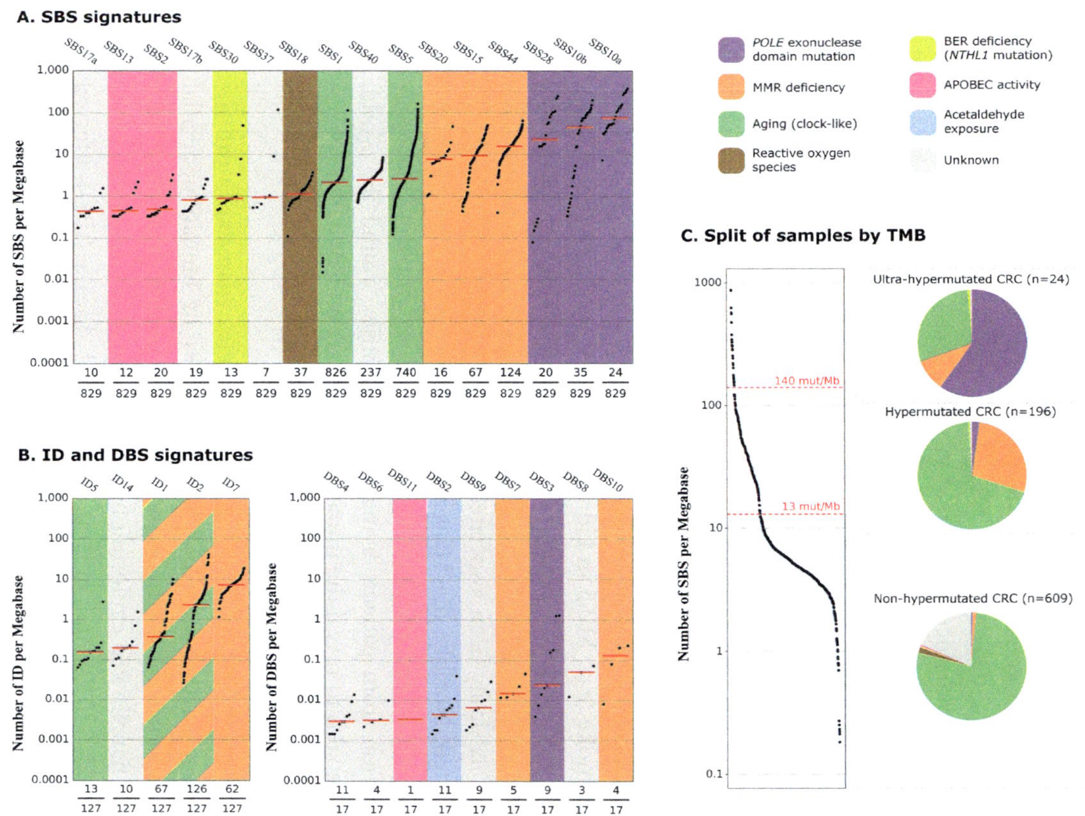

Fig. 2 Tumor mutational burden of mutational signatures found in CRC samples from the PCAWG consortium. Mutational signatures were separated by types of somatic mutations: (A) single base substitutions (SBSs), (B) small insertions and deletions (IDs), and (C) doublet base substitutions (DBSs). Only samples with a minimum reconstruction accuracy of 0.90 and a minimum of 15 SBSs were considered. In the case of SBS signatures, only signatures operative in more than 5 samples and contributing at least 10 mutations were included in the plots. (D) Classification of CRC samples by their SBS burden (thresholds were defined by fitting a Gaussian mixture model).

signature SBS30 (Grolleman et al., 2019). However, the potential of this methodology is even broader, since it has proved useful both to ascertain the pathogenicity of previously unknown significant mutations in high-risk genes such as *POLE* (Castellsagué et al., 2019), and also to identify DNA repair deficiencies in cases lacking mutations in genes commonly evaluated as part of standard clinical practice, *e.g.*, homologous recombination deficient breast cancer patients lacking mutations in *BRCA1*, *BRCA2*, *RAD51*, or *PALB2* (Davies et al., 2017).

Despite the importance of different environmental mutagens in CRC risk, including alcohol consumption (Bagnardi et al., 2015), tobacco smoking (Botteri et al., 2008) or high intake of red and processed meats (Bouvard et al., 2015), no mutational signature found in colorectal cancer has been associated with any of these risk factors. The lack of direct signature relationship may be due to poorly annotated genomic datasets that lack information for most risk factors. Nevertheless, the lack of association with known smoking signatures is particularly striking as mutational signatures have been attributed to tobacco smoking in both tissues directly exposed to tobacco smoke (*e.g.*, SBS4 in lung cancer) and tissues indirectly exposed to tobacco smoke (*e.g.*, SBS5 in kidney cancer) (Alexandrov et al., 2016). Although aging and DNA repair processes currently underpin the most common mutational signatures found in CRC, different environmental risk factors have classically been associated with colorectal cancer and a future link is expected. Notably, the *Mutographs of cancer* project is currently generating 1000 whole-genome sequenced CRC with extensive epidemiological information from patients living in 11 countries around the world. *Mutographs*' goal is to reveal the mutational processes underlying the previously reported epidemiological associations (Ginsburg, Ashton-Prolla, Cantor, Mariosa, & Brennan, 2021).

In the last decade, thousands of cancer genomes were sequenced allowing the exploration of mutational signatures in human neoplasia. Recently, sequencing data from laser-capture microdissection of colonic crypt clones has allowed exploring mutational signatures in normal colon (Lee-Six et al., 2019) and inflammatory bowel disease (IBD)-affected colon (Olafsson et al., 2020). Accordingly, the main goal of this chapter will be to summarize the current knowledge about mutational signatures both in normal colorectal tissue and colorectal cancer as well as to discuss the potential clinical applications and future directions of this methodology.

2. Mutational signatures of DNA repair and polymerase deficiencies

DNA repair mechanisms act as a physiological barrier against tumor progression, limiting genome instability and the accumulation of somatic mutations, which could benefit neoplastic cells by the acquisition of certain cancer hallmarks (Hanahan & Weinberg, 2011). Different classes of DNA repair are activated in response to specific DNA lesions, which can originate from endogenous or exogenous sources (Jeggo, Pearl, & Carr, 2016). Genetic and epigenetic alterations in the different DNA repair pathways can result in their failure, thus, increasing the accumulation of mutations with specific patterns that can be captured by analysis of mutational signatures (Ma et al., 2018).

2.1 Mutational signatures of failed mismatch repair (MMR)

Impairment of the mismatch repair (MMR) pathway has been most closely associated with both sporadic and hereditary CRC cases (Cancer Genome Atlas Network et al., 2012; Kuipers et al., 2015); the latter related to the well-known Lynch syndrome, characterized by germline mutations in the genes belonging to the MMR pathway and accounting for up to 4% of all CRC cases (Jasperson, Tuohy, Neklason, & Burt, 2010). MMR is the mechanism used by cells to correct the mismatches generated during DNA replication and not fixed by the intrinsic proofreading activity of the DNA polymerases (Kunkel & Erie, 2015). Post-replicative MMR repairs both mis-paired bases, as well as insertion/deletion loops arising from strand slippage during DNA replication (Peña-Diaz & Jiricny, 2012). In consequence, MMR deficiency leads to a large number of mutations, also referred to as high tumor mutational burden or hypermutation, including single base substitutions and, especially, small insertions and deletions. Mismatch repair plays an essential role in microsatellite regions, repetitive sequences normally shorter than five base-pairs (Helleday et al., 2014), prone to replication slippage which are normally stabilized by MMR. Thus, the well-known microsatellite instability (MSI) phenotype is almost always a direct consequence of malfunctioning MMR (Grilley, Holmes, Yashar, & Modrich, 1990; Shibata, Peinado, Ionov, Malkhosyan, & Perucho, 1994), and has been used as a biomarker for this DNA repair deficiency in CRC throughout the past three decades (Ionov, Peinado, Malkhosyan, Shibata, & Perucho, 1993).

The number of mutational signatures associated with MMR has increased with the number of examined samples. Originally signature SBS6, characterized by a large number of C>T transitions at Np<u>C</u>pG and Gp<u>C</u>pN trinucleotide contexts and associated with a large number of indels, was found in MSI samples and attributed to failed MMR (Alexandrov, Nik-Zainal, Wedge, Aparicio, et al., 2013). A recent characterization of large numbers of cancer patients (Alexandrov et al., 2020) has revealed a number of signatures putatively attributed to deficiency in the MMR pathway: seven SBS signatures (SBS6, SBS14, SBS15, SBS20, SBS21, SBS26 and SBS44), two DBS signatures (DBS7 and DBS10), and three different ID signatures (ID1, ID2 and ID7). Two of these signatures (SBS14 and SBS20) were linked to concurrent defects on MMR and polymerase proofreading ability by DNA polymerases (Haradhvala et al., 2018). Interestingly, a recent report suggests that MMR deficiency could also be the underlying cause of signature SBS11 when present in conjunction with temozolomide treatment (Touat et al., 2020). SBS11 was previously thought to be originated directly from the exposure to temozolomide (Alexandrov, Nik-Zainal, Wedge, Aparicio, et al., 2013) but its pattern differed significantly from the one arising from direct exposure to temozolomide in human induced pluripotent stem cells after *in vitro* exposure (Kucab et al., 2019).

Among SBS signatures, other than SBS14, all other signatures associated with MMR deficiency were found in the CRC PCAWG dataset, although only three of them affected more than five samples, namely SBS15, SBS20 and SBS44 (Fig. 2A). Their presence is significantly enriched in the group of hypermutated CRCs (mean TMB of 46 mutations per megabase), where they represent 28% of all contributions of the known SBS signatures (Fig. 2C). Interestingly, the mutational profile of SBS44, the most prevalent in CRC, was almost perfectly reconstructed by a CRISPR-Cas9-generated *MLH1* knockout performed in human colon organoids, therefore highlighting the importance of this specific mutational signature in CRC and its link with this particular MMR gene deficiency (Drost et al., 2017). Experimental validation of other MMR-related signatures was also performed in *C. elegans* and human cell lines, even though mixed profiles were observed in these studies, encompassing different contributions of the mutational signatures of failed MMR (Meier et al., 2018; Zou et al., 2018). On the other hand, SBS20 was previously associated with a concurrent MMR and polymerase proofreading deficiency driven by mutations in the exonuclease domain of polymerase delta (*POLD1*) (Haradhvala et al., 2018), a combination that was already reported in CRC (Jansen et al., 2016). Interestingly, in the PCAWG dataset, SBS20 was not detected in any ultra-hypermutated colorectal tumor or in combination with any *POLE* mutation-associated signatures. Rather, SBS20 was almost exclusively found in hypermutated MSI samples without any known *POLE* deficiencies (Fig. 2).

Considering the distribution of mutations across the cell cycle, in general, SBSs tend to accumulate in heterochromatic late replicating regions (Woo & Li, 2012). However, recent studies suggested that this is not the case for somatic mutations generated by some MMR-related signatures (Morganella et al., 2016; Tomkova, Tomek, Kriaucionis, & Schuster-Bockler, 2018). In fact, MMR activity has been proposed to be the cause of the enrichment of mutations in late replicating regions (Supek & Lehner, 2015). Interestingly, two of the MMR SBS signatures (SBS21 and SBS26, both dominated by T > C transitions) present transcriptional strand bias (TSB), with more mutated T than A on the untranscribed strand of the coding genome compatible with damage to thymine, a feature commonly associated with transcription-coupled nucleotide excision repair (Alexandrov et al., 2015; Nik-Zainal et al., 2016). A link between transcription-coupled nucleotide excision repair and MMR was historically observed in human cell lines (Mellon, Rajpal, Koi, Boland, & Champe, 1996; Zhao, Jain, Iyer, Modrich, & Vasquez, 2009) and recently suggested as the mechanism behind the TSB found in indels across different types of human cancer (Georgakopoulos-Soares et al., 2020). Further research is needed in order to evaluate this interaction between DNA repair pathways and why TSB is affecting only two of the mutational signatures associated with MMR deficiency (none of them prevalent in CRC).

In the case of indel (ID) signatures, three different signatures were associated with MMR repair defects in CRC (ID1, ID2 and ID7). All of these signatures were dominated by one base-pair insertions or deletions at long mononucleotide repeats (Fig. 1). Interestingly, both ID1 and ID2 were also found in most other cancer types, presenting a correlation with the age of diagnosis in non-hypermutated samples, and are potentially due to strand slippage during DNA replication (Alexandrov et al., 2020). However, in MMR-deficient cases, both of these signatures have a highly increased number of mutations, which is accompanied by high contributions of one or more of the SBS signatures also linked with MMR deficiency (Fig. 2B). This increment would be in agreement with the impaired repair of strand slippages caused by a malfunctioning MMR (Peña-Diaz & Jiricny, 2012).

The number of available samples with sufficient reconstruction accuracy was considerably reduced for doublet base substitutions (DBS) as these types of somatic mutations are rare and, generally, cannot be detected from whole-exome sequencing; note that the majority of CRC samples in the PCAWG dataset are whole-exome sequenced (94%). Nevertheless, two DBS signatures associated with MMR deficiency (DBS7 and DBS10; dominated by TpT > ApA and CpG > TpA substitutions respectively) (Fig. 1) were detected in CRC whole-genome sequenced samples with these samples harboring high numbers of other MMR SBS signatures (Alexandrov et al., 2020). Furthermore, signature DBS10 presented the highest median number of mutations of all DBS signatures identified in CRC (Fig. 2B).

2.2 Mutational signatures of deficient polymerase proofreading

The main role of DNA polymerases *POLE* and *POLD1* is to accurately replicate the nuclear DNA from the leading and lagging strands, respectively (Georgescu et al., 2014). In addition, both enzymes present exonuclease proofreading activity, constituting the first defense barrier of cells against the intrinsic slight infidelity of the DNA replication process and followed by the aforementioned post-replicative MMR (Rayner et al., 2016). Interestingly, there is also evidence supporting *POLD1* as the responsible for both DNA strands synthesis, whereas *POLE* being in charge of the leading strand proofreading (Johnson, Klassen, Prakash, & Prakash, 2015). Although both polymerase proofreading and MMR are responsible for the DNA replication integrity, the role of MMR failures in human cancer was established much earlier (Ionov et al., 1993). In contrast, the evidence for the role of *POLE* and *POLD1* in cancer mutagenesis has been accumulating over the last three decades including experimental reports showing that mutated polymerases lead to a large increase in the number of mutations in yeast (Morrison, Johnson, Johnston, & Sugino, 1993) as well as to the appearance of tumors in mice (Albertson et al., 2009; Goldsby et al., 2001). The link to human cancer was only recently uncovered, connecting colorectal and endometrial cancers to specific missense mutations targeting the exonuclease proofreading domain of *POLE* and *POLD1* both in sporadic (Cancer Genome Atlas Network et al., 2012, 2013) and in hereditary cases (Palles et al., 2013). This proofreading domain is in charge of the repair activity produced immediately after DNA synthesis during replication (Jain, Aggarwal, & Rechkoblit, 2018; Rayner et al., 2016). Missense mutations affecting the exonuclease domain impair the proofreading capabilities of these two polymerases, while maintaining the essential DNA replication abilities. Lack of truncating driver mutations in these two genes may be also a consequence of their involvement in this critical cellular mechanism (Bellido et al., 2016).

Similar to signatures of failed MMR, mutational signatures associated with deficiencies of polymerase proofreading have increased as the number of examined CRC samples have grown. The first pan-cancer analysis associated signature SBS10 with mutations in the exonuclease domain of *POLE* (Alexandrov, Nik-Zainal, Wedge, Aparicio, et al., 2013), a signature that was subsequently divided into SBS10a and SBS10b, after the decomposition of its two main components, TpCpT > TpApT and TpCpG > TpTpG substitutions (Alexandrov et al., 2020). Interestingly, an additional signature, SBS28, was also found in most samples harboring SBS10a and SBS10b mutations, potentially explaining the third minor component of the original SBS10 signature related to T > G transversions (particularly in TpTpT trinucleotide contexts) (Fig. 1). CRC data from the PCAWG consortium showed the interplay among these three signatures, with a special relevance in the ultra-hypermutator group of samples, defined by a TMB larger than 140 mutations/megabase, and characterized by somatic *POLE* exonuclease domain mutations and predominantly microsatellite stable (MSS) phenotype (Fig. 2A and C). Both the number of mutations and the distribution of samples correlate with the intensity of the different characteristic features previously observed in the original SBS10 signature, with a highest prevalence of the C > A component (captured by SBS10a), followed by C > T (SBS10b) and T > G (SBS28). Interestingly, this subset of ultra-hypermutated CRC was found to have a better prognosis, in comparison with MSS *POLE* wild-type cases (Domingo et al., 2016). This pattern is also observed in other cancer types, including endometrial cancer (Church et al., 2015) and glioblastoma (Erson-Omay et al., 2015). The better prognosis is likely attributable to the enrichment in neoepitopes generated by the high volume of mutations and the proposed associated increase in the immunogenic response (van Gool et al., 2015). The consequences of the *POLE* exonuclease domain mutations on the mutational spectrum and their association with mutational signatures were also experimentally validated in mice (Li et al., 2018), yeast (Xing et al., 2019), and human cell lines (Hodel et al., 2020). Interestingly, a replication strand bias was found for these signatures, with more mutations attributed to the leading strand (Shinbrot et al., 2014; Tomkova et al., 2018), which is in agreement with *POLE* being the polymerase in charge of at least the leading strand proofreading and, potentially, also its synthesis (Johnson et al., 2015). Apart from purely

POLE-specific mutational signatures, SBS14 and SBS20 signatures were observed in samples with a combination of failure of MMR and polymerase proofreading deficiencies in either *POLE* or *POLD1* (Haradhvala et al., 2018). It should be noted that no individual mutational signature has been associated with deficiencies in *POLD1* in MSS CRC. In addition to the signatures of single base substitutions, a doublet base substitution signature, namely, DBS3, has been attributed to *POLE* deficiency (Fig. 2B). No indel signatures have been associated with polymerase proofreading deficiency in CRC or any other cancer type (Alexandrov et al., 2020).

2.3 Mutational signatures of failed base excision repair (BER)

Apart from failure of MMR or failure of polymerase proofreading, a deficiency in base excision repair (BER) has classically been linked to CRC, especially in hereditary cases (Weren et al., 2018). BER is in charge of the removal and replacement of different types of aberrant bases in the DNA sequence, with a predominant role in correcting oxidative damage (Chatterjee & Walker, 2017). Correction of oxidative damage requires prior lesion recognition, which is preferentially performed by DNA glycosylases (Odell, Wallace, & Pederson, 2013). Biallelic germline mutations in one of these glycosylases, *MUTYH*, are known to cause an inherited adenomatous polyposis syndrome (namely, MAP) presenting an 80% lifetime risk of CRC (Al-Tassan et al., 2002; Kanth, Grimmett, Champine, Burt, & Samadder, 2017). However, a moderate increase (1.5- to 2-fold) of CRC risk was also found in cases carrying monoallelic germline *MUTYH* mutations, particularly in those patients with a first-degree relative diagnosed with CRC (Win et al., 2014). Recently, a pattern of somatic mutations was identified in MAP patients and linked to a specific mutational signature, that was initially identified as SBS18 (Pilati et al., 2017) and remapped to SBS36 in subsequent studies (Alexandrov et al., 2020; Viel et al., 2017). Interestingly, the current proposed etiology for SBS18 is damage by reactive oxygen species (ROS), which is the specific target of *MUTYH* repair activity. Notably, signatures SBS18 and SBS36 present similar mutational profiles, dominated by C>A transversions, with a cosine similarity of 0.91 and they likely represent mutations due to slightly different combinations of oxidative damage. *MUTYH* avoids potential C>A mutations, which may be produced when a ROS product (*e.g.*, 8-oxo-guanine) is mis-paired with adenine (Poetsch, 2020). Similarly, signature SBS45, an artifactual signature also due to oxidative damage, is characterized by C>A mutations but, in the case of signature

SBS45, these mutations are introduced during DNA acoustic shearing in the context of library preparation for next generation sequencing (Alexandrov et al., 2020; Costello et al., 2013).

Additionally, a unique mutational signature, SBS30, was found in samples with deficiency in the *NTHL1* glycosylase with a profile dominated by C>T transitions at NpCpA and NpCpC contexts (Fig. 1). Recent studies showed the link between this BER gene deficiency and cancer predisposition in the form of an autosomal recessive polyposis syndrome (namely, NAP) giving rise to both CRC (Weren et al., 2015) and potentially affecting 13 other cancer types (Grolleman, de Voer, et al., 2019). However, unlike *MUTYH*, no increased CRC risk has been found for monoallelic germline carriers of *NTHL1* mutations (Elsayed et al., 2020). Subsequent functional studies using CRISPR-Cas9 in human colon organoids were performed in order to validate the association between *NTHL1* biallelic mutations and signature SBS30 (Drost et al., 2017). SBS30 signature is found in only a very small subset of CRC cases, which is in agreement with the very low prevalence of NAP (Weren et al., 2018). Surprisingly, SBS30 has a median TMB significantly reduced in comparison to other signatures related to DNA repair defects (Fig. 2A). As in the case of some MMR-related mutational signatures, the common enrichment in late replicating regions for SBSs was not found for those mutations derived from SBS30 activity (Morganella et al., 2016).

2.4 Deficiencies in other DNA repair pathways

Malfunctioning of other DNA repair pathways has also been associated with specific mutational signatures despite these signatures not being commonly present in CRC, including signatures due to malfunctioning of nucleotide excision repair (NER) or signatures due to failure of repair of DNA double strand breaks by homologous recombination (HR). NER is a well-known mechanism in charge of removing bulky DNA lesions, recently linked to signature SBS8 in case of its global genome sub-pathway (Jager et al., 2019). In addition, transcription-coupled NER activity is proposed as the underlying mechanism for the TSB observed in several of the COSMIC mutational signatures (Alexandrov et al., 2020). On the other hand, HR deficiency favors more error-prone alternative pathways, including non-homologous end joining (NHEJ), that generates a large number of structural variants and microhomology-mediated indels, as well as specific patterns of single base substitutions (Ceccaldi, Rondinelli, & D'Andrea,

2016). The current mutational signature framework captured these patterns, mainly as signatures SBS3 and ID6. Signatures SBS3 and ID6 are strongly associated with germline and somatic mutations in *BRCA1* and *BRCA2* genes in breast, pancreatic and ovarian cancers (Alexandrov et al., 2020; Nik-Zainal et al., 2016) as well as, in the case of SBS3, with mutations in other genes in the HR pathway (Polak et al., 2017). Additionally, to a lesser extent, signatures SBS8, ID8 and two rearrangement signatures (one specifically linked to *BRCA1* alterations) were also proposed to be related to HR failure and increased NHEJ activity in regard to double strand breaks, particularly in breast cancer (Alexandrov et al., 2020; Nik-Zainal et al., 2016). None of these signatures were found to be significantly active in the CRC PCAWG cohort (only five cases affected by signature SBS3), which contradicts prior observations of an increase in CRC risk found in patients harboring *BRCA1* mutations (Oh et al., 2018) as well as studies reporting both *BRCA1* and *BRCA2* as moderate-risk CRC susceptibility genes (Valle, Vilar, Tavtigian, & Stoffel, 2019). Interestingly, certain mutational signatures related to HR or MMR deficiencies were found enriched in brain metastasis, which are rare but extremely lethal in CRC patients (Sun et al., 2019). Lastly, an increase in the risk of CRC has been established for patients with Li-Fraumeni syndrome (Valle, de Voer, et al., 2019). Li-Fraumeni syndrome has historically been associated with a number of cancer types and is caused by germline mutations in the tumor suppressor gene *TP53*, responsible for the modulation of DNA repair activities (Kratz et al., 2017). Thus far, no specific mutational signature has been associated with Li-Fraumeni syndrome or with mutations in *TP53*.

3. Signatures of clock-like mutational processes

Certain mutational processes are permanently active throughout life due to normal cellular activities (Alexandrov et al., 2015). These age-associated processes give rise to specific patterns of mutations that are accumulating in a steady clock-like manner since the fertilized egg (Alexandrov et al., 2015) and are observed not only in cancers but also in normal somatic cells. Clock-like signatures exhibit a strong correlation with age, in contrast to all other processes that may have acted in an episodic manner (Alexandrov & Stratton, 2014). SBS1 and SBS5 were the first signatures observed to follow this clock-like pattern in certain tissues, being responsible from a sustained increase in mutations correlating with age at diagnosis in patients across the complete spectrum of human cancer types (Alexandrov et al., 2015). SBS1 and SBS5 were consistently identified in normal cells from different tissue types of healthy individuals exhibiting a strong correlation with age (Blokzijl et al., 2016; Brunner et al., 2019; Lee-Six et al., 2019; Martincorena et al., 2018; Moore et al., 2020; Yoshida et al., 2020). Remarkably, clock-like properties were tissue-dependent for both signatures, with SBS1 found significantly associated with age both in colorectal cancer and normal colon, but not SBS5 (Alexandrov et al., 2015; Blokzijl et al., 2016).

SBS1 has been attributed to the spontaneous deamination of 5-methylcytosine at NpCpG trinucleotides leading to T-G mismatches, which, if not repaired before DNA replication, result in C>T transitions (Bell & Dutta, 2002). While SBS1 is found in almost every cancer genome, it is responsible for a different number of mutations depending on the cancer type. Since those differences cannot be attributed to variations in CpG methylation across cell types (Horvath, 2013), they reflect differences in the mitotic rates among tissues, thus, establishing SBS1 as a molecular clock recording the number of mitoses completed by a particular cell during its lifetime (Alexandrov et al., 2015; Tomasetti, Li, & Vogelstein, 2017). In contrast, the etiology of SBS5 remains unknown, although the TSB observed in the dominant T>C transitions at ApTpN trinucleotides suggests a potential involvement of transcription-coupled NER activity (Alexandrov et al., 2015). In urothelial cancer, high elevation of SBS5 was associated with mutations in the NER gene *ERCC2* (Kim et al., 2016). However, in certain tissues, SBS5 was not found to correlate with age despite accounting for a large number of mutations, therefore suggesting the hypothesis of being activated by external factors during life or as part of the tumor evolution process in certain cancer types (Alexandrov et al., 2015). Interestingly, an association with tobacco smoking was previously reported for SBS5, with larger numbers of mutations originated by this signature found in smokers (Alexandrov et al., 2016; Kim et al., 2016). In addition, in the case of kidney cancer, the SBS5 mutation rate was found to be elevated in both renal clear cell carcinoma and renal papillary carcinoma but not in kidney chromophobe tumors. This distinction could potentially link SBS5 with a continuous exposure to certain metabolic mutagens, since the first two cancer types originate from kidney proximal tubular epithelium, which is responsible for the absorption of metabolites (Alexandrov et al., 2015).

Other mutational signatures were also found to correlate with the age of cancer diagnosis in certain cancer types. This is the case of SBS40, with an unknown etiology and a mutational profile closely resembling the one from SBS5, and DBS signatures DBS2 and DBS4 (Alexandrov et al., 2020), which aggregated mutational profile matches the one found in normal mouse cells (Hill, Wang, Farwell, & Sommer, 2003). In the case of indels, a clear age correlation was found for ID5 and ID8, whereas in the case of ID1 and ID2, this correlation is only significant in non-hypermutated cancers because of the aforementioned large accumulation of indels arising from both ID1 and ID2 signatures in MMR-deficient cases (Alexandrov et al., 2020). Signatures ID1, ID2 and ID5 were found in the CRC PCAWG data; ID1 and ID2 were found to contribute high numbers of indels in MSI hypermutated colorectal tumors (Fig. 2B).

Recently, the sequencing of colonic crypts obtained by laser-capture microdissection allowed the analysis of mutational signatures in normal (Lee-Six et al., 2019) and inflammatory bowel disease (IBD)-affected colon (Olafsson et al., 2020). These studies showed a ubiquitous activity of clock-like mutational signatures in normal colon cells. IBD is a well-known risk factor for CRC, as a consequence of the associated chronic inflammation, increasing the risk in parallel with the duration and extent of the disease (Beaugerie & Itzkowitz, 2015; Jess, Rungoe, & Peyrin-Biroulet, 2012; Lutgens et al., 2013). Interestingly, IBD-affected colon was found to harbor a 2.4-fold increase in TMB *vs* healthy tissue, with most additional mutations linked to clock-like mutational processes (Olafsson et al., 2020).

Mutational signatures in both normal and IBD datasets were evaluated using the state-of-the-art computational method for the analysis of mutational signatures (Alexandrov et al., 2020). Interestingly, at least one of the clock-like SBS mutational signatures (SBS1 and SBS5) was found active in all samples. The repertoire of signatures of normal and IBD-affected colon was found dominated by the aging signatures, with an average contribution over 80% in both cases (Fig. 3A and B), in contrast with the decrease in contribution in CRC along with the increase in TMB (77% in non-hypermutated CRC, 69% in hypermutated and only 29% in ultra-hypermutated) (Fig. 2C). In terms of the total number of generated mutations, a significant increase was identified in CRC patients in comparison with normal and IBD-affected cases (Fig. 3C). Additionally, in CRC, a correlation was observed between the number of total mutations and the ones linked to the clock-like mutational signatures, with ultra-hypermutated CRC cases harboring the largest numbers of mutations despite their reduced average contribution (Fig. 3C). This observation indicates that additional factors could be influencing the underlying mechanisms of these signatures in cancer,

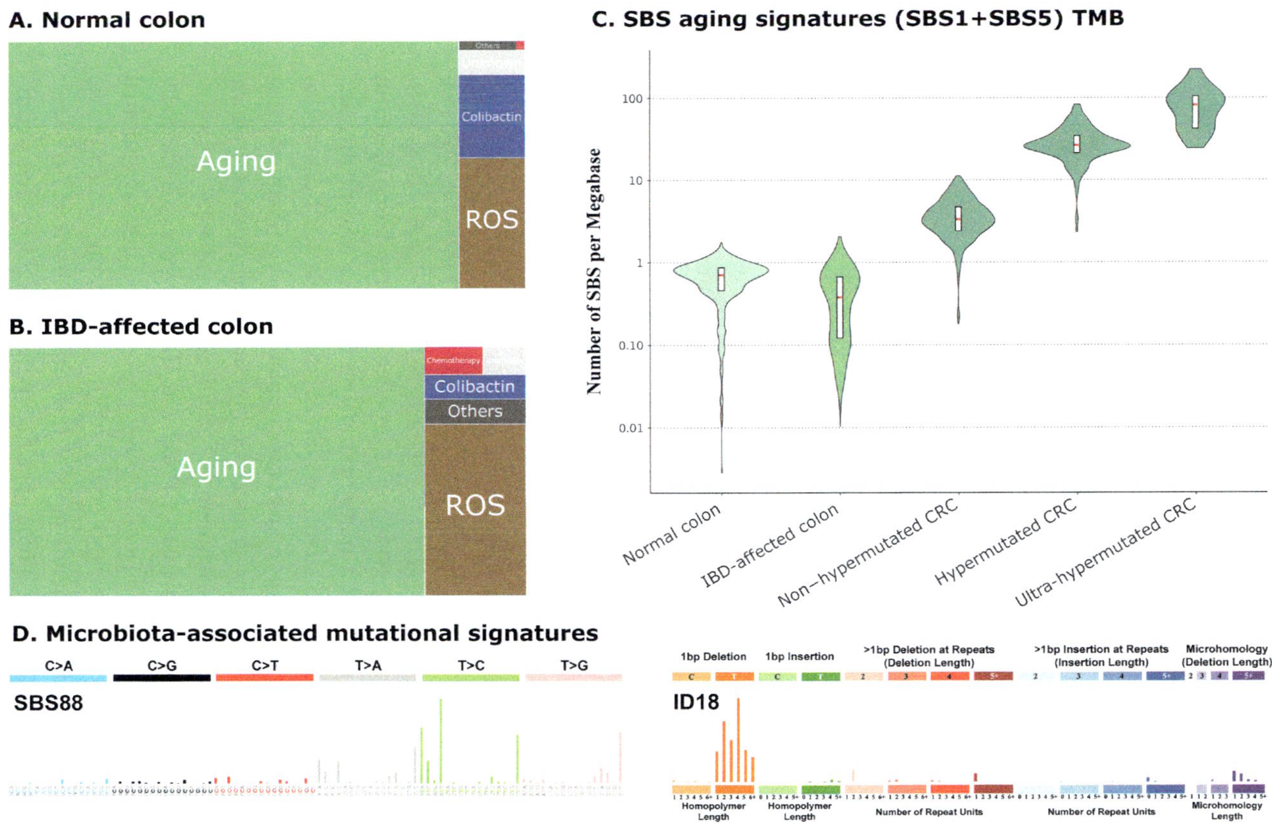

Fig. 3 Repertoire of mutational signatures in normal and inflammatory bowel disease (IBD)-affected colonic tissue. Average signature contributions in (A) normal colon and (B) IBD-affected colonic tissue. (C) Comparison of the mutation burden arising from SBS age-related signatures in normal colon, IBD-affected colon, and different subtypes of CRC. (D) Mutational profiles of colibactin-associated signatures SBS88 and ID88, found in normal and IBD-affected colon.

accelerating the aging mechanisms found active in normal cells. In the hypermutated CRC subgroup, which is characterized by MMR deficiency, clock-like signatures were also found to be dominant (69% of all single base substitutions) (Fig. 2C), raising the question of whether MMR deficiency or other non-clock-like factors could be responsible for the increased mutational burden generated by these signatures. Similarly, clock-like signatures were also identified as the major contributors in non-hypermutated CRC, whereas SBS40, which correlates with age in other tissue types, was predominant among the signatures with unknown etiology (Fig. 2C) (Alexandrov et al., 2020).

4. Mutational signatures associated with gut microbiota

Recent analyses of mutational signatures in both normal (Lee-Six et al., 2019) and IBD-affected colon (Olafsson et al., 2020) have allowed the identification of the catalog of somatic mutations in non-cancerous colonic tissue. This new large set of whole-genome sequencing data also facilitated the deciphering of novel mutational signatures which had not been identified in the CRC PCAWG data, likely due to the lack of statistical power. Interestingly, the first two mutational signatures attributed to an interplay with the gut microbiota, single base substitution signature SBS88 and indel signature ID18 (initially named SBSA and IDA) were identified in these datasets. Both signatures were first extracted in normal colon (Lee-Six et al., 2019) and then replicated in IBD patients (Olafsson et al., 2020). Signature SBS88 is dominated by T>C transitions at Ap$\underline{\text{T}}$pA, Ap$\underline{\text{T}}$pT and Tp$\underline{\text{T}}$pT as well as T>G transversions at Tp$\underline{\text{T}}$pT (Fig. 3D). SBS88 also presents a strong enrichment in adenine three bases upstream of the mutated thymine and a transcriptional strand bias toward thymine mutations on the transcribed strand, which is compatible with damage to adenine and activity of transcription-coupled NER (Pleguezuelos-Manzano et al., 2020). ID18 is characterized by single deletions at thymine homopolymers, with an enrichment in adenines immediately upstream (Fig. 3D). Subsequently, signatures SBS88 and ID18 were experimentally proven to be the result of colibactin exposure by utilizing human intestinal organoids (Pleguezuelos-Manzano et al., 2020). Colibactin is a genotoxic compound generated by certain strains of *Escherichia coli* bacteria harboring the pathogenicity island *pks*. These specific strains were identified significantly increased in IBD and CRC patients, and were found to contribute to tumorigenesis in mice by inducing inflammation (Arthur et al., 2012).

Interestingly, colibactin was found to generate adenine adducts in mammalian cells (Wilson et al., 2019; Xue et al., 2019), which are commonly repaired by transcription-coupled NER. These results are consistent with the observed TSB found in colibactin-associated signature SBS88 (Pleguezuelos-Manzano et al., 2020).

Mutational signatures SBS88 and ID18 were also identified in human tumors. Although most cases corresponded to CRC patients, other cancer types were found harboring both colibactin signatures contributions, including head-and-neck cancers as well as tumors of the urinary track (Boot et al., 2020; Pleguezuelos-Manzano et al., 2020). *E. coli* infection was previously described for both tissues (Banerjee et al., 2017; McLellan & Hunstad, 2016; Zawadzki et al., 2016). Remarkably, SBS88 mutations were detected as early events in normal and IBD-affected colon (Lee-Six et al., 2019; Olafsson et al., 2020), estimated to be produced in the first decade of life (Lee-Six et al., 2019). This early enrichment is in agreement with CRC data, where SBS88 mutations were also identified more significantly active among clonal than subclonal mutational events (Pleguezuelos-Manzano et al., 2020).

Recently, a recurrent *APC* splicing mutation was found to be caused by the colibactin-associated mutational signature SBS88 in unexplained colorectal polyposis patients (Terlouw et al., 2020). Although germline *APC* mutations are known to be the cause of familial adenomatous polyposis (Bodmer et al., 1987; Leppert et al., 1987), which is responsible for up to 1% of all CRCs (Kanth et al., 2017), some polyposis cases remain unexplained and are supposed to be caused by *de novo* mutations (Ripa, Bisgaard, Bulow, & Nielsen, 2002). The c.835-8A > G mutation was found enriched in a cohort of 201 unexplained polyposis patients and was proposed to be derived from signature SBS88 activity, according to its extended context and potential origin from adenine adducts on the untranscribed strand, compatible with the observed TSB (Terlouw et al., 2020).

SBS88 and ID18 were included in the latest update of COSMIC database (version 3.1) released in June 2020, along with the inclusion of four additional SBS mutational signatures: SBS86, SBS87, SBS89 and SBS90. Signature SBS89 was also found in normal and IBD-affected colonic tissue (named SBSB in both studies and renamed to SBS89 when included in COSMIC), although its etiology remains unknown (Lee-Six et al., 2019; Olafsson et al., 2020). As in the case of SBS88, mutations related to SBS89 were found present mostly in early stages during the evolution of the normal somatic colon tissue. SBS89 activity in normal colon correlated with DBS8, a doublet base substitution signature of unknown etiology, and a newly reported indel signature, known as IDB and dominated by deletions of isolated thymines (Lee-Six et al., 2019).

Signatures SBS86, SBS87 and SBS90 were found associated with different antineoplastic treatments. Signatures SBS86 and SBS87 were linked to chemotherapy in a recent study considering relapsed acute lymphoblastic leukemia patients (Li et al., 2020). More specifically, an association between signature SBS87 and thiopurine treatment was found and experimentally validated by exposing human cell lines to this chemotherapy agent (Li et al., 2020). Signature SBS90 was identified in human liver cancer cell lines after exposure to duocarmycin, a cytotoxic compound naturally produced by different species of *Streptomyces* bacteria which is used as a cancer therapy (Boot et al., 2020). Remarkably, a similar mutational profile to signature SBS90 was found in an independent study (Priestley et al., 2019) for two patients of breast and esophagus metastatic cancer treated with a duocarmycin-based HER2-targeting antibody-drug conjugate (Black et al., 2016). A very strong enrichment of sequence context was found for SBS90 beyond the common trinucleotide framework, with a prevalence of thymine up to four bases upstream of its characteristic T>A transversions (Boot et al., 2020). Other mutational signatures have been previously associated with DNA-damaging chemotherapy agents: SBS31, SBS35 and DBS5, all linked to platinum treatments (Alexandrov et al., 2020; Boot et al., 2018); SBS25, found in cell lines derived from Hodgkin lymphomas treated with chemotherapy (Alexandrov et al., 2015); and SBS17b, with an association to 5-fluorouracil (5-FU) (Pich et al., 2019) recently observed using *in vitro* intestinal organoids (Christensen et al., 2019). Although in most tumors where signature SBS17b is active its etiology remains unknown (Alexandrov et al., 2020), this recent finding suggests that for some cases SBS17b may represent the first treatment-related signature implicated in CRC. The association of SBS17b with 5-FU would be in agreement with this compound being used as a primary chemotherapy neoadjuvant treatment for CRC (Kuipers et al., 2015).

5. Other mutational processes involved in CRC

In addition to mutational signatures associated with DNA repair deficiencies, aging and microbiota, 19 other mutational signatures have also been found in CRC (12 SBS, 6 DBS and 1 ID). Most of these signatures were identified in a very small proportion of samples, and the molecular mechanisms of their underlying mutational processes remain mysterious.

Reactive oxygen species (ROS)-related signature SBS18 was found in a number of CRC cases contributing modest numbers of somatic mutations (Fig. 2A). Experimental validation for signature SBS18's etiology has been recently performed by treating human stem cells with potassium bromate, which is a source of ROS, obtaining its characteristic mutational profile dominated by C>A transversions (Fig. 1) (Kucab et al., 2019). Interestingly, SBS18 was also found as the most prevalent signature in normal and IBD-affected colon tissue besides aging signatures (Fig. 3A and B). This high incidence suggests a role of SBS18-associated mutations in early events prior to the neoplastic development in CRC, which is in agreement with recent findings in the evolutionary trajectory analysis performed using the PCAWG dataset (Gerstung et al., 2020). SBS18 was also ubiquitously present in cell lines grown in culture due to the higher levels of oxygen used in most incubator environments (Kucab et al., 2019).

Mutational signatures attributed to the activity of the APOBEC family of cytidine deaminases were also found in CRC (Fig. 2). Interestingly, the number of affected samples and mutations attributed to these APOBEC signatures were much lower than in most other cancer types (Alexandrov et al., 2020). APOBEC signatures SBS2, SBS13, and DBS11 were present in a small number of CRC samples in the PCAWG dataset. Originally, the mutational profiles of signatures SBS2 and SBS13 (Fig. 1) were attributed to single strand DNA hypermutation by APOBEC deaminases (Nik-Zainal et al., 2012). Recently, detailed mutational profiles were extracted for signatures SBS2 and SBS13, with specific forms of both signatures preferentially attributed to APOBEC3A and APOBEC3B by using extended mutational contexts beyond the trinucleotide context encompassing the immediate 5′ and 3′ bases adjacent to the mutation (Alexandrov et al., 2020; Chan et al., 2015; Petljak et al., 2019). Replication strand bias was found for both SBS2 and SBS13 with an enrichment of mutations on the lagging strand, a potential source of the single strand DNA needed for APOBEC mutagenesis (Morganella et al., 2016; Nik-Zainal et al., 2016). Mutational signature DBS11 is characterized by a combination of predominant CpC>TpT dinucleotide substitutions accompanied by other CpC>NpN mutations (Fig. 1), making this signature different from the pure CpC>TpT mutational profile well-known to be a consequence of ultraviolet light exposure (Miller, 1985) which is captured in signature DBS1 (Alexandrov et al., 2020; Chen, Ferec, & Cooper, 2013).

The etiology link of DBS11 with APOBEC activity was inferred by its association with signature SBS2, especially in cancer types not associated with ultraviolet light exposure (Alexandrov et al., 2020).

A very small proportion of CRC PCAWG samples (~0.5%) exhibited signature SBS9, which is caused by error-prone polymerase eta replication as part of somatic hypermutation, commonly affecting hematological cancers (Alexandrov, Nik-Zainal, Wedge, Aparicio, et al., 2013; Puente et al., 2011). Signature DBS2 was found contributing hundreds to thousands of mutations in cancers associated with tobacco smoking, encompassing the expected CpC>ApA doublet base substitutions (Chen et al., 2013). DBS2 was also identified in many other cancer types unrelated to smoking, including CRC (Alexandrov et al., 2020). Interestingly, an age correlation was found for signature DBS2 (Alexandrov et al., 2020), in agreement with its predominant role in normal mouse cells (Hill et al., 2003). In addition, experimental studies identified a similar mutational profile to DBS2 in cell lines exposed to acetaldehyde, an oxidation product of alcohol which is also present in tobacco smoke and automotive exhaust gases (Matsuda, Kawanishi, Yagi, Matsui, & Takebe, 1998). Although the number of samples tested is small, interestingly, DBS2 was the most prevalent DBS signature in the CRC PCAWG cohort (11/17 tumors; 65%) along with DBS4, which has an unknown etiology (Fig. 2C).

Approximately one quarter of all mutational signatures found in CRC have unknown etiologies with mutations attributed to these signatures accounting for almost 20% of all mutations in non-hypermutated CRC (Fig. 2C). Among signatures with unknown etiology, SBS40 is the one accounting for a greater number of mutations, affecting 29% of CRC PCAWG samples (237/829) with a similar mutation burden as the clock-like signatures SBS1 and SBS5 (Fig. 2A). A flat featureless mutational profile is observed for SBS40 (Fig. 1), similar to the one observed in SBS5 although lacking the TSB found in SBS5 for T>C transitions (Alexandrov et al., 2020). Although a correlation between SBS40 exposure and age of diagnosis was found for certain cancer types, including kidney and prostate cancers, this was not the case for CRC (Alexandrov et al., 2020). Interestingly, SBS40 as well as signatures SBS17a and SBS17b (all with unknown etiology) were recently found to be associated with late clonal mutations in CRC, and also late subclonal mutations in the case of SBS40 (Gerstung et al., 2020). In contrast, age-correlated signatures SBS1 and SBS5 were found predominant in early mutational events, as well as in non-neoplastic colonic tissue (Lee-Six et al., 2019). Mutational signatures SBS17a and SBS17b were found in a small number of CRC PCAWG samples. Despite the aforementioned link of SBS17b with 5-FU chemotherapy, in most cases the etiology of this mutational signature is still unknown. Interestingly, signature SBS17, subsequently split in current SBS17a and SBS17b (Secrier et al., 2016), was proposed to be linked to ROS damage, by incorporation of 8-oxo-guanine nucleotides from the nucleotide pool (Dulak et al., 2013; Nones et al., 2014; Tomkova et al., 2018). This association with oxidative damage is in agreement with the strong replication strand bias toward the lagging strand observed in the dominant T > G transversions (Tomkova et al., 2018) after being experimentally observed in yeast (Mudrak, Welz-Voegele, & Jinks-Robertson, 2009; Pavlov, Mian, & Kunkel, 2003). Other signatures with unknown etiology found in CRC PCAWG samples were SBS signatures SBS37 and SBS41, indel signature ID14 and DBS signatures DBS4, DBS6, DBS8 and DBS9 (Fig. 2). Interestingly, DBS8 was found in highly mutated CRCs, as well as other DBS signatures linked to MMR (DBS7 and DBS10) or polymerase proofreading (DBS3) deficiencies. However, tumors presenting high levels of DBS8 signature were apparently not harboring any of these DNA repair defects but were rather linked to clock-like mutational signatures SBS1 and SBS5 (Alexandrov et al., 2020). Signatures SBS43, SBS45 and SBS49, putatively related to sequencing artifacts, were also found in a very small proportion of the analyzed samples from the CRC PCAWG cohort (~0.1%).

Lastly, it is worth noting that some differences observed in mutational signatures may be due to the utilized sequencing approach. Sequencing the whole genome of a tumor provides a greater resolution to decipher the mutational signatures operative in a cancer genome when compared to sequencing just the coding regions of a cancer (*i.e.*, whole-exome sequencing). This difference in resolution may result in important differences, both in the profiles and activities of the identified mutational signatures (Alexandrov et al., 2020), giving the differences observed in mutation rate and DNA repair activity in coding *vs* non-coding regions of the genome (Frigola et al., 2017). This is especially important in the PCAWG Consortium data since, in contrast to other cancer types, the CRC dataset contained a high number (94%) of whole-exome sequenced samples and only a few whole-genome sequenced tumors (6%). Thus, prior analyses have likely captured all highly potent mutational signatures (*i.e.*, ones that generate large numbers of somatic mutations making them observable from whole-exome sequencing) but these analyses have likely missed more subtle signatures due to the lack of whole-genome sequencing data.

6. Conclusions and potential clinical applications

Significant advances have been made in molecularly characterizing colorectal cancer including the derivation of mutational signatures from the CRC PCAWG dataset. In principle, mutational signatures marked a major turning point in regard to the identification of the underlying mechanisms giving rise to somatic mutations. In CRC, mutational signatures were first validated by identifying well-known processes involved in CRC tumorigenesis, such as aging or deficiencies in MMR (Alexandrov, Nik-Zainal, Wedge, Aparicio, et al., 2013). More recently, mutational signatures have improved our understanding of the molecular landscape of colorectal neoplasia, by uncovering novel associations with previously unrelated mechanisms, including deficiency in base excision repair due to mutations in *NTHL1* (Grolleman, de Voer, et al., 2019) or an interplay with the gut microbiota (Pleguezuelos-Manzano et al., 2020), among others. However, despite the large epidemiological evidence accumulated over many decades, thus far, the analysis of mutational signatures was unable to associate specific signatures to any of the known environmental risk factors for CRC. The lack of a sufficient number of CRC samples and, in particular, the lack of whole-genome sequenced CRC samples with detailed epidemiological information, is likely contributing to the absence of associations since the analysis of mutational signatures requires a large number of well-annotated samples to perform proper statistical analysis (Alexandrov, Nik-Zainal, Wedge, Campbell, & Stratton, 2013). Further, it should be noted that most CRC data currently available is derived from patients located in either North America or in Europe. In order to address this bias, the *Mutographs of cancer* project is currently generating whole-genome sequencing data from colorectal tumors from patients located in 11 different countries around the world; these countries were selected based on large epidemiological differences in the risk factors for developing CRC of their general populations (Ginsburg et al., 2021).

Importantly, the molecular mechanisms behind a significant proportion of mutational signatures found in colorectal tumors remain unknown. This lack of knowledge may be due to the signatures being active in small sets of samples or having a small effect size, therefore hampering a reliable computational extraction and subsequent associations with known risk factors. Different strategies are currently emerging in order to look for the underlying mutational processes generating these signatures, including both further research on their associated mutational profiles and experimental evaluations of known carcinogens (Hu, Xu, & De, 2020).

Significant associations with DNA structure, replication time, multiple histone modifications, several transcription factors' binding sites and biases affecting the accumulation of mutations in both replication and transcription strands have already been found for a number of mutational signatures (Georgakopoulos-Soares, Morganella, Jain, Hemberg, & Nik-Zainal, 2018; Morganella et al., 2016; Singh, Rastogi, Hu, Wang, & De, 2020; Tomkova et al., 2018). Different patterns of local hypermutation were also identified and associated with particular mutational signatures. Clusters of high numbers of mutations in small genomic regions were identified following a mutational profile resembling signatures SBS2 and SBS13, both attributed to the activity of APOBEC deaminases (Nik-Zainal et al., 2012; Roberts et al., 2012; Supek & Lehner, 2017). This focal hypermutation was denotated as *kataegis* (Greek for thunderstorm) and is thought to be initiated by DNA double strand break repair by homologous recombination or break-induced replication pathways. Both repair pathways generate large stretches of single stranded DNA, which is a known source for APOBEC mutagenesis (Lei et al., 2018; Sakofsky et al., 2014; Taylor et al., 2013). Two additional clustered mutational signatures, SBS84 and SBS85, were identified in lymphoid cancers affecting the immunoglobulin gene and other regions. Signatures SBS84 and SBS85 are potentially encompassing the activity of activation-induced cytidine deaminases and translesion DNA synthesis by error-prone polymerases (Kasar et al., 2015). Recently, a different type of nonrecurrent diffuse hypermutation mediated by APOBEC deaminases, denotated as *omikli* (Greek for fog), has also been identified. *Omikli* clusters have been linked to mismatch repair activity, which can generate the single strand DNA substrate required for APOBEC-related mutagenesis (Mas-Ponte & Supek, 2020).

Experimental models have also been utilized in the last 5 years in order to gain knowledge of the etiology of unexplained mutational signatures (Koh, Zou, & Nik-Zainal, 2020). Different model organisms have been used, including: human cell lines (Boot et al., 2018; Kucab et al., 2019; Zou et al., 2018), human organoids (chapter "Modeling colorectal cancers using multidimensional organoids" by Sayed et al.) (Christensen et al., 2019; Drost et al., 2017; Pleguezuelos-Manzano et al., 2020), yeast (Chan et al., 2015; Loeillet et al., 2020; Xing et al., 2019), *C. elegans* (Meier et al., 2014, 2018), mouse models and mouse cell lines (Huang et al., 2017; Li et al., 2018; Nik-Zainal et al., 2015; Riva et al., 2020), and even chicken cell lines

(Poti et al., 2019; Zamborszky et al., 2017). Large-scale experimental screens were conducted in order to characterize the patterns of mutations generated by well-known and prospective mutagenic compounds and to find correlations with mutational signatures of unknown etiology previously derived from the analysis of cancer genomes (Kucab et al., 2019; Riva et al., 2020).

Mutational signatures have also contributed to public policies in regard to the use of certain mutagenic agents, especially in the case of those compounds related to signatures previously found in human cancers. A paradigmatic example is the case of aristolochic acid, found in *Aristolochia* plants, commonly used for traditional herbal remedies and which exposure has been associated with a high risk of developing kidney cancer since the early 1990s (Cosyns, Jadoul, Squifflet, Wese, & van Ypersele de Strihou, 1999; Nortier et al., 2000). Recently, the mutational signature of aristolochic acid, termed, signature SBS22, was experimentally confirmed (Nik-Zainal et al., 2015; Poon et al., 2013) as well as found in a number of different cancer types (Hoang et al., 2013; Poon et al., 2015). Another example is a study that found a very specific mutational signature in Japanese printing workers exposed to haloalkanes. The printing workers developed high incidence of early-onset cholangiocarcinoma and their tumors presented with high levels of signature SBS42 (Mimaki et al., 2016). Aflatoxin, a known mutagen compound produced by molds growing in grains, peanuts or other food was also found related to a particular mutational signature (mainly, SBS24) after the analysis of exposed liver cancer patients (Zhang et al., 2017) and further experimental validation in human cell lines and mouse models (Huang et al., 2017). Additional mutagenic exposures leading to particular mutational signatures were found for several medications, including signature SBS32 which was shown to be caused by the immunosuppressive agent azathioprine (Inman et al., 2018) as well as several SBS and DBS signatures which were imprinted on the genomes of secondary or recurrent cancers by different antineoplastic chemotherapy treatments (Boot et al., 2018; Pich et al., 2019).

Uncovering the molecular mechanisms involved in mutagenesis could also be achieved by elaborating the mutational signature framework by including additional classes of somatic mutations or by extending the currently examined classes of single base substitutions, doublet base substitutions, and small insertions and deletions. Additional classes of somatic mutations may include copy number and structural variants, that have already been used in several tissue-specific analyses (Macintyre et al., 2018; Nik-Zainal et al., 2016). Moreover, an extension of the original

SBS trinucleotide context has also shown potential to benefit mutational signatures specificity, by including additional peripheral nucleotides (*e.g.*, pentanucleotides), or topography features (*e.g.*, transcriptional or replicational strand bias) (Alexandrov et al., 2020). Interestingly, mutational signatures methodology could also be adjusted for non-discrete data, such as gene promoter methylation data, as it was suggested by SigFit, one of the latest tools developed for the analysis of mutational signatures (Baez-Ortega et al., 2019).

Mutational signatures were highlighted by the American Society of Clinical Oncology as a major achievement in clinical cancer research, emphasizing their potential applications in preventive strategies, prognosis estimates and treatment selection for different cancer types, including CRC (Masters et al., 2015). However, any potential future application of mutational signatures in a clinical setting would require the adoption of tumor sequencing procedures as a routine technique in most cancer cases, which is still a work in progress in most countries around the world. Hitherto, the use of mutational signatures as clinical biomarkers has been focused on breast cancer. For this cancer type, a computational predictor based on mutational signatures was developed to identify those patients harboring homologous recombination DNA repair pathway deficiency, known as HRDetect (Davies et al., 2017). Interestingly, benefitting from mutational signatures, this computational model is able to detect tumors lacking *BRCA1* and *BRCA2* mutations but harboring a similar mutational profile, therefore suggesting that the underlying molecular pathway is affected in a different way. Thus, previously unidentified patients are susceptible to receive PARP inhibitors, an effective treatment for patients harboring this DNA repair deficiency (Fong et al., 2009).

Importantly, cancer predisposition syndromes related to DNA repair deficiencies, many of which are common in CRC, may also benefit from clinical assessment based on mutational signatures in regard to early diagnosis (Ma et al., 2018; Van Hoeck et al., 2019). Base excision repair deficiencies due to biallelic *MUTYH* and *NTHL1* mutations, leading to inherited colorectal polyposis syndromes, have been recently associated with a strong activity of mutational signatures SBS36 (Pilati et al., 2017; Viel et al., 2017) and SBS30 (Drost et al., 2017; Grolleman, de Voer, et al., 2019), respectively. In addition, up to seven SBS, two DBS and three ID mutational signatures have been associated with different MMR deficiency (Alexandrov et al., 2020), leading to well-known Lynch syndrome; and four different signatures were linked to *POLE* exonuclease domain mutations (Alexandrov et al., 2020), which, if found in a person's germline DNA, are responsible for polymerase proofreading-associated polyposis syndrome (Palles et al., 2013). Thus, mutational signature analysis could become useful for an early diagnosis of these inherited CRC syndromes, benefitting genetic counseling in the affected families by an accurate identification of carriers. Mutational signatures could also benefit the pathogenicity assessment of previously unknown mutations affecting certain cancer-associated genes, as it has been already shown for *POLE* (Castellsagué et al., 2019).

Interestingly, mutational signatures have also been proposed as biomarkers for immunotherapy treatment, which is commonly deployed based on the total number of somatic mutations in a cancer (Chan et al., 2019; Yarchoan, Hopkins, & Jaffee, 2017), with certain signatures associated with improved patient response and better survival rates (chapter "Immunotherapy in colorectal cancer" by Agarwal et al.). In bladder cancer, APOBEC mutagenesis was suggested to lead the generation of tumor-specific neoantigens, that may benefit the effectiveness of these treatments (Chan et al., 2015). Thus, selection of immunotherapy treatment could be improved by specifically considering the burden of certain mutational signatures (*e.g.*, APOBEC-related signatures SBS2 and SBS13), since not all mutations are likely to induce immunogenic neoantigens (Chalmers et al., 2017).

In summary, during the last decade, genomics analyses have revealed the mutational signatures that have shape the genomic landscape of colorectal cancer, providing both validation of previous findings and deciphering novel molecular mechanisms of carcinogenesis. Future experimental studies and additional improvements of the computational methodologies are expected to further increase the knowledge about the genomic complexities of colorectal cancer and to provide significant opportunities for utilizing mutational signatures for both treating and preventing colorectal cancer.

Acknowledgments

The authors are grateful to the patients, their families and all physicians and researchers who laboriously gathered, processed, sequenced and analyzed all tumor samples considered in this report. This work was supported by Cancer Research UK Grand Challenge Award C98/A24032. L.B.A. is an Abeloff V scholar and he is personally supported by an Alfred P. Sloan Research Fellowship and a Packard Fellowship for Science and Engineering.

References

Albertson, T. M., Ogawa, M., Bugni, J. M., Hays, L. E., Chen, Y., Wang, Y., et al. (2009). DNA polymerase epsilon and delta proofreading suppress discrete mutator and cancer phenotypes in mice. *Proceedings of the National Academy of Sciences of the United States of America*, *106*, 17101–17104.

Alexandrov, L. B., Jones, P. H., Wedge, D. C., Sale, J. E., Campbell, P. J., Nik-Zainal, S., et al. (2015). Clock-like mutational processes in human somatic cells. *Nature Genetics*, *47*, 1402–1407.

Alexandrov, L. B., Ju, Y. S., Haase, K., Van Loo, P., Martincorena, I., Nik-Zainal, S., et al. (2016). Mutational signatures associated with tobacco smoking in human cancer. *Science*, *354*, 618–622.

Alexandrov, L. B., Kim, J., Haradhvala, N. J., Huang, M. N., Ng, A. W. T., Wu, Y., et al. (2020). The repertoire of mutational signatures in human cancer. *Nature*, *578*, 94–101.

Alexandrov, L. B., Nik-Zainal, S., Wedge, D. C., Aparicio, S. A., Behjati, S., Biankin, A. V., et al. (2013). Signatures of mutational processes in human cancer. *Nature*, *500*, 415–421.

Alexandrov, L. B., Nik-Zainal, S., Wedge, D. C., Campbell, P. J., & Stratton, M. R. (2013). Deciphering signatures of mutational processes operative in human cancer. *Cell Reports*, *3*, 246–259.

Alexandrov, L. B., & Stratton, M. R. (2014). Mutational signatures: The patterns of somatic mutations hidden in cancer genomes. *Current Opinion in Genetics & Development*, *24*, 52–60.

Al-Tassan, N., Chmiel, N. H., Maynard, J., Fleming, N., Livingston, A. L., Williams, G. T., et al. (2002). Inherited variants of MYH associated with somatic G:C→T:A mutations in colorectal tumors. *Nature Genetics*, *30*, 227–232.

Arthur, J. C., Perez-Chanona, E., Muhlbauer, M., Tomkovich, S., Uronis, J. M., Fan, T. J., et al. (2012). Intestinal inflammation targets cancer-inducing activity of the microbiota. *Science*, *338*, 120–123.

Baez-Ortega, A., & Gori, K. (2019). Computational approaches for discovery of mutational signatures in cancer. *Briefings in Bioinformatics*, *20*, 77–88.

Baez-Ortega, A., Gori, K., Strakova, A., Allen, J. L., Allum, K. M., Bansse-Issa, L., et al. (2019). Somatic evolution and global expansion of an ancient transmissible cancer lineage. *Science*, *365*, eaau9923.

Bagnardi, V., Rota, M., Botteri, E., Tramacere, I., Islami, F., Fedirko, V., et al. (2015). Alcohol consumption and site-specific cancer risk: A comprehensive dose-response meta-analysis. *British Journal of Cancer*, *112*, 580–593.

Banerjee, S., Tian, T., Wei, Z., Peck, K. N., Shih, N., Chalian, A. A., et al. (2017). Microbial signatures associated with oropharyngeal and oral squamous cell carcinomas. *Scientific Reports*, 7, 4036.

Beaugerie, L., & Itzkowitz, S. H. (2015). Cancers complicating inflammatory bowel disease. *The New England Journal of Medicine*, *372*, 1441–1452.

Bell, S. P., & Dutta, A. (2002). DNA replication in eukaryotic cells. *Annual Review of Biochemistry*, *71*, 333–374.

Bellido, F., Pineda, M., Aiza, G., Valdes-Mas, R., Navarro, M., Puente, D. A., et al. (2016). POLE and POLD1 mutations in 529 kindred with familial colorectal cancer and/or polyposis: Review of reported cases and recommendations for genetic testing and surveillance. *Genetics in Medicine*, *18*, 325–332.

Black, J., Menderes, G., Bellone, S., Schwab, C. L., Bonazzoli, E., Ferrari, F., et al. (2016). SYD985, a novel duocarmycin-based HER2-targeting antibody-drug conjugate, shows antitumor activity in uterine serous carcinoma with HER2/Neu expression. *Molecular Cancer Therapeutics*, *15*, 1900–1909.

Blokzijl, F., de Ligt, J., Jager, M., Sasselli, V., Roerink, S., Sasaki, N., et al. (2016). Tissue-specific mutation accumulation in human adult stem cells during life. *Nature*, *538*, 260–264.

Bodmer, W. F., Bailey, C. J., Bodmer, J., Bussey, H. J., Ellis, A., Gorman, P., et al. (1987). Localization of the gene for familial adenomatous polyposis on chromosome 5. *Nature*, *328*, 614–616.

Boot, A., Huang, M. N., Ng, A. W. T., Ho, S. C., Lim, J. Q., Kawakami, Y., et al. (2018). In-depth characterization of the cisplatin mutational signature in human cell lines and in esophageal and liver tumors. *Genome Research*, *28*, 654–665.

Boot, A., Ng, A. W. T., Chong, F. T., Ho, S. C., Yu, W., Tan, D. S. W., et al. (2020). Characterization of colibactin-associated mutational signature in an Asian oral squamous cell carcinoma and in other mucosal tumor types. *Genome Research*, *30*, 803–813.

Botteri, E., Iodice, S., Bagnardi, V., Raimondi, S., Lowenfels, A. B., & Maisonneuve, P. (2008). Smoking and colorectal cancer: A meta-analysis. *JAMA*, *300*, 2765–2778.

Bouvard, V., Loomis, D., Guyton, K. Z., Grosse, Y., Ghissassi, F. E., Benbrahim-Tallaa, L., et al. (2015). Carcinogenicity of consumption of red and processed meat. *The Lancet Oncology*, *16*, 1599–1600.

Brunner, S. F., Roberts, N. D., Wylie, L. A., Moore, L., Aitken, S. J., Davies, S. E., et al. (2019). Somatic mutations and clonal dynamics in healthy and cirrhotic human liver. *Nature*, *574*, 538–542.

Campbell, B. B., Light, N., Fabrizio, D., Zatzman, M., Fuligni, F., de Borja, R., et al. (2017). Comprehensive analysis of hypermutation in human cancer. *Cell*, *171*, 1042–1056 e1010.

Cancer Genome Atlas Network, Muzny, D. M., Bainbridge, M. N., Chang, K., Dinh, H. H., Drummond, J. A., et al. (2012). Comprehensive molecular characterization of human colon and rectal cancer. *Nature*, *487*, 330–337.

Cancer Genome Atlas Research Network, Kandoth, C., Schultz, N., Cherniack, A. D., Akbani, R., Liu, Y., et al. (2013). Integrated genomic characterization of endometrial carcinoma. *Nature*, *497*, 67–73.

Carethers, J. M., & Jung, B. H. (2015). Genetics and genetic biomarkers in sporadic colorectal cancer. *Gastroenterology*, *149*, 1177–1190 e1173.

Castellsagué, E., Li, R., Aligué, R., González, S., Sanz, J., Martin, E., et al. (2019). Novel POLE pathogenic germline variant in a family with multiple primary tumors results in distinct mutational signatures. *Human Mutation*, *40*, 36–41.

Ceccaldi, R., Rondinelli, B., & D'Andrea, A. D. (2016). Repair pathway choices and consequences at the double-strand break. *Trends in Cell Biology*, *26*, 52–64.

Chalmers, Z. R., Connelly, C. F., Fabrizio, D., Gay, L., Ali, S. M., Ennis, R., et al. (2017). Analysis of 100,000 human cancer genomes reveals the landscape of tumor mutational burden. *Genome Medicine*, *9*, 34.

Chan, K., Roberts, S. A., Klimczak, L. J., Sterling, J. F., Saini, N., Malc, E. P., et al. (2015). An APOBEC3A hypermutation signature is distinguishable from the signature of background mutagenesis by APOBEC3B in human cancers. *Nature Genetics*, *47*, 1067–1072.

Chan, T. A., Yarchoan, M., Jaffee, E., Swanton, C., Quezada, S. A., Stenzinger, A., et al. (2019). Development of tumor mutation burden as an immunotherapy biomarker: Utility for the oncology clinic. *Annals of Oncology*, *30*, 44–56.

Chatterjee, N., & Walker, G. C. (2017). Mechanisms of DNA damage, repair, and mutagenesis. *Environmental and Molecular Mutagenesis*, *58*, 235–263.

Chen, J. M., Ferec, C., & Cooper, D. N. (2013). Patterns and mutational signatures of tandem base substitutions causing human inherited disease. *Human Mutation*, *34*, 1119–1130.

Christensen, S., Van der Roest, B., Besselink, N., Janssen, R., Boymans, S., Martens, J. W. M., et al. (2019). 5-Fluorouracil treatment induces characteristic T>G mutations in human cancer. *Nature Communications*, *10*, 4571.

Church, D. N., Stelloo, E., Nout, R. A., Valtcheva, N., Depreeuw, J., ter Haar, N., et al. (2015). Prognostic significance of POLE proofreading mutations in endometrial cancer. *Journal of the National Cancer Institute*, *107*, 402.

Costello, M., Pugh, T. J., Fennell, T. J., Stewart, C., Lichtenstein, L., Meldrim, J. C., et al. (2013). Discovery and characterization of artifactual mutations in deep coverage targeted capture sequencing data due to oxidative DNA damage during sample preparation. *Nucleic Acids Research*, *41*, e67.

Cosyns, J.-P., Jadoul, M., Squifflet, J.-P., Wese, F.-X., & van Ypersele de Strihou, C. (1999). Urothelial lesions in Chinese-herb nephropathy. *American Journal of Kidney Diseases*, *33*, 1011–1017.

Davies, H., Glodzik, D., Morganella, S., Yates, L. R., Staaf, J., Zou, X., et al. (2017). HRDetect is a predictor of BRCA1 and BRCA2 deficiency based on mutational signatures. *Nature Medicine*, *23*, 517–525.

Díaz-Gay, M., Vila-Casadesus, M., Franch-Exposito, S., Hernandez-Illan, E., Lozano, J. J., & Castellvi-Bel, S. (2018). Mutational signatures in cancer (MuSiCa): A web application to implement mutational signatures analysis in cancer samples. *BMC Bioinformatics*, *19*, 224.

Dienstmann, R., Vermeulen, L., Guinney, J., Kopetz, S., Tejpar, S., & Tabernero, J. (2017). Consensus molecular subtypes and the evolution of precision medicine in colorectal cancer. *Nature Reviews. Cancer*, *17*, 79–92.

Domingo, E., Freeman-Mills, L., Rayner, E., Glaire, M., Briggs, S., Vermeulen, L., et al. (2016). Somatic POLE proofreading domain mutation, immune response, and prognosis in colorectal cancer: A retrospective, pooled biomarker study. *The Lancet Gastroenterology & Hepatology*, *1*, 207–216.

Drost, J., van Boxtel, R., Blokzijl, F., Mizutani, T., Sasaki, N., Sasselli, V., et al. (2017). Use of CRISPR-modified human stem cell organoids to study the origin of mutational signatures in cancer. *Science*, *358*, 234–238.

Dulak, A. M., Stojanov, P., Peng, S., Lawrence, M. S., Fox, C., Stewart, C., et al. (2013). Exome and whole-genome sequencing of esophageal adenocarcinoma identifies recurrent driver events and mutational complexity. *Nature Genetics*, *45*, 478–486.

Elsayed, F. A., Grolleman, J. E., Ragunathan, A., NTHL1 Study Group, Buchanan, D. D., van Wezel, T., et al. (2020). Monoallelic NTHL1 loss-of-function variants and risk of polyposis and colorectal cancer. *Gastroenterology*, *159*, 2241–2243.

Erson-Omay, E. Z., Caglayan, A. O., Schultz, N., Weinhold, N., Omay, S. B., Ozduman, K., et al. (2015). Somatic POLE mutations cause an ultramutated giant cell high-grade glioma subtype with better prognosis. *Neuro-Oncology*, *17*, 1356–1364.

Fong, P. C., Boss, D. S., Yap, T. A., Tutt, A., Wu, P., Mergui-Roelvink, M., et al. (2009). Inhibition of poly(ADP-ribose) polymerase in tumors from BRCA mutation carriers. *The New England Journal of Medicine*, *361*, 123–134.

Frigola, J., Sabarinathan, R., Mularoni, L., Muinos, F., Gonzalez-Perez, A., & Lopez-Bigas, N. (2017). Reduced mutation rate in exons due to differential mismatch repair. *Nature Genetics*, *49*, 1684–1692.

Georgakopoulos-Soares, I., Koh, G., Momen, S. E., Jiricny, J., Hemberg, M., & Nik-Zainal, S. (2020). Transcription-coupled repair and mismatch repair contribute towards preserving genome integrity at mononucleotide repeat tracts. *Nature Communications*, *11*, 1980.

Georgakopoulos-Soares, I., Morganella, S., Jain, N., Hemberg, M., & Nik-Zainal, S. (2018). Noncanonical secondary structures arising from non-B DNA motifs are determinants of mutagenesis. *Genome Research*, *28*, 1264–1271.

Georgescu, R. E., Langston, L., Yao, N. Y., Yurieva, O., Zhang, D., Finkelstein, J., et al. (2014). Mechanism of asymmetric polymerase assembly at the eukaryotic replication fork. *Nature Structural & Molecular Biology*, *21*, 664–670.

Gerstung, M., Jolly, C., Leshchiner, I., Dentro, S. C., Gonzalez, S., Rosebrock, D., et al. (2020). The evolutionary history of 2,658 cancers. *Nature*, *578*, 122–128.

Ginsburg, O., Ashton-Prolla, P., Cantor, A., Mariosa, D., & Brennan, P. (2021). The role of genomics in global cancer prevention. *Nature Reviews. Clinical Oncology*, *18*, 116–128.

Goldsby, R. E., Lawrence, N. A., Hays, L. E., Olmsted, E. A., Chen, X., Singh, M., et al. (2001). Defective DNA polymerase-delta proofreading causes cancer susceptibility in mice. *Nature Medicine*, *7*, 638–639.

Grilley, M., Holmes, J., Yashar, B., & Modrich, P. (1990). Mechanisms of DNA-mismatch correction. *Mutation Research: DNA Repair, 236*, 253–267.

Grolleman, J. E., de Voer, R. M., Elsayed, F. A., Nielsen, M., Weren, R. D. A., Palles, C., et al. (2019). Mutational signature analysis reveals NTHL1 deficiency to cause a multi-tumor phenotype. *Cancer Cell, 35*, 256–266 e255.

Grolleman, J. E., Díaz-Gay, M., Franch-Exposito, S., Castellvi-Bel, S., & de Voer, R. M. (2019). Somatic mutational signatures in polyposis and colorectal cancer. *Molecular Aspects of Medicine, 69*, 62–72.

Hanahan, D., & Weinberg, R. A. (2011). Hallmarks of cancer: The next generation. *Cell, 144*, 646–674.

Haradhvala, N. J., Kim, J., Maruvka, Y. E., Polak, P., Rosebrock, D., Livitz, D., et al. (2018). Distinct mutational signatures characterize concurrent loss of polymerase proofreading and mismatch repair. *Nature Communications, 9*, 1746.

Helleday, T., Eshtad, S., & Nik-Zainal, S. (2014). Mechanisms underlying mutational signatures in human cancers. *Nature Reviews. Genetics, 15*, 585–598.

Hill, K. A., Wang, J., Farwell, K. D., & Sommer, S. S. (2003). Spontaneous tandem-base mutations (TBM) show dramatic tissue, age, pattern and spectrum specificity. *Mutation Research, Genetic Toxicology and Environmental Mutagenesis, 534*, 173–186.

Hoang, M. L., Chen, C. H., Sidorenko, V. S., He, J., Dickman, K. G., Yun, B. H., et al. (2013). Mutational signature of aristolochic acid exposure as revealed by whole-exome sequencing. *Science Translational Medicine, 5*, 197ra102.

Hodel, K. P., Sun, M. J. S., Ungerleider, N., Park, V. S., Williams, L. G., Bauer, D. L., et al. (2020). POLE mutation spectra are shaped by the mutant allele identity, its abundance, and mismatch repair status. *Molecular Cell, 78*, 1166–1177 e1166.

Horvath, S. (2013). DNA methylation age of human tissues and cell types. *Genome Biology, 14*, R115.

Hu, X., Xu, Z., & De, S. (2020). Characteristics of mutational signatures of unknown etiology. *NAR Cancer, 2*, zcaa026.

Huang, M. N., Yu, W., Teoh, W. W., Ardin, M., Jusakul, A., Ng, A. W. T., et al. (2017). Genome-scale mutational signatures of aflatoxin in cells, mice, and human tumors. *Genome Research, 27*, 1475–1486.

ICGC/TCGA Pan-Cancer Analysis of Whole Genomes Consortium. (2020). Pan-cancer analysis of whole genomes. *Nature, 578*, 82–93.

Inman, G. J., Wang, J., Nagano, A., Alexandrov, L. B., Purdie, K. J., Taylor, R. G., et al. (2018). The genomic landscape of cutaneous SCC reveals drivers and a novel azathioprine associated mutational signature. *Nature Communications, 9*, 3667.

Ionov, Y., Peinado, M. A., Malkhosyan, S., Shibata, D., & Perucho, M. (1993). Ubiquitous somatic mutations in simple repeated sequences reveal a new mechanism for colonic carcinogenesis. *Nature, 363*, 558–561.

Jager, M., Blokzijl, F., Kuijk, E., Bertl, J., Vougioukalaki, M., Janssen, R., et al. (2019). Deficiency of nucleotide excision repair is associated with mutational signature observed in cancer. *Genome Research, 29*, 1067–1077.

Jain, R., Aggarwal, A. K., & Rechkoblit, O. (2018). Eukaryotic DNA polymerases. *Current Opinion in Structural Biology, 53*, 77–87.

Jansen, A. M., van Wezel, T., van den Akker, B. E., Ventayol Garcia, M., Ruano, D., Tops, C. M., et al. (2016). Combined mismatch repair and POLE/POLD1 defects explain unresolved suspected Lynch syndrome cancers. *European Journal of Human Genetics, 24*, 1089–1092.

Jasperson, K. W., Tuohy, T. M., Neklason, D. W., & Burt, R. W. (2010). Hereditary and familial colon cancer. *Gastroenterology, 138*, 2044–2058.

Jeggo, P. A., Pearl, L. H., & Carr, A. M. (2016). DNA repair, genome stability and cancer: A historical perspective. *Nature Reviews. Cancer, 16*, 35–42.

Jess, T., Rungoe, C., & Peyrin-Biroulet, L. (2012). Risk of colorectal cancer in patients with ulcerative colitis: A meta-analysis of population-based cohort studies. *Clinical Gastroenterology and Hepatology*, *10*, 639–645.

Johnson, R. E., Klassen, R., Prakash, L., & Prakash, S. (2015). A major role of DNA polymerase delta in replication of both the leading and lagging DNA strands. *Molecular Cell*, *59*, 163–175.

Kanth, P., Grimmett, J., Champine, M., Burt, R., & Samadder, N. J. (2017). Hereditary colorectal polyposis and cancer syndromes: A primer on diagnosis and management. *The American Journal of Gastroenterology*, *112*, 1509–1525.

Kasar, S., Kim, J., Improgo, R., Tiao, G., Polak, P., Haradhvala, N., et al. (2015). Whole-genome sequencing reveals activation-induced cytidine deaminase signatures during indolent chronic lymphocytic leukaemia evolution. *Nature Communications*, *6*, 8866.

Kim, J., Mouw, K. W., Polak, P., Braunstein, L. Z., Kamburov, A., Kwiatkowski, D. J., et al. (2016). Somatic ERCC2 mutations are associated with a distinct genomic signature in urothelial tumors. *Nature Genetics*, *48*, 600–606.

Koh, G., Zou, X., & Nik-Zainal, S. (2020). Mutational signatures: Experimental design and analytical framework. *Genome Biology*, *21*, 37.

Kratz, C. P., Achatz, M. I., Brugieres, L., Frebourg, T., Garber, J. E., Greer, M. C., et al. (2017). Cancer screening recommendations for individuals with Li-Fraumeni syndrome. *Clinical Cancer Research*, *23*, e38–e45.

Kucab, J. E., Zou, X., Morganella, S., Joel, M., Nanda, A. S., Nagy, E., et al. (2019). A compendium of mutational signatures of environmental agents. *Cell*, *177*, 821–836 e816.

Kuipers, E. J., Grady, W. M., Lieberman, D., Seufferlein, T., Sung, J. J., Boelens, P. G., et al. (2015). Colorectal cancer. *Nature Reviews. Disease Primers*, *1*, 15065.

Kunkel, T. A., & Erie, D. A. (2015). Eukaryotic mismatch repair in relation to DNA replication. *Annual Review of Genetics*, *49*, 291–313.

Lappalainen, T., Scott, A. J., Brandt, M., & Hall, I. M. (2019). Genomic analysis in the age of human genome sequencing. *Cell*, *177*, 70–84.

Lee-Six, H., Olafsson, S., Ellis, P., Osborne, R. J., Sanders, M. A., Moore, L., et al. (2019). The landscape of somatic mutation in normal colorectal epithelial cells. *Nature*, *574*, 532–537.

Lei, L., Chen, H., Xue, W., Yang, B., Hu, B., Wei, J., et al. (2018). APOBEC3 induces mutations during repair of CRISPR-Cas9-generated DNA breaks. *Nature Structural & Molecular Biology*, *25*, 45–52.

Leppert, M., Dobbs, M., Scambler, P., O'Connell, P., Nakamura, Y., Stauffer, D., et al. (1987). The gene for familial polyposis coli maps to the long arm of chromosome 5. *Science*, *238*, 1411–1413.

Li, B., Brady, S. W., Ma, X., Shen, S., Zhang, Y., Li, Y., et al. (2020). Therapy-induced mutations drive the genomic landscape of relapsed acute lymphoblastic leukemia. *Blood*, *135*, 41–55.

Li, H. D., Cuevas, I., Zhang, M., Lu, C., Alam, M. M., Fu, Y. X., et al. (2018). Polymerase-mediated ultramutagenesis in mice produces diverse cancers with high mutational load. *The Journal of Clinical Investigation*, *128*, 4179–4191.

Loeillet, S., Herzog, M., Puddu, F., Legoix, P., Baulande, S., Jackson, S. P., et al. (2020). Trajectory and uniqueness of mutational signatures in yeast mutators. *Proceedings of the National Academy of Sciences of the United States of America*, *117*, 24947–24956.

Lutgens, M. W., van Oijen, M. G., van der Heijden, G. J., Vleggaar, F. P., Siersema, P. D., & Oldenburg, B. (2013). Declining risk of colorectal cancer in inflammatory bowel disease: An updated meta-analysis of population-based cohort studies. *Inflammatory Bowel Diseases*, *19*, 789–799.

Lynch, H. T., & Krush, A. J. (1971). Cancer family "G" revisited: 1895-1970. *Cancer*, *27*, 1505–1511.

Lynch, H. T., Snyder, C. L., Shaw, T. G., Heinen, C. D., & Hitchins, M. P. (2015). Milestones of Lynch syndrome: 1895-2015. *Nature Reviews. Cancer*, *15*, 181–194.

Ma, J., Setton, J., Lee, N. Y., Riaz, N., & Powell, S. N. (2018). The therapeutic significance of mutational signatures from DNA repair deficiency in cancer. *Nature Communications*, *9*, 3292.

Macintyre, G., Goranova, T. E., De Silva, D., Ennis, D., Piskorz, A. M., Eldridge, M., et al. (2018). Copy number signatures and mutational processes in ovarian carcinoma. *Nature Genetics*, *50*, 1262–1270.

Martincorena, I., Fowler, J. C., Wabik, A., Lawson, A. R. J., Abascal, F., Hall, M. W. J., et al. (2018). Somatic mutant clones colonize the human esophagus with age. *Science*, *362*, 911–917.

Mas-Ponte, D., & Supek, F. (2020). DNA mismatch repair promotes APOBEC3-mediated diffuse hypermutation in human cancers. *Nature Genetics*, *52*, 958–968.

Masters, G. A., Krilov, L., Bailey, H. H., Brose, M. S., Burstein, H., Diller, L. R., et al. (2015). Clinical cancer advances 2015: Annual report on progress against cancer from the American Society of Clinical Oncology. *Journal of Clinical Oncology*, *33*, 786–809.

Matsuda, T., Kawanishi, M., Yagi, T., Matsui, S., & Takebe, H. (1998). Specific tandem GG to TT base substitutions induced by acetaldehyde are due to intra-strand crosslinks between adjacent guanine bases. *Nucleic Acids Research*, *26*, 1769–1774.

McLellan, L. K., & Hunstad, D. A. (2016). Urinary tract infection: Pathogenesis and outlook. *Trends in Molecular Medicine*, *22*, 946–957.

Meier, B., Cooke, S. L., Weiss, J., Bailly, A. P., Alexandrov, L. B., Marshall, J., et al. (2014). C. elegans whole-genome sequencing reveals mutational signatures related to carcinogens and DNA repair deficiency. *Genome Research*, *24*, 1624–1636.

Meier, B., Volkova, N. V., Hong, Y., Schofield, P., Campbell, P. J., Gerstung, M., et al. (2018). Mutational signatures of DNA mismatch repair deficiency in C. elegans and human cancers. *Genome Research*, *28*, 666–675.

Mellon, I., Rajpal, D. K., Koi, M., Boland, C. R., & Champe, G. N. (1996). Transcription-coupled repair deficiency and mutations in human mismatch repair genes. *Science*, *272*, 557–560.

Miller, J. H. (1985). Mutagenic specificity of ultraviolet light. *Journal of Molecular Biology*, *182*, 45–65.

Mimaki, S., Totsuka, Y., Suzuki, Y., Nakai, C., Goto, M., Kojima, M., et al. (2016). Hypermutation and unique mutational signatures of occupational cholangiocarcinoma in printing workers exposed to haloalkanes. *Carcinogenesis*, *37*, 817–826.

Moore, L., Leongamornlert, D., Coorens, T. H. H., Sanders, M. A., Ellis, P., Dentro, S. C., et al. (2020). The mutational landscape of normal human endometrial epithelium. *Nature*, *580*, 640–646.

Morganella, S., Alexandrov, L. B., Glodzik, D., Zou, X., Davies, H., Staaf, J., et al. (2016). The topography of mutational processes in breast cancer genomes. *Nature Communications*, 7, 11383.

Morrison, A., Johnson, A. L., Johnston, L. H., & Sugino, A. (1993). Pathway correcting DNA replication errors in Saccharomyces cerevisiae. *The EMBO Journal*, *12*, 1467–1473.

Mudrak, S. V., Welz-Voegele, C., & Jinks-Robertson, S. (2009). The polymerase eta translesion synthesis DNA polymerase acts independently of the mismatch repair system to limit mutagenesis caused by 7,8-dihydro-8-oxoguanine in yeast. *Molecular and Cellular Biology*, *29*, 5316–5326.

Nguyen, L. H., Goel, A., & Chung, D. C. (2020). Pathways of colorectal carcinogenesis. *Gastroenterology*, *158*, 291–302.

Nik-Zainal, S., Alexandrov, L. B., Wedge, D. C., Van Loo, P., Greenman, C. D., Raine, K., et al. (2012). Mutational processes molding the genomes of 21 breast cancers. *Cell, 149*, 979–993.

Nik-Zainal, S., Davies, H., Staaf, J., Ramakrishna, M., Glodzik, D., Zou, X., et al. (2016). Landscape of somatic mutations in 560 breast cancer whole-genome sequences. *Nature, 534*, 47–54.

Nik-Zainal, S., Kucab, J. E., Morganella, S., Glodzik, D., Alexandrov, L. B., Arlt, V. M., et al. (2015). The genome as a record of environmental exposure. *Mutagenesis, 30*, 763–770.

Nones, K., Waddell, N., Wayte, N., Patch, A. M., Bailey, P., Newell, F., et al. (2014). Genomic catastrophes frequently arise in esophageal adenocarcinoma and drive tumorigenesis. *Nature Communications, 5*, 5224.

Nortier, J. L., Martinez, M. C., Schmeiser, H. H., Arlt, V. M., Bieler, C. A., Petein, M., et al. (2000). Urothelial carcinoma associated with the use of a Chinese herb (Aristolochia fangchi). *The New England Journal of Medicine, 342*, 1686–1692.

Odell, I. D., Wallace, S. S., & Pederson, D. S. (2013). Rules of engagement for base excision repair in chromatin. *Journal of Cellular Physiology, 228*, 258–266.

Oh, M., McBride, A., Yun, S., Bhattacharjee, S., Slack, M., Martin, J. R., et al. (2018). BRCA1 and BRCA2 gene mutations and colorectal cancer risk: Systematic review and meta-analysis. *Journal of the National Cancer Institute, 110*, 1178–1189.

Olafsson, S., McIntyre, R. E., Coorens, T., Butler, T., Jung, H., Robinson, P. S., et al. (2020). Somatic evolution in non-neoplastic IBD-affected colon. *Cell, 182*, 672–684 e611.

Omichessan, H., Severi, G., & Perduca, V. (2019). Computational tools to detect signatures of mutational processes in DNA from tumours: A review and empirical comparison of performance. *PLoS One, 14*, e0221235.

Palles, C., Cazier, J. B., Howarth, K. M., Domingo, E., Jones, A. M., Broderick, P., et al. (2013). Germline mutations affecting the proofreading domains of POLE and POLD1 predispose to colorectal adenomas and carcinomas. *Nature Genetics, 45*, 136–144.

Pavlov, Y. I., Mian, I. M., & Kunkel, T. A. (2003). Evidence for preferential mismatch repair of lagging strand DNA replication errors in yeast. *Current Biology, 13*, 744–748.

Peña-Diaz, J., & Jiricny, J. (2012). Mammalian mismatch repair: Error-free or error-prone? *Trends in Biochemical Sciences, 37*, 206–214.

Petljak, M., Alexandrov, L. B., Brammeld, J. S., Price, S., Wedge, D. C., Grossmann, S., et al. (2019). Characterizing mutational signatures in human cancer cell lines reveals episodic APOBEC mutagenesis. *Cell, 176*, 1282–1294 e1220.

Pich, O., Muinos, F., Lolkema, M. P., Steeghs, N., Gonzalez-Perez, A., & Lopez-Bigas, N. (2019). The mutational footprints of cancer therapies. *Nature Genetics, 51*, 1732–1740.

Pilati, C., Shinde, J., Alexandrov, L. B., Assie, G., Andre, T., Helias-Rodzewicz, Z., et al. (2017). Mutational signature analysis identifies MUTYH deficiency in colorectal cancers and adrenocortical carcinomas. *The Journal of Pathology, 242*, 10–15.

Pleguezuelos-Manzano, C., Puschhof, J., Rosendahl Huber, A., van Hoeck, A., Wood, H. M., Nomburg, J., et al. (2020). Mutational signature in colorectal cancer caused by genotoxic pks(+) E. coli. *Nature, 580*, 269–273.

Poetsch, A. R. (2020). The genomics of oxidative DNA damage, repair, and resulting mutagenesis. *Computational and Structural Biotechnology Journal, 18*, 207–219.

Polak, P., Kim, J., Braunstein, L. Z., Karlic, R., Haradhavala, N. J., Tiao, G., et al. (2017). A mutational signature reveals alterations underlying deficient homologous recombination repair in breast cancer. *Nature Genetics, 49*, 1476–1486.

Poon, S. L., Huang, M. N., Choo, Y., McPherson, J. R., Yu, W., Heng, H. L., et al. (2015). Mutation signatures implicate aristolochic acid in bladder cancer development. *Genome Medicine*, *7*, 38.

Poon, S. L., Pang, S. T., McPherson, J. R., Yu, W., Huang, K. K., Guan, P., et al. (2013). Genome-wide mutational signatures of aristolochic acid and its application as a screening tool. *Science Translational Medicine*, *5*, 197ra101.

Poti, A., Gyergyak, H., Nemeth, E., Rusz, O., Toth, S., Kovacshazi, C., et al. (2019). Correlation of homologous recombination deficiency induced mutational signatures with sensitivity to PARP inhibitors and cytotoxic agents. *Genome Biology*, *20*, 240.

Priestley, P., Baber, J., Lolkema, M. P., Steeghs, N., de Bruijn, E., Shale, C., et al. (2019). Pan-cancer whole-genome analyses of metastatic solid tumours. *Nature*, *575*, 210–216.

Puente, X. S., Pinyol, M., Quesada, V., Conde, L., Ordonez, G. R., Villamor, N., et al. (2011). Whole-genome sequencing identifies recurrent mutations in chronic lymphocytic leukaemia. *Nature*, *475*, 101–105.

Rayner, E., van Gool, I. C., Palles, C., Kearsey, S. E., Bosse, T., Tomlinson, I., et al. (2016). A panoply of errors: Polymerase proofreading domain mutations in cancer. *Nature Reviews. Cancer*, *16*, 71–81.

Ripa, R., Bisgaard, M. L., Bulow, S., & Nielsen, F. C. (2002). De novo mutations in familial adenomatous polyposis (FAP). *European Journal of Human Genetics*, *10*, 631–637.

Riva, L., Pandiri, A. R., Li, Y. R., Droop, A., Hewinson, J., Quail, M. A., et al. (2020). The mutational signature profile of known and suspected human carcinogens in mice. *Nature Genetics*, *52*, 1189–1197.

Roberts, S. A., Sterling, J., Thompson, C., Harris, S., Mav, D., Shah, R., et al. (2012). Clustered mutations in yeast and in human cancers can arise from damaged long single-strand DNA regions. *Molecular Cell*, *46*, 424–435.

Rogozin, I. B., Pavlov, Y. I., Goncearenco, A., De, S., Lada, A. G., Poliakov, E., et al. (2018). Mutational signatures and mutable motifs in cancer genomes. *Briefings in Bioinformatics*, *19*, 1085–1101.

Rosenthal, R., McGranahan, N., Herrero, J., Taylor, B. S., & Swanton, C. (2016). DeconstructSigs: Delineating mutational processes in single tumors distinguishes DNA repair deficiencies and patterns of carcinoma evolution. *Genome Biology*, *17*, 31.

Sakofsky, C. J., Roberts, S. A., Malc, E., Mieczkowski, P. A., Resnick, M. A., Gordenin, D. A., et al. (2014). Break-induced replication is a source of mutation clusters underlying kataegis. *Cell Reports*, *7*, 1640–1648.

Secrier, M., Li, X., de Silva, N., Eldridge, M. D., Contino, G., Bornschein, J., et al. (2016). Mutational signatures in esophageal adenocarcinoma define etiologically distinct subgroups with therapeutic relevance. *Nature Genetics*, *48*, 1131–1141.

Shibata, D., Peinado, M. A., Ionov, Y., Malkhosyan, S., & Perucho, M. (1994). Genomic instability in repeated sequences is an early somatic event in colorectal tumorigenesis that persists after transformation. *Nature Genetics*, *6*, 273–281.

Shinbrot, E., Henninger, E. E., Weinhold, N., Covington, K. R., Goksenin, A. Y., Schultz, N., et al. (2014). Exonuclease mutations in DNA polymerase epsilon reveal replication strand specific mutation patterns and human origins of replication. *Genome Research*, *24*, 1740–1750.

Singh, V. K., Rastogi, A., Hu, X., Wang, Y., & De, S. (2020). Mutational signature SBS8 predominantly arises due to late replication errors in cancer. *Communications Biology*, *3*, 421.

Stratton, M. R. (2011). Exploring the genomes of cancer cells: Progress and promise. *Science*, *331*, 1553–1558.

Stratton, M. R., Campbell, P. J., & Futreal, P. A. (2009). The cancer genome. *Nature*, *458*, 719–724.

Sun, J., Wang, C., Zhang, Y., Xu, L., Fang, W., Zhu, Y., et al. (2019). Genomic signatures reveal DNA damage response deficiency in colorectal cancer brain metastases. *Nature Communications*, *10*, 3190.

Supek, F., & Lehner, B. (2015). Differential DNA mismatch repair underlies mutation rate variation across the human genome. *Nature*, *521*, 81–84.

Supek, F., & Lehner, B. (2017). Clustered mutation signatures reveal that error-prone DNA repair targets mutations to active genes. *Cell*, *170*, 534–547 e523.

Tate, J. G., Bamford, S., Jubb, H. C., Sondka, Z., Beare, D. M., Bindal, N., et al. (2019). COSMIC: The catalogue of somatic mutations in cancer. *Nucleic Acids Research*, *47*, D941–D947.

Taylor, B. J., Nik-Zainal, S., Wu, Y. L., Stebbings, L. A., Raine, K., Campbell, P. J., et al. (2013). DNA deaminases induce break-associated mutation showers with implication of APOBEC3B and 3A in breast cancer kataegis. *eLife*, *2*, e00534.

Terlouw, D., Suerink, M., Boot, A., van Wezel, T., Nielsen, M., & Morreau, H. (2020). Recurrent APC splice variant c.835-8A>G in patients with unexplained colorectal polyposis fulfilling the colibactin mutational signature. *Gastroenterology*, *159*, 1612–1614.

Tomasetti, C., Li, L., & Vogelstein, B. (2017). Stem cell divisions, somatic mutations, cancer etiology, and cancer prevention. *Science*, *355*, 1330–1334.

Tomkova, M., Tomek, J., Kriaucionis, S., & Schuster-Bockler, B. (2018). Mutational signature distribution varies with DNA replication timing and strand asymmetry. *Genome Biology*, *19*, 129.

Touat, M., Li, Y. Y., Boynton, A. N., Spurr, L. F., Iorgulescu, J. B., Bohrson, C. L., et al. (2020). Mechanisms and therapeutic implications of hypermutation in gliomas. *Nature*, *580*, 517–523.

Valle, L., de Voer, R. M., Goldberg, Y., Sjursen, W., Forsti, A., Ruiz-Ponte, C., et al. (2019). Update on genetic predisposition to colorectal cancer and polyposis. *Molecular Aspects of Medicine*, *69*, 10–26.

Valle, L., Vilar, E., Tavtigian, S. V., & Stoffel, E. M. (2019). Genetic predisposition to colorectal cancer: Syndromes, genes, classification of genetic variants and implications for precision medicine. *The Journal of Pathology*, *247*, 574–588.

van Gool, I. C., Eggink, F. A., Freeman-Mills, L., Stelloo, E., Marchi, E., de Bruyn, M., et al. (2015). POLE proofreading mutations elicit an antitumor immune response in endometrial cancer. *Clinical Cancer Research*, *21*, 3347–3355.

Van Hoeck, A., Tjoonk, N. H., van Boxtel, R., & Cuppen, E. (2019). Portrait of a cancer: Mutational signature analyses for cancer diagnostics. *BMC Cancer*, *19*, 457.

Viel, A., Bruselles, A., Meccia, E., Fornasarig, M., Quaia, M., Canzonieri, V., et al. (2017). A specific mutational signature associated with DNA 8-oxoguanine persistence in MUTYH-defective colorectal cancer. *eBioMedicine*, *20*, 39–49.

Vogelstein, B., Fearon, E. R., Hamilton, S. R., Kern, S. E., Preisinger, A. C., Leppert, M., et al. (1988). Genetic alterations during colorectal-tumor development. *New England Journal of Medicine*, *319*, 525–532.

Volkova, N. V., Meier, B., Gonzalez-Huici, V., Bertolini, S., Gonzalez, S., Vohringer, H., et al. (2020). Mutational signatures are jointly shaped by DNA damage and repair. *Nature Communications*, *11*, 2169.

Weren, R. D., Ligtenberg, M. J., Geurts van Kessel, A., De Voer, R. M., Hoogerbrugge, N., & Kuiper, R. P. (2018). NTHL1 and MUTYH polyposis syndromes: Two sides of the same coin? *The Journal of Pathology*, *244*, 135–142.

Weren, R. D., Ligtenberg, M. J., Kets, C. M., de Voer, R. M., Verwiel, E. T., Spruijt, L., et al. (2015). A germline homozygous mutation in the base-excision repair gene NTHL1 causes adenomatous polyposis and colorectal cancer. *Nature Genetics*, *47*, 668–671.

Wilson, M. R., Jiang, Y., Villalta, P. W., Stornetta, A., Boudreau, P. D., Carra, A., et al. (2019). The human gut bacterial genotoxin colibactin alkylates DNA. *Science*, *363*, eaar7785.

Win, A. K., Dowty, J. G., Cleary, S. P., Kim, H., Buchanan, D. D., Young, J. P., et al. (2014). Risk of colorectal cancer for carriers of mutations in MUTYH, with and without a family history of cancer. *Gastroenterology*, *146*(1208–1211), e1201–e1205.

Woo, Y. H., & Li, W. H. (2012). DNA replication timing and selection shape the landscape of nucleotide variation in cancer genomes. *Nature Communications*, *3*, 1004.

Xing, X., Kane, D. P., Bulock, C. R., Moore, E. A., Sharma, S., Chabes, A., et al. (2019). A recurrent cancer-associated substitution in DNA polymerase epsilon produces a hyperactive enzyme. *Nature Communications*, *10*, 374.

Xue, M., Kim, C. S., Healy, A. R., Wernke, K. M., Wang, Z., Frischling, M. C., et al. (2019). Structure elucidation of colibactin and its DNA cross-links. *Science*, *365*, eaax2685.

Yarchoan, M., Hopkins, A., & Jaffee, E. M. (2017). Tumor mutational burden and response rate to PD-1 inhibition. *The New England Journal of Medicine*, *377*, 2500–2501.

Yates, L. R., & Campbell, P. J. (2012). Evolution of the cancer genome. *Nature Reviews. Genetics*, *13*, 795–806.

Yoshida, K., Gowers, K. H. C., Lee-Six, H., Chandrasekharan, D. P., Coorens, T., Maughan, E. F., et al. (2020). Tobacco smoking and somatic mutations in human bronchial epithelium. *Nature*, *578*, 266–272.

Zamborszky, J., Szikriszt, B., Gervai, J. Z., Pipek, O., Poti, A., Krzystanek, M., et al. (2017). Loss of BRCA1 or BRCA2 markedly increases the rate of base substitution mutagenesis and has distinct effects on genomic deletions. *Oncogene*, *36*, 746–755.

Zawadzki, P. J., Perkowski, K., Starosciak, B., Baltaza, W., Padzik, M., Pionkowski, K., et al. (2016). Identification of infectious microbiota from oral cavity environment of various population group patients as a preventive approach to human health risk factors. *Annals of Agricultural and Environmental Medicine*, *23*, 566–569.

Zhang, W., He, H., Zang, M., Wu, Q., Zhao, H., Lu, L. L., et al. (2017). Genetic features of aflatoxin-associated hepatocellular carcinoma. *Gastroenterology*, *153*(249–262), e242.

Zhao, J., Jain, A., Iyer, R. R., Modrich, P. L., & Vasquez, K. M. (2009). Mismatch repair and nucleotide excision repair proteins cooperate in the recognition of DNA interstrand crosslinks. *Nucleic Acids Research*, *37*, 4420–4429.

Zou, X., Owusu, M., Harris, R., Jackson, S. P., Loizou, J. I., & Nik-Zainal, S. (2018). Validating the concept of mutational signatures with isogenic cell models. *Nature Communications*, *9*, 1744.

CHAPTER ELEVEN

Epigenetic alterations in the gastrointestinal tract: Current and emerging use for biomarkers of cancer

William M. Grady[a,b,*]

[a]Clinical Research Division, Fred Hutchinson Cancer Research Center, Seattle, WA, United States
[b]Division of Gastroenterology, University of Washington School of Medicine, Seattle, WA, United States
*Corresponding author: e-mail address: wgrady@fredhutch.org

Contents

Advances in Cancer Research, Volume 151
ISSN 0065-230X
https://doi.org/10.1016/bs.acr.2021.02.006

Abstract

Colorectal cancer is a leading cause of cancer related deaths worldwide. One of the hallmarks of cancer and a fundamental trait of virtually all gastrointestinal cancers is genomic and epigenomic DNA alterations. Cancer cells acquire genetic and epigenetic alterations that drive the initiation and progression of the cancers by altering the molecular and cell biological process of the cells. These alterations, as well as other host and microenvironment factors, ultimately mediate the initiation and progression of cancers, including colorectal cancer. Epigenetic alterations, which include changes affecting DNA methylation, histone modifications, chromatin structure, and noncoding RNA expression, have emerged as a major class of molecular alteration in colon polyps and colorectal cancer. The classes of epigenetic alterations, their status in colorectal polyps and cancer, their effects on neoplasm biology, and their application to clinical care will be discussed.

1. Introduction

1.1 DNA alterations in cancer

Cancer develops as a consequence of disruption of the mechanisms that regulate fundamental processes, such as cellular proliferation, cell metabolism, angiogenesis, cell death, invasion, metastasis, as well as other hallmark behaviors of cancer (Hanahan & Weinberg, 2000, 2011) (Fig. 1). These disruptions arise through an evolutionary process that stably encodes acquired oncogenic alterations, which can then accumulate in clonal lineages (Sievers, Grady, Halberg, et al., 2017). Genetic alterations including gene sequence mutations, gene copy number alterations, insertions, deletions, and recombination events were the first and most clearly demonstrated of these oncogenic alterations of DNA, leading to the prominence of the model of cancer genetics in cancer biology, which proposes that somatic mutations in oncogenes and tumor suppressor genes drive the initiation and progression of cancer by providing progressive growth advantages to cells, thus driving cancer initiation and progression.

More recently epigenetic alterations have been established as another fundamental oncogenic mechanism and appear to play a prominent role in the pathogenesis of cancer by inducing hallmark cancer cell behaviors. In its most broad sense, epigenetics refers the phenomenon of heritable phenotype changes that do not involve alterations in the DNA sequence. At a molecular level, epigenetics alterations involve functionally relevant changes to the genome that do not involve a change in the nucleotide sequence, which includes alterations in DNA methylation, histone modifications, and the chromatin structure of the DNA. Genome-wide genomic

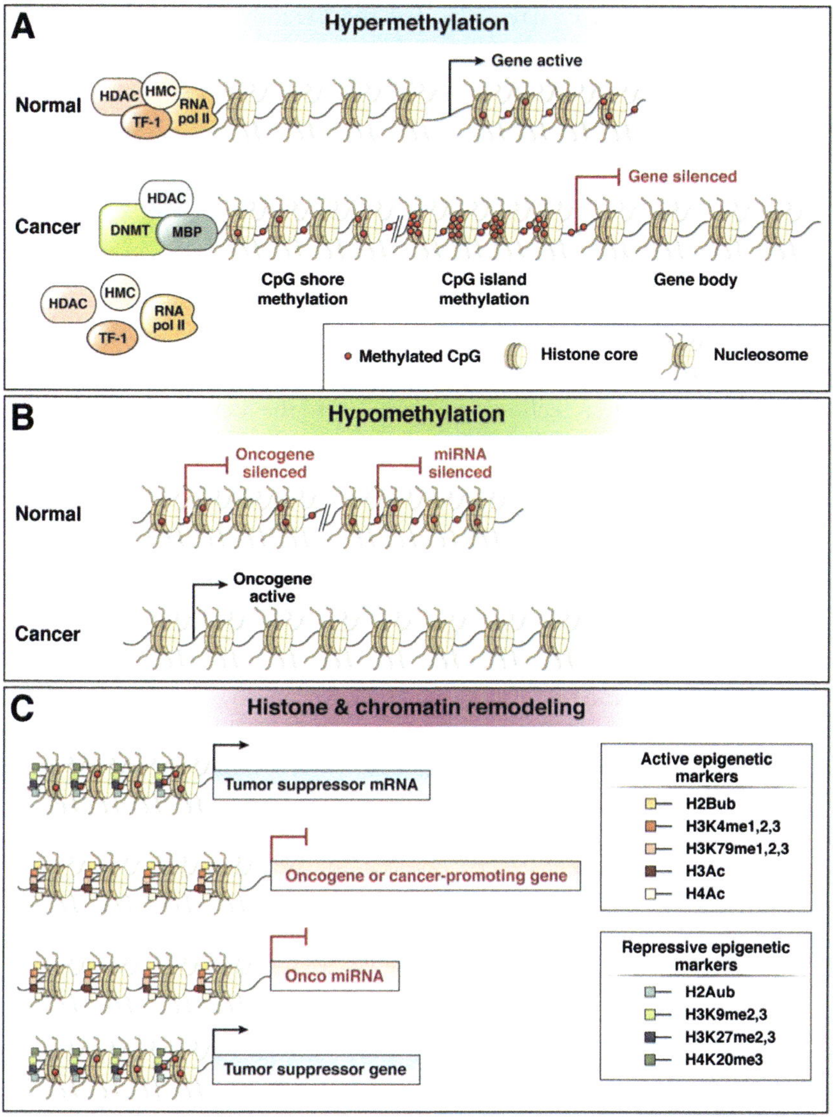

Fig. 1 (A, B) Schematic diagram of DNA hypermethylation and DNA hypomethylation in cancer. (TF-1, transcription factor 1; RNA pol II, RNA polymerase II; HDAC, histone deacetylase complex; DNMT, DNA methyltransferase; MBP, methyl binding protein). (C) Schematic diagram of DNA chromatin structure and histone modification states that affect conformation of chromatin. (H2A, histone 2A; H2Aub, ubiquityl histone H2A; H2B, histone 2B; H2Bub, ubiquityl histone H2B; H3, histone 3; H4, histone 4; ub, ubiquitylated; me, methylated; Ac, acetylated; K, lysine. Onco miR is an oncogenic microRNA.) *Adapted from Grady, W. M., Yu, M., Markowitz, S. D., (2020). Epigenetic alterations in the gastrointestinal tract: Current and emerging use for biomarkers of cancer.* Gastroenterology, *S0016-5085(20)35511-6. PMID: 33279516.*

and epigenomic analyses have revealed the breadth of alterations to the epigenome in cancer cells as well as the widespread occurrence of mutations in genes that encode epigenetic regulators (Liu, Sethi, Hinoue, et al., 2018; You & Jones, 2012). It is clear that genetic and epigenetic mechanisms influence and cooperate with each other to enable the acquisition of the hallmarks of cancer (Hanahan & Weinberg, 2011).

The prevailing consensus suggests that epigenetic alterations in cancer occur early and are more common than genetic alterations. A subset of these alterations play a role in mediating cancer formation. This has been well demonstrated for colorectal cancer (CRC) (Kim, Petko, Dzieciatkowski, et al., 2006; Luo, Wong, Kaz, et al., 2014). In addition, advances in genomic and epigenomic analysis technologies have led to the identification of a variety of specific epigenetic alterations that can and are being used as clinical biomarkers for gastrointestinal (GI) cancer patients.

The use of these recently developed technologies has led to insights into cancer epidemiology and hereditary cancer syndromes as well as cancer biology. Because of the potential for environmental factors to modify epigenetic states, the association of cancer risk factors, such as tobacco exposure, with cancer relevant epigenetic alterations has also been assessed recently. There is modestly robust evidence to date suggesting such factors have significant effects in humans (Herceg, 2007; Holland, 2017). It is also notable that GI cancer syndromes, such as Lynch syndrome, can arise from germline epigenetic alterations (Cerrato, Sparago, & Ariani, 2020; Hitchins, 2013). However, familial epigenetic cancer syndromes are rare and appear to have a low inheritance risk to offspring (Morak, Ibisler, Keller, et al., 2018).

This chapter briefly describes epigenetic modifications, outlines the fundamental basis of epigenetic alterations in cancer, discusses the role of these alterations in the molecular pathogenesis of CRC, and details the current state of the field regarding the promise and clinical usefulness of various epigenetic alterations as biomarkers for the early detection, diagnosis, prognosis and management of GI pre-cancer and cancer, with a primary focus of colorectal and esophageal cancer. The classes of cancer epigenetic alterations that will be the focus of this review will be predominantly (1) DNA methylation and (2) chromatin and histone structure and function because the quality of data on these topics is robust. Noncoding RNAs and microRNAs will also be discussed but in less detail due to the lack of high quality studies to guide our understanding of this class of molecular alteration in CRC. The interested reader is directed to recent reviews of this subject for more information (Dragomir, Kopetz, Ajani, et al., 2020; Okugawa, Grady, & Goel, 2015).

1.2 Cancer epigenetics

Epigenetic mechanisms stably regulate cell behavior by controlling the transcriptional availability of various parts of the genome through differential DNA methylation and chromatin marking as well as by regulating DNA packaging and compaction via histone modifications. The best studied of these mechanisms involve direct DNA modifications, primarily CpG cytosine-5′ methylation (Baylin, Herman, Graff, et al., 1998; Liu et al., 2018), as well as CpG cytosine-5′ hydroxylation, formylation, and carboxylation (Meldi & Figueroa, 2015; Paluch, Naqash, Brumberger, et al., 2016). The role of DNA hydroxymethylation, formylation and carboxylation in cancer is being assessed and is not well defined at the time this review is being written (Eleftheriou & Ruzov, 2021). Other epigenetic alterations observed in cancer include changes in nucleosome occupancy and positioning (Brahma & Henikoff, 2020; Gaffney, McVicker, Pai, et al., 2012; Valouev, Johnson, Boyd, et al., 2011); nucleosome alterations (e.g. histone variants, different histone modifications) (Tan & Davey, 2011); and in noncoding RNAs (Calin, Pekarsky, & Croce, 2007; Fabbri, 2010; Wolin, Sim, & Chen, 2012). Cancer related epigenetic alterations cooperate through a network of mutually reinforcing or counteracting signals to regulate the expression of genes or activity state of noncoding DNA. Genome-scale projects charting the human epigenome are rapidly extending our understanding of epigenetic marks and how they interact (Shen, Wang, Zhang, et al., 2013). (The interested reader can visit the NIH Roadmap Epigenomics Project for more details and updated information.) Importantly, epigenetic mechanisms induce stable cell phenotypes but can adapt to changing developmental or environmental needs through the activities of proteins (e.g. chromatin readers, enzymes that regulate DNA methylation, etc) and noncoding RNAs (ncRNA), that also affect the epigenetic state of the genome and its effects on transcription. Recent studies of the aberrant regulation of noncoding DNA, primarily gene enhancers and super-enhancers, in cancer have further extended our understanding of the effects of epigenetic alterations on tumor formation and have shown how epigenetic alterations in noncoding DNA elements can promote cancer formation (Sengupta & George, 2017; Sur & Taipale, 2016).

The epigenome regulators can be classified as: (1) initiators (e.g. long noncoding RNAs, transcription factors); (2) histone writers, which establish the epigenetic marks; (3) readers, which interpret the epigenetic marks; (4) erasers, which remove the epigenetic marks; (5) remodelers, which can alter chromatin states; and (6) insulators, which form boundaries between epigenetic domains. Another class of epigenetic regulators includes enzymes that methylate and demethylate DNA as well as other direct DNA modifying enzymes. The establishment, maintenance, and modification of epigenetic marks are intricately regulated, with crosstalk among the marks and writers to shape and alter the epigenetic landscape (Shen et al., 2013).

1.3 DNA methylation

DNA methylation is an epigenetic DNA modification that can mediate a variety of processes in cells, including the maintenance of genome integrity, genomic imprinting, transcriptional regulation, and also can regulate embryonic developmental processes (Kanwal & Gupta, 2010). As noted previously, DNA methylation in eukaryotic cells primarily occurs at the 5-prime position of the cytosine ring within CpG dinucleotides and regulates gene transcription via effects on promoters and noncoding DNA elements, such as enhancers. The modification at 5-methyl cytosine is catalyzed by a family of DNA methyltransferases (DNMTs) that include DNMT1, a DNA methylation maintenance enzyme, and DNMT3A and DNMT3B, which are de novo enzymes that target unmethylated CpGs to initiate methylation at previously unmethylated sites. DNMT3A and DNMT3B are highly expressed during embryogenesis and only minimally in adult tissues; whereas DNMT1 is constitutively active in all adult replicating tissues (Kanwal & Gupta, 2010). Another family member is DNMT-3L, which lacks intrinsic methyltransferase activity and interacts with DNMT3A to facilitate DNA methylation (Zhang & Xu, 2017). In normal cells, DNA methylation occurs predominantly in repetitive genomic regions, including satellite DNA and parasitic elements such as long interspersed transposable elements (LINEs) and short interspersed transposable elements (SINEs). This is believed to help maintain genomic integrity (Ambrosi, Manzo, & Baubec, 2017; Robertson, 2005) (Fig. 2A). Methylated cytosines account for approximately 1% of total nucleotides and is found in about 75% of all CpG dinucleotides in the human genome. (There are approximately 28 million CpGs/genome.) The CpG dinucleotides are unevenly distributed across the human genome and are concentrated in areas called CpG islands (CGIs). CpG islands are often defined as regions >200 basepairs in length with at least 50% cytosine composition and an observed-to-expected CpG ratio greater than 60%. Roughly 50–60% of gene promoters lie within CpG islands, and it is estimated that the human genome contains approximately 29,000 CGI sequences (Kim, Samaranayake, & Pradhan, 2009).

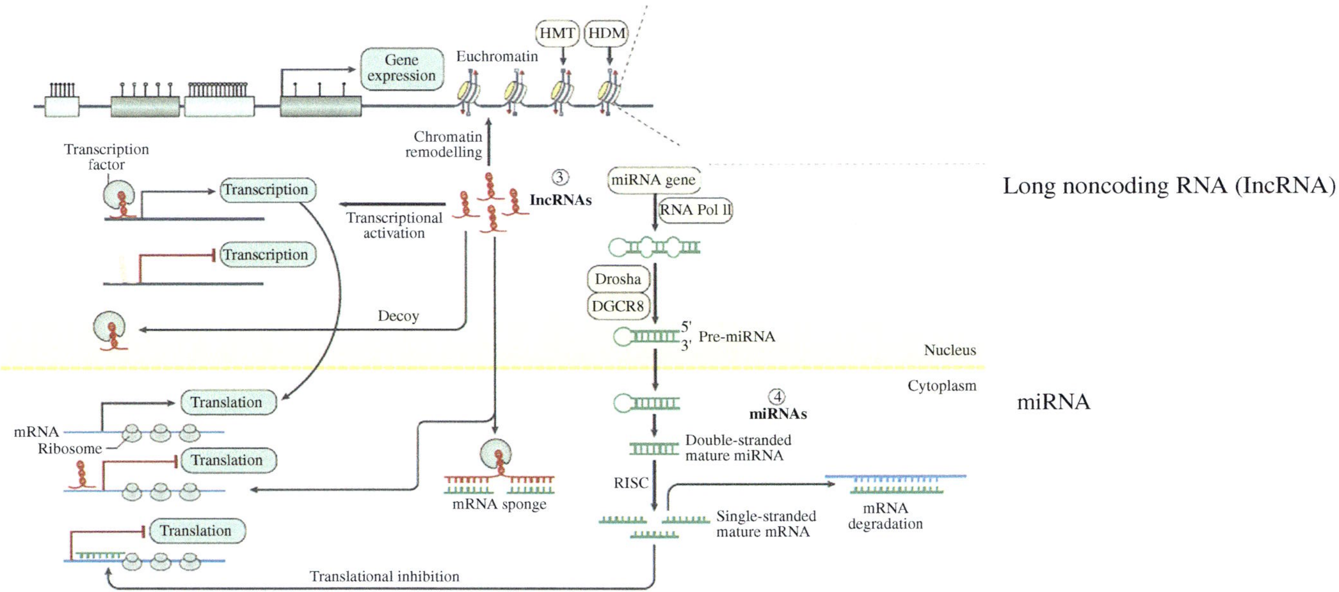

Fig. 2 Schematic diagram of long non-coding RNAs (lncRNAs) and microRNAs (miRNAs) processing and gene expression regulatory mechanisms is shown. lncRNAs influence gene and protein expression through different molecular mechanisms. They can enhance transcription by recruiting transcription factors or repress transcription by decoying transcription factors and preventing their recruitment to transcriptional start sites. lncRNAs can also restore translation by binding and deactivating miRNAs that would otherwise prevent translation of their corresponding mRNA. lncRNAs can also directly inhibit translation. The biogenesis of miRNAs starts with the transcription of the miRNA gene by RNA polymerase II (RNA Pol II). The double-stranded, hairpin-formed pri-miRNA is processed to the pre-miRNA by the Drosha–DGCR8 complex and then translocated to the cytoplasm by exportin 5. The RNase III enzyme DICER cuts the hairpin loop, resulting in a double-stranded miRNA–miRNA molecule. The RNA induced silencing complex (RISC) incorporates one of the strands and mediates its interaction with the target mRNA, leading either to translational inhibition or mRNA degradation. *Adapted from Jung, G., Hernández-Illán, E., Moreira, L., Balaguer, F., Goel, A., (2020). Epigenetics of colorectal cancer: Biomarker and therapeutic potential.* Nature Reviews Gastroenterology & Hepatology, 17 *(2), 111–130. doi: 10.1038/s41575-019-0230-y. Epub 2020 Jan 3.PMID: 31900466.*

Unlike the rest of the genome, CpG islands, particularly those associated with promoters, are generally unmethylated in normal cells, providing access to transcription factors and chromatin-associated proteins for the expression of most housekeeping genes and several other regulated genes. Some of the CpGs (~6%) become methylated in a tissue-specific manner during early development or in differentiated tissues (Kanwal & Gupta, 2010; Portela & Esteller, 2010).

DNA methylation can inhibit gene expression directly, by inhibiting the binding of specific transcription factors, and indirectly, by recruiting methyl-CpG-binding domain (MBD) proteins. The associated MBD family members in turn recruit histone-modifying and chromatin-remodeling complexes to methylated sites (Leighton & Williams Jr., 2019). To date, six methyl-CpG-binding proteins, including methylcytosine binding protein 2 (MECP2), MBD1, MBD2, MBD3, MBD4 and Kaiso, have been identified in mammals. Furthermore, it has been shown that a nucleosome remodeling complex (NuRD) can methylate DNA by interacting with the DNA methylation binding protein MBD2, which directs the NuRD complex to methylate DNA. DNA can also be actively demethylated by processes involving base excision repair (BER) enzymes (Dehan, Kustermans, Guenin, et al., 2009; Kohli & Zhang, 2013; Torchy, Hamiche, & Klaholz, 2015). These and other recent findings have established that DNA cytosine methylation is a critical component of epigenetic gene regulation and gene expression.

In cancer, including all gastrointestinal cancers, global DNA hypomethylation is observed as well as aberrant regional DNA hypermethylation. DNA hypermethylation has been extensively studied in virtually all cancers and is believed to contribute to cancer formation via the repression of tumor suppressor gene expression (Grady, Willis, Guilford, et al., 2000; Klutstein, Nejman, Greenfield, et al., 2016; Veigl, Kasturi, Olechnowicz, et al., 1998). In contrast, the role of DNA hypomethylation, which is also hallmark feature of cancer, in cancer formation is less well understood. It is believed to influence CRC development by inducing chromosomal instability, causing global loss of imprinting and activating super-enhancers (Eden, Gaudet, Waghmare, et al., 2003; Gaudet, Hodgson, Eden, et al., 2003; Heyn, Vidal, Ferreira, et al., 2016; Suzuki, Suzuki, Leodolter, et al., 2006). Genome-wide hypomethylation generally occurs within repetitive transposable DNA elements such as the LINE-1 or short interspersed nucleotide elements (SINE, or Alu) sequences (Lander, Linton, Birren, et al., 2001; Ogino, Kawasaki, Nosho, et al., 2008; Yamamoto, Toyota, Suzuki, et al., 2008). LINE-1 hypomethylation inversely associates with MSI and/or CIMP (Estecio, Gharibyan, Shen, et al., 2007; Ogino, Kawasaki, et al., 2008).

Furthermore, a number of studies have demonstrated that a high degree of LINE-1 hypomethylation correlates with worse patient survival (Ahn, Chung, Maeda, et al., 2011; Antelo, Balaguer, Shia, et al., 2012; Ogino, Nosho, Kirkner, et al., 2008; Rhee, Kim, Bae, et al., 2012). One hypothesis is that hypomethylation of LINE/SINE sequences may induce inadvertent activation of potential proto-oncogenes (Xiao-Jie, Hui-Ying, Qi, et al., 2016), which implies that LINE-1 hypomethylation has a functional role in CRC formation (Hur, Cejas, Feliu, et al., 2014). However, definitive evidence for a cancer-causing impact of DNA hypomethylation in primary human cancer is lacking at this time (Dokun, Florl, Seifert, et al., 2008; Van Tongelen, Loriot, & De Smet, 2017; Vilain, Vogt, Dutrillaux, et al., 1999; Wilson, Power, & Molloy, 1775)

1.4 Mechanism of DNA hypermethylation

The molecular mechanisms responsible for the causing aberrant DNA methylation in the majority of cancers are not clear, but several putative candidate mechanisms for DNA hypermethylation and hypomethylation have been identified. The first set of candidate mechanisms is related to the activity of the DNMTs. Some studies have shown increased DNMT1 activity in cancers but it is unclear if this is a primary phenomenon or secondary to increased proliferation. DNMT1 activity is cell cycle dependent and increased in S phase. DNMT1 also has an error prone nature and does not copy with 100% fidelity the DNA methylation patterns from parent DNA strand to daughter DNA strand (Tse, Jenkins, Chionh, et al., 2017). Consistent with DNMT activity and function derangement being the mechanism of cancer related to DNA methylation alterations, hypomethylated CpG dinucleotide rich regions (i.e., CpG-islands) have been shown to undergo methylation because of sporadic de novo methylation errors that occur when stem cells replicate their genome hundreds to thousands of times during an individual's lifetime (Issa, 2014; Kim, Kang, Koo, et al., 2004; Shibata, 2008, 2009). A second and more definitively proven mechanism is related to the effects of oncogenic mutations of *IDH1* and *TET*. Mutations in these enzymes can alter the activity of the TET enzymes. The *ten-eleven translocation* (TET) enzymes oxidize 5-methylcytosines (5mCs) and promote the locus-specific reversal of DNA methylation through base excision repair (Kohli & Zhang, 2013). *IDH1* mutations are found in subsets of some cancer types, like glioma and cholangiocarcinoma, and through perturbing alpha ketoglutarate metabolism create a metabolic state in the cell that disrupts TET function (Dawson, 2017). A third mechanism is related to the effect of environmental factors, like nutritional elements such as folate, copper; drugs (e.g. alcohol, etc) and the effect these factors have on the activity of enzymes that regulate DNA methylation (Zhang, 2015). These will be discussed further later.

With regards to the mechanism responsible for cancer associated DNA hypomethylation, the prevailing belief is that it usually results from a passive process secondary to inadequate maintenance of methylation during DNA replication (Auclair & Weber, 2012). However, recent studies have also demonstrated that DNA hypomethylation in cancer can be secondary to an active process in some tumors. The ten-eleven translocation enzymes (TET) can catalyze the formation of 5-hydroxymethylcytosine (5-hmC) from 5-methylcytosine, which is recognized by the Base Excision Repair (BER) proteins for removal and replacement with unmethylated cytosine. The ten-eleven translocation (TET) family members TET1 and TET2 are mutated in leukemia and a biochemically associated enzyme, isocitrate dehydrogenase (IDH1) is mutated in some gliomas. As noted above, it has also been proposed that mutant *TET1* and *TET2* and mutant *IDH1* are the mechanisms for DNA hypermethylation in leukemias and gliomas, respectively (Cairns & Mak, 2013; Kriaucionis & Heintz, 2009; Tahiliani, Koh, Shen, et al., 2009).

Of interest, global loss of hydroxymethylated DNA (5-hmC) is also reported in several types of solid tumors, including CRC, which may be due to deregulated TET enzyme activity (Kudo, Tateishi, Yamamoto, et al., 2012). Possible biological consequences of altered 5-hmC levels are suggested by studies that have shown downregulation of *TET1* expression in early stages of colon cancers, and loss of TET expression associating with inhibition of the WNT signaling pathway and suppression of tumor proliferation (Neri, Dettori, Incarnato, et al., 2015; Wu, Li, Shi, et al., 2019). However, the biological significance of 5-hmC and the TET proteins with regards to LINE-1 hypomethylation in CRC still remains unclear, and requires further investigation. Of note, *IDH1* and *TET1 or TET2* mutations do not appear to be a common cause of aberrant DNA methylation in CRC.

1.5 Epigenetic drift

It is of particular interest that accumulating evidence supports that the landscape of DNA methylation can be modified as a "function of age." DNA methylation has been proposed to result from a gradual stochastic age-dependent dysregulation caused by a combination of external environmental factors and internal spontaneous random errors in the maintenance of methylation. This process of age-dependent alterations in methylation is defined as "epigenetic drift" (Issa, 2014). Interestingly, such age-associated DNA methylation often targets the promoters of tumor-suppressive genes (Ahuja & Issa, 2000; Ahuja, Li, Mohan, et al., 1998). In monozygotic twins, epigenetic divergence with age suggests the underlying epigenetic drift may in part help explain the disease discordance (Bell & Spector, 2012; Bell, Wilson, Butcher, et al., 2012; Fraga, Ballestar, Paz, et al., 2005). Since this new concept closely relates to the "field effect," identification of biomarkers that overlap both the "epigenetic drift" and "field effect" in colorectal mucosa may allow development of next-generation biomarkers for determining risk for CRC development.

It is also worth noting that there is another process involving age related DNA methylation changes that involves epigenetic clocks. Epigenetic clocks appear to be based on a different mechanism from epigenetic drift. There several methods for assessing epigenetic age by the epigenetic clocks and many different epigenetic clocks described at this time (Horvath & Raj, 2018; Yu, Hazelton, Luebeck, et al., 2020). A number of recent studies have found associations of epigenetic drift and epigenetic clock status with aging, mortality, and a variety of diseases (Horvath & Raj, 2018; Yu et al., 2020).

1.6 Chromatin alterations and histone modifications

Histones are a family of small basic proteins that include H2A, H2B, H3, and H4 and that have a globular domain and a flexible charged NH2 terminus known as the histone tail, which protrudes from a protein complex called a nucleosome. A nucleosome is an octomer of histone proteins, which consists of members of each of H2A, H2B, H3, and H4, and encompasses ~146 bp of DNA wrapped around this octomer. These octamers consist of two subunits of each of the core histone proteins (Gilbert, 2019). Histone modifications influence chromatin structure, which plays an important role in gene regulation and carcinogenesis (Cedar & Bergman, 2009; Du, Luu, Stirzaker, et al., 2015). In addition to histone modifications, there are histone variants that provide an additional layer of regulation. Histone variants are characterized by a distinct protein sequence and a selection of designated chaperone systems and chromatin remodeling complexes that regulate their localization in the genome. They include H2A.Z, MacroH2A, H2A-Bbd, H2AvD, H2A.X, H3.3, CenH3, and H3.4 (Kamakaka & Biggins, 2005). The main histone genes have multiple copies in the genome and are expressed during S phase. Single-copy variants are also expressed at other phases of the cell cycle and have distinct functions and/or locations. H2A has the largest number of variants including H2A.Z, MacroH2A, H2A-Bbd, H2AvD, and H2A.X 44. The H3 variants include H3.3 and centromeric H3 (CenH3 or CENP-A), as well as a mammalian testis-specific histone H3 variant called H3.4. Histone variants have the capacity to endow specific regions of chromatin with unique character and function in a regulated manner. Emerging data is uncovering their role in cancer (Martire & Banaszynski, 2020).

Chromatin is a complex molecular structure that consists of DNA and protein. It has a primary function to package long DNA molecules into more compact, denser structures and is involved in DNA compaction, DNA replication, and gene regulation. It is a highly ordered B-form structure consisting of repeats of nucleosomes connected by linker DNA and exists in two distinct conformation states: heterochromatin, which is densely compacted and transcriptionally inert, and euchromatin, which is decondensed and transcriptionally active (Bersaglieri & Santoro, 2019) (Fig. 2B). Chromatin consists of DNA, histones, and non-histone proteins condensed into nucleoprotein complexes and functions as the physiological template of all eukaryotic genetic information (Suganuma & Workman, 2011). Regulation of gene expression occurs through posttranslational modifications of the histone tails provided by covalent modifications, such as acetylation, methylation, phosphorylation, ubiquitination, sumoylation, proline isomerization, and ADP ribosylation (Bartke & Kouzarides, 2011; Cohen, Poreba, Kamieniarz, et al., 2011; Kouzarides, 2007a; Tessarz & Kouzarides, 2014). Posttranslational modifications to histone tails govern the structural status of chromatin and resulting transcriptional status of genes within a particular region of DNA. Euchromatin is characterized by high levels of acetylation and trimethylated H3K4, H3K36 and H3K79. On the other hand, heterochromatin is characterized by low levels of acetylation and high levels of H3K9, H3K27 and H4K20 methylation (Bartova, Krejci, Harnicarova, et al., 2008; Hadley, Noonepalle, Banik, et al., 1983). These modifications are reversible and are controlled by a group of enzymes that include histone acetyltransferases (HATs) and deacetylases (HDACs), methyltransferases (HMTs) and demethylases (HDMs), kinases, phosphatases, ubiquitin ligases and deubiquitinases, SUMO ligases and proteases, which function to add and remove the respective modifications (Kouzarides, 2007b; Tessarz & Kouzarides, 2014). These histone modifiers generally act in complexes, such as the repressive Polycomb (PcG) and activating Trithorax (TrxG) group complexes, which counterbalance each other in the regulation of genes important for development. They are often mutated in cancers, including CRC, and have been implicated in the pathogenesis of cancer (Feinberg, Koldobskiy, & Gondor, 2016; Mills, 2010; Tahara, Yamamoto, Madireddi, et al., 2014).

The most thoroughly studied post-translational histone modifications that alter gene expression in the context of cancer, including CRC, are histone acetylation and methylation (Jung, Hernandez-Illan, & Moreira, 2020). This is partly as a result of the assessment of histone modifications being limited due to the availability of reliable reagents. It is expected that our understanding of this area will advance more quickly as new methods are developed.

With regards to histone acetylation, the acetylation and deacetylation of histones is catalyzed by histone acetyl transferases (HATs) and histone deacetylases (HDACs), respectively. Histone acetylation is believed to affect chromatin structure by neutralizing the positive charge on the histone tails, which weakens the electrostatic interaction between the DNA and histones (Portela & Esteller, 2010). Hyperacetylation of specific lysine residues on H3 and H4 associates with gene expression changes, and can permit increased gene expression, whereas hypoacetylation of these residues is believed to contribute to silencing the genes by blocking promoter access to transcription factors. Thus, there is a dual role of histone acetylation status in cancer development and progression through oncogene activation and tumor suppressor gene silencing (Leighton & Williams Jr., 2019). Unsurprisingly, in view of these key cancer related activities, these enzymes are viewed as potential therapeutic targets in various malignancies, including CRC (Kohli & Zhang, 2013), although the clinical utility of HDAC inhibitors in the treatment of CRC is still under investigation (Suraweera, O'Byrne, & Richard, 2018).

In contrast to histone acetylation, histone methylation not only changes the compaction status of the DNA, but also can create docking sites in the chromatin that can be recognized by various proteins, such as those comprising transcriptional complexes (Dehan et al., 2009; Torchy et al., 2015). As with acetylation, the lysine and arginine residues in the histone tails are the most common sites for methylation, which is added to specific sites with one to three methyl groups per site. This modification can lead to the activation or repression of gene expression, depending on the residue that is methylated. Histone methylation regulates many biological functions that are crucial for normal cell differentiation and likely has a central role in carcinogenesis and CRC progression (Grady et al., 2000). Histone methylation and demethylation are regulated by histone methyltransferases (HMTs) and histone demethylases (HDMs), respectively, the overexpression or under-expression of which might change the global histone methylation status, alter the expression of hundreds of oncogenes or tumor suppressor genes and, eventually, promote cancer development or progression (Dehan et al., 2009).

ATP-dependent chromatin-remodeling complexes are responsible for sliding of the nucleosomes, as well as insertion and ejection of histone octamers, processes that are important for transcriptional repression and activation, and other important cellular functions such as DNA replication and repair. There are also remodeling complexes that can be divided into four families: SWI/SNF, CHD (chromodomain and helicase-like domain), ISWI, and INO80.

With regards to cancer and as mentioned above, histone modification alterations, alterations in chromatin structure and deregulation of the genes that regulate histone modifications and chromatin structure have been found to occur commonly in many cancer types, including gastrointestinal cancers, and have been extensively characterized (Corces, Granja, & Shams, 2018; Feinberg et al., 2016; Okugawa et al., 2015). These alterations appear to have cancer specific signature patterns and have potential to be used as cancer biomarkers. There is also interest in developing epigenetic therapies for cancer, although these have not been shown to be particularly effective in the treatment of gastrointestinal cancers to date (Abdelfatah, Kerner, Nanda, et al., 2016; Baretti & Azad, 2018).

1.7 Noncoding RNA (ncRNA)

The central dogma of molecular biology, which describes the sequential transfer of genetic information and the concept that "DNA makes RNA and RNA makes protein", was developed in 1956 and provided a fundamental framework for modern molecular biology for 40 years (Crick, Barnett, Brenner, et al., 1961). Recently, advances in our understanding of the regulation of gene expression has been provided by the Encyclopedia of DNA Elements Consortium (ENCODE) transcriptome project, which revealed that protein-coding genes represent less than 2% of the total genome and approximately 80% of the genome is actively transcribed into non-coding RNA (ncRNA) (Derrien, Johnson, Bussotti, et al., 2012). Although ncRNAs were previously believed to be "background transcriptional noise" related to "junk DNA," ncRNAs are now recognized to play a significant role in many biological processes, including the regulation of oncogenes and tumor suppressor genes in cancer (Costa, 2005). NcRNAs are broadly categorized into two groups based upon their size: small ncRNAs, that are shorter than <200 nucleotides, composed of miRNAs (miRNAs), piwi-interacting RNAs (piRNAs), and small nucleolar RNAs (snoRNAs); and long ncRNAs (lncRNA) that are longer than 200 nucleotides (Pauli, Rinn, & Schier, 2011). Recent studies have begun to characterize the functional role of ncRNAs, particularly miRNAs, in human cancers and have also provided important insights into the feasibility of their use as clinically relevant biomarkers or therapeutic targets in cancer.

1.8 Overview of miRNAs

In the family of ncRNAs, the miRNAs are the class that has been most extensively studied and best characterized in cellular processes, including in the context of carcinogenesis. MiRNAs are endogenous single-stranded small RNAs that are 18–25 nucleotides in length that were first discovered in 1993 as repressive post-transcriptional regulators in *Caenorhabditis elegans* (Lee, Feinbaum, & Ambros, 1993; Wightman, Ha, & Ruvkun, 1993). During their biogenesis, premature-miRNAs are exported from the nucleus to the cytoplasm. Subsequent processing of the pre-miRNA generates mature-miRNA, which binds to 3'UTR "seed sequences" of target mRNAs, a process that is catalyzed by the RNA-induced silencing complex (RISC). The binding of miRNA to the target mRNA can result in degradation of the target mRNA or inhibition of its translation into protein, with the degree of sequence complementarity between the miRNA and mRNA determining which mechanism is employed (He & Hannon, 2004; Mendell, 2005; Vasudevan, Tong, & Steitz, 2007) (Fig. 2C). The role of miRNAs in cancer was first shown in 2002 by Croce and colleagues, who demonstrated decreased expression of miR-15 and miR-16 in patients with chronic lymphocytic leukemia (Calin, Dumitru, Shimizu, et al., 2002). Since then, hundreds of miRNAs have been shown to be deregulated in other human malignancies. Some of these deregulated miRNAs appear to have a functional role in tumorigenesis, including CRC formation, by regulating the expression of important oncogenes and tumor-suppressor genes (Amirkhah, Schmitz, Linnebacher, et al., 2015; Slaby, Svoboda, Michalek, et al., 2009).

2. Epigenetic alterations and colorectal polyps and cancer

The current paradigm of the molecular pathogenesis of CRC is that colon neoplasms can arise through a variety of discrete molecular pathways. These pathways can be characterized based on fundamental features of the resulting CRCs. One popular characterization is based on the form of genomic or epigenomic instability present. The major classes of CRC recognized with this scheme are chromosomal unstable (CIN), microsatellite unstable (MSI), and epigenetically unstable or hypermethylated (CpG Island Methylator Phenotype (CIMP) CRC (Carethers & Jung, 2015; Grady & Carethers, 2008; Komor, Bosch, Bounova, et al., 2018). Another system for defining these pathways is based on patterns of altered gene expression and includes the CMS classes of CMS1 (microsatellite instability immune), CMS2 (canonical), CMS3 (metabolic) and CMS4 (mesenchymal) (Guinney, Dienstmann, Wang, et al., 2015). Importantly, regardless of the pathway through which CRCs develop, epigenetic alterations are present in essentially all CRCs, although the number of alterations varies between subclasses with some having high proportions of methylated genes and others having low numbers (Hinoue, Weisenberger, Lange, et al., 2012; Zhang, Yang, Chen, et al., 2014).

With specific regards to the epigenetic alterations found in colorectal polyps and CRC, aberrant DNA methylation is the best understood class of epigenetic alterations in CRC and is present in colon polyps as well as established CRCs (Kim et al., 2006; Luo et al., 2014). Few studies of alterations in chromatin structure or histone modification states have been conducted to date. These studies have shown increased expression of HDAC2 and increased acetylation of H4K12 and H3K18 in adenomas compared to normal colon but little else is known (Ashktorab, Belgrave, Hosseinkhah, et al., 2009). Consequently, in this chapter, the majority of the discussion of epigenetic alterations in colon polyps will focus on DNA methylation.

2.1 Epigenetic alterations in CRC initiation and polyp formation

Aberrant DNA methylation appears to have a functional role in both the formation of colon polyps as well as their progression to CRC (Carragher, Snell, & Giblett, 2010; Kim et al., 2006). The discovery of a molecular subclass of CRCs defined as having a "CpG island methylator phenotype (or CIMP)" in 1999 was a significant advance in our understanding of the molecular mechanisms that orchestrate colorectal tumor formation and revealed the role of aberrant DNA methylation in particular as an important class of epigenetic alteration in CRC (Toyota, Ahuja, Ohe-Toyota, et al., 1999; Weisenberger, Siegmund, Campan, et al., 2006). Interestingly, although not initially recognized as precancerous until approximately 2008, sessile serrated polyps commonly have CIMP (Bettington, Walker, Clouston, et al., 2013). It is now appreciated that CIMP CRC arise almost exclusively from sessile serrated lesions, which include sessile serrated adenomas, sessile serrated polyps, and traditional serrated adenomas, and exhibit unique clinicopathological and molecular features, including a predilection for proximal location in the colon, female gender, poor and mucinous histology, the presence of frequent *KRAS* and *BRAF* mutations and frequent MSI due to biallelic *MLH1* methylation (Hampel, Frankel, Martin, et al., 2005; Pai, Bettington, Srivastava, et al., 2019; Samowitz, Albertsen, Herrick, et al., 2005; Toyota et al., 1999; Toyota, Ohe-Toyota, Ahuja, et al., 2000; Valle, Serena-Acedo, Liyanarachchi, et al., 2008; Veigl et al., 1998; Weisenberger et al., 2006). In fact, biallelic methylation of *MLH1* is the underlying mechanism responsible for the majority of sporadic MSI CRCs (Veigl et al., 1998).

2.2 Environmental factors and the methylome

DNA methylation depends on the intrinsic activity state of the DNMTs and is also affected by the availability of methyl-donors, such as those provided by the one-carbon cycle, and on cofactors that modulate the enzymatic activity of DNA methyltransferases (DNMTs) (Zhang, 2015). *S*-adenosylmethionine (SAM) is the only known physiological donor of methyl groups, allowing their transfer to cytosine (which becomes methylated, 5mC) (Feil & Fraga, 2012). SAM is synthesized, via methionine, from different dietary precursors (folate, choline, and betaine), using homocysteine as methylation substrate (Zeisel, 2009). Some studies have shown reduced availability of methyl donors results in low SAM synthesis, which results in DNA methylation alterations (either hypermethylation or hypomethylation, depending on the tissue type and developmental stage), while increased availability of methyl donors is usually associated with DNA hypermethylation (Niculescu & Lupu, 2011).

One of the first studies in people to suggest diet and nutritional factor effect DNA methylation status, and possibly cancer, is a study in people who lived through the Dutch Hunger Winter (1944–1945). The cohort of children and adolescents who survived the severe famine have significant modifications of the DNA methylation status for several cancer-related genes, and to have a lower risk for CRC (Mathe, Nguyen, Bowman, et al., 2009). It has also been shown that perinatal maternal undernutrition correlates with a lower methylation state of the insulin-like growth factor 2 (IGF2) (Heijmans, Tobi, Stein, et al., 2008). These studies raise the possibility that early nutritional exposures during fetal and childhood development can permanently alter the DNA methylation pattern of an individual, leading to modifications that could be related with an increased risk for diseases like cancer (Bishop & Ferguson, 2015).

Specific nutritional factors that could plausibly effect DNA methylation include one carbon metabolism methyl donors, such as folate, methionine, cobalamin, pyridoxine, and riboflavin, which have been associated with cancer-related alterations of DNA methylation. The role of folate is still controversial, as various epidemiological studies have shown contradictory effects. Most studies have shown either that increased dietary availability of folate can modulate the DNA methylation status and have anticarcinogenic effects or that there is no effect of folate on DNA methylation and cancer risk (Mahmoud & Ali, 2019; Maruti, Ulrich, & White, 2009). Some studies have shown low folate intakes associate with increased risk for cancer development through DNA damage, abnormal methylation and inhibition of DNA methyltransferases (DNMTs) (Mahmoud & Ali, 2019). A study of low intakes of methionine, pyridoxine, cobalamin, niacin, and riboflavin showed an association with an increased risk for breast cancer, while supplementation with folate associated with decreased breast cancer risk in premenopausal women (Bhowmick, Ghiassi, Aakre, et al., 2003; Maruti et al., 2009). Pyridoxine is a cofactor for the serine hydroxymethyltransferase involved in 5,10-MTHF synthesis and plays an important role in glutathione synthesis from homocysteine (Maruti et al., 2009). Low levels of pyridoxine may inhibit DNMT activity, cause DNA hypomethylation and an increase the risk for cancer (Mahmoud & Ali, 2019).

Beyond one carbon metabolism related nutrients, phytoestrogens, such as resveratrol and genistein, have been shown to affect DNA methylation. Genistein (4′,5,7-trihydroxyisoflavone), one of the polyphenols from soybean, has been shown to decrease promoter methylation in renal carcinoma cell lines and prostate cancer cells and to have anti-cancer effects similar to 5-aza-cytidine (Majid, Kikuno, Nelles, et al., 2008). Bioflavonoids (such as quercetin), tea catechins {epigallocatechin-3-gallate (EGCG)} and coffee polyphenols (caffeic acid, chlorogenic acid) have inhibitory effects on DNA methylation and the activity of DNMTs (Yang, Landau, Huang, et al., 2001). A high intake of chlorogenic acid was associated with elevated concentrations of homocysteine, probably due to its inhibitory role upon DNA methylation reactions. The tea catechins and bioflavonoids have also been reported to inhibit DNA methyltransferase activity by interfering with the substrates for the methylation reactions mediated by catechol-O-methyltransferase (COMT). EGCG can exert a dual inhibitory effect on DNMT-mediated DNA methylation and methyltransferase activity by increasing SAH in the reaction catalyzed by COMT (Stefanska, Karlic, Varga, et al., 2012). Other nutritional factors that have been shown to affect DNMT activity and DNA methylation include curcumin and lycopenes (Zhang, 2015). The mechanisms though which they operate are not understood at this time. In aggregate, these studies suggest that diet and nutritional factors can modify DNA and influence the pathogenesis of CRC.

2.3 DNA methylation and the gut microbiome

The effect of the gut microbiome on DNA methylation in the intestines is an area of relatively recent investigation. Studies of the gut microbiome and intestinal DNA methylation have demonstrated associations of DNA methylation patterns of intestinal epithelial cells and bacterial species and a potential role in intestinal development and homeostasis (Ansari, Raddatz, Gutekunst, et al., 2020). Furthermore, studies of health-promoting probiotic species, such as *Lactobacillus acidophilus*, have shown a common methylation signature in colon epithelial cells that is distinct from the signature induced by pathogenic bacteria. Several studies have shown that butyrate, which is commonly metabolized by gut bacteria, can affect both the methylation of DNA globally (Wippermann, Klausing, Rupp, et al., 2014; Wippermann, Rupp, Brinkrolf, et al., 2017) and the expression of genes that function in DNA methylation or demethylation pathways (Sarkar, Abujamra, Loew, et al., 2011; Wei, Melas, & Wegener, 2014). With regards to CRC formation, studies of gut microbes and DNA methylation in transformed cells have shown that Enterotoxigenic *Bacteroides fragilis* (ETBF)-induced tumors in mice contained more hypermethylated differentially methylated regions (DMRs) and fewer hypomethylated DMRs compared to spontaneous tumors. Many of the hypermethylated DMRs were associated with the CpG islands of genes with known tumor-suppressive functions, such as *Hoxa5*, *Polg*, *Runx1*, *Runx3*, *CD37*, *Stx11*, *Tceb2*, *Lgr6*, *Cdx1*, and *Fut4* (Sobhani, Bergsten, Couffin, et al., 2019). More recent studies of

CRC-associated gut dysbiosis and altered DNA methylation in colon neoplasms in humans and in mice have shown in germ-free mice receiving fecal samples from subjects with normal colonoscopy or from CRC patients that CRC-associated microbiota induced higher numbers of hypermethylated genes, including *Sept9, Penk, Sfrp1, 2, 3,* and *Wif1* in murine colonic mucosa. These studies suggest that CRC-related dysbiosis induces methylation of host genes and may promote CRC formation. However, more studies are needed to better understand how methylation changes that are induced by specific microbes and their toxins contribute to CRC development. Many of the studies to date have been done using in vitro CRC or non-intestinal cell lines demonstrating the need for further studies in this field (Cristiano, Leal, Phallen, et al., 2019).

2.4 Epigenetics and "field cancerization" in the colon

The concept of "field cancerization" (or field effect) was first proposed in 1953 by Slaughter, Southwick, and Smejkal (1953). Field cancerization is characterized by the occurrence of genetic and epigenetic alterations in histologically normal-appearing tissues, and is believed to lead to an increased risk for synchronous or metachronous primary tumors. While genetic alterations are common in CRC cells, these are believed to be rare in normal cells (Lee-Six, Olafsson, Ellis, et al., 2019). In contrast, some studies suggest somatic epigenetic dysregulation occurs not only in cancer tissues, but also in non-cancerous and pre-neoplastic tissues. Considering that epigenetic alterations could contribute to the early events predisposing to malignant transformation, these studies suggest epigenetic events are potentially more promising somatic CRC risk markers (aka field cancerization markers) than are gene mutations. Methylation changes in tumor-suppressor genes occurs more frequently in the normal colonic mucosa of CRC patients than healthy controls (Luo et al., 2014), suggesting that they may be one of the earliest events that predispose normal mucosa to tumorigenic transformation in CRC (Issa, Ottaviano, Celano, et al., 1994). Furthermore, loss of the insulin-like growth factor-II (*IGF2*) gene imprinting occurs at a higher frequency in the normal mucosa adjacent to cancer tissue, compared to normal mucosa in patients without CRC (Cui, Onyango, Brandenburg, et al., 2002), emphasizing the potential of loss of *IGF2* imprinting to be a biomarker to identify patients at greater risk for CRC development. Other studies have revealed that both hypermethylation of tumor-suppressive genes such as *SFRP, ESR1, MYOD, EVL,* and *MGMT,* as well as LINE-1 hypomethylation in normal colonic mucosa correlates with an increased risk of CRC compared to patients without these traits (W.M. Grady, personal communication; Grady, Parkin, Mitchell, et al., 2008; Kamiyama, Suzuki, Maeda, et al., 2012; Kawakami, Ruszkiewicz, Bennett, et al., 2006; Shen, Kondo, Rosner, et al., 2005; Suzuki, Watkins, Jair, et al., 2004; Yu et al., 2020). Further information on this topic is provided in the chapter by Rebello et al. (2021).

2.5 Aberrant DNA methylation in the adenomatous polyp and in the serrated polyp pathways

As stated earlier, there are multiple molecular pathways that can give rise to colon polyps and CRC. The two most common pathways to CRC formation are initiated by either sessile serrated lesions (SSLs) or by adenomas (Bettington et al., 2013; Bettington, Rosty, Whitehall, et al., 2018; Chang, Willis, Reumers, et al., 2018; Komor et al., 2018). There are at least two or three recognized SSL pathways that begin with different precursor lesions: hyperplastic polyps, serrated polyps, and serrated adenomas. The malignant potential of these precursors is believed to vary significantly, but appears to be less than that of adenomas (Pai et al., 2019). It is unclear what initiates SSLs rather than adenomas. Recent findings suggest that in addition to microenvironment factors, genetic and epigenetic alterations that occur at low frequency in the normal colon mucosa are a major factor that determine whether the initial lesion in CRC formation is an adenoma or SSL (Kaz, Wong, Dzieciatkowski, et al., 2014; Lee-Six et al., 2019; Luo et al., 2014). For both SSLs and adenomas, the majority of DNA alterations that are found are epigenetic alterations rather than DNA mutations. Most epigenetic alterations appear to arise during polyp formation, which is not true of genetic alterations, which occur predominantly after the polyp initiation (Luo et al., 2014). There is also a clear difference in the proportion of methylated loci that are found in most SSLs compared to adenomas, which suggests that factors driving the epigenetic alterations may play a role in determining whether the initial neoplastic lesion in the colon is an SSL vs. an adenoma.

The adenoma to CRC pathway is initiated by alterations in the WNT signaling pathway, most commonly *APC* mutations, and subsequent progression is associated with alterations that affect the MAPK and TP53 pathways (e.g. *KRAS* and *TP53* mutations and methylation of genes that regulate these pathways). The SSL to CRC sequence is primarily characterized by having a activation of the mitogen activated protein kinase (MAPK) pathway by oncogenic mutations in *BRAF* and *KRAS*. Serrated polyps that carry mutant *BRAF* are also commonly CIMP and can progress to microsatellite stable or unstable CRC, depending on whether epigenetic inactivation of the mismatch repair protein MLH1 occurs (Bettington et al., 2013). In contrast, traditional serrated adenomas more commonly activate the MAPK pathway via *KRAS* mutations, carry *RSPO* fusion transcripts, and typically have a low CIMP status. Unlike classic adenomatous polyps, sessile serrated polyps and traditional serrated adenomas do not typically have genetic alterations in *APC* or *CTNNB1* (Bettington et al., 2013; Laiho, Kokko, Vanharanta, et al., 2007), and appear to activate WNT signaling late in the polyp to CRC sequence via aberrant DNA methylation of *SFRP* family members, *CDX2*, or *MCC* (Dhir, Yachida, Van Neste, et al., 2011; Kohonen-Corish, Sigglekow, Susanto, et al., 2007). SSLs also more commonly employ epigenetic alterations to disrupt other CRC pathways and processes, including the p53 signaling pathway (via alterations in *IGFBP7)* (Suzuki, Igarashi, Nojima, et al., 2010) and cell cycle control proteins (via alterations in *CDKN2A*) (Kriegl, Neumann, Vieth, et al., 2011).

3. Clinical applications of DNA methylation in the prevention and management of CRC

There is significant potential for the use of cancer related molecular alterations as biomarkers for the prevention and management of a variety of cancers, including the majority of GI cancers. Aberrantly methylated DNA biomarkers have proven the most robust and successful clinically used molecular markers to date. Another epigenetic feature, chromatin alterations, is an emerging class of promising biomarkers that will likely be in clinical use in the near future (Snyder, Kircher, Hill, et al., 2016). There are many DNA-methylation based biomarkers in the prevention and management of CRC now in clinical use (Fig. 3, Table 1). The earliest application of methylated gene biomarkers in CRC clinical care occurred in the late 1990s and was the use of methylated *MLH1* for determining the likelihood of MSI CRC being sporadic vs. hereditary in origin (Pritchard & Grady, 2011). Methylated *MLH1* and *BRAF* mutations arise almost exclusively in sporadic MSI CRC and are used in the clinic to identify CRC patients who should be considered for genetic testing.

3.1 Screening/early detection

During the early stages of colorectal carcinogenesis epigenetic alterations are much more frequent than genetic mutations throughout the polyps

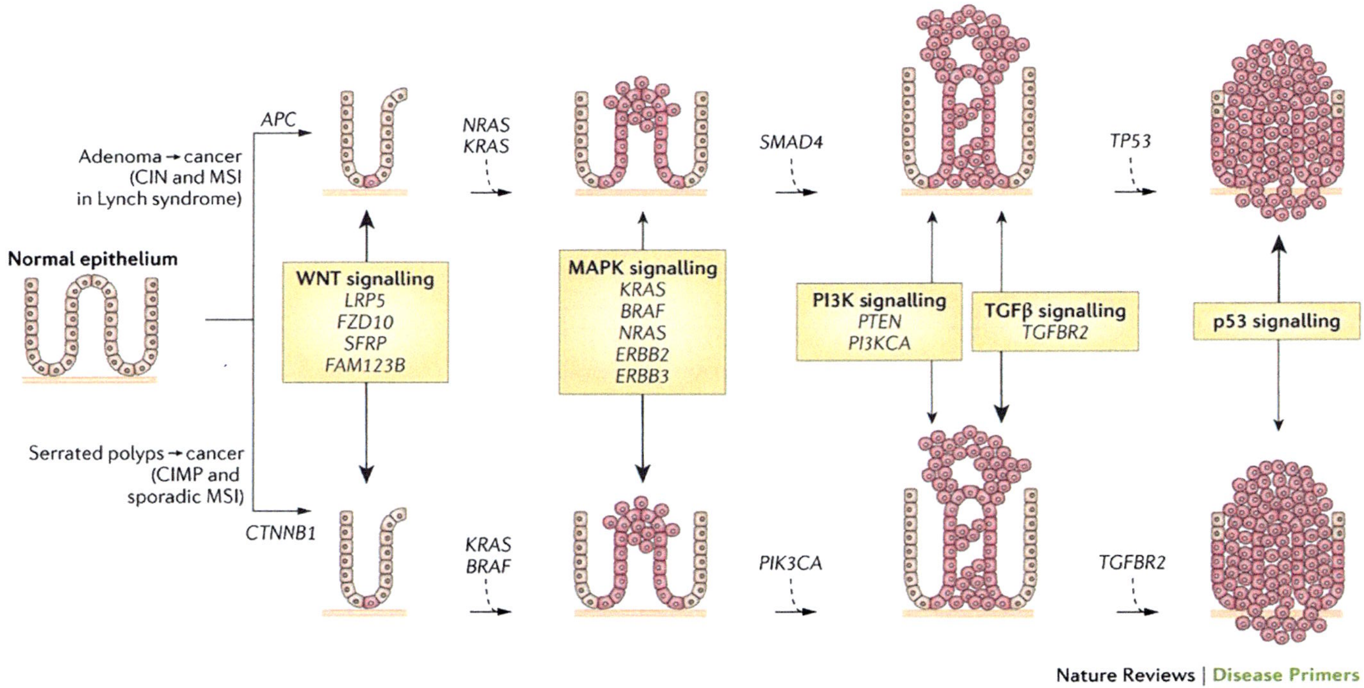

Fig. 3 Diagram showing the polyp-to-CRC progression sequence and associated methylated gene alterations and mutations. *Adapted from Kuipers, E. J., et al. (2015). Colorectal cancer.* Nature Reviews. Disease Primers, 1, *15065.*

Table 1 Validated DNA-methylation biomarkers for colorectal cancer.

Clinical use	Biomarkers	Available commercial assays	Study design of major trials with assay	Ref.
Stool-based CRC screening	*mVIM,*	ColoSure™	Case ($N=42$) control ($N=241$) study	Itzkowitz, Jandorf, Brand, et al. (2007)
	*m*BMP3 and m*NDRG4*	Cologuard® (detects mutant *KRAS*, and includes a FIT test)	Prospective cohort based clinical trial in screening population ($N=9989$)	Imperiale, Ransohoff, Itzkowitz, et al. (2014)
Blood-based diagnostic marker	*mSEPT9*	EpiproColon® 1.0; ColoVantage®; RealTi*me m*S9	Multiple trials: (1) Prospective cohort based clinical trial in screening population ($N=7941$) (Church); (2) Case-Control study ($N=269$) (deVos); (3) Case-Control study ($N=312$) (Lofton-Day, Model, Devos, et al., 2008)	Nikolaou, Qiu, Fiorentino, et al. (2018), Church, Wandell, Lofton-Day, et al. (2014), Lofton-Day et al. (2008), Grutzmann, Molnar, Pilarsky, et al. (2008)
	mBCAT1 *mIKZF1*	Colvera	Cross-sectional study ($N=220$)	Young, Pedersen, Mansfield, et al. (2016)
Tissue-based prognostic markers	*CIMP* panel	NA	Multiple trials: (1) Case-Control study from 2 phase I/II clinical trials ($N=31$) (Ogino, Kawasaki, Kirkner, et al., 2007; Ogino, Meyerhardt, Kawasaki, et al., 2007): (2) Case-Control study from phase 3 clinical trial ($N=615$) (Shiovitz, Bertagnolli, Renfro, et al., 2014); (3) Observational cohort study ($N=2050$) (Phipps, Limburg, Baron, et al., 2015)	Shen, Catalano, Benson, et al. (2007), Shiovitz et al. (2014), Ogino, Meyerhardt, et al. (2007), Phipps et al. (2015)
Diagnostic tool to screen for Lynch syndrome	*mMLH1*	MLH1 hypermethylation analysis	(1) Cross sectional study ($N=1066$)	Hampel et al. (2005)

(Myint, Verma, Fernandez-Garcia, et al., 2018). Consequently, they have great potential as diagnostic biomarkers for the detection of colonic polyps and cancers, which is evident by the fact that the currently clinically approved screening assays for CRC, the ColoGuard assay (Exact Sciences) and EpiProcolon (Epigenomics) are based on methylated DNA. The potential for methylated DNA based alterations and perhaps other epigenetic alterations to detect colon polyp and early stage CRC in circulating DNA has also led to the development of blood-based assays for CRC screening that are currently under clinical investigation or that are available clinically. (See section below.)

3.1.1 Stool-based biomarkers

Specific biofluids, such as blood (plasma or serum) and feces are the most common analytes used in CRC screening tests. Since the initial discovery by Sidransky and colleagues of mutant *KRAS* in fecal specimens from patients with CRC (Sidransky, Tokino, Hamilton, et al., 1992), numerous studies have shown fecal DNA in potential screening assays for the early detection of CRC.

To date, a large number of hypermethylated genes including *APC, ATM, BMP3, CDKN2A, SFRP2, GATA4, GSTP1, HLTF, MLH1, MGMT, NDRG4, RASSF2A, SFRP2, TFPI2, VIM, WIF1* as well as others, have been analyzed in fecal DNA for the early detection of CRC (Chen, Han, Skoletsky, et al., 2005; Glockner, Dhir, Yi, et al., 2009; Hellebrekers, Lentjes, van den Bosch, et al., 2009; Lee, Lee, Jung, et al., 2009; Leung et al., 2007; Lidgard, Domanico, Bruinsma, et al., 2013; Petko, Ghiassi, Shuber, et al., 2005; Wang & Tang, 2008). Of this large list of potential epigenetic based CRC screening markers, methylated *VIM, BMP3,* and *NDRG4* have been shown to be robust and accurate enough to be approved for clinical use. Methylated *VIM* was the first stool-based epigenetic biomarker approved for the early detection of CRC and was marketed under the name ColoSure™, (Lab Corp, Burlington, NC) (Itzkowitz et al., 2007; Itzkowitz, Brand, Jandorf, et al., 2008).

After the ColoSure assay, a next generation stool based multi target (MT) stool DNA assay was developed and later FDA approved (in 2014). This assay is a stool DNA based assay that detects methylated *BMP3*, methylated *NDRG4*, mutant *KRAS*, and occult hemoglobin (FIT test) (Cologuard® (Exact Sciences Corporation). In a large clinical trial of average risk individuals (the Deep C trial), this MT stool DNA assay was compared to the FIT assay and to colonoscopy ($N=9989$) (Imperiale et al., 2014), and showed an overall sensitivity of 92% (95% CI, 83–97.5%) for CRC and

93% (95% CI 83.8–98.2%) for stage I-III CRC, compared to sensitivity of FIT at 74% (95% CI, 61.5–84%) and 73% (95% CI, 60.3–83.9%), respectively ($P=0.002$). For advanced adenomas and sessile serrated polyps, the sensitivity of the test increased proportionately with lesion size and grade. The assay was significantly more sensitive than FIT for advanced adenomas: 42% (95% CI, 38.9–46%) vs 24% (95% CI, 20.8–27%), respectively, for those ≥1 cm and 66% vs 43% for those ≥2 cm ($P<0.001$). Sessile serrated polyps ≥1 cm were detected at a rate of 42% for the molecular assay compared 5% for FIT ($P<0.001$). In this study the test specificity for detection of CRC and advanced pre-cancers was 87% (Imperiale et al., 2014).

Since its FDA approval, >3 million MT stool DNA assays have been conducted, leading to a real-world use experience of this assay, which has similar performance to that seen in the Deep C trial (Eckmann, Ebner, & Kisiel, 2020). In addition, recent studies have assessed the performance of the MT stool DNA assay in noncompliant patient populations and found improved compliance in this group (Prince, Lester, Chiniwala, et al., 2017). The assay has been also assessed in African Americans, who have a higher risk for CRC than the general population, and was found to have equivalent performance to that seem in Caucasian patients (Cooper, Markowitz, Chen, et al., 2018).

3.1.2 Blood-based biomarkers

Due to the convenience of blood based tests and their high patient acceptance, blood is invariably regarded to be the most ideal analyte for cancer biomarkers. The majority of blood-based biomarkers for cancer to date have been proteins or glycoproteins (e.g. PSA, CEA, CA-125, etc.). Recent studies have demonstrated the potential of circulating tumor DNA (ctDNA) for cancer detection and management, and there are now "liquid biopsy" assays being used in the clinical care of cancer patients (Mouliere & Rosenfeld, 2015; Thomas & Marcato, 2018). For CRC, ctDNA somatic tumor-derived mutations are promising markers for the early detection of recurrent cancer, for monitoring treatment response, and for prognosis (Bi, Wang, Dong, et al., 2020; El Messaoudi, Mouliere, Du Manoir, et al., 2016; Reinert, Scholer, Thomsen, et al., 2016; Siravegna, Mussolin, Buscarino, et al., 2015). However, they have not been shown to be detected at a high frequency in patients with advanced polyps and early stage CRC using current technology, which likely reflects the very low frequency of these mutations in the plasma (6.6 ng/mL blood, <0.1–0.01% ct DNA) (Myint et al., 2018; Reinert, Henriksen, Christensen, et al., 2019; Tie, Cohen,

Wang, et al., 2019; Wang, Li, Cohen, et al., 2019). Methylated DNA and chromatin fragmentation patterns, in contrast can be found in patients with advanced polyps and early stage CRC more commonly. This is likely a consequence of the high frequency of epigenetic alterations in colon polyps and early CRC. Consequently, methylated DNA has been detected in higher proportions of patients with early stage CRC patients compared to DNA mutations, which has led to their assessment as blood-based early detection assays (Bartak, Kalmar, Peterfia, et al., 2017; Nikolaou et al., 2018).

To date, several potential blood-based diagnostic methylation biomarkers have been identified for CRC detection, including *ALX4* (Ebert, Model, Mooney, et al., 2006), *APC* (Lee et al., 2009), *CDKN2A* (Petko et al., 2005), *HLTF* (Leung et al., 2005), *HPP1* (Wallner, Herbst, Behrens, et al., 2006), *MLH1* (Leung et al., 2005), *MGMT* (Lee et al., 2009), *NEUROG1* (Herbst, Rahmig, Stieber, et al., 2011), *NGFR*, *RASSF2A* (Lee et al., 2009), *SFRP2*, *VIM* (Lee et al., 2009), and *WIF1* (Lee et al., 2009), *B4GAT1* (Picardo, Romanelli, & Muinelo-Romay, 2019), *BCAT1*, *IKZF1* (Aad, Abbott, Abdallah, et al., 2015); *SFRP1*, *SDC2*, and *PRIMA1* (Bartak et al., 2017). The best studied methylated DNA blood-based biomarkers to date are *mSEPT9*, *mSDC2*, and a combination of *mBCAT1* and m*IKZF1*, which is marketed under the name Colvera (Clinical Genomics). They have been shown to consistently detect CRC in serum or plasma samples and have high potential to eventually be used in the clinic for monitoring for recurrent CRC (Church et al., 2014; Nikolaou et al., 2018; Oh, Kim, Moon, et al., 2013; Young et al., 2016). Other methylated DNA biomarkers are under active investigation for the early detection of recurrent CRC and for screening for CRC.

Currently, the most established methylated DNA blood biomarker is methylated Septin 9 (*SEPT9*), which belongs to the gene family that encodes a group of GTP-binding (Finger, 2002) and filament-forming proteins (Sheffield, Oliver, Kremer, et al., 2003) involved in cytoskeletal formation. Lofton-Day and colleagues first identified methylated *SEPT9* *(mSEPT9)* as a non-invasive diagnostic biomarker for CRC with 69% sensitivity and 86% specificity (Lofton-Day et al., 2008). However, a subsequent prospective CRC screening trial (PRESEPT) showed lower sensitivity for CRC (48.2% at 91.5% specificity) (Grutzmann et al., 2008). Subsequent studies have validated the clinical significance of m*SEPT9* as a potential biomarker for CRC screening, but it is only recommended in people who refuse to undergo other CRC screening tests, because of its low sensitivity for CRC (52–72%) compared to other CRC screening assays (Nikolaou et al., 2018). It is commercially-offered as a blood-based screening test in various assays including EpiproColon® 1.0 (Epigenomics, Seattle, WA), ColoVantage® (Quest Diagnostics, Madison, NJ) and RealTi*me m*S9 (Abbott Laboratories, Des Plaines, IL) and has been approved for CRC screening by the Chinese FDA. A major issue of m*SEPT9* as a CRC screening assay, is its low sensitivity for the detection of advanced adenomas (11%), underscoring the need for further improvement of this test if it is to be used for population-based screening of colorectal neoplasia. A recent study demonstrated that the methylated *SEPT9* assay was superior to fecal immunochemical (FIT) at detecting CRC neoplasms, but in this study both approaches were suboptimal for diagnosing patients with advanced adenomas (Jin, Kang, Wang, et al., 2015).

In summary, although the quality of the experimental data to support the clinical use of CRC screening biomarkers for the early detection of recurrent CRC, methylated DNA plasma-based markers have high potential to eventually be used in the clinic for the early detection of recurrent CRC and possibly for CRC screening. The most promising CRC epigenetic alterations under evaluation at this time for CRC screening are based on a combination of biomarkers including methylated DNA and chromatin fragmentation patterns and are being used in ongoing CRC screening trials at the time this chapter is being written {Circulating Cell-free Genome Atlas study (NCT02889978; GRAIL); Session VCTPLO2, CT021—Prediction of cancer and tissue of origin in individuals with suspicion of cancer using a cell-free DNA multi-cancer early detection test, and the ECLIPSE trial of the Lunar-2 assay (Kim, AACR 2020 abstract #916; Guardant Health)}. Thus, there appears to be potential for a robust biomarker panel of methylated genes to be a clinically accurate CRC screening method.

3.2 Epigenetic prognostic biomarkers

Currently, the most accurate means for assessing CRC patient prognosis is based on the pathological staging and specific histologic features of the tumor. However, the heterogeneity of survival times in patients with the same stage of CRC is well known, which highlights the need for a more accurate system for determining CRC patient prognosis. Multiple large and sufficiently powered clinical studies with independent external validation cohorts have demonstrated the feasibility of using specific methylated DNA signatures for developing prognostic biomarkers in CRC (Okugawa et al., 2015).

Among all epigenetic biomarker candidates, CIMP status is the most promising prognostic indicator for CRC patients to date. CIMP-positive cancers correlate with an overall unfavorable prognosis (Kim, Colangelo, Wieand, et al., 2007; Ogino, Kawasaki, et al., 2007; Rhee, Jair, Yen, et al., 2000; Shen et al., 2007; Zlobec, Minoo, Baumhoer, et al., 2008). Rijnsoever and colleagues showed in a cohort of 206 stage III CRC patients that CIMP-positive status associated with poor survival (Van Rijnsoever, Elsaleh, Joseph, et al., 2003). Another independent study analyzed more than 600 CRC patients and also found that CIMP associated with poor prognosis in MSS (microsatellite stable) CRC patients (Lee, Cho, Yoo, et al., 2008). Some studies suggested that poor prognosis in CIMP-positive CRCs is from co-existing V600E *BRAF* mutations (Lochhead, Kuchiba, Imamura, et al., 2013; Pai, Jayachandran, Koong, et al., 2012), however, in addition to CIMP, MSI status remains an important confounding factor that likely underlies the difference in prognosis of CIMP-positive MSS vs. MSI cancers (Juo, Johnston, Zhang, et al., 2014). CIMP positive MSI CRCs have MSI due to biallelic methylation and inactivation of MLH1, which presumably alters the biology of the CRC to make it less lethal. These data highlight that the prognosis of patients with CIMP CRCs is affected by the MSI status of the tumor.

In addition to hypermethylation of various genes/loci, growing evidence suggests that DNA hypomethylation status associates with the prognosis of CRC patients. Ogino and colleagues have reported a correlation between LINE-1 hypomethylation and poor survival in prospective cohort studies of CRC patients (Ogino, Nosho, et al., 2008). Subsequent studies not only validated this association for LINE-1 hypomethylation and CRC prognosis (Ahn et al., 2011; Antelo et al., 2012; Rhee et al., 2012), but also identified other potential genes that correlate with adverse outcomes (Baba, Nosho, Shima, et al., 2010; Draht, Smits, Tournier, et al., 2014; Nagasaka, Sharp, Notohara, et al., 2003; Nilsson, Lof-Ohlin, & Sun, 2013; Park, Kim, Hong, et al., 2015; Zhang, Kang, Piao, et al., 2014).

In aggregate, these studies provide evidence that aberrantly methylated DNA loci have potential for use as prognostic biomarkers for CRC; however, further investigation is required to develop clinically reliable, standardized assays in order for these assays to be used in clinical care.

3.3 Epigenetic predictive biomarkers for response to treatment

Despite recent advances in the development of cancer therapeutics, the currently used chemotherapeutic drugs have modest efficacy for advanced CRCs, especially when used without consideration for molecular subtypes. It is now well recognized that molecular markers in CRC, such as mutant

RAS (*KRAS, NRAS*) family genes and microsatellite instability, have clinical utility for targeting anti-EGFR (predict resistance to EGFR blockade) and immune checkpoint blockade therapies, respectively (Dienstmann, Salazar, & Tabernero, 2015). There is an unmet need for predictive biomarkers that can be used to target cytotoxic and targeted therapies to those CRC patients most likely to benefit.

Over the last decade, a number of aberrantly methylated genes have been assessed as predictive biomarkers for CRC patients undergoing various chemotherapeutic regimens. The majority of these studies have not progressed beyond Phase I/II discovery phase and thus will not be discussed further. CIMP as a predictive marker has been intensively studied for more than a decade; however, the best done studies to date have yielded conflicting results on its use for predicting response to 5-fluorouracil (5-FU) (Jover, Nguyen, & Perez-Carbonell, 2011; Van Rijnsoever et al., 2003). Thus, at this time, CIMP is not used clinically for directing 5-FU based therapy. More recently, prospective studies assessing CIMP as a predictive marker for adjuvant irinotecan and oxaliplatin have shown modest prognostic effects for overall survival in stage III, MSS, and CIMP-positive CRCs with the addition of irinotecan to adjuvant 5FU and leucovorin and in stage III CIMP-positive CRCs treated with 5FU, leucovorin, and oxaliplatin (FOLFOX-4) (Gallois, Taieb, Le Corre, et al., 2018; Shiovitz et al., 2014; Min, Bae, & Lee, 2011). These studies suggest promise for the use of CIMP as a potential prognostic marker and also highlight the need for additional studies of the interaction between CIMP status and therapeutic response to various treatments.

Although mostly still in the early phase of development, some promising single gene pharmaco-epigenetic biomarkers have been identified in various cancers with methylated *MGMT* for directing temozolamide treatment of gliomas being the best established to date (Weller & Stupp, 2012). A recent study showed the feasibility of using hypermethylated Transcription Factor AP-2 Epsilon (*TFAP2E*) as a predictive biomarker for response to 5-FU based chemotherapy in CRC patients (Ebert, Tanzer, Balluff, et al., 2012). Furthermore, DNA methylation microarray profiling of oxaliplatin sensitive vs resistant CRC cell lines revealed that oxaliplatin-resistant cells exhibited hypermethylation of the BRCA1 interactor *SRBC* gene; which was subsequently shown to associate with poor progression free survival (PFS) in CRC cohorts treated with oxaliplatin although the results of the initial study have not been replicated to date (Moutinho, Martinez-Cardus, Santos, et al., 2014). The results of these early phase studies using methylated genes as predictive markers and the ability to develop reliable assays for methylated genes is expected to continue to drive the investigation of methylated genes as response predictors for CRC therapy.

3.4 Histone modification alterations in CRC: Potential for use as biomarkers

Altered histone modifications are commonly found in CRC (Fraga & Esteller, 2005) and associate with altered chromatin activity states and gene expression through effects on gene promoters and noncoding regulatory elements (Audia & Campbell, 2016; Cohen, Saiakhova, Corradin, et al., 2017). Some studies have demonstrated potential for histone modifications to be CRC biomarkers, however, due to the technical limitations of assays that assess the post-translational histone modification state, no reliable tests for use in clinical care have been created. Global alterations of specific-histones in primary tissues have been the focus for biomarker development in CRC. Studies of H3K4me2, H3K9ac, and H3K9me2 alterations detected by immunohistochemical staining in liver metastases suggest that high H3K4me2 expression levels inversely correlate with poor prognosis (Tamagawa, Oshima, Shiozawa, et al., 2012). Additionally, other studies in CRC suggest that histone modifications, such as acetylation of H3 lysine 56 and di- or tri-methylation of H3 lysine 9 and 27, have potential to be prognostic markers in CRC (Baylin & Jones, 2011; Benard, Goossens-Beumer, van Hoesel, et al., 2014, 2015; Fraga, Ballestar, Villar-Garea, et al., 2005; Tamagawa et al., 2012; Tamagawa, Oshima, Numata, et al., 2013). Similarly, studies of histone modifications in circulating nucleosomes have identified reduced levels of H3K9me3- and H4K20me3 as potential diagnostic biomarkers for CRC (Gezer, Ustek, Yoruker, et al., 2013; Leszinski, Gezer, Siegele, et al., 2012). However, thus far, these studies are all proof of concept Phase I biomarker studies.

4. Conclusions

In summary, epigenetic alterations, including DNA methylation alterations, histone modification and chromatin perturbations and dysregulated non-coding RNAs, are frequently found in pre-malignant and malignant tumors of the colorectum and have been found in virtually all CRCs to date. Although initially believed to be a nonfunctional molecular alteration that merely accompanied CRC formation, epigenetic alterations are now known to not only effect the initiation and progression of the polyp-to-CRC progression sequence by altering the expression of tumor suppressor genes and oncogenes. They also appear to be robust biomarkers that can be used for the early detection of colon polyps as well as early stage cancers. Thus, epigenetic alterations in CRC are a central mechanism affecting CRC formation and have significant potential to be used to enhance the care of people with colon polyps and CRC.

Acknowledgments

These studies were supported by funding from the NIH:UO1CA152756, RO1CA194663, P30CA015704, U54CA163060, UO1CA086402, UO1CA182940 to WMG, P50CA150964 and UO1CA152756 to SDM, R50CA233042 to MY. Funding is also provided by the Cottrell Family Fund, R.A.C.E. Charities and the Listwin Foundation to WMG.

We also wish to acknowledge the ColoCare study team (Fred Hutchinson Cancer Research Center) and GICaRes study team (University of Washington, Seattle, WA) for their efforts and support of biomarker assay development.

Author contributions

WM Grady: study concept and design; drafting of the manuscript; critical revision of the manuscript for important intellectual content.

Grant support

NIH grants (P30CA15704, R01CA194663, RO1CA220004, RO1CA189184, U54CA143862, P01CA077852, R01CA207371, U01CA206110), R.A.C.E. Charities, Cottrell Family Fund, U01152756, Rodger Haggitt Endowed Chair, Listwin Family Foundation, Seattle Translational Tumor Research program, Fred Hutchinson Cancer Research Center to WMG; NCI (R50CA233042) to MY, and P50CA150964 and UO1CA152756 to SDM.

Conflict of interests

WM Grady is on the advisory boards for Freenome, Guardant Health, and SEngine and consults for Diacarta. He is also an investigator for a clinical trial sponsored by Janssen and receives services for investigator initiated research from Tempus.

References

Aad, G., Abbott, B., Abdallah, J., et al. (2015). Combined measurement of the Higgs boson mass in pp collisions at sqrt[s]=7 and 8 TeV with the ATLAS and CMS experiments. *Physical Review Letters*, *114*, 191803.

Abdelfatah, E., Kerner, Z., Nanda, N., et al. (2016). Epigenetic therapy in gastrointestinal cancer: The right combination. *Therapeutic Advances in Gastroenterology*, *9*, 560–579.

Ahn, J. B., Chung, W. B., Maeda, O., et al. (2011). DNA methylation predicts recurrence from resected stage III proximal colon cancer. *Cancer*, *117*, 1847–1854.

Ahuja, N., & Issa, J. P. (2000). Aging, methylation and cancer. *Histology and Histopathology*, *15*, 835–842.

Ahuja, N., Li, Q., Mohan, M., et al. (1998). Aging and DNA methylation in colorectal mucosa and cancer. *Cancer Research*, *58*, 5489–5494.

Ambrosi, C., Manzo, M., & Baubec, T. (2017). Dynamics and context-dependent roles of DNA methylation. *Journal of Molecular Biology*, *429*, 1459–1475.

Amirkhah, R., Schmitz, U., Linnebacher, M., et al. (2015). MicroRNA-mRNA interactions in colorectal cancer and their role in tumor progression. *Genes, Chromosomes & Cancer*, *54*, 129–141.

Ansari, I., Raddatz, G., Gutekunst, J., et al. (2020). The microbiota programs DNA methylation to control intestinal homeostasis and inflammation. *Nature Microbiology*, *5*, 610–619.

Antelo, M., Balaguer, F., Shia, J., et al. (2012). A high degree of LINE-1 hypomethylation is a unique feature of early-onset colorectal cancer. *PLoS One*, *7*, e45357.

Ashktorab, H., Belgrave, K., Hosseinkhah, F., et al. (2009). Global histone H4 acetylation and HDAC2 expression in colon adenoma and carcinoma. *Digestive Diseases and Sciences*, *54*, 2109–2117.

Auclair, G., & Weber, M. (2012). Mechanisms of DNA methylation and demethylation in mammals. *Biochimie*, *94*, 2202–2211.

Audia, J. E., & Campbell, R. M. (2016). Histone modifications and Cancer. *Cold Spring Harbor Perspectives in Biology*, *8*, a019521.

Baba, Y., Nosho, K., Shima, K., et al. (2010). HIF1A overexpression is associated with poor prognosis in a cohort of 731 colorectal cancers. *The American Journal of Pathology*, *176*, 2292–2301.

Baretti, M., & Azad, N. S. (2018). The role of epigenetic therapies in colorectal cancer. *Current Problems in Cancer*, *42*, 530–547.

Bartak, B. K., Kalmar, A., Peterfia, B., et al. (2017). Colorectal adenoma and cancer detection based on altered methylation pattern of SFRP1, SFRP2, SDC2, and PRIMA1 in plasma samples. *Epigenetics*, *12*, 751–763.

Bartke, T., & Kouzarides, T. (2011). Decoding the chromatin modification landscape. *Cell Cycle*, *10*, 182.

Bartova, E., Krejci, J., Harnicarova, A., et al. (2008). Histone modifications and nuclear architecture: A review. *The Journal of Histochemistry and Cytochemistry*, *56*, 711–721.

Baylin, S. B., Herman, J. G., Graff, J. R., et al. (1998). Alterations in DNA methylation: A fundamental aspect of neoplasia. *Advances in Cancer Research*, *72*, 141–196.

Baylin, S. B., & Jones, P. A. (2011). A decade of exploring the cancer epigenome—Biological and translational implications. *Nature Reviews. Cancer*, *11*, 726–734.

Bell, J. T., & Spector, T. D. (2012). DNA methylation studies using twins: What are they telling us? *Genome Biology*, *13*, 172.

Bell, C. G., Wilson, G. A., Butcher, L. M., et al. (2012). Human-specific CpG "beacons" identify loci associated with human-specific traits and disease. *Epigenetics*, *7*, 1188–1199.

Benard, A., Goossens-Beumer, I. J., van Hoesel, A. Q., et al. (2014). Histone trimethylation at H3K4, H3K9 and H4K20 correlates with patient survival and tumor recurrence in early-stage colon cancer. *BMC Cancer*, *14*, 531.

Benard, A., Goossens-Beumer, I. J., van Hoesel, A. Q., et al. (2015). Nuclear expression of histone deacetylases and their histone modifications predicts clinical outcome in colorectal cancer. *Histopathology*, *66*, 270–282.

Bersaglieri, C., & Santoro, R. (2019). Genome organization in and around the nucleolus. *Cell*, *8*, 579.

Bettington, M., Rosty, C., Whitehall, V., et al. (2018). A morphological and molecular study of proposed early forms of traditional serrated adenoma. *Histopathology*, *73*, 1023–1029.

Bettington, M., Walker, N., Clouston, A., et al. (2013). The serrated pathway to colorectal carcinoma: Current concepts and challenges. *Histopathology*, *62*, 367–386.

Bhowmick, N. A., Ghiassi, M., Aakre, M., et al. (2003). TGF-ß-induced RhoA and p160ROCK activation is involved in the inhibition of Cdc25A with resultant cell-cycle arrest. *Proceedings of the National Academy of Sciences of the United States of America*, *100*, 15548–15553.

Bi, F., Wang, Q., Dong, Q., et al. (2020). Circulating tumor DNA in colorectal cancer: Opportunities and challenges. *American Journal of Translational Research*, *12*, 1044–1055.
Bishop, K. S., & Ferguson, L. R. (2015). The interaction between epigenetics, nutrition and the development of cancer. *Nutrients*, *7*, 922–947.
Brahma, S., & Henikoff, S. (2020). Epigenome regulation by dynamic nucleosome unwrapping. *Trends in Biochemical Sciences*, *45*, 13–26.
Cairns, R. A., & Mak, T. W. (2013). Oncogenic isocitrate dehydrogenase mutations: Mechanisms, models, and clinical opportunities. *Cancer Discovery*, *3*, 730–741.
Calin, G. A., Dumitru, C. D., Shimizu, M., et al. (2002). Frequent deletions and down-regulation of micro- RNA genes miR15 and miR16 at 13q14 in chronic lymphocytic leukemia. *Proceedings of the National Academy of Sciences of the United States of America*, *99*, 15524–15529.
Calin, G. A., Pekarsky, Y., & Croce, C. M. (2007). The role of microRNA and other non-coding RNA in the pathogenesis of chronic lymphocytic leukemia. *Best Practice & Research. Clinical Haematology*, *20*, 425–437.
Carethers, J. M., & Jung, B. H. (2015). Genetics and genetic biomarkers in sporadic colorectal cancer. *Gastroenterology*, *149*, 1177–1190e3.
Carragher, L. A., Snell, K. R., Giblett, S. M., et al. (2010). V600EBraf induces gastrointestinal crypt senescence and promotes tumour progression through enhanced CpG methylation of p16INK4a. *EMBO Molecular Medicine*, *2*, 458–471.
Cedar, H., & Bergman, Y. (2009). Linking DNA methylation and histone modification: Patterns and paradigms. *Nature Reviews. Genetics*, *10*, 295–304.
Cerrato, F., Sparago, A., Ariani, F., et al. (2020). DNA methylation in the diagnosis of monogenic diseases. *Genes (Basel)*, *11*, 355.
Chang, K., Willis, J. A., Reumers, J., et al. (2018). Colorectal premalignancy is associated with consensus molecular subtypes 1 and 2. *Annals of Oncology*, *29*, 2061–2067.
Chen, W. D., Han, Z. J., Skoletsky, J., et al. (2005). Detection in fecal DNA of colon cancer-specific methylation of the nonexpressed vimentin gene. *Journal of the National Cancer Institute*, *97*, 1124–1132.
Church, T. R., Wandell, M., Lofton-Day, C., et al. (2014). Prospective evaluation of methylated SEPT9 in plasma for detection of asymptomatic colorectal cancer. *Gut*, *63*, 317–325.
Cohen, I., Poreba, E., Kamieniarz, K., et al. (2011). Histone modifiers in cancer: Friends or foes? *Genes & Cancer*, *2*, 631–647.
Cohen, A. J., Saiakhova, A., Corradin, O., et al. (2017). Hotspots of aberrant enhancer activity punctuate the colorectal cancer epigenome. *Nature Communications*, *8*, 14400.
Cooper, G. S., Markowitz, S. D., Chen, Z., et al. (2018). Performance of multitarget stool DNA testing in African American patients. *Cancer*, *124*, 3876–3880.
Corces, M. R., Granja, J. M., Shams, S., et al. (2018). The chromatin accessibility landscape of primary human cancers. *Science*, *362*, 420.
Costa, F. F. (2005). Non-coding RNAs: New players in eukaryotic biology. *Gene*, *357*, 83–94.
Crick, F. H., Barnett, L., Brenner, S., et al. (1961). General nature of the genetic code for proteins. *Nature*, *192*, 1227–1232.
Cristiano, S., Leal, A., Phallen, J., et al. (2019). Genome-wide cell-free DNA fragmentation in patients with cancer. *Nature*, *570*, 385–389.
Cui, H., Onyango, P., Brandenburg, S., et al. (2002). Loss of imprinting in colorectal cancer linked to hypomethylation of H19 and IGF2. *Cancer Research*, *62*, 6442–6446.
Dawson, M. A. (2017). The cancer epigenome: Concepts, challenges, and therapeutic opportunities. *Science*, *355*, 1147–1152.
Dehan, P., Kustermans, G., Guenin, S., et al. (2009). DNA methylation and cancer diagnosis: New methods and applications. *Expert Review of Molecular Diagnostics*, *9*, 651–657.

Derrien, T., Johnson, R., Bussotti, G., et al. (2012). The GENCODE v7 catalog of human long noncoding RNAs: Analysis of their gene structure, evolution, and expression. *Genome Research*, *22*, 1775–1789.

Dhir, M., Yachida, S., Van Neste, L., et al. (2011). Sessile serrated adenomas and classical adenomas: An epigenetic perspective on premalignant neoplastic lesions of the gastrointestinal tract. *International Journal of Cancer*, *129*, 1889–1898.

Dienstmann, R., Salazar, R., & Tabernero, J. (2015). Personalizing colon cancer adjuvant therapy: Selecting optimal treatments for individual patients. *Journal of Clinical Oncology*, *33*, 1787–1796.

Dokun, O. Y., Florl, A. R., Seifert, H. H., et al. (2008). Relationship of SNCG, S100A4, S100A9 and LCN2 gene expression and DNA methylation in bladder cancer. *International Journal of Cancer*, *123*, 2798–2807.

Dragomir, M. P., Kopetz, S., Ajani, J. A., et al. (2020). Non-coding RNAs in GI cancers: From cancer hallmarks to clinical utility. *Gut*, *69*, 748–763.

Draht, M. X., Smits, K. M., Tournier, B., et al. (2014). Promoter CpG island methylation of RET predicts poor prognosis in stage II colorectal cancer patients. *Molecular Oncology*, *8*, 679–688.

Du, Q., Luu, P. L., Stirzaker, C., et al. (2015). Methyl-CpG-binding domain proteins: Readers of the epigenome. *Epigenomics*, 7, 1051–1073.

Ebert, M. P., Model, F., Mooney, S., et al. (2006). Aristaless-like homeobox-4 gene methylation is a potential marker for colorectal adenocarcinomas. *Gastroenterology*, *131*, 1418–1430.

Ebert, M. P., Tanzer, M., Balluff, B., et al. (2012). TFAP2E-DKK4 and chemoresistance in colorectal cancer. *The New England Journal of Medicine*, *366*, 44–53.

Eckmann, J. D., Ebner, D. W., & Kisiel, J. B. (2020). Multi-target stool DNA testing for colorectal cancer screening: Emerging learning on real-world performance. *Current Treatment Options in Gastroenterol*, *18*, 109–119.

Eden, A., Gaudet, F., Waghmare, A., et al. (2003). Chromosomal instability and tumors promoted by DNA hypomethylation. *Science*, *300*, 455.

El Messaoudi, S., Mouliere, F., Du Manoir, S., et al. (2016). Circulating DNA as a strong multimarker prognostic tool for metastatic colorectal cancer patient management care. *Clinical Cancer Research*, *22*, 3067–3077.

Eleftheriou, M., & Ruzov, A. (2021). Modified forms of cytosine in eukaryotes: DNA (De) methylation and beyond. *Methods in Molecular Biology*, *2198*, 3–13.

Estecio, M. R., Gharibyan, V., Shen, L., et al. (2007). LINE-1 hypomethylation in cancer is highly variable and inversely correlated with microsatellite instability. *PLoS One*, *2*, e399.

Fabbri, M. (2010). miRNAs as molecular biomarkers of cancer. *Expert Review of Molecular Diagnostics*, *10*, 435–444.

Feil, R., & Fraga, M. F. (2012). Epigenetics and the environment: Emerging patterns and implications. *Nature Reviews. Genetics*, *13*, 97–109.

Feinberg, A. P., Koldobskiy, M. A., & Gondor, A. (2016). Epigenetic modulators, modifiers and mediators in cancer aetiology and progression. *Nature Reviews. Genetics*, *17*, 284–299.

Finger, F. P. (2002). One ring to bind them. Septins and actin assembly. *Developmental Cell*, *3*, 761–763.

Fraga, M. F., Ballestar, E., Paz, M. F., et al. (2005). Epigenetic differences arise during the lifetime of monozygotic twins. *Proceedings of the National Academy of Sciences of the United States of America*, *102*, 10604–10609.

Fraga, M. F., Ballestar, E., Villar-Garea, A., et al. (2005). Loss of acetylation at Lys16 and trimethylation at Lys20 of histone H4 is a common hallmark of human cancer. *Nature Genetics*, *37*, 391–400.

Fraga, M. F., & Esteller, M. (2005). Towards the human cancer epigenome: A first draft of histone modifications. *Cell Cycle*, *4*, 1377–1381.

Gaffney, D. J., McVicker, G., Pai, A. A., et al. (2012). Controls of nucleosome positioning in the human genome. *PLoS Genetics*, *8*, e1003036.

Gallois, C., Taieb, J., Le Corre, D., et al. (2018). Prognostic value of methylator phenotype in stage III colon cancer treated with oxaliplatin-based adjuvant chemotherapy. *Clinical Cancer Research*, *24*, 4745–4753.

Gaudet, F., Hodgson, J. G., Eden, A., et al. (2003). Induction of tumors in mice by genomic hypomethylation. *Science*, *300*, 489–492.

Gezer, U., Ustek, D., Yoruker, E. E., et al. (2013). Characterization of H3K9me3- and H4K20me3-associated circulating nucleosomal DNA by high-throughput sequencing in colorectal cancer. *Tumour Biology*, *34*, 329–336.

Gilbert, N. (2019). Biophysical regulation of local chromatin structure. *Current Opinion in Genetics & Development*, *55*, 66–75.

Glockner, S. C., Dhir, M., Yi, J. M., et al. (2009). Methylation of TFPI2 in stool DNA: A potential novel biomarker for the detection of colorectal cancer. *Cancer Research*, *69*, 4691–4699.

Grady, W. M., & Carethers, J. M. (2008). Genomic and epigenetic instability in colorectal cancer pathogenesis. *Gastroenterology*, *135*, 1079–1099.

Grady, W. M., Parkin, R. K., Mitchell, P. S., et al. (2008). Epigenetic silencing of the intronic microRNA hsa-miR-342 and its host gene EVL in colorectal cancer. *Oncogene*, *27*, 3880–3888.

Grady, W. M., Willis, J., Guilford, P. J., et al. (2000). Methylation of the CDH1 promoter as the second genetic hit in hereditary diffuse gastric cancer [in process citation]. *Nature Genetics*, *26*, 16–17.

Grutzmann, R., Molnar, B., Pilarsky, C., et al. (2008). Sensitive detection of colorectal cancer in peripheral blood by septin 9 DNA methylation assay. *PLoS One*, *3*, e3759.

Guinney, J., Dienstmann, R., Wang, X., et al. (2015). The consensus molecular subtypes of colorectal cancer. *Nature Medicine*, *21*, 1350–1356.

Hadley, M., Noonepalle, S., Banik, D., et al. (1983). Functional analysis of HDACs in tumorigenesis. *Methods in Molecular Biology*, *2019*, 279–307.

Hampel, H., Frankel, W. L., Martin, E., et al. (2005). Screening for the lynch syndrome (hereditary nonpolyposis colorectal cancer). *The New England Journal of Medicine*, *352*, 1851–1860.

Hanahan, D., & Weinberg, R. A. (2000). The hallmarks of cancer. *Cell*, *100*, 57–70.

Hanahan, D., & Weinberg, R. A. (2011). Hallmarks of cancer: The next generation. *Cell*, *144*, 646–674.

He, L., & Hannon, G. J. (2004). MicroRNAs: Small RNAs with a big role in gene regulation. *Nature Reviews. Genetics*, *5*, 522–531.

Heijmans, B. T., Tobi, E. W., Stein, A. D., et al. (2008). Persistent epigenetic differences associated with prenatal exposure to famine in humans. *Proceedings of the National Academy of Sciences of the United States of America*, *105*, 17046–17049.

Hellebrekers, D. M., Lentjes, M. H., van den Bosch, S. M., et al. (2009). GATA4 and GATA5 are potential tumor suppressors and biomarkers in colorectal cancer. *Clinical Cancer Research*, *15*, 3990–3997.

Herbst, A., Rahmig, K., Stieber, P., et al. (2011). Methylation of NEUROG1 in serum is a sensitive marker for the detection of early colorectal cancer. *The American Journal of Gastroenterology*, *106*, 1110–1118.

Herceg, Z. (2007). Epigenetics and cancer: Towards an evaluation of the impact of environmental and dietary factors. *Mutagenesis*, *22*, 91–103.

Heyn, H., Vidal, E., Ferreira, H. J., et al. (2016). Epigenomic analysis detects aberrant super-enhancer DNA methylation in human cancer. *Genome Biology*, *17*, 11.

Hinoue, T., Weisenberger, D. J., Lange, C. P., et al. (2012). Genome-scale analysis of aberrant DNA methylation in colorectal cancer. *Genome Research*, *22*, 271–282.

Hitchins, M. P. (2013). The role of epigenetics in lynch syndrome. *Familial Cancer, 12*, 189–205.
Holland, N. (2017). Future of environmental research in the age of epigenomics and exposomics. *Reviews on Environmental Health, 32*, 45–54.
Horvath, S., & Raj, K. (2018). DNA methylation-based biomarkers and the epigenetic clock theory of ageing. *Nature Reviews. Genetics, 19*, 371–384.
Hur, K., Cejas, P., Feliu, J., et al. (2014). Hypomethylation of long interspersed nuclear element-1 (LINE-1) leads to activation of proto-oncogenes in human colorectal cancer metastasis. *Gut, 63*, 635–646.
Imperiale, T. F., Ransohoff, D. F., Itzkowitz, S. H., et al. (2014). Multitarget stool DNA testing for colorectal-cancer screening. *The New England Journal of Medicine, 370*, 1287–1297.
Issa, J. P. (2014). Aging and epigenetic drift: A vicious cycle. *The Journal of Clinical Investigation, 124*, 24–29.
Issa, J. P., Ottaviano, Y. L., Celano, P., et al. (1994). Methylation of the oestrogen receptor CpG island links ageing and neoplasia in human colon. *Nature Genetics*, 7, 536–540.
Itzkowitz, S., Brand, R., Jandorf, L., et al. (2008). A simplified, noninvasive stool DNA test for colorectal cancer detection. *The American Journal of Gastroenterology, 103*, 2862–2870.
Itzkowitz, S. H., Jandorf, L., Brand, R., et al. (2007). Improved fecal DNA test for colorectal cancer screening. *Clinical Gastroenterology and Hepatology, 5*, 111–117.
Jin, P., Kang, Q., Wang, X., et al. (2015). Performance of a second-generation methylated SEPT9 test in detecting colorectal neoplasm. *Journal of Gastroenterology and Hepatology, 30*, 830–833.
Jover, R., Nguyen, T. P., Perez-Carbonell, L., et al. (2011). 5-Fluorouracil adjuvant chemotherapy does not increase survival in patients with CpG island methylator phenotype colorectal cancer. *Gastroenterology, 140*, 1174–1181.
Jung, G., Hernandez-Illan, E., Moreira, L., et al. (2020). Epigenetics of colorectal cancer: Biomarker and therapeutic potential. *Nature Reviews. Gastroenterology & Hepatology, 17*(2), 111–130. https://doi.org/10.1038/s41575-019-0230-y. Epub 2020 Jan 3. PMID: 31900466.
Juo, Y. Y., Johnston, F. M., Zhang, D. Y., et al. (2014). Prognostic value of CpG island methylator phenotype among colorectal cancer patients: A systematic review and meta-analysis. *Annals of Oncology, 25*, 2314–2327.
Kamakaka, R. T., & Biggins, S. (2005). Histone variants: Deviants? *Genes & Development, 19*, 295–310.
Kamiyama, H., Suzuki, K., Maeda, T., et al. (2012). DNA demethylation in normal colon tissue predicts predisposition to multiple cancers. *Oncogene, 31*, 5029–5037.
Kanwal, R., & Gupta, S. (2010). Epigenetics and cancer. *Journal of Applied Physiology (1985), 109*, 598–605.
Kawakami, K., Ruszkiewicz, A., Bennett, G., et al. (2006). DNA hypermethylation in the normal colonic mucosa of patients with colorectal cancer. *British Journal of Cancer, 94*, 593–598.
Kaz, A. M., Wong, C. J., Dzieciatkowski, S., et al. (2014). Patterns of DNA methylation in the normal colon vary by anatomical location, gender, and age. *Epigenetics, 9*, 492–502.
Kim, G. P., Colangelo, L. H., Wieand, H. S., et al. (2007). Prognostic and predictive roles of high-degree microsatellite instability in colon cancer: A National Cancer Institute-National Surgical Adjuvant Breast and bowel project collaborative study. *Journal of Clinical Oncology, 25*, 767–772.
Kim, S. H., Kang, Y. K., Koo, D. B., et al. (2004). Differential DNA methylation reprogramming of various repetitive sequences in mouse preimplantation embryos. *Biochemical and Biophysical Research Communications, 324*, 58–63.
Kim, Y. H., Petko, Z., Dzieciatkowski, S., et al. (2006). CpG island methylation of genes accumulates during the adenoma progression step of the multistep pathogenesis of colorectal cancer. *Genes, Chromosomes & Cancer, 45*, 781–789.

Kim, J. K., Samaranayake, M., & Pradhan, S. (2009). Epigenetic mechanisms in mammals. *Cellular and Molecular Life Sciences*, *66*, 596–612.

Klutstein, M., Nejman, D., Greenfield, R., et al. (2016). DNA methylation in Cancer and aging. *Cancer Research*, *76*, 3446–3450.

Kohli, R. M., & Zhang, Y. (2013). TET enzymes, TDG and the dynamics of DNA demethylation. *Nature*, *502*, 472–479.

Kohonen-Corish, M. R., Sigglekow, N. D., Susanto, J., et al. (2007). Promoter methylation of the mutated in colorectal cancer gene is a frequent early event in colorectal cancer. *Oncogene*, *26*, 4435–4441.

Komor, M. A., Bosch, L. J., Bounova, G., et al. (2018). Consensus molecular subtype classification of colorectal adenomas. *The Journal of Pathology*, *246*, 266–276.

Kouzarides, T. (2007a). Chromatin modifications and their function. *Cell*, *128*, 693–705.

Kouzarides, T. (2007b). SnapShot: Histone-modifying enzymes. *Cell*, *131*, 822.

Kriaucionis, S., & Heintz, N. (2009). The nuclear DNA base 5-hydroxymethylcytosine is present in Purkinje neurons and the brain. *Science*, *324*, 929–930.

Kriegl, L., Neumann, J., Vieth, M., et al. (2011). Up and downregulation of p16(Ink4a) expression in BRAF-mutated polyps/adenomas indicates a senescence barrier in the serrated route to colon cancer. *Modern Pathology*, *24*, 1015–1022.

Kudo, Y., Tateishi, K., Yamamoto, K., et al. (2012). Loss of 5-hydroxymethylcytosine is accompanied with malignant cellular transformation. *Cancer Science*, *103*, 670–676.

Laiho, P., Kokko, A., Vanharanta, S., et al. (2007). Serrated carcinomas form a subclass of colorectal cancer with distinct molecular basis. *Oncogene*, *26*, 312–320.

Lander, E. S., Linton, L. M., Birren, B., et al. (2001). Initial sequencing and analysis of the human genome. *Nature*, *409*, 860–921.

Lee, S., Cho, N. Y., Yoo, E. J., et al. (2008). CpG island methylator phenotype in colorectal cancers: Comparison of the new and classic CpG island methylator phenotype marker panels. *Archives of Pathology & Laboratory Medicine*, *132*, 1657–1665.

Lee, R. C., Feinbaum, R. L., & Ambros, V. (1993). The C. Elegans heterochronic gene lin-4 encodes small RNAs with antisense complementarity to lin-14. *Cell*, *75*, 843–854.

Lee, B. B., Lee, E. J., Jung, E. H., et al. (2009). Aberrant methylation of APC, MGMT, RASSF2A, and Wif-1 genes in plasma as a biomarker for early detection of colorectal cancer. *Clinical Cancer Research*, *15*, 6185–6191.

Lee-Six, H., Olafsson, S., Ellis, P., et al. (2019). The landscape of somatic mutation in normal colorectal epithelial cells. *Nature*, *574*, 532–537.

Leighton, G., & Williams, D. C., Jr. (2019). The methyl-CpG-binding domain 2 and 3 proteins and formation of the nucleosome remodeling and deacetylase complex. *Journal of Molecular Biology*, *432*, 1624–1639.

Leszinski, G., Gezer, U., Siegele, B., et al. (2012). Relevance of histone marks H3K9me3 and H4K20me3 in cancer. *Anticancer Research*, *32*, 2199–2205.

Leung, W. K., To KF, Man, E. P., et al. (2005). Quantitative detection of promoter hypermethylation in multiple genes in the serum of patients with colorectal cancer. *The American Journal of Gastroenterology*, *100*, 2274–2279.

Leung, W. K., To KF, Man, E. P., et al. (2007). Detection of hypermethylated DNA or cyclooxygenase-2 messenger RNA in fecal samples of patients with colorectal cancer or polyps. *The American Journal of Gastroenterology*, *102*, 1070–1076.

Lidgard, G. P., Domanico, M. J., Bruinsma, J. J., et al. (2013). Clinical performance of an automated stool DNA assay for detection of colorectal neoplasia. *Clinical Gastroenterology and Hepatology*, *11*, 1313–1318.

Liu, Y., Sethi, N. S., Hinoue, T., et al. (2018). Comparative molecular analysis of gastrointestinal adenocarcinomas. *Cancer Cell*, *33*, 721–735e8.

Lochhead, P., Kuchiba, A., Imamura, Y., et al. (2013). Microsatellite instability and BRAF mutation testing in colorectal cancer prognostication. *Journal of the National Cancer Institute*, *105*, 1151–1156.

Lofton-Day, C., Model, F., Devos, T., et al. (2008). DNA methylation biomarkers for blood-based colorectal cancer screening. *Clinical Chemistry, 54*, 414–423.

Luo, Y., Wong, C. J., Kaz, A. M., et al. (2014). Differences in DNA methylation signatures reveal multiple pathways of progression from adenoma to colorectal cancer. *Gastroenterology, 147*, 418–29e8.

Mahmoud, A. M., & Ali, M. M. (2019). Methyl donor micronutrients that modify DNA methylation and cancer outcome. *Nutrients, 11*.

Majid, S., Kikuno, N., Nelles, J., et al. (2008). Genistein induces the p21WAF1/CIP1 and p16INK4a tumor suppressor genes in prostate cancer cells by epigenetic mechanisms involving active chromatin modification. *Cancer Research, 68*, 2736–2744.

Martire, S., & Banaszynski, L. A. (2020). The roles of histone variants in fine-tuning chromatin organization and function. *Nature Reviews. Molecular Cell Biology, 21*, 522–541.

Maruti, S. S., Ulrich, C. M., & White, E. (2009). Folate and one-carbon metabolism nutrients from supplements and diet in relation to breast cancer risk. *The American Journal of Clinical Nutrition, 89*, 624–633.

Mathe, E. A., Nguyen, G. H., Bowman, E. D., et al. (2009). MicroRNA expression in squamous cell carcinoma and adenocarcinoma of the esophagus: Associations with survival. *Clinical Cancer Research, 15*, 6192–6200.

Meldi, K. M., & Figueroa, M. E. (2015). Cytosine modifications in myeloid malignancies. *Pharmacology & Therapeutics, 152*, 42–53.

Mendell, J. T. (2005). MicroRNAs: Critical regulators of development, cellular physiology and malignancy. *Cell Cycle, 4*, 1179–1184.

Mills, A. A. (2010). Throwing the cancer switch: Reciprocal roles of polycomb and trithorax proteins. *Nature Reviews. Cancer, 10*, 669–682.

Min, B. H., Bae, J. M., Lee, E. J., et al. (2011). The CpG island methylator phenotype may confer a survival benefit in patients with stage II or III colorectal carcinomas receiving fluoropyrimidine-based adjuvant chemotherapy. *BMC Cancer, 11*, 344.

Morak, M., Ibisler, A., Keller, G., et al. (2018). Comprehensive analysis of the MLH1 promoter region in 480 patients with colorectal cancer and 1150 controls reveals new variants including one with a heritable constitutional MLH1 epimutation. *Journal of Medical Genetics, 55*, 240–248.

Mouliere, F., & Rosenfeld, N. (2015). Circulating tumor-derived DNA is shorter than somatic DNA in plasma. *Proceedings of the National Academy of Sciences of the United States of America, 112*, 3178–3179.

Moutinho, C., Martinez-Cardus, A., Santos, C., et al. (2014). Epigenetic inactivation of the BRCA1 interactor SRBC and resistance to oxaliplatin in colorectal cancer. *Journal of the National Cancer Institute, 106*, djt322.

Myint, N. N. M., Verma, A. M., Fernandez-Garcia, D., et al. (2018). Circulating tumor DNA in patients with colorectal adenomas: Assessment of detectability and genetic heterogeneity. *Cell Death & Disease, 9*, 894.

Nagasaka, T., Sharp, G. B., Notohara, K., et al. (2003). Hypermethylation of O6-methylguanine-DNA methyltransferase promoter may predict nonrecurrence after chemotherapy in colorectal cancer cases. *Clinical Cancer Research, 9*, 5306–5312.

Neri, F., Dettori, D., Incarnato, D., et al. (2015). TET1 is a tumour suppressor that inhibits colon cancer growth by derepressing inhibitors of the WNT pathway. *Oncogene, 34*, 4168–4176.

Niculescu, M. D., & Lupu, D. S. (2011). Nutritional influence on epigenetics and effects on longevity. *Current Opinion in Clinical Nutrition and Metabolic Care, 14*, 35–40.

Nikolaou, S., Qiu, S., Fiorentino, F., et al. (2018). Systematic review of blood diagnostic markers in colorectal cancer. *Techniques in Coloproctology, 22*, 481–498.

Nilsson, T. K., Lof-Ohlin, Z. M., & Sun, X. F. (2013). DNA methylation of the p14ARF, RASSF1A and APC1A genes as an independent prognostic factor in colorectal cancer patients. *International Journal of Oncology, 42*, 127–133.

Ogino, S., Kawasaki, T., Kirkner, G. J., et al. (2007). Evaluation of markers for CpG Island methylator phenotype (CIMP) in colorectal cancer by a large population-based sample. *The Journal of Molecular Diagnostics*, *9*, 305–314.

Ogino, S., Kawasaki, T., Nosho, K., et al. (2008). LINE-1 hypomethylation is inversely associated with microsatellite instability and CpG island methylator phenotype in colorectal cancer. *International Journal of Cancer*, *122*, 2767–2773.

Ogino, S., Meyerhardt, J. A., Kawasaki, T., et al. (2007). CpG island methylation, response to combination chemotherapy, and patient survival in advanced microsatellite stable colorectal carcinoma. *Virchows Archiv*, *450*, 529–537.

Ogino, S., Nosho, K., Kirkner, G. J., et al. (2008). A cohort study of tumoral LINE-1 hypomethylation and prognosis in colon cancer. *Journal of the National Cancer Institute*, *100*, 1734–1738.

Oh, T., Kim, N., Moon, Y., et al. (2013). Genome-wide identification and validation of a novel methylation biomarker, SDC2, for blood-based detection of colorectal cancer. *The Journal of Molecular Diagnostics*, *15*, 498–507.

Okugawa, Y., Grady, W. M., & Goel, A. (2015). Epigenetic alterations in colorectal cancer: Emerging biomarkers. *Gastroenterology*, *149*, 1204–1225 (e12).

Pai, R. K., Bettington, M., Srivastava, A., et al. (2019). An update on the morphology and molecular pathology of serrated colorectal polyps and associated carcinomas. *Modern Pathology*, *32*, 1390–1415.

Pai, R. K., Jayachandran, P., Koong, A. C., et al. (2012). BRAF-mutated, microsatellite-stable adenocarcinoma of the proximal colon: An aggressive adenocarcinoma with poor survival, mucinous differentiation, and adverse morphologic features. *The American Journal of Surgical Pathology*, *36*, 744–752.

Paluch, B. E., Naqash, A. R., Brumberger, Z., et al. (2016). Epigenetics: A primer for clinicians. *Blood Reviews*, *30*, 285–295.

Park, S. J., Kim, S. M., Hong, Y. S., et al. (2015). TFAP2E methylation status and prognosis of patients with radically resected colorectal cancer. *Oncology*, *88*, 122–132.

Pauli, A., Rinn, J. L., & Schier, A. F. (2011). Non-coding RNAs as regulators of embryogenesis. *Nature Reviews. Genetics*, *12*, 136–149.

Petko, Z., Ghiassi, M., Shuber, A., et al. (2005). Aberrantly methylated CDKN2A, MGMT, and MLH1 in colon polyps and in fecal DNA from patients with colorectal polyps. *Clinical Cancer Research*, *11*, 1203–1209.

Phipps, A. I., Limburg, P. J., Baron, J. A., et al. (2015). Association between molecular subtypes of colorectal cancer and patient survival. *Gastroenterology*, *148*, 77–87e2.

Picardo, F., Romanelli, A., Muinelo-Romay, L., et al. (2019). Diagnostic and prognostic value of B4GALT1 hypermethylation and its clinical significance as a novel circulating cell-free DNA biomarker in colorectal cancer. *Cancers (Basel)*, *11*, 1598.

Portela, A., & Esteller, M. (2010). Epigenetic modifications and human disease. *Nature Biotechnology*, *28*, 1057–1068.

Prince, M., Lester, L., Chiniwala, R., et al. (2017). Multitarget stool DNA tests increases colorectal cancer screening among previously noncompliant medicare patients. *World Journal of Gastroenterology*, *23*, 464–471.

Pritchard, C. C., & Grady, W. M. (2011). Colorectal cancer molecular biology moves into clinical practice. *Gut*, *60*, 116–129.

Rebello, D., Rebello, E., Custodio, M., Xu, X., Gandhi, S., & Hemant K. Roy (2021) Field carcinogenesis for risk stratification of colorectal cancer. *Advances in Cancer Research*, *151*, 305–344.

Reinert, T., Henriksen, T. V., Christensen, E., et al. (2019). Analysis of plasma cell-free DNA by ultradeep sequencing in patients with stages I to III colorectal cancer. *JAMA Oncology*, *5*, 1124–1131.

Reinert, T., Scholer, L. V., Thomsen, R., et al. (2016). Analysis of circulating tumour DNA to monitor disease burden following colorectal cancer surgery. *Gut*, *65*, 625–634.

Rhee, I., Jair, K. W., Yen, R. W., et al. (2000). CpG methylation is maintained in human cancer cells lacking DNMT1. *Nature*, *404*, 1003–1007.

Rhee, Y. Y., Kim, M. J., Bae, J. M., et al. (2012). Clinical outcomes of patients with microsatellite-unstable colorectal carcinomas depend on L1 methylation level. *Annals of Surgical Oncology*, *19*, 3441–3448.

Robertson, K. D. (2005). DNA methylation and human disease. *Nature Reviews. Genetics*, *6*, 597–610.

Samowitz, W. S., Albertsen, H., Herrick, J., et al. (2005). Evaluation of a large, population-based sample supports a CpG island methylator phenotype in colon cancer. *Gastroenterology*, *129*, 837–845.

Sarkar, S., Abujamra, A. L., Loew, J. E., et al. (2011). Histone deacetylase inhibitors reverse CpG methylation by regulating DNMT1 through ERK signaling. *Anticancer Research*, *31*, 2723–2732.

Sengupta, S., & George, R. E. (2017). Super-enhancer-driven transcriptional dependencies in cancer. *Trends Cancer*, *3*, 269–281.

Sheffield, P. J., Oliver, C. J., Kremer, B. E., et al. (2003). Borg/septin interactions and the assembly of mammalian septin heterodimers, trimers, and filaments. *The Journal of Biological Chemistry*, *278*, 3483–3488.

Shen, L., Catalano, P. J., Benson, A. B., 3rd, et al. (2007). Association between DNA methylation and shortened survival in patients with advanced colorectal cancer treated with 5-fluorouracil based chemotherapy. *Clinical Cancer Research*, *13*, 6093–6098.

Shen, L., Kondo, Y., Rosner, G. L., et al. (2005). MGMT promoter methylation and field defect in sporadic colorectal cancer. *Journal of the National Cancer Institute*, *97*, 1330–1338.

Shen, J., Wang, S., Zhang, Y. J., et al. (2013). Exploring genome-wide DNA methylation profiles altered in hepatocellular carcinoma using Infinium HumanMethylation 450 BeadChips. *Epigenetics*, *8*, 34–43.

Shibata, D. (2008). Stem cells as common ancestors in a colorectal cancer ancestral tree. *Current Opinion in Gastroenterology*, *24*, 59–63.

Shibata, D. (2009). Inferring human stem cell behaviour from epigenetic drift. *The Journal of Pathology*, *217*, 199–205.

Shiovitz, S., Bertagnolli, M. M., Renfro, L. A., et al. (2014). CpG island methylator phenotype is associated with response to adjuvant irinotecan-based therapy for stage III colon cancer. *Gastroenterology*, *147*, 637–645.

Sidransky, D., Tokino, T., Hamilton, S., et al. (1992). Identification of *ras* oncogene mutations in the stool of patients with curable colorectal tumors. *Science (Washington, DC)*, *256*, 102–105.

Sievers, C. K., Grady, W. M., Halberg, R. B., et al. (2017). New insights into the earliest stages of colorectal tumorigenesis. *Expert Review of Gastroenterology & Hepatology*, *11*, 723–729.

Siravegna, G., Mussolin, B., Buscarino, M., et al. (2015). Clonal evolution and resistance to EGFR blockade in the blood of colorectal cancer patients. *Nature Medicine*, *21*, 827.

Slaby, O., Svoboda, M., Michalek, J., et al. (2009). MicroRNAs in colorectal cancer: Translation of molecular biology into clinical application. *Molecular Cancer*, *8*, 102.

Slaughter, D. P., Southwick, H. W., & Smejkal, W. (1953). Field cancerization in oral stratified squamous epithelium; clinical implications of multicentric origin. *Cancer*, *6*, 963–968.

Snyder, M. W., Kircher, M., Hill, A. J., et al. (2016). Cell-free DNA comprises an in vivo nucleosome footprint that informs its tissues-of-origin. *Cell*, *164*, 57–68.

Sobhani, I., Bergsten, E., Couffin, S., et al. (2019). Colorectal cancer-associated microbiota contributes to oncogenic epigenetic signatures. *Proceedings of the National Academy of Sciences of the United States of America, 116*, 24285–24295.

Stefanska, B., Karlic, H., Varga, F., et al. (2012). Epigenetic mechanisms in anti-cancer actions of bioactive food components—The implications in cancer prevention. *British Journal of Pharmacology, 167*, 279–297.

Suganuma, T., & Workman, J. L. (2011). Signals and combinatorial functions of histone modifications. *Annual Review of Biochemistry, 80*, 473–499.

Sur, I., & Taipale, J. (2016). The role of enhancers in cancer. *Nature Reviews. Cancer, 16*, 483–493.

Suraweera, A., O'Byrne, K. J., & Richard, D. J. (2018). Combination therapy with histone deacetylase inhibitors (HDACi) for the treatment of cancer: Achieving the full therapeutic potential of HDACi. *Frontiers in Oncology, 8*, 92.

Suzuki, H., Igarashi, S., Nojima, M., et al. (2010). IGFBP7 is a p53-responsive gene specifically silenced in colorectal cancer with CpG island methylator phenotype. *Carcinogenesis, 31*, 342–349.

Suzuki, K., Suzuki, I., Leodolter, A., et al. (2006). Global DNA demethylation in gastrointestinal cancer is age dependent and precedes genomic damage. *Cancer Cell, 9*, 199–207.

Suzuki, H., Watkins, D. N., Jair, K. W., et al. (2004). Epigenetic inactivation of SFRP genes allows constitutive WNT signaling in colorectal cancer. *Nature Genetics, 36*, 417–422.

Tahara, T., Yamamoto, E., Madireddi, P., et al. (2014). Colorectal carcinomas with CpG island methylator phenotype 1 frequently contain mutations in chromatin regulators. *Gastroenterology, 146*, 530–538 (e5).

Tahiliani, M., Koh, K. P., Shen, Y., et al. (2009). Conversion of 5-methylcytosine to 5-hydroxymethylcytosine in mammalian DNA by MLL partner TET1. *Science, 324*, 930–935.

Tamagawa, H., Oshima, T., Numata, M., et al. (2013). Global histone modification of H3K27 correlates with the outcomes in patients with metachronous liver metastasis of colorectal cancer. *European Journal of Surgical Oncology, 39*, 655–661.

Tamagawa, H., Oshima, T., Shiozawa, M., et al. (2012). The global histone modification pattern correlates with overall survival in metachronous liver metastasis of colorectal cancer. *Oncology Reports, 27*, 637–642.

Tan, S., & Davey, C. A. (2011). Nucleosome structural studies. *Current Opinion in Structural Biology, 21*, 128–136.

Tessarz, P., & Kouzarides, T. (2014). Histone core modifications regulating nucleosome structure and dynamics. *Nature Reviews. Molecular Cell Biology, 15*, 703–708.

Thomas, M. L., & Marcato, P. (2018). Epigenetic modifications as biomarkers of tumor development, therapy response, and recurrence across the cancer care continuum. *Cancers (Basel), 10*.

Tie, J., Cohen, J. D., Wang, Y., et al. (2019). Circulating tumor DNA analyses as markers of recurrence risk and benefit of adjuvant therapy for stage III colon cancer. *JAMA Oncology, 5*, 1710–1717.

Torchy, M. P., Hamiche, A., & Klaholz, B. P. (2015). Structure and function insights into the NuRD chromatin remodeling complex. *Cellular and Molecular Life Sciences, 72*, 2491–2507.

Toyota, M., Ahuja, N., Ohe-Toyota, M., et al. (1999). CpG island methylator phenotype in colorectal cancer. *Proceedings of the National Academy of Sciences of the United States of America, 96*, 8681–8686.

Toyota, M., Ohe-Toyota, M., Ahuja, N., et al. (2000). Distinct genetic profiles in colorectal tumors with or without the CpG island methylator phenotype. *Proceedings of the National Academy of Sciences of the United States of America, 97*, 710–715.

Tse, J. W. T., Jenkins, L. J., Chionh, F., et al. (2017). Aberrant DNA methylation in colorectal Cancer: What should we target? *Trends Cancer*, *3*, 698–712.

Valle, L., Serena-Acedo, T., Liyanarachchi, S., et al. (2008). Germline allele-specific expression of TGFBR1 confers an increased risk of colorectal cancer. *Science*, *321*, 1361–1365.

Valouev, A., Johnson, S. M., Boyd, S. D., et al. (2011). Determinants of nucleosome organization in primary human cells. *Nature*, *474*, 516–520.

Van Rijnsoever, M., Elsaleh, H., Joseph, D., et al. (2003). CpG island methylator phenotype is an independent predictor of survival benefit from 5-fluorouracil in stage III colorectal cancer. *Clinical Cancer Research*, *9*, 2898–2903.

Van Tongelen, A., Loriot, A., & De Smet, C. (2017). Oncogenic roles of DNA hypomethylation through the activation of cancer-germline genes. *Cancer Letters*, *396*, 130–137.

Vasudevan, S., Tong, Y., & Steitz, J. A. (2007). Switching from repression to activation: MicroRNAs can up-regulate translation. *Science*, *318*, 1931–1934.

Veigl, M. L., Kasturi, L., Olechnowicz, J., et al. (1998). Biallelic inactivation of hMLH1 by epigenetic gene silencing, a novel mechanism causing human MSI cancers. *Proceedings of the National Academy of Sciences of the United States of America*, *95*, 8698–8702.

Vilain, A., Vogt, N., Dutrillaux, B., et al. (1999). DNA methylation and chromosome instability in breast cancer cell lines. *FEBS Letters*, *460*, 231–234.

Wallner, M., Herbst, A., Behrens, A., et al. (2006). Methylation of serum DNA is an independent prognostic marker in colorectal cancer. *Clinical Cancer Research*, *12*, 7347–7352.

Wang, Y., Li, L., Cohen, J. D., et al. (2019). Prognostic potential of circulating tumor DNA measurement in postoperative surveillance of nonmetastatic colorectal cancer. *JAMA Oncology*, *5*, 1118–1123.

Wang, D. R., & Tang, D. (2008). Hypermethylated SFRP2 gene in fecal DNA is a high potential biomarker for colorectal cancer noninvasive screening. *World Journal of Gastroenterology*, *14*, 524–531.

Wei, Y., Melas, P. A., Wegener, G., et al. (2014). Antidepressant-like effect of sodium butyrate is associated with an increase in TET1 and in 5-hydroxymethylation levels in the Bdnf gene. *The International Journal of Neuropsychopharmacology*, *18*, 1–10.

Weisenberger, D. J., Siegmund, K. D., Campan, M., et al. (2006). CpG island methylator phenotype underlies sporadic microsatellite instability and is tightly associated with BRAF mutation in colorectal cancer. *Nature Genetics*, *38*, 787–793.

Weller, M., & Stupp, R. (2012). Prime time for molecular marker diagnostics in neurooncology. *Current Opinion in Neurology*, *25*, 743–744.

Wightman, B., Ha, I., & Ruvkun, G. (1993). Posttranscriptional regulation of the heterochronic gene lin-14 by lin-4 mediates temporal pattern formation in C. elegans. *Cell*, *75*, 855–862.

Wilson, A. S., Power, B. E., & Molloy, P. L. (1775). DNA hypomethylation and human diseases. *Biochimica et Biophysica Acta*, *2007*, 138–162.

Wippermann, A., Klausing, S., Rupp, O., et al. (2014). Establishment of a CpG island microarray for analyses of genome-wide DNA methylation in Chinese hamster ovary cells. *Applied Microbiology and Biotechnology*, *98*, 579–589.

Wippermann, A., Rupp, O., Brinkrolf, K., et al. (2017). Integrative analysis of DNA methylation and gene expression in butyrate-treated CHO cells. *Journal of Biotechnology*, *257*, 150–161.

Wolin, S. L., Sim, S., & Chen, X. (2012). Nuclear noncoding RNA surveillance: Is the end in sight? *Trends in Genetics*, *28*, 306–313.

Wu, J., Li, H., Shi, M., et al. (2019). TET1-mediated DNA hydroxymethylation activates inhibitors of the Wnt/beta-catenin signaling pathway to suppress EMT in pancreatic tumor cells. *Journal of Experimental & Clinical Cancer Research*, *38*, 348.

Xiao-Jie, L., Hui-Ying, X., Qi, X., et al. (2016). LINE-1 in cancer: Multifaceted functions and potential clinical implications. *Genetics in Medicine, 18*, 431–439.

Yamamoto, E., Toyota, M., Suzuki, H., et al. (2008). LINE-1 hypomethylation is associated with increased CpG island methylation in Helicobacter pylori-related enlarged-fold gastritis. *Cancer Epidemiology, Biomarkers & Prevention, 17*, 2555–2564.

Yang, C. S., Landau, J. M., Huang, M. T., et al. (2001). Inhibition of carcinogenesis by dietary polyphenolic compounds. *Annual Review of Nutrition, 21*, 381–406.

You, J. S., & Jones, P. A. (2012). Cancer genetics and epigenetics: Two sides of the same coin? *Cancer Cell, 22*, 9–20.

Young, G. P., Pedersen, S. K., Mansfield, S., et al. (2016). A cross-sectional study comparing a blood test for methylated BCAT1 and IKZF1 tumor-derived DNA with CEA for detection of recurrent colorectal cancer. *Cancer Medicine, 5*, 2763–2772.

Yu, M., Hazelton, W. D., Luebeck, G. E., et al. (2020). Epigenetic aging: More than just a clock when it comes to Cancer. *Cancer Research, 80*, 367–374.

Zeisel, S. H. (2009). Epigenetic mechanisms for nutrition determinants of later health outcomes. *The American Journal of Clinical Nutrition, 89*, 1488S–1493S.

Zhang, N. (2015). Epigenetic modulation of DNA methylation by nutrition and its mechanisms in animals. *Animal Nutrition, 1*, 144–151.

Zhang, R., Kang, K. A., Piao, M. J., et al. (2014). Epigenetic alterations are involved in the overexpression of glutathione S-transferase pi-1 in human colorectal cancers. *International Journal of Oncology, 45*, 1275–1283.

Zhang, W., & Xu, J. (2017). DNA methyltransferases and their roles in tumorigenesis. *Biomarker Research, 5*, 1.

Zhang, W. P., Yang, H., Chen, H., et al. (2014). Gene expression analysis of potential genes and pathways involved in the pathogenic mechanisms of parvovirus B19 in human colorectal cancer. *Oncology Letters, 8*, 523–532.

Zlobec, I., Minoo, P., Baumhoer, D., et al. (2008). Multimarker phenotype predicts adverse survival in patients with lymph node-negative colorectal cancer. *Cancer, 112*, 495–502.

9780128240786